Stem Cells and Regenerative Medicine

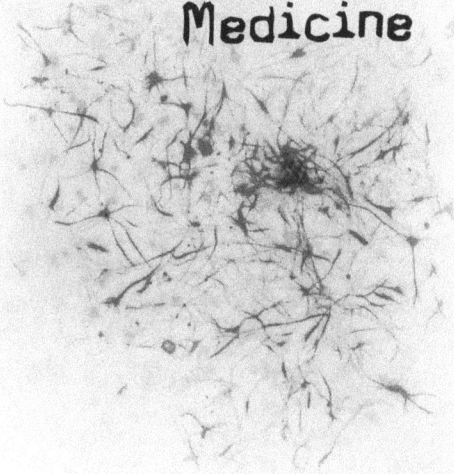

Stem Cells and Regenerative Medicine

editors

Walter C. Low
University of Minnesota, USA

Catherine M. Verfaillie
University of Minnesota, USA & Catholic University of Leuven, Belgium

World Scientific

NEW JERSEY · LONDON · SINGAPORE · BEIJING · SHANGHAI · HONG KONG · TAIPEI · CHENNAI

Published by

World Scientific Publishing Co. Pte. Ltd.

5 Toh Tuck Link, Singapore 596224

USA office: 27 Warren Street, Suite 401-402, Hackensack, NJ 07601

UK office: 57 Shelton Street, Covent Garden, London WC2H 9HE

Library of Congress Cataloging-in-Publication Data
Stem cells and regenerative medicine / editors, Walter C. Low, Catherine M. Verfaillie.
    p. ; cm.
  Includes bibliographical references and index.
  ISBN-13: 978-981-277-576-4 (hardcover : alk. paper)
  ISBN-10: 981-277-576-5 (hardcover : alk. paper)
  1. Stem cells. 2. Regeneration (Biology) I. Low, Walter C. II. Verfaillie, Catherine M.
  [DNLM: 1. Stem Cells. 2. Regenerative Medicine--methods. 3. Stem Cell Transplantation.
  QU 325 S82345 2008]
  QH588.S83S745 2008
  616'.02774--dc22

      2008009898

British Library Cataloguing-in-Publication Data
A catalogue record for this book is available from the British Library.

First published 2008 (Hardcover)
Reprinted 2016 (in paperback edition)
ISBN 978-981-3203-45-7

Typeset by Stallion Press
Email: enquiries@stallionpress.com

Preface

Stem cells represent an exciting new technology for regenerative medicine and repair of the human body in disease and aging. The study of stem cells in laboratory animals has a long and rich history. This field of study was greatly accelerated by the work of James Thomson and his colleagues in their isolation and characterization of human embryonic stem cells. Several chapters in this book are devoted to recent advances in human embryonic stem cells in relation to the development of the immune system, as a model for the development of islet cells in the pancreas, and in the identification of nuclear reprogramming factors to achieve an embryonic stem cell state.

Other chapters in this book focus on the isolation and characterization of stem cells derived from adult and neonatal tissue. These include stem cells derived from bone marrow, umbilical cord blood, brain, muscle, and other organ systems of the body. The use of these stem cells to produce differentiated cells such as heart cells, skeletal muscle cells, kidney, liver, and cells of the nervous system is discussed in the chapters.

The application of stem cells for regenerative medicine is critically dependent on the ability to generate sufficient quantities of clinically relevant cells. These issues are discussed in the chapters that focus on the maintenance and large scale manufacturing of stem cells.

The compilation of this book on the latest advances in the field of stem cell research required the participation of many individuals. However, we would like to acknowledge, in particular, the contribution of Elizabeth Hedin for her careful and meticulous editing of the chapters in this book.

Walter C. Low, Ph.D.
Dept. of Neurosurgery
Stem Cell Institute
University of Minnesota
Minneapolis, MN, U.S.A.

Catherine M. Verfaillie, M.D.
Stem Cell Institute Leuven
Catholic University of Leuven
Leuven, Belgium

Contents

Preface   v

List of Contributors   xi

EMBRYONIC STEM CELLS, NUCLEAR REPROGRAMMING,
AND STEM CELL FATE

Chapter 1
Development of Diverse Hematopoietic Cell   3
Populations from Human Embryonic Stem Cells
Rizwan Romee, Ketan Doshi and Dan S. Kaufman

Chapter 2
Using Embryonic Stem Cells as a Model   27
of Pancreatic Development
Zhaohui Geng, Lucas Chase and Meri T. Firpo

Chapter 3
Approaches to Identifying Nuclear Reprogramming   45
Factors from Embryonic Stem Cell Biology
and Somatic Cell Nuclear Cloning
Nobuaki Kikyo

Chapter 4
Cell Cycle Control of Stem Cell Fate   63
Lucas Nacusi and Robert Sheaff

ADULT STEM CELLS

Chapter 5
Multipotent Adult Progenitor Cells   95
Yuehua Jiang, April Breyer and Catherine Verfaillie

Chapter 6
Ocular Surface Epithelial Stem Cells   111
De-Quan Li, Stephen C. Pflugfelder and Andrew J. W. Huang

STEM CELLS FOR MUSCLE AND HEART

Chapter 7
Stem Cells in Skeletal Muscle Regeneration   145
Atsushi Asakura

Chapter 8
Myogenic Precursor Cells in the Extraocular Muscles   177
Kristen M. Kallestad and Linda K. McLoon

Chapter 9
Treating the Continuum of Coronary Heart Disease   191
with Progenitor-Cell-based Repair: The University
of Minnesota Experience
Doris A. Taylor, Jonathan D. McCue and Andrey G. Zenovich

Chapter 10
Postnatal Stem Cells for Myocardial Repair   221
Mohammad Nurulqadr Jameel, Peter Eckman,
Arthur From, Catherine Verfaillie and Jianyi Zhang

STEM CELLS FOR THE NERVOUS SYSTEM

Chapter 11
Use of a *β*-Galactosidase Reporter Coupled   265
to Cell-Specific Promoters to Examine
Differentiation of Neural Progenitor
Cells *In Vivo* and *In Vitro*
Dale S. Gregerson

Chapter 12
From Neural Stem Cells to Neuroregeneration   291
Terry C. Burns, Walter C. Low and Catherine M. Verfaillie

Chapter 13
Cochlear Stem Cells/Progenitors   327
Jizhen Lin, Water Low and Catherine Verfaillie

Chapter 14
Intravascular Delivery Systems for Stem Cell   355
Transplantation in Neurologic Disorders
Vallabh Janardhan, Adnan I. Qureshi and Walter C. Low

Chapter 15
Stem Cell Strategies for Treating Inner Ear Dysfunction   397
John H. Anderson and Steven K. Juhn

STEM CELLS FOR LIVER AND KIDNEY

Chapter 16
Renal Stem Cells   419
Sandeep Gupta and Mark E. Rosenberg

Chapter 17
Liver Stem Cells in Regenerative Medicine   445
Xin Wang, Yiping Hu, Zongyu Chen and Xiaoyan Ding

TECHNOLOGY FOR PRODUCTION OF STEM CELLS

Chapter 18
Clinical Manufacture of Stem Cells   487
and Derivative Cell Populations
David H. McKenna and John Wagner

Chapter 19
Stem Cell Culture Engineering   513
Kartik Subramanian and Wei-Shou Hu

Index   543

List of Contributors

John H. Anderson
Department of Otolaryngology
Graduate Program in Neuroscience
University of Minnesota
Minneapolis, MN 55455
USA

Atsushi Asakura
Stem Cell Institute
University of Minnesota Medical School
McGuire Translational Research Facility
Minneapolis, MN 55455
USA

April Breyer
Department of Medicine and Stem Cell Institute
University of Minnesota
Minneapolis, MN 55455
USA

Terry C. Burns
Department of Neurosurgery
Graduate Program in Neuroscience
University of Minnesota Medical School
Minneapolis, MN 55455
USA

Lucas Chase
University of Minnesota Stem Cell Institute
Department of Medicine and Division of Endocrinology
Minneapolis, Minnesota 55455
USA

Zongyu Chen
Department of Medicine
University of Minnesota
Minneapolis, MN 55455
USA

Xiaoyan Ding
Key Laboratory of Molecular and Cell Biology
Institutes of Biochemistry and Cell Biology
Shanghai Institute of Biological Sciences
Shanghai 200031
P. R. China

Ketan Doshi
Stem Cell Institute and Department of Medicine
University of Minnesota
Minneapolis, MN 55455
USA

Peter Eckman
Department of Medicine
University of Minnesota
Minneapolis, MN 55455
USA

Meri T. Firpo
University of Minnesota Stem Cell Institute
Department of Medicine and Division of Endocrinology
Minneapolis, Minnesota 55455
USA

Arthur From
Center for Magnetic Resonance Research
University of Minnesota
Minneapolis, MN 55455
USA

Zhaohui Geng
University of Minnesota Stem Cell Institute
Department of Medicine and Division of Endocrinology
Minneapolis, Minnesota 55455
USA

Dale S. Gregerson
Department of Ophthalmology
University of Minnesota
Minneapolis, MN 55431-2404
USA

Sandeep Gupta
Division of Renal Diseases and Hypertension
Department of Medicine
University of Minnesota
Minneapolis, MN 55455
USA

Yiping Hu
Department of Laboratory Medicine and Pathology
University of Minnesota
Minneapolis, MN 55455
USA

Wei-Shou Hu
Department of Chemical Engineering and Materials Science
Stem Cell Institute
University of Minnesota
Minneapolis, MN 55455-0132
USA

Andrew J. W. Huang
Department of Ophthalmology and Visual Sciences
Washington University
St. Louis, MO 63110
USA

Mohammad Nurulqadr Jameel
Department of Medicine
University of Minnesota
Minneapolis, MN 55455
USA

Vallabh Janardhan
Department of Neurology, Radiology and Neurosurgery
Stroke Center
University of Minnesota
Minneapolis, MN 55455
USA

Yuehua Jiang
Department of Medicine and Stem Cell Institute
University of Minnesota
Minneapolis, MN 55455
USA

Steven K. Juhn
Department of Otolaryngology and Stem Cell Institute
University of Minnesota
Minneapolis, MN 55455
USA

Kristen M. Kallestad
Department of Ophthalmology and Neuroscience
University of Minnesota
Minneapolis, MN 55455
USA

Dan S. Kaufman
Stem Cell Institute and Department of Medicine
University of Minnesota
Minneapolis, MN 55455
USA

Nobuaki Kikyo
Stem Cell Institute, Department of Medicine, and
Division of Hematology, Oncology, and Transplantation
University of Minnesota
Minneapolis, MN 55455
USA

De-Quan Li
Department of Ophthalmology
Center for Cell and Gene Therapy
Baylor College of Medicine
Houston, TX
USA

Jizhen Lin
Auditory of Molecular Biology Laboratory
Department of Otolaryngology
Stem Cell Institute
University of Minnesota
Minneapolis, MN 55455
USA

Walter C. Low
Department of Neurosurgery
Graduate Program in Neuroscience
Stem Cell Institute
University of Minnesota Medical School
Minneapolis, MN 55455
USA

Jonathan D. McCue
Department of Thoracic and Cardiovascular Surgery
University of Minnesota Medical School
Minneapolis, MN 55455
USA

David H. McKenna
Clinical Cell Therapy Laboratory
University of Minnesota Medical Center
Minneapolis, MN 55455
USA

Linda K. McLoon
Department of Ophthalmology
University of Minnesota
Minneapolis, MN 55455
USA

Lucas Nacusi
Department of Biochemistry, Molecular Biology, and Biophysics
University of Minnesota
Minneapolis, MN 55455
USA

Stephen C. Pflugfelder
The Cullen Eye Institute
Department of Ophthalmology
Baylor College of Medicine
Houston, Texas 77030
USA

Adnan I. Qureshi
Department of Neurology, Radiology, and Neurosurgery
Stroke Center
University of Minnesota
Minneapolis, MN 55455
USA

Rizwan Romee
Stem Cell Institute and Department of Medicine
University of Minnesota
Minneapolis, MN 55455
USA

Mark E. Rosenberg
Department of Medicine and Stem Cell Institute
University of Minnesota
Minneapolis, MN 55455
USA

Robert Sheaff
Department of Chemistry and Biochemistry
University of Tulsa
Tulsa, OK
USA

Kartik Subramanian
Department of Chemical Engineering and Materials Science
Stem Cell Institute
University of Minnesota
Minneapolis, MN 55455-0132
USA

Doris A. Taylor
Center for Cardiovascular Repair
University of Minnesota
Minneapolis, MN 55455
USA

Catherine M. Verfaillie
Department of Medicine and Stem Cell Institute
University of Minnesota
Minneapolis, MN 55455
USA

and

Stem Cell Institute Leuven
Catholic University of Leuven
Leuven, Belgium

John Wagner
Molecular and Cellular Therapeutics
Department of Pediatrics
University of Minnesota
1900 Fitch Avenue, Saint Paul
MN 55108
USA

Xin Wang
Stem Cell Institute
University of Minnesota
Minneapolis, MN 55455
USA

Andrey G. Zenovich
Center for Cardiovascular Repair
Department of Medicine
University of Minnesota
Minneapolis, MN 55455
USA

Jianyi Zhang
Department of Medicine/Cardiology
University of Minnesota Medical School
Minneapolis, MN 55455
USA

EMBRYONIC STEM CELLS, NUCLEAR REPROGRAMMING, AND STEM CELL FATE

1

Development of Diverse Hematopoietic Cell Populations from Human Embryonic Stem Cells

Rizwan Romee, Ketan Doshi and Dan S. Kaufman*

1. Introduction

Mouse and human embryonic stem cells (hereafter called mESCs and hESCs respectively) are pluripotent cells derived from the inner cell mass of preimplantation blastocyst stage embryos. mESCs and hESCs are defined by their ability to self-renew and maintain the normal karyotype after prolonged periods in culture, while retaining the capacity to generate cells of all three germ cell layers when induced or supported to differentiate.[1–3] While there are important differences between mESCs and hESCs in terms of morphology, growth characteristics, and culture conditions required for maintenance of the undifferentiated state,[2–4] these differences will not be discussed in further detail here.

Hematopoiesis (blood cell development) has been one of the most intensively studied cell lineages using not only mESCs and hESCs but also other model systems such as the zebrafish[5] *Xenopus*,[6,7] developing mouse embryos,[8–11] mouse and human bone marrow, and human umbilical cord blood (UCB). With all these more established model systems available, there are at least three main reasons why studies of hematopoiesis from hESCs are important:

(i) To understand the basic human developmental biology of self-renewal, differentiation, and commitment to specific lineage. hESCs

*Correspondence: Stem Cell Institute and Department of Medicine, University of Minnesota, Room 2-220 Translational Research Facility, 2001 6th Street SE, Mail Code 2873, Minneapolis, MN 55455, USA. E-mail: kaufm020@umn.edu

3

allow easy access to a homogeneous cell population to define the earliest developmental events that occur during human embryogenesis, a stage that is difficult to obtain from primary human tissue.

(ii) hESC-derived cells can be used to substitute and/or supplement other cell populations now typically utilized for hematopoietic stem cell transplantation (HCT) or other hematopoietic cell therapies. Much of the clinical interest in studying hESC-derived hematopoiesis comes from the fact that many malignant hematological disorders can be treated and possibly cured by HCT. Indeed, HCT is an example of a "stem cell therapy" that has been used clinically for over thirty years.[12,13] At present there are three sources of transplantable hematopoietic stem cells (HSCs): adult bone marrow, growth-factor-mobilized peripheral blood, and UCB. Limitations to further expanding clinical use of HSCs include scarce supply, inability to expand HSCs *in vitro* without loss of stem cell function, risk of pathogen contamination, risk to donor (pain, infection, bleeding), and risk of graft-versus-host disease. These limitations could potentially be circumvented if HSCs derived from hESCs are used. Additionally, hESCs can potentially be manipulated to match or mismatch human leukocyte antigen (HLA) types for therapeutic advantage. In the situation of graft failure, patients could be given a supplemental dose of hESC-derived HSCs. Additionally, hESC-derived hematopoietic cells may be used to induce tolerance to subsequent transplantation of other hESC-derived cells and tissues (such as neurons, pancreatic islets, or cardiomyocytes) while minimizing the need for prolonged heavy immunosuppressant therapy.[14–16] Eventually, with somatic cell nuclear transfer, we may be able to have designer isogenic cell lines for each patient. An example of this has been done to successfully create hematopoietic cells from mESCs suitable for correcting a mouse model of immunodeficiency.[17]

(iii) hESC-derived cells can be used to generate other pure cell populations on a larger scale, such as red blood cells, platelets, granulocytes, or lymphocytes. These could be made potentially safer, without concern about contamination, with infectious agents such as human immunodeficiency virus (HIV), hepatitis, or prions that currently can contaminate our donor blood supply. In this manner, hESCs could also be used to potentially grow clinical grade hematopoietic cells on a large scale to create a virtually endless source of these cells for the purpose of transfusion medicine.

2. Hematopoietic Differentiation from mESCs

Studies of mESCs demonstrate that differentiation into hematopoietic cells closely mimics the ordered gene expression and development of specific cell populations that occur during normal embryonic development.[2,18,19] During murine embryonic development, hematopoiesis develops in distinct waves with the yolk sac, giving rise to primitive hematopoietic cells followed by cells of the aorta gonads and mesonephros (AGM) region, which in turn gives rise to definitive hematopoietic cells. However, the yolk sac can potentially also generate cells of the definitive hematopoietic program, including HSCs, macrophage, definitive erythroid, and mast cells.[8,10,11] This can become important when one is analyzing the development of HSCs and other hematopoietic cells from mESCs and hESCs, as it may not be readily apparent whether the ESC-derived cells are representative of the yolk sac (primitive hematopoiesis) or AGM (definitive hematopoiesis).

Development of specific hematopoietic lineages from mESCs has been well described. These hematopoietic cell lineages include erythrocytes,[18,20] megakaryocytes,[21] granulocytes,[22] mast cells,[18,23] eosinophils,[24] T and B lymphocytes[25-28] macrophages, dendritic cells,[29,30] and natural killer (NK)[31-33] cells. Methods to generate specific lineage type cells in relatively pure populations now include addition of supplemental cytokines, chemicals (such as dexamethasone), and use of various stromal cell lines, such as S17, OP9, and MS-5. Stromal cells are thought to influence differentiation and commitment of mESCs to particular cell types by mechanisms which could include cell–cell interaction, secretion of soluble or cell-bound growth/differentiation or lineage commitment influencing factors, and/or inhibition of factors maintaining mESCs in the undifferentiated state.[25,34-36] Embryoid body (EB) formation can also be used to induce or support differentiation of mESCs without use of stromal cell lines (37).

One example of the utility of mESCs has been to characterize the common development of the hematopoietic and vascular systems. Initially based on observation that vascular and blood islands appear in close proximity within the early developing mouse embryo, it was thought that a common progenitor cell, the "hemangioblast," gave rise to both tissues. This hypothesis was supported by observations like common shared gene expressions [brachyury, vascular endothelial growth factor (VEGF), and Flk-1][38,39] in progenitors of both lineages. Subsequently, carefully timed studies of mESC-derived EBs were able to characterize the phenotype and developmental requirements of hemangioblast cells.[40,41] This work using mESCs was then translated into isolating a similar population of putative hemangioblasts in mouse embryos.[42] More recently, hemangioblasts from hESCs have also been described, demonstrating the ability to extend mESC-based analyses into the human system.[43]

3. Hematopoietic Differentiation from hESCs

An overview of hematopoietic development and analysis from hESCs is given in Fig. 1. In the initial studies of hematopoietic development from hESCs, CD34[+] hematopoietic precursor cells expressing hematopoietic

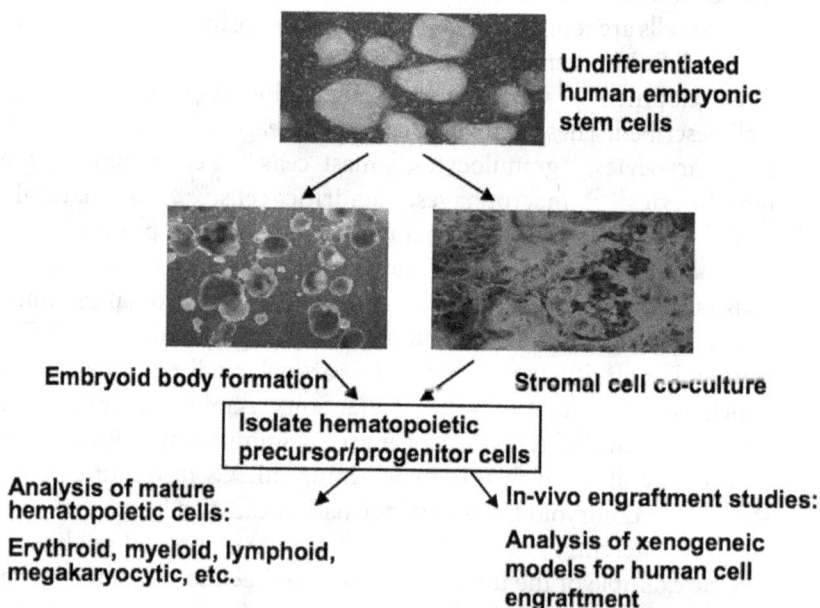

Figure 1. Overview of hematopoietic development from hESCs. Undifferentiated hESCs are typically induced to differentiate via either suspension culture or embryoid body formation, or as adherent cells by coculture on stromal cell lines. Either method of differentiation can be employed, either using FBS alone or with addition of defined cytokines and growth factors. hESCs allowed to differentiate under these conditions can be analyzed and sorted for hematopoietic cell populations based on specific phenotypes. These hESC-derived hematopoietic cell populations can be either induced to further differentiate *in vitro* into defined mature blood cell populations, or analyzed with *in vivo* models for ability to sustain long term, multilineage engraftment.

transcription factors were derived by utilizing the coculture method with the murine bone marrow cell line S17 or the yolk sac endothelial cell line C166 in the presence of fetal bovine serum (FBS), without any added cytokines.[44] Importantly, CD34+ selection led to enrichment of the hematopoietic progenitor cells, as confirmed by the colony-forming cell (CFC) assay giving rise to characteristic myeloid, erythroid, and megakaryocytic colonies. More mature hematopoietic cells derived from these colonies also expressed normal surface antigens corresponding to various cell types appropriately.

Subsequently, the EB method of differentiation was used for hESCs.[45] Here, treatment of hESCs during EB development with a combination of FBS and defined cytokines like stem cell factor (SCF), Flt-3 ligand, interleukin (IL)3, IL-6, granulocyte colony-stimulating factor (G-CSF), and bone morphogenetic protein 4 (BMP-4) strongly promoted differentiation into CD45+ hematopoietic cells. Despite the removal of cytokines at day 10, hematopoietic differentiation of hESCs continued, suggesting that the cytokines act on hematopoietic precursors as opposed to more differentiated hematopoietic cells.

Another study using EBs demonstrated that erythromyelopoiesis from hESCs in serum-free clonogenic assays parallels erythromyelopoiesis from the human embryo/yolk sac by stepwise passage through embryonic and definitive hematopoiesis.[46] Here, erythropoietin (EPO) and granulocyte-macrophage colony-stimulating factor (GM-CSF), in addition to SCF, G-CSF, IL-3, and IL-6, were used. Initially, EB-derived CD45+ cells differentiated into semiadherent mesodermal–hematoendothelial (MHE) colonies that formed organized, yolk sac-like structures, and subsequently led to generation of multipotent primitive hematopoietic cells, erythroblasts, and macrophages. A first wave of hematopoiesis follows MHE colony emergence, giving rise to primitive erythrocytes with brilliant red embryonic/fetal hemoglobin, expressing CD71/CD325a (glycophorin A). A second wave then gives rise to definitive-type erythroid, granulocyte, macrophage CFCs, and multilineage CFCs.

Other stromal-cell-based studies were able to obtain up to 20% of CD34+ cells and isolate up to 10^7 CD34+ cells with more than 95% purity from a similar number of initially plated hESCs after 8–9 days of coculture with the OP9 cell line.[47] Again, as found in the previous reports,[44,45] these hESC-derived CD34+ cells were expressing typical hematopoiesis-associated genes, highly enriched in CFCs, and retained clonogenic potential after *in vitro* expansion. Phenotypically these CD34+ hematopoietic progenitor cells derived here were consistent with primitive hematopoietic progenitors as defined by coexpression of CD90, CD117, and CD164, along with a lack of CD38 expression. Moreover, this population contained aldehyde dehydrogenase-positive cells, as well as cells with verapamil-sensitive ability to efflux rhodamine 123.[48–50] Additionally, when cocultured on MS-5 stromal cells with appropriate cytokines, these hESC/OP9 coculture-derived CD34+ cells gave rise to phenotypic lymphoid (B and NK cells) as well as myeloid (macrophages and granulocytes) lineages. However, these lymphoid cells were not tested for functional capacity, as was done in some of the lineage-specific studies described below.

Additional studies by the same group using the OP9 system demonstrated that early progenitors committed to hematopoietic development could be identified by surface expression of leukosialin (CD43).[51] The appearance of CD43 was found to precede that of CD45 on all types of emerging clonogenic progenitors, and CD43 can reliably separate the hematopoietic CD34+ population from CD34+CD43- endothelial and mesenchymal cells. Furthermore, these studies demonstrated that multipotent lymphohematopoietic progenitors followed precommitted erythro-megakaryocytic progenitors in appearance, and their gene/transcription factor profile was representative of initial stages of definitive hematopoiesis originating from endothelium-like precursors. Once these cells acquired CD34, CD43, and CD45 expression, they were largely devoid of VE-cadherin and KDR (also termed Flk-1 or VEGF receptor-2) expression (endothelial markers) and had a distinct gene expression profile consistent with commitment to lymphohematopoietic fate.

Two studies have evaluated the effect of overexpression of the homeobox B4 gene (HOXB4) on hematopoiesis from hESCs. Several previous studies using mESCs or mouse and human HSCs have found HOXB4 to improve expansion and *in vivo* engraftment of HSCs.[52–54] One study found that overexpression of HOXB4 considerably augmented hematopoietic development of hESCs.[55] Here, differentiation of hESCs as EBs in a serum-containing medium without the use of additional cytokines led to sequential expansion of first erythroid and then myeloid progenitor cells. However, interpretation of these results is complicated by the minimal hematopoietic development from hESC-derived control cell populations that did not overexpress HOXB4, suggesting that differentiation of these cells was not optimized. Additionally, the second study to overexpress HOXB4 in hESC-derived hematopoietic cells did not find increased hematopoietic engraftment,[56] as is the case for mESC-derived blood cells.

However, stable HOXB4 expression was not demonstrated in the transplanted cells, again complicating this analysis.

Finally, as Keller and colleagues have done for mESCs,[40,41] they have recently reported isolation of two separate populations of cells from hESCs that meets the criteria for definition of hemangioblasts, i.e. blast colonies having both hematopoietic and vascular potential. In their experiment, 72–96 hours of BMP-4–stimulated EB differentiation gave rise to a KDR[+] population of cells with both hematopoietic and vascular potential before appearance of the primitive erythroid cells.[43] Two distinct types of hemangioblasts were identified: those that give rise to primitive erythroid cells, macrophages, and endothelial cells, and those that generate only the primitive erythroid population and endothelial cells.

4. *In Vitro* Production of Specific Hematopoietic Lineages from hESCs

In addition to the studies described above that evaluated hematopoietic development from hESCs based on derivation of hematopoietic precursor and progenitor cell populations, several other studies have now characterized development of specific mature blood cell lineages from hESCs.

4.1. Erythrocytes

VEGF is well known to support hematopoietic and endothelial cell growth and development in many systems. Indeed, it was found to selectively promote erythropoietic development from hESCs.[57] Here, when hESC-derived EBs were treated with VEGF, authors found a higher frequency of cells coexpressing CD34 and KDR (hematoendothelial phenotype). These cells also expressed erythroid markers and increased expression of embryonic zeta (ζ) and epsilon (ε) globins, though no change in fetal/adult hemoglobin was noticed. These effects of VEGF were dependent on the presence of other hematopoietic cytokines (Flt-3 ligand, IL-3, IL-6, G-CSF, and BMP-4), and were augmented by addition of EPO. It was also noticed that the *in vitro* self-renewal potential of primitive hematopoietic cells with erythroid progenitor capacity was enhanced by the presence of VEGF.

More specific analysis of globin gene expression in erythroid cells derived from hESCs used hESCs cocultured on FH-B-hTERT (an immortalized fetal liver cell line) or S17 cells.[58] Analysis of mRNA expression from the β-globin locus revealed that hESC-derived erythroid cells produced ε- and γ-globin mRNAs but no or very little amounts of β-globin expression. Over time in culture, the mean ratio of γ/ε increased by more than 10-fold, recapitulating the ε-globin to γ-globin switch but not the γ-globin to β-globin switch that occurs around birth. Subsequent studies by this group have also described an experimental protocol to produce large numbers of primitive erythroid cells from hESCs.[59] Here, a more than 5000-fold increase in primitive erythroid cells was obtained by differentiating hESC-derived CD34+ cells into liquid cultures. Again, these erythroid cells were morphologically and functionally similar to primitive erythroid cells present in the yolk sac of early human embryos, i.e. they did not enucleate, were fully hemoglobinized, and expressed a mixture of embryonic and fetal globins but no β-globin.

Another study also evaluated the generation of erythroid cells from hESCs, with similar results to those described above. Here, morphologically definitive erythroid cells, coexpressing high levels of embryonic (ε) and fetal (γ) globins at mRNA and protein levels with little or no adult globin (β), were generated from human EB-derived CD45+ hematopoietic development.[60] This globin expression pattern was not altered by factors like culture duration, FBS, VEGF, Flt3-Ligand, or coculture with OP-9 during erythroid differentiation. This coexpression of both embryonic and fetal globins by definitive-type erythroid cells did not faithfully recapitulate either yolk sac embryonic globins or their fetal liver counterparts. Therefore, it remains unclear if these results are spurious due to *in vitro* culture conditions, or if there is some abnormal promiscuity of globin gene expression that may be normal, but not typically observed during early human development. Regardless, this ability to achieve erythroid cells coexpressing embryonic and fetal globins generated from hESCs at high frequency can help us to understand and explore molecular mechanisms of hematopoiesis.

4.2. Megakaryocytes

The OP9 stromal cell coculture system was also used to generate megakaryocytes from hESCs.[61] These platelets were functionally similar to adult human platelets with submaximal adenosine diphosphate responses augmented by epinephrine. Moreover, hESC-derived megakaryocytes undergo lamellipodium formation, actin filament assembly, and vinculin localization at focal adhesions when plated on a fibrinogen-coated surface, characteristic of $\alpha_{IIb}\beta_3$ outside-in signaling typically seen with these cells.

Together, these studies of hESC-derived erythroid and megakaryocytic cells highlight the potential of hESCs to serve as the starting point for cells suitable for transfusion medicine therapies. While red blood cells and platelets can be readily obtained from the Red Cross and similar agencies, use of hESC-derived cells for transfusion medicine would be known to start from a homogeneous, well-characterized cell population that could be screened and guaranteed to be free of infectious blood-borne pathogens such as the HIV, hepatitis, prions, and other infectious agents. Moreover, hESC-derived cells for transfusions could be useful for patients with rare blood types or incompatibility with erythroid cell antigens.

4.3. Lymphocytes and other immune cells

Initial studies to demonstrate development of functional lymphocytes from hESCs used a two-step culture method by first sorting CD34$^+$ cells derived from S17 stromal cell coculture, then using the AFT024 stromal cell line with defined cytokines known to support NK cell development.[62] This work demonstrated that these CD56$^+$CD45$^+$ hESC-derived lymphocytes expressed typical markers and functioned like mature NK cells, i.e. expressed killer cell Ig-like receptors, natural cytotoxicity receptors, and CD16, and were able to lyse human tumor cells by both direct cell-mediated cytotoxicity and antibody-dependent cellular cytotoxicity.

Next, T cells were derived from hESCs by using OP9-mediated differentiation into CD34$^+$ and CD133$^+$ cells that were transplanted into human thymic tissues that had been engrafted into immunodeficient mice.[63] These hESC-derived T cells expressed surface antigens such as CD4, CD8, CD1a, and CD7 typical of T cells. Some functional capacity was demonstrated by expression of activation markers upon costimulation of T lineage cells by CD3 and CD28. Additionally, a lentiviral vector expressing enhanced green fluorescent protein (GFP) under the control of the elongation factor 1α promoter, introduced at the hESC stage, continued to express the reporter gene at high frequency throughout thymopoiesis. These results suggest that genetically manipulated hESCs may hold promise for treatment or modeling disorders of the T cell lineage, including HIV infection.

Other cells of the immune system, such as macrophages and dendritic cells, have also been derived from hESCs. CD34$^+$ cells derived by

S17-stromal cell-mediated culture of hESCs were sorted and cultured with defined cytokines [FBS, GM-CSF, and macrophage colony-stimulating factor A(M-CSF)] to derive macrophages.[64] Phenotypic and functional analyses carried out on these hESC-derived macrophages confirmed the typical phenotype of macrophages. They also displayed normal functionality by efficient phagocytosis, by up-regulation of the costimulatory molecule B7.1, and by cytokine secretion in response to lipopolysaccharide stimulation. As in the other studies described above, lentiviral transduced hESCs expressing the transgene GFP also differentiated into functionally and phenotypically normal macrophages, indicating no adverse effects due to lentivirus infection.

Dendritic cells (DCs) are another hematopoietic cell lineage that serves as principal antigen-presenting cells (APCs) in triggering and supporting immune responses. One study used EB-mediated differentiation to derive both functional DCs and macrophages. These APCs expressed high levels of HLA class II molecules and were able to stimulate mixed leukocyte reactions (MLRs) as an *in vitro* measure of immune activity.[65] A second study used the OP9 coculture system to promote differentiation of hESCs into myeloid cells.[66] First, myeloperoxidase-expressing $CD4^+CD11b^+CD11c^+$ $CD16^+CD123^{low}HLA-DR^-$ myeloid cells were generated and expanded by sequentially coculturing them first with OP9, followed by feeder free conditions and GM-CSF stimulation. When transferred to a serum-free medium with GM-CSF and IL-4, these myeloid cells led to generation of cells expressing high levels of major histocompatibility complex class I and II molecules, CD1a, CD11c, CD80, CD86, and CD40, had typical dendritic morphology, and were functional as they were capable of antigen processing and triggering T cell responses.

These studies to derive functional DCs from hESCs under appropriate conditions will facilitate future studies to better understand DC biology and potential novel DC vaccines or DC-mediated induction of immune tolerance.

4.4. Engraftment of hESC-derived hematopoietic cells

HSCs are defined by their ability to successfully support long term, multilineage development of all hematopoietic cell lineages *in vivo*. Since it is not feasible to test transplantation and engraftment of still poorly characterized hematopoietic cell populations such as those derived from hESCs in patients, immune-deficient mice are typically used for these analyses.

So-called SCID-repopulating cells (SRCs) are considered to be a close surrogate for HSCs, and this assay has been commonly used to test other populations of putative HSCs, such as those derived or isolated from human umbilical cord blood.

To date, three published studies have evaluated the engraftment potential of hESC-derived hematopoietic cells.[56,67,68] All these studies were done using the H1 and H9 cell lines. Two studies utilized NOD/SCID mice, whereas one report utilized the Dorset Marino Sheep fetal sheep model (< 65 days old) for engraftment. Using an intra-bone-marrow transplantation (IBMT) technique where the hematopoietic cells differentiated cells are injected directly into the bone marrow (usually femur) of a mouse, one study was able to show successful engraftment of hESC-derived cells in 11 out of 19 mice.[56] The engrafted cells included lymphoid (CD45$^+$CD19$^+$), myeloid (CD45$^+$CD133$^+$), and erythroid (glycophorinA$^+$) cells. However, the levels of human reconstitution and the frequency of detection were limited compared with UCB-derived cells. Interestingly, this same study was unable to demonstrate successful engraftment after IV (tail vein) injection of the mice with hESC-derived hematopoietic cells. Indeed, there was actually a decrease in survival of the mice after IV injection due to aggregation of the cells postinjection, resulting in pulmonary emboli. No secondary transplantation studies were done here.

The other NOD/SCID transplantation study,[68] using both IBMT and the IV injection of hESC-derived hematopoietic cells, did find successful engraftment in the mice without any decreased survival or pulmonary emboli after IV injection. This difference probably reflects variations in the cell populations derived by alternative methods (by coculture on S17 stromal cells versus by EB formation). In this study, bone marrow analyzed three or more months after IV injection showed on average 0.69% human CD45$^+$ cells (compared to 2.98% of human CD45$^+$ cells seen in mice injected with cells derived from UCB). In mice where IBMT was used, the level of engraftment was seen to be 1.88% in the femur directly injected with the cells and 1.79% in the contra lateral femur. Analysis of the engrafted cells found most to be CD45$^+$CD33$^+$ myeloid cells; however, some CD34$^+$ cells were also seen, suggesting some HSC survival. Secondary transplantation studies were done to demonstrate successful long term engraftment in the secondary recipients, though at a level only detectable by very sensitive PCR analysis.

Although NOD/SCID mice are reported to have little NK cell activity, several analyses of peripheral blood mononuclear cells demonstrated that these mice retain some NK cell activity.[69,70] In this regard, the second study described above also evaluated mice treated with anti-asialo GM1 (ASGM1) antiserum (which depletes NK cells) the day before injection of hESC-derived cells, and subsequently every 11 days posttransplantation. This treatment led to an enhanced level of engraftment, to 1.74%. This increase in the engraftment is likely related to the fact that hESC-derived progenitors have a lower HLA class I molecule expression as compared to UCB-derived HSCs,[68] which would predispose them to NK-cell-mediated lysis.

In the fetal sheep transplantation study, hESC-derived hematopoietic cells were injected *in utero* into the peritoneal cavity of sheep at less than 65 days' gestation. Five to seventeen months after birth, approximately 0.1% human CD34$^+$ or CD45$^+$ cells were seen in the BM and/or peripheral blood (PB). Also, both myeloid (CD15$^+$CD36$^+$) and lymphoid (CD2$^+$) lineage cells were identified. Overall, this level of engraftment is deceased compared to use of UCB-derived cells, where 2–3% chimerism is typical.[71] However, the low level engraftment was confirmed in the hESC studies by PCR for human DNA; BM samples were chimeric in 6 out of 8 animals with 0.001% to 0.09% of cells, analyzed at 33–39 months after transplantation. Furthermore, human hematopoiesis in secondary transplanted sheep was followed for up to 22 months, proving actual engraftment rather than just transfer of the hematopoietic progenitors. Overall, successful secondary engraftment helps to confirm that the recipient animals provide an appropriate hematopoietic niche for the development of the engrafted stem cells, and that the transplanted population contains true HSCs capable of long term engraftment.

In the three published studies, one[68] used unsorted cells, though the other[67] used sorted CD34$^+$lin$^-$ and CD34$^+$CD38$^-$ cells. It is not clear whether the cells used in the other study[56] were sorted or unsorted. The hESC-derived CD34$^+$ cells consist of a heterogeneous cell population that includes endothelial cells in addition to hematopoietic cells. This mixed population with fewer SRCs may account for the difficulty of demonstrating high levels of long term multilineage engraftment. Further studies are needed to directly compare engraftment and survival of mice that receive unsorted versus sorted populations of different hESC-derived blood cells in order to evaluate the strengths and weaknesses of various phenotypic populations.

The age of the recipient is likely to be an important factor for successful engraftment of hematopoietic cells. It has been shown in mouse models that engraftment of HSCs derived from the embryonic (yolk sac) stage fails in the sublethally irradiated mice but is successful if injected into the newborn pups.[72] It has been hypothesized that the fetal liver (which is still active in hematopoiesis in the newborn pups) may provide an age-appropriate environment for the maturation of developmentally naïve HSCs derived from the embryonic stem cells. Interestingly, these transplanted cells in the newborn pups are able to reconstitute the sublethally immune-deficient adult mice upon secondary transplantation.

As seen in the above studies, compared to UCB-derived HSCs, the levels of engraftment in using HSCs derived from hESCs are relatively poor. Studies have shown that HSCs derived from hESCs have a distinct molecular signature that more closely resembles the HSCs derived from an early developmental stage at which the yolk sac and the fetal liver are the primary site of the hematopoiesis.[73] It has been noted that major gene families are expressed differentially in hESCs- and UB/PB-derived HSCs.[56] They include genes associated with cell adhesion, cell migration and cell transcription, and transcriptional regulation. This differential expression of various gene families may explain unique *in vivo* properties of the hESC-derived HSCs. For example, higher expression of adhesion proteins like CKLF-1 and β_3-integrin may reduce the ability of hESC-derived hematopoietic cells to migrate beyond the injection site and enter the circulation.

In contrast, compared to hESC-derived cells, UCB- and BM-isolated cells express higher levels of CXCR4, CD44, and l-selectin, proteins known to mediate the homing and engraftment behavior of HSCs.[56] Similarly, proteins like ADAM8[74] and ADAM17[75] that are involved in establishing HSC residence within the BM niche also have higher expression in UCB- and BM-isolated hematopoietic cells than in hESC-derived hematopoietic cells.

In addition, hESC-derived HSCs differentially express genes involved with accelerated cell cycle progression and loss of stem cell self-renewal ability, whereas somatic cells express genes required for maintenance and control of the quiescent cell cycle status, essential for the function of the transplantable HSCs.[56] As there is no single gene family which is expressed differently by hESC-derived HSCs, it is unlikely that ectopic expression of single genes like HoxB4 will make the behavior of hESC-derived hematopoietic cells similar to that of HSCs derived from somatic sources. Ectopic expression of the HoxB4 gene in mESC-derived HSCs does lead to markedly improved hematopoietic engraftment potential.[16] However, ectopic expression of the same gene was unable to improve engraftment of hESC-derived HSCs,[56] though it was difficult to determine if there was stable expression of HoxB4 in the hESC-derived blood cells. These results suggest intrinsic differences in the basic biology of mESC-derived HSCs from the hESC-derived HSCs cells. Also, the limitations in the ability of hESC-derived HSCs to activate a genetic program similar to the spectrum of genes expressed by somatic HSCs may account for their limited proliferation and migratory capacity *in vivo*.[56]

To produce HSCs with better long term multilineage engraftment potential from hESCs, it is likely necessary to develop culture techniques that more closely resemble the *in vivo* microenvironment required to stimulate a genetic program needed for not only the hematopoietic specification of the hESCs, but also the transition from primitive to definitive hematopoiesis. In this regard, we need a better understanding of the pathways involved in this complex process. Several signaling pathways, like Wnt, Notch, and Hedgehog, are likely to play a prominent role in this genetic program.[56]

One important but often-overlooked outcome of these three studies concerns the safety of hESC-based therapies due to the ability of the undifferentiated hESCs to form teratomas upon injection into animals. However, no teratoma formation has been seen in any of these engraftment studies done so far using hematopoietic progenitors derived from hESCs. This is despite using immune-compromised mice that were also sublethally irradiated, some of which received additional anti-NK-cell treatment. While a more rigorous study would be needed to definitively prove that transplantation of hESC-derived hematopoietic cells poses no risk of teratoma development, these studies seem to suggest the safety of this cellular therapy.

5. Hematopoiesis and Strategies to Overcome the Immune Barriers

Any future therapies derived from hESC-derived cells would typically result in the use of allogeneic cells and be subject to immune-mediated rejection. Potential means of preventing rejection of these cells have previously been outlined.[15,16,76] These methods include the use of immune-suppressant drugs such as cyclosporin A, modification of the hESCs to make them more tolerant, and creation of "banks" of hESC lines that would provide appropriate histocompatibility matching.[16] Potentially, the most efficacious strategy would be to use somatic cell nuclear transfer (SCNT) to derive HLA-matched patient-specific hESC lines. These patient-specific hESCs could then be induced to differentiate into the lineage of choice, as needed for a particular therapy. Indeed, this type of strategy was used to demonstrate effective correction of genetic immunodeficiency in a mouse model.[17]

However, in addition to the unfortunate ethical/political debate and concerns, to date it has not been possible to derive hESCs via this technique, though this has been effective in other species.[77-79] Many questions remain unsolved regarding the biology of the ES cells derived from SCNT. They include epigenetic influences, genomic imprinting and genomic stability of these cell lines. Furthermore, there has been an increased incidence of abnormalities in the animals derived from the technique. Therefore, it is crucial to fully understand the basic biology behind SCNT before this technique can be put to practical use.

Hematopoietic chimerism is also a very intriguing means of overcoming immune rejection of hESC-derived cells. This method is especially relevant to discussion of hESC-derived blood cells, as development of this lineage will be necessary for eventual chimerism-induced tolerance. This principle comes out of both animal experiments and human clinical experiences. Several animal models clearly demonstrate that successful immune tolerance can be achieved to the donor graft tissue by inducing mixed hematopoietic chimerism.[80,81] This mixed chimerism results from engraftment of pluripotent HSCs, which leads to stable long term coexistence of multilineage hematopoietic cells of the donor within the host. Chimeric recipients have specific immunological tolerance to the alloantigens that are expressed by hematopoietic cells of the same donor, and do not reject a tissue or organ allograft of the donor haplotype.[81] Several cases have been reported in which successful immune tolerance to solid organ transplantation was induced by prior bone marrow transplantation.[82,83]

Similarly, it may be possible to induce mixed chimerism using hESCs differentiated into hematopoietic cells.[76] In this approach a recipient requiring a particular organ would first be transplanted with blood-group-ABO-matched (to prevent hyperacute rejection) but not necessarily completely HLA-matched HSCs to induce chimerism. After inducing stable mixed chimerism, the patient would be transplanted with a cell population or tissue (i.e. neurons, pancreatic islets, or cardiomyocytes) that the patient requires, derived from the same hESC line used to induce the hematopoietic chimerism. One study using rat-ES-like cells suggested that such a strategy might prove more effective.[84] In this study stable chimerism was induced, and the host readily accepted a heart transplant from the same donor rat strain without any evidence of rejection.

6. Future Directions for Studies of Hematopoiesis from Human ES Cells

NOD/SCID mice have been the primary strain used to evaluate human SRCs isolated from sources such as bone marrow or UCB. However, this strain may not be optimal for engraftment of hESC-derived hematopoietic cells. NOD/SCID mice do have some innate immunity, and effectors such as mouse NK cells may recognize and kill hESC-derived blood cells.[66] Indeed, we have found that hESC-derived blood cells express lower levels of HLA class I molecules than hematopoietic cells obtained from more mature sources.[85] NK cells are especially attuned to kill cells with low class I expression. Indeed, studies of mESC-derived blood cells also have low class I expression and are better able to engraft in mice that lack NK cell activity.[86]

Other strains of mice that are "more immunodeficient" have recently been utilized to demonstrate more effective development of a functional immune system when engrafted with human hematopoietic cells. These more effective mouse strains include Rag2$^{-/-}$/γc$^{-/-}$ mice that have defective expression of both the recombinase activating gene (Rag) which results in a lack of adaptive (T and B cell) immunity and the common g chain of the IL2, IL4, IL7, and other cytokine receptors leading to a lack of NK cell activity. Also, the the NOD/SCID/γc$^{-/-}$ mouse which combines the NOD/SCID defects with the gc deletion results in more complete immuno-suppression and increased multilineage engraftment of human hematopoietic cells. Other novel mouse strains suitable for engraftment of human cells have also been described.[87] It will be of interest to now utilize these strains for studies of human-ES-cell-derived hematopoietic precursor cells. Use of neonatal mice for these transplantation studies may also be advantageous.

Additionally, nonhuman primates can be used in developing and testing several novel therapies prior to being tested in humans. Nonhuman primate ES cells differ from mESCs and resemble hESCs in several aspects, including colony morphology, growth requirements, and developmental molecular signature.[3,88,89] Therefore, they would serve as an excellent preclinical model for hESC engraftment studies. Indeed, one study has used genetically labeled (GFP+) cynomolgus ES cells and transplanted them *in utero* into the cynomolgus fetus in the abdominal cavity or into the liver.[90] While there was successful engraftment of these cells, it was surprising to also see teratoma formation in some of the animals.[90] Another group has induced mesodermal differentiation of cynomolgus embryonic stem cells and then transplanted them into the fetal sheep liver.[91] Here, cynomolgus hematopoietic progenitor cells were detected in bone marrow at a level of 1–2%, with no teratoma formation.

Genetic modification of hESCs is another technology essential for better characterization of how specific genes regulate development of specific human cell lineages. While a variety of methodologies have been utilized to obtain stable transgene expression in hESCs, most of them suffer from low efficiency and gene silencing.[92] Use of lentivirus-based vectors has allowed the most stable expression of foreign genes in hESCs, though these vectors are relatively complicated and time-consuming to produce. Recently, we have demonstrated the use of transposon-based genetic systems that allow efficient and stable transgene expression in hESCs.[93] We anticipate that transposons may develop into an optimal method for genetic modification of hESCs to allow future gain-and-loss-of-function analyses.

Summary

This chapter has provided a comprehensive review of studies to derive hematopoietic cells from hESCs, beginning soon after hESCs were first isolated and described in 1998. In these few years, essentially all blood cell types of the adult body and bone marrow have demonstrated feasibility in HSCs being produced from hESCs using methods that are defined to varying degrees. While it is difficult to know for sure, it is likely that more work has been published on hematopoietic development from hESCs than any other cell lineage. Indeed, as described, the strong 50-plus-year history of research on blood cell development has greatly facilitated progress in this new developmental model.

Characterization of putative HSCs capable of long term, multilineage engraftment that can be derived from hESCs remains less than satisfactory. While hematopoietic engraftment in the immunodeficient mouse and fetal sheep models shows some intriguing potential,[56,67,68] more work is clearly needed. The level of engraftment in these studies is reminiscent of work using mESCs, though there syngeneic mice could be utilized to reduce immune barriers. mESC-derived hematopoietic engraftment is facilitated by overexpression of genes such as HOXB4. However, similar studies using hESCs have not been as successful or clear-cut.[55,56] This highlights the possibility that there are distinct differences in HSC development between the two models, as well as technical barriers with hESCs, that remain to be resolved.

Finally, ESC and adult stem cell research are often described as two separate and distinct systems. This is clearly not the case, since studies of isolated adult stem cells such as HSCs definitely aid in analysis of similar populations that may be derived from hESCs. Additionally, it is likely that the hESC developmental model will have positive feedback on adult stem cell models. Future progress in both the basic science and clinical applications of hematopoietic cell development will require optimal use of all available resources.

References

1. Thomson JA, Itskovitz-Eldor J, Shapiro SS, *et al.* (1998) Embryonic stem cell lines derived from human blastocysts. *Science* **282:** 1145–1147.
2. Keller G. (2005) Embryonic stem cell differentiation: emergence of a new era in biology and medicine. *Genes Dev* **19:** 1129–1155.
3. Pera MF, Trounson AO. (2004) Human embryonic stem cells: prospects for development. *Development* **131:** 5515–5525.
4. Smith AG. (2001) Embryo-derived stem cells: of mice and men. *Annu Rev Cell Dev Biol* **17:** 435–462.
5. de Jong JL, Zon LI. (2005) Use of the zebrafish system to study primitive and definitive hematopoiesis. *Annu Rev Genet* **39:** 481–501.
6. Walmsley M, Ciau-Uitz A, Patient R. (2005) Tracking and programming early hematopoietic cells in *Xenopus* embryos. *Methods Mol Med* **105:** 123–136.
7. Turpen JB. (1998) Induction and early development of the hematopoietic and immune systems in *Xenopus. Dev Comp Immunol* **22:** 265–278.
8. McGrath KE, Palis J. (2005) Hematopoiesis in the yolk sac: more than meets the eye. *Exp Hematol* **33:** 1021–1028.
9. Baron MH. (2005) Early patterning of the mouse embryo: implications for hematopoietic commitment and differentiation. *Exp Hematol* **33:** 1015–1020.
10. Godin I, Cumano A. (2002) The hare and the tortoise: an embryonic haematopoietic race. *Nat Rev Immunol* **2:** 593–604.

11. Dzierzak E. (2002) Hematopoietic stem cells and their precursors: developmental diversity and lineage relationships. *Immunol Rev* **187**: 126–138.

12. Thomas ED, Blume KG. (1999) Historical markers in the development of allogeneic hematopoietic cell transplantation. *Biol Blood Marrow Transplant* **5**: 341–346.

13. Thomas ED. (1999) Bone marrow transplantation: a review. *Semin Hematol* **36**: 95–103.

14. Rifle G, Mousson C. (2003) Donor-derived hematopoietic cells in organ transplantation: a major step toward allograft tolerance? *Transplantation* **75**: 3S–7S.

15. Kaufman DS, Thomson JA. (2002) Human ES cells — haematopoiesis and transplantation strategies. *J Anat* **200**: 243–248.

16. Bradley JA, Bolton EM, Pedersen RA. (2002) Stem cell medicine encounters the immune system. *Nat Rev Immunol* **2**: 859–871.

17. Rideout WM, Hochedlinger K, Kyba M, *et al.* (2002) Correction of a genetic defect by nuclear transplantation and combined cell and gene therapy. *Cell* **109**: 17–27.

18. Keller G, Kennedy M, Papayannopoulou T, Wiles MV. (1993) Hematopoietic commitment during embryonic stem cell differentiation in culture. *Mol Cell Biol* **13**: 473–486.

19. Kyba M, Daley GQ. (2003) Hematopoiesis from embryonic stem cells: lessons from and for ontogeny. *Exp Hematol* **31**: 994–1006.

20. Carotta S, Pilat S, Mairhofer A, *et al.* (2004) Directed differentiation and mass cultivation of pure erythroid progenitors from mouse embryonic stem cells. *Blood* **104**: 1873–1880.

21. Fujimoto TT, Kohata S, Suzuki H, *et al.* (2003) Production of functional platelets by differentiated embryonic stem (ES) cells *in vitro. Blood* **102**: 4044–4051.

22. Lieber JG, Webb S, Suratt BT, *et al.* (2004) The *in vitro* production and characterization of neutrophils from embryonic stem cells. *Blood* **103**: 852–859.

23. Tsai M, Wedemeyer J, Ganiatsas S, *et al.* (2000) *In vivo* immunological function of mast cells derived from embryonic stem cells: an approach for the rapid analysis of even embryonic lethal mutations in adult mice *in vivo. Proc Natl Acad Sci USA* **97**: 9186–9190.

24. Hamaguchi-Tsuru E, Nobumoto A, Hirose N, *et al.* (2004) Development and functional analysis of eosinophils from murine embryonic stem cells. *Br J Haematol* **124**: 819–827.

25. Nakano T, Kodama H, Honjo T. (1994) Generation of lymphohematopoietic cells from embryonic stem cells in culture. *Science* **265**: 1098–1101.

26. Cho SK, Webber TD, Carlyle JR, *et al.* (1999) Functional characterization of B lymphocytes generated *in vitro* from embryonic stem cells. *Proc Natl Acad Sci USA* **96**: 9797–9802.

27. de Pooter RF, Cho SK, Carlyle JR, Zuniga-Pflucker JC. (2003) *In vitro* generation of T lymphocytes from embryonic stem cell-derived prehematopoietic progenitors. *Blood* 102: 1649–1653.
28. Schmitt TM, de Pooter RF, Gronski MA, *et al.* (2004) Induction of T cell development and establishment of T cell competence from embryonic stem cells differentiated *in vitro. Nat Immunol* 5: 410–417.
29. Senju S, Hirata S, Matsuyoshi H, *et al.* (2003) Generation and genetic modification of dendritic cells derived from mouse embryonic stem cells. *Blood* 101: 3501–3508.
30. Fairchild PJ, Nolan KF, Cartland S, *et al.* (2003) Stable lines of genetically modified dendritic cells from mouse embryonic stem cells. *Transplantation* 76: 606–608.
31. Takei F, McQueen KL, Maeda M, *et al.* (2001) Ly49 and CD94/NKG2: developmentally regulated expression and evolution. *Immunol Rev* 181: 90–103.
32. Lian RH, Kumar V. (2002) Murine natural killer cell progenitors and their requirements for development. *Semin Immunol* 14: 453–460.
33. Lian RH, Maeda M, Lohwasser S, *et al.* (2002) Orderly and nonstochastic acquisition of CD94/NKG2 receptors by developing NK cells derived from embryonic stem cells *in vitro. J Immunol* 168: 4980–4987.
34. Krassowska A, Gordon-Keylock S, Samuel K, *et al.* (2006) Promotion of haematopoietic activity in embryonic stem cells by the aorta-gonad-mesonephros microenvironment. *Exp Cell Res* 312: 3595–3603.
35. Zhang WJ, Park C, Arentson E, Choi K. (2005) Modulation of hematopoietic and endothelial cell differentiation from mouse embryonic stem cells by different culture conditions. *Blood* 105: 111–114.
36. Berthier R, Prandini MH, Schweitzer A, *et al.* (1997) The MS-5 murine stromal cell line and hematopoietic growth factors synergize to support the megakaryocytic differentiation of embryonic stem cells. *Exp Hematol* 25: 481–490.
37. Keller GM. (1995) *In vitro* differentiation of embryonic stem cells. *Curr Opin Cell Biol* 7: 862–869.
38. Watt SM, Gschmeissner SE, Bates PA. (1995) PECAM-1: its expression and function as a cell adhesion molecule on hemopoietic and endothelial cells. *Leuk Lymphoma* 17: 229–244.
39. Shalaby F, Rossant J, Yamaguchi TP, *et al.* (1995) Failure of blood-island formation and vasculogenesis in Flk-1-deficient mice. *Nature* 376: 62–66.
40. Kennedy M, Firpo M, Choi K, *et al.* (1997) A common precursor for primitive erythropoiesis and definitive haematopoiesis. *Nature* 386: 488–493.
41. Choi K, Kennedy M, Kazarov A, *et al.* (1998) A common precursor for hematopoietic and endothelial cells. *Development* 125: 725–732.

42. Huber TL, Kouskoff V, Fehling HJ, *et al.* (2004) Haemangioblast commitment is initiated in the primitive streak of the mouse embryo. *Nature* **432:** 625–630.

43. Kennedy M, D'Souza SL, Lynch-Kattman M, *et al.* (2007) Development of the hemangioblast defines the onset of hematopoiesis in human ES cell differentiation cultures. *Blood* **109**(7): 2679–2687

44. Kaufman DS, Hanson ET, Lewis RL, *et al.* (2001) Hematopoietic colony-forming cells derived from human embryonic stem cells. *Proc Natl Acad Sci USA* **98:** 10716–10721.

45. Chadwick K, Wang L, Li L, *et al.* (2003) Cytokines and BMP-4 promote hematopoietic differentiation of human embryonic stem cells. *Blood* **102:** 906–915.

46. Zambidis ET, Peault B, Park TS, *et al.* (2005) Hematopoietic differentiation of human embryonic stem cells progresses through sequential hematoendothelial, primitive, and definitive stages resembling human yolk sac development. *Blood* **106:** 860–870.

47. Vodyanik MA, Bork JA, Thomson JA, Slukvin, II. (2005) Human embryonic stem cell-derived CD34⁺ cells: efficient production in the coculture with OP9 stromal cells and analysis of lymphohematopoietic potential. *Blood* **105:** 617–626.

48. Baum CM, Weissman IL, Tsukamoto AS, *et al.* (1992) Isolation of a candidate human hematopoietic stem-cell population. *Proc Natl Acad Sci USA* **89:** 2804–2808.

49. Goodell MA, Rosenzweig M, Kim H, *et al.* (1997) Dye efflux studies suggest that hematopoietic stem cells expressing low or undetectable levels of CD34 antigen exist in multiple species. *Nat Med* **3:** 1337–1345.

50. Uchida N, Combs J, Chen S, *et al.* (1996) Primitive human hematopoietic cells displaying differential efflux of the rhodamine 123 dye have distinct biological activities. *Blood* **88:** 1297–1305.

51. Vodyanik MA, Thomson JA, Slukvin, II. (2006) Leukosialin (CD43) defines hematopoietic progenitors in human embryonic stem cell differentiation cultures. *Blood* **108:** 2095–2105.

52. Kyba M, Perlingeiro RC, Daley GQ. (2002) HoxB4 confers definitive lymphoid–myeloid engraftment potential on embryonic stem cell and yolk sac hematopoietic progenitors. *Cell* **109:** 29–37.

53. Sauvageau G, Thorsteinsdottir U, Eaves CJ, *et al.* (1995) Overexpression of HOXB4 in hematopoietic cells causes the selective expansion of more primitive populations *in vitro* and *in vivo*. *Genes Dev* **9:** 1753–1765.

54. Antonchuk J, Sauvageau G, Humphries RK. (2002) HOXB4-induced expansion of adult hematopoietic stem cells *ex vivo*. *Cell* **109:** 39–45.

55. Bowles KM, Vallier L, Smith JR, *et al.* (2006) HOXB4 overexpression promotes hematopoietic development by human embryonic stem cells. *Stem Cells* **24**: 1359–1369.

56. Wang L, Menendez P, Shojaei F, *et al.* (2005) Generation of hematopoietic repopulating cells from human embryonic stem cells independent of ectopic HOXB4 expression. *J Exp Med* **201**: 1603–1614.

57. Cerdan C, Rouleau A, Bhatia M. (2004) VEGF-A165 augments erythropoietic development from human embryonic stem cells. *Blood* **103**: 2504–2512.

58. Qiu C, Hanson E, Olivier E, *et al.* (2005) Differentiation of human embryonic stem cells into hematopoietic cells by coculture with human fetal liver cells recapitulates the globin switch that occurs early in development. *Exp Hematol* **33**: 1450–1458.

59. Olivier EN, Qiu C, Velho M, *et al.* (2006) Large-scale production of embryonic red blood cells from human embryonic stem cells. *Exp Hematol* **34**: 1635–1642.

60. Chang KH, Nelson AM, Cao H, *et al.* (2006) Definitive-like erythroid cells derived from human embryonic stem cells coexpress high levels of embryonic and fetal globins with little or no adult globin. *Blood* **108**: 1515–1523.

61. Gaur M, Kamata T, Wang S, *et al.* (2006) Megakaryocytes derived from human embryonic stem cells: a genetically tractable system to study megakaryocytopoiesis and integrin function. *J Thromb Haemost* **4**: 436–442.

62. Woll PS, Martin CH, Miller JS, Kaufman DS. (2005) Human embryonic stem cell-derived NK cells acquire functional receptors and cytolytic activity. *J Immunol* **175**: 5095–5103.

63. Galic Z, Kitchen SG, Kacena A, *et al.* (2006) T lineage differentiation from human embryonic stem cells. *Proc Natl Acad Sci USA* **103**: 11742–11747.

64. Anderson JS, Bandi S, Kaufman DS, Akkina R. (2006) Derivation of normal macrophages from human embryonic stem (hES) cells for applications in HIV gene therapy. *Retrovirology* **3**: 24.

65. Zhan X, Dravid G, Ye Z, *et al.* (2004) Functional antigen-presenting leucocytes derived from human embryonic stem cells *in vitro*. *Lancet* **364**: 163–171.

66. Slukvin, II, Vodyanik MA, Thomson JA, *et al.* (2006) Directed differentiation of human embryonic stem cells into functional dendritic cells through the myeloid pathway. *J Immunol* **176**: 2924–2932.

67. Narayan AD, Chase JL, Lewis RL, *et al.* (2006) Human embryonic stem cell-derived hematopoietic cells are capable of engrafting primary as well as secondary fetal sheep recipients. *Blood* **107**: 2180–2183.

68. Tian X, Woll PS, Morris JK, *et al.* (2006) Hematopoietic engraftment of human embryonic stem cell-derived cells is regulated by recipient innate immunity. *Stem Cells* **24**: 1370–1380.

69. Yoshino H, Ueda T, Kawahata M, *et al.* (2000) Natural killer cell depletion by anti-asialo GM1 antiserum treatment enhances human hematopoietic stem cell engraftment in NOD/Shi-scid mice. *Bone Marrow Transplant* **26**: 1211–1216.

70. Ito M, Hiramatsu H, Kobayashi K, *et al.* (2002) NOD/SCID/γ_c^{null} mouse: an excellent recipient mouse model for engraftment of human cells. *Blood* **100**: 3175–3182.

71. McNiece IK, Almeida-Porada G, Shpall EJ, Zanjani E. (2002) *Ex vivo* expanded cord blood cells provide rapid engraftment in fetal sheep but lack long-term engrafting potential. *Exp Hematol* **30**: 612–616.

72. Yoder MC, Hiatt K. (1997) Engraftment of embryonic hematopoietic cells in conditioned newborn recipients. *Blood* **89**: 2176–2183.

73. Lu SJ, Li F, Vida L, Honig GR. (2004) CD34$^+$CD38$^-$ hematopoietic precursors derived from human embryonic stem cells exhibit an embryonic gene expression pattern. *Blood* **103**: 4134–4141.

74. Higuchi Y, Yasui A, Matsuura K, Yamamoto S. (2002) CD156 transgenic mice: different responses between inflammatory types. *Pathobiology* **70**: 47–54.

75. Boissy P, Lenhard TR, Kirkegaard T, *et al.* (2003) An assessment of ADAMs in bone cells: absence of TACE activity prevents osteoclast recruitment and the formation of the marrow cavity in developing long bones. *FEBS Lett* **553**: 257–261.

76. Odorico JS, Kaufman DS, Thomson JA. (2001) Multilineage differentiation from human embryonic stem cell lines. *Stem Cells* **19**: 193–204.

77. Wilmut I, Beaujean N, de Sousa PA, *et al.* (2002) Somatic cell nuclear transfer. *Nature* **419**: 583–586.

78. Gurdon JB, Byrne JA, Simonsson S. (2003) Nuclear reprogramming and stem cell creation. *Proc Natl Acad Sci USA* **100** **Suppl 1**: 11819–11822.

79. Briggs R, King TJ. (1952) Transplantation of living nuclei from blastula cells into enucleated frogs' eggs. *Proc Natl Acad Sci USA* **38**: 455–463.

80. Ildstad ST, Sachs DH. (1984) Reconstitution with syngeneic plus allogeneic or xenogeneic bone marrow leads to specific acceptance of allografts or xenografts. *Nature* **307**: 168–170.

81. Sykes M. (2001) Mixed chimerism and transplant tolerance. *Immunity* **14**: 417–424.

82. Dey B, Sykes M, Spitzer TR. (1998) Outcomes of recipients of both bone marrow and solid organ transplants: a review. *Medicine (Baltimore)* **77**: 355–369.

83. Millan MT, Shizuru JA, Hoffmann P, *et al.* (2002) Mixed chimerism and immunosuppressive drug withdrawal after HLA-mismatched kidney and hematopoietic progenitor transplantation. *Transplantation* **73**: 1386–1391.

84. Fandrich F, Lin X, Chai GX, *et al.* (2002) Preimplantation-stage stem cells induce long-term allogeneic graft acceptance without supplementary host conditioning. *Nat Med* **8**: 171–178.

85. Drukker M, Katz G, Urbach A, *et al.* (2002) Characterization of the expression of MHC proteins in human embryonic stem cells. *Proc Natl Acad Sci USA* **99**: 9864–9869.

86. Hochedlinger K, Rideout WM, Kyba M, *et al.* (2004) Nuclear transplantation, embryonic stem cells and the potential for cell therapy. *Hematol J* **5 Suppl 3**: S114–117.

87. Shultz LD, Ishikawa F, Greiner DL. (2007) Humanized mice in translational biomedical research. *Nat Rev Immunol* **7**: 118–130.

88. Smith AG. (2001) Embryo-derived stem cells: of mice and men. *Annu Rev Cell Dev Biol* **17**: 435–462.

89. Bhattacharya B, Miura T, Brandenberger R, *et al.* (2004) Gene expression in human embryonic stem cell lines: unique molecular signature. *Blood* **103**: 2956–2964.

90. Asano T, Ageyama N, Takeuchi K, *et al.* (2003) Engraftment and tumor formation after allogeneic *in utero* transplantation of primate embryonic stem cells. *Transplantation* **76**: 1061–1067.

91. Sasaki K, Nagao Y, Kitano Y, *et al.* (2005) Hematopoietic microchimerism in sheep after *in utero* transplantation of cultured cynomolgus embryonic stem cells. *Transplantation* **79**: 32–37.

92. Papapetrou EP, Zoumbos NC, Athanassiadou A. (2005) Genetic modification of hematopoietic stem cells with nonviral systems: past progress and future prospects. *Gene Ther* **12 Suppl 1**: S118–130.

93. Wilber A, Linehan JL, Tian X, *et al.* (2007) Use of the sleeping beauty transposon system for genetic engineering of human embryonic stem cell-derived hematopoietic cells. *Stem Cells* **25**: 2919–2927.

2

Using Embryonic Stem Cells as a Model of Pancreatic Development

Zhaohui Geng, Lucas Chase and Meri T. Firpo*

Embryonic stem cells, which can differentiate into insulin-producing cells and express pancreas specific markers during development *in vitro*, provide a model for studying pancreatic development. Current models of pancreatic development are based on animal models, including zebrafish, chicken and mouse. Genetic manipulations in the mouse have been a major source of information to date. More recently, *in vitro* differentiation of mouse embryonic stem cells has been used to allow both environmental and genetic manipulation during differentiation. While gene expression during *in vitro* differentiation appears to reproduce patterns found *in vivo*, comparison of mouse and human embryonic stem cell differentiation has demonstrated differences between regulation of mouse and human development. In this chapter, we review several studies on ES cell differentiation into insulin-producing cells *in vitro* and analysis of pancreatic-specific gene expression during differentiation, and discuss the potential of using human ES cells as a model of human pancreatic development.

1. Introduction

The mammalian pancreas is a complex organ composed of endocrine and exocrine tissues, and plays important roles in metabolism. The main function of the exocrine pancreas is the release of digestive enzymes and salts, while the endocrine pancreas, located within the islets of Langerhans, secretes hormones which are responsible for maintaining blood glucose homeostasis.[1,2] Defective pancreatic function can result in severe disorders,

*Correspondence: University of Minnesota Stem Cell Institute, Department of Medicine and Division of Endocrinology, Minneapolis, Minnesota, 55455, USA. E-mail: firpo001@umn.edu

including diabetes mellitus, a condition characterized by hyperglycemia. This disease affects an estimated 170 million people worldwide[3] and is the most common and devastating pancreatic disorder. Type 1 diabetes is an autoimmune disorder that destroys more than 80%[4] of the insulin-producing β cells in the pancreatic islets, and subsequently reduces or eliminates insulin production. Type 2 diabetes is also characterized by an up to 65% decrease in β cell mass.[5,6] Islet cell preservation is an important target in management of diabetes and its related complications, such as cardiovascular disease, blindness and kidney failure. Treatments range from restricted diet and pharmaceutical metabolic intervention for mild type 2 diabetes to insulin injections for type 1 and severe type 2 diabetes. Insulin therapy is the most common treatment for type 1 diabetes. For patients who require kidney transplants, pancreatic transplant has been found to be an effective treatment,[7–9] and in some patients, transplantation of isolated cadaveric islets have been used for regulation of blood glucose for up to several years.[10–13] However, these therapies are not without problems. Insulin therapy carries the risk of severe hypoglycemia, and limited availability of donor organs limits the use of transplantation therapies. The continued need for an alternative means of replacing β cells in patients with diabetes has fostered scientific and public interest in the potential of stem cells as a source of transplanted islets.

There is an ongoing controversy regarding the existence of stem cells for new β cells in adult exocrine pancreatic parenchyma, pancreatic ducts, pancreatic islets, liver, spleen and bone marrow from mouse development, regeneration models and clinical observations.[14] Evidence exists from one mouse model that new insulin-producing β cells in the pancreas are not generated by stem cells within the pancreas, but by the existing adult β cells.[15] Thus, the debate continues whether the pancreas or other distant organs contain stem cells that could regenerate the insulin-producing β cells. The introduction of human embryonic stem (hES) cells opened the possibility of obtaining enough mature β cells for transplantation therapies from an *in vitro* source.[16–20]

In the vertebrate embryo, the process of pancreatic development takes place in distinct stages. The pancreas develops from the fusion of endoderm-derived dorsal and ventral diverticula, known as pancreatic buds.[21] The dorsal bud arises first and ultimately generates most of the pancreas. The ventral bud arises from the bile duct and forms only part of the head and uncinate process of the pancreas. The exocrine pancreas, which represents 98% of the pancreatic mass, is composed of acinar cells that produce digestive enzymes and duct cells that secrete a bicarbonate solution to carry the digestive enzymes to the small intestine. Endocrine progenitors, expanding from the budding ducts, then form aggregates of differentiated cells known as the islets of Langerhans. The islets are made of four main cell populations, which produce hormones, in which the α cells produce glucagon, the β cells produce insulin, the PP cells make pancreatic polypeptide, and the δ cells produce somatostatin.[22,23] The endocrine tissue come to reside embedded within the exocrine tissues and comprises 1–2% of the cellular mass. The secreted hormones function together to regulate blood glucose levels in the fed and fasted state.

In humans, the ventral pancreatic bud comes into contact and eventually fuses with the dorsal pancreatic bud during the sixth week of development.[24] Restricted access to the developing human embryo limits the direct investigation of human pancreatic development. Furthermore, significant differences exist between human development and animal models.[25] Therefore, a human model system offers an important experimental approach to the study of genes and signals regulating human pancreatic differentiation. The regulation of pancreatic development has, however, been studied extensively in mouse, zebrafish and chicken.[26–32] Pancreatic organogenesis is regulated by a cascade of transcription factors expressed at specific times and in sequential order during development.[27,28] Table 1 summarizes genes with a role in the regulation of mouse pancreatic development, and Fig. 1 reviews the pattern of expression of these genes.

Table 1. Transcription Factors Function in Mouse Pancreas Development

Gene	Embryonic Expression	Knockout Phenotype	Reference
Sox17	E5.5–6.5	Required for pancreas specification	33
Hex	E8.5	Failure in ventral pancreatic specification	34
Pdx1	E8–10	Pancreatic agenesis	35, 36
Isl1	E9.0	Lack of islet cells; lack of dorsal mesenchyme formation	37
Nkx2.2	E8.75–9.0	Decreased α, β and PP cells	38
Ngn3	E9–9.5	Absence of endocrine cells	39
Nkx6.1	E9.0–9.5	Inhibition of β cell formation	40
Pax6	E9.0–9.5	Absence of α cells	41
Pax4	E9.5	α and δ cell agenesis	42
NeuroD	E9.5	Reduction of endocrine cells	43

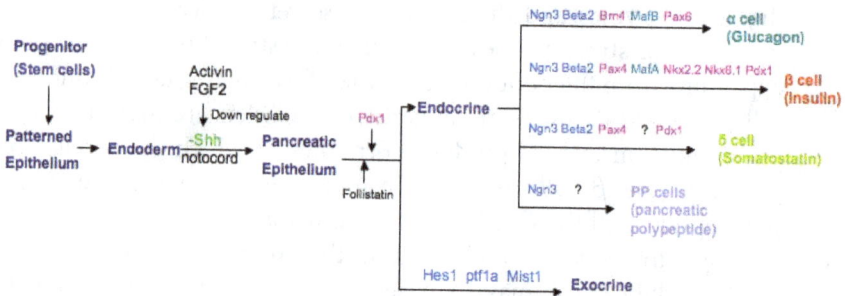

Figure 1. Gene expression patterns of pancreatic genes during development of the mouse pancreas. The roles of the listed genes in mouse development are summarized in Table 1. These genes can be used as markers to monitor pancreatic differentiation from ES cells *in vitro*.

Embryonic stem (ES) cells are pluripotent stem cells derived from the inner cell mass (ICM) of the blastocyst, displaying the properties of self-renewal and multilineage differentiation potential.[44–46] *In vivo*, cells within the ICM can give rise to all the differentiated cell types in the adult. Unlike ICM cells, which undergo a limited number of self-renewal cell divisions, ES cells can be continually propagated *in vitro* without differentiation. In theory, ES cells can be expanded indefinitely *in vitro* in an undifferentiated state and differentiated into all cell types when exposed to the appropriate signals. As a true measure of pluripotency, mouse ES cells have been shown to contribute to all fetal lineages plus yolk sac mesoderm, allantois, and amnion after being injected back to the host blastocyst.[47] Since the first establishment of mES cells lines,[44,45] they have been used as a model for studying the development of many tissues *in vitro*, including differentiation into insulin-producing cells.[48–53] Although the use of ES cells has been greatly expanded using the mouse model, many aspects of embryogenesis, including pancreatic development, differ greatly between mice and humans.[54,55] With the recent ability to generate human ES cells,[46] an opportunity has arisen by which one can establish an *in vitro* model of human development, and therefore a better understanding of the unique aspects of human development.

When cultured in suspension, ES cells form cell aggregates comprising three lineages (mesoderm, ectoderm and endoderm), known as embryoid bodies (EB). Evidence suggests that the sequential expression of endoderm genes in mouse EB during differentiation in culture correlates closely with that of normal embryo development.[56] Many studies demonstrated that both mES and hES cells have the capacity for differentiation into endoderm, pancreatic precursor cells and islet endocrine cell types. Under specific culture conditions, ES cells differentiate into insulin-producing cells, in which transcripts for pancreas development markers such as Pdx1, Ngn3, Isl1, Glut-2 and GK are expressed.[57] Blocking the activity of Sonic Hedgehog (Shh), known as an inhibitor of pancreas development by treatment with anti-Shh antibody[49] or the Shh inhibitor cyclopamine[58] increased insulin expression. Supplementing differentiation cultures with Activin A, a key factor during endoderm formation, also increased the generation insulin-producing cells.[53,58] Thus, evidence indicates that ES cells can be used as a model in studying pancreas development.

2. Strategies for Differentiating ES Cells into Insulin-producing Cells as a Model of Pancreatic Development

The ability to obtain multiple cell lineages from ES cells opens opportunities to model embryonic development *in vitro* and to study the events regulating the stages of pancreatic induction and specification. In order to study the mechanisms regulating pancreatic development and to get mature insulin-producing cells from embryonic stem cells, several methods have been employed. One strategy used a selection method that introduced an antibiotic resistance gene under the control of promoters specifically expressed in β cells, such as insulin or Nkx6.1. This allowed the selection of insulin-expressing cells that had spontaneously differentiated within EB.[48,49] Another strategy manipulated culture conditions in a multistep protocol that resulted in differentiation of ES cells into insulin-producing cells. This strategy has been conducted with both mouse and human ES cells.[50–53,58–61] Overexpression of pancreatic-specific transcription factors in ES cells provides a method to investigate the regulatory mechanisms in ES cell differentiation to the pancreatic fate.[62–64] In recent years, modifications of the differentiation protocols have been made in efforts to improve the yield and maturation of insulin-producing cells. Here we summarize the main strategies of differentiation of ES cells into insulin-producing cells, and the advancement of an *in vitro* pancreatic differentiation model. To evaluate the maturation of insulin-producing cells differentiated from ES cells, several aspects of detection are included.

Several methods are routinely used to evaluate differentiation of insulin-producing cells from ES cells. First, differentiation *in vitro* is monitored by tracking the cascade of transcription factors known to be expressed sequentially during *in vivo* development.[27,28] Figure 1 summarizes the patterns of genes expressed during pancreatic development. Several key genes known to be involved in this process include *Pdx1, Ngn3, Isl1, Nkx6-1, Pax6*, and *Glut2*. Secondly, ES cell differentiation into β cells results in insulin production that can be detected by mRNA or detection of mature insulin protein in the cells or medium by immunological means. Alternatively, because it is released in the processing of insulin, c-peptide is an indicator of β cell differentiation. Measuring c-peptide release can confirm that insulin is actually being produced and processed by the cells, and not merely a result of taking up insulin from the medium. Insulin storage granule formation and structure is another useful marker of mature β cells that can be monitored by electron microscopy. The capacity of insulin secretion in response to increased glucose concentration can be assessed. Finally, assessment of the function of insulin produced by differentiated ES cells can be conducted. *In vivo* experiments can be used to check the insulin function in diabetic mouse models such as streptozoticin (STZ)-treated or genetic models of diabetic mice. Treated mice can be monitored for blood glucose levels, glucose tolerance and changes in weight.

2.1. Differentiation of mouse ES cells into pancreatic cells

Because mES cells have been available longer, more studies have been done with mouse than human ES cells. The main approach to differentiating mES cells has been to modify culture conditions to support differentiation into pancreatic endocrine cells, including insulin-producing cells. Table 2 summarizes the studies on mouse pancreas development from mouse ES cells.

In order to investigate ES cells as a model for the complex process of the development of islet cells *in vitro*, Kahan *et al.*[65] established a method for differentiation of mouse ES cell by plating cells in adherent culture following seven days of EB differentiation. Using RT-PCR and immunostaining analysis, they evaluated the expression of pancreas-specific genes and concluded that *Pdx1, Pax6*, Isl1, *Pax4* and *Nkx6.1* could be detected. *Nkx2.2* expression increased significantly, and pancreatic progenitor cell markers peptide YY (YY) and islet amyloid polypeptide (IAPP) were expressed in many cells clustered in discrete foci after differentiation. Early EB cells co-expressed *Pdx1* with YY, and late-stage EB expressed *Pdx1* in

Table 2. Summary of the Studies on Mouse Pancreatic Development and Insulin-producing Cells Differentiated from mES Cells

Publication	Lumelsky, 2001 (50)	Hori, 2002 (51)	Blyszczuk, 2004 (52)	Shi, 2005 (53)	Miyazaki, 2004 (62)	Leon-Quinto, 2004 (49)
mES Cell Line	R1, E14.1, B5	JM1, ROSA	R1, Pax4+	R1	RTF-pdx-1	D3
Differentiation	Nestin Selection	Induced	Spontaneous	Induced	Pdx1 Overexpression	Nkx6.1 Selection
Assay						
Insulin detection by immunohistochemistry	Positive	Positive	Positive	Positive	Positive	Positive
Insulin detection in medium	Positive	Positive	Positive	Positive	Positive	Positive
Insulin secretion response to glucose level	N/A	Positive	Positive	Positive	Positive	Positive
C-peptide	N/A	Positive	Positive	Positive	Positive	N/A
Granule formation	N/A	N/A	N/A	N/A	N/A	N/A
Pancreas markers assayed by RT-PCR						
Insulin	Positive	Positive	Positive	Positive	Positive	Positive
Pdx1	Positive	Positive	Positive	Positive	Positive	Positive
Isl1	N/A	N/A	N/A	Positive	N/A	N/A
IAPP	Positive	N/A	Positive	N/A	Positive	N/A
CK19	N/A	N/A	Positive	N/A	N/A	N/A
Glut2	Positive	Positive	N/A	Positive	Negative	Positive
GK	N/A	Positive	N/A	N/A	Positive	Positive

N/A: Not available.

nearly all insulin-positive and many somatostatin-positive cells. Evidence from this study indicated that the gene regulation program and hormone expression in mES cell differentiation into islet-like cell types were consistent with mouse embryonic pancreas development, although the percentage of islet-like cells was low under these conditions.

In 2000, Soria *et al.*[48] developed an antibiotic resistance method in selecting mES cells differentiating into insulin-producing cells, in which undifferentiated mES cells were transduced with a plasmid containing the human insulin promoter driving the neomycin resistance (*βgeo*) gene and a phosphoglycerate kinase promoter driving a hygromycin resistance gene. After G418 selection, cells were detected for insulin synthesis by radioimmunoassay and for insulin function by maintaining a stable *in vivo* glucose response in STZ-treated mice after implantation of the cells in the spleen. Later in 2004,[49] the same group constructed another vector containing an Nkx6.1 promoter followed by a neomycin resistance gene. After selection, transfected cells were differentiated into islet precursors through culturing of EB with whole mouse fetal (E17.5) pancreas. Selection of *Nkx6.1*-expressing cells was performed by neomycin selection to enrich for the insulin-producing cell population. After selection in culture for 20–30 days, enhanced expression of insulin and other *β* cell-related genes was observed in the culture. Transplanted cells in STZ-treated mice normalized blood glucose levels for 3 weeks before removal of the transplant.

Rather than using drug selection, a multistep method to selectively differentiate mES cells into insulin-producing cells was reported by Lumelsky *et al.*[50] Although the insulin production demonstrated in this study is significantly lower than pancreatic *β* cells, it yielded improved production of insulin over spontaneous differentiation and selection. Further, this study began to establish an *in vitro* ES model for studying the development of pancreas based on information gained from *in vivo* studies. This multistep method included EB formation and selected expansion of nestin-positive cells. Progress was monitored by measuring expression levels of several pancreas-related genes. During the first differentiation stage (2–3 days), *Ins1*, *Ins2*, *Gluc*, and *Iapp* were negative, while *Glut2* and *Pdx-1* were positive. In later stages (24–26 days), *Ins1*, *Ins2* and *Iapp* expression became detectable. *Glut2* and *Pdx-1* expression was weaker, but at this stage was still detectable.

Later, Hori *et al.*[51] established a multistep culture method to promote mouse ES cell differentiation into insulin-producing cell by adding growth inhibitors, based on data suggesting that *in vitro* treatment of human fetal pancreas with nicotinamide and LY294002, an inhibitor of phosphoinositide 3-kinase (PI3K), could increase the total endocrine cell number and insulin content. Using this method, after two days of EB formation, cells were plated in adherent cultures in an insulin-transferrin-selenium-fibronectin (ITSF) medium for six days and then transferred to an N2 medium: insulin, transferrin, progesterone, putrascine and selenite plus basic fibroblast growth factor (bFGF). For the last stage, the cells were cultured in an N2 medium supplemented with nicotinamide and LY294002 for at least six days. Immunohistochemistry staining indicated that 95–97% of harvested cells at the end of differentiation were insulin positive cells. Insulin positive cells expressed numerous pancreatic β cell markers, including Pdx1, Glut2, and GK by RT-PCR. This group was also able to detect c-peptide by immunostaining. Cultured cells released insulin in response to increased glucose concentrations *in vitro*. The transplantation of insulin-positive cells enabled the survival of NOD *scid* mice with STZ-induced diabetes, improving both glucose regulation and circulating insulin levels.

However, some evidence has been presented suggesting that insulin detected in differentiated ES cultures could have been taken up from the medium, rather than being produced in the cultured ES-derived cells.[66,67] Subsequently, studies on a modification of the multistep differentiation method were performed with additional functional analyses. Without selection of nestin-positive cells, Pax4 positive mES cells were used in a fourth differentiation strategy.[52] First, mES cells were subjected to five days of spontaneous differentiation in EB, which were then plated in adherent cultures in a medium supplemented with 20% fetal bovine serum. After another nine days in culture, the cells were transferred to a pancreatic differentiation medium N2 plus nicotinamide to induce pancreatic differentiation for up to 28 days in culture. By RT-PCR analysis, *Pdx-1, Pax4,* insulin, *IAPP* and *CK19* were detected. In addition, by immunofluorescence analysis, c-peptide-positive cells reached 26% of the total population. After transplantation of differentiated cells by injection under the kidney capsule and into the spleen of STZ-treated mice, blood glucose levels were reduced for two weeks following transplantation, and attained normoglycemia by five weeks.

Later, more complex strategies for differentiation into insulin-producing cells were established based on the regulation of mouse pancreas development *in vivo*, with evidence indicating activin and retinoic acid (RA) are important factors in endoderm formation and pancreas bud formation being cited.[53] A three-step method demonstrated that combined treatment with activin and RA improved the efficiency of differentiation into insulin-producing cells. In this study, mES cells were subjected to: 1) EB formation, followed by adherence of EB and treatment with Activin A and RA; 2) expansion of insulin-producing precursors by culturing with bFGF; and 3) maturation of the insulin-producing cells with the supplement of N2, B27, laminin, bFGF and nicotinamide. At the end of the two-week process, gene expression for insulin, *Pdx-1*, *Glut2*, *Foxa2* and *Isl1* were detected by RT-PCR; c-peptide was detected with immunohistochemistry; and insulin release in high-glucose medium (27.7 mM) was shown to be nearly six times higher than that in low-glucose medium (5.5 mM). Finally, the insulin-producing cells sustained the STZ-induced diabetic mice after transplantation into the left renal capsule for 40 days. The survival rate of differentiated cell-treated mice was 70%, while in the PBS transplanted mice, survival was limited to 25%.

Overexpression of transcription factors related to pancreatic differentiation, including Pdx1, Nkx2.2 and Ngn3, in ES cells offered another strategy for studying gene regulation during ES cell differentiation into pancreatic tissues, and improving hormone production. In 2004, Miyazaki *et al.*[62] designed a transgenic mouse ES cell line (RTF-pdx-1) from parental EB3 cells, in which *Pdx1* gene expression was overexpressed upon withdrawal of doxycycline (Dox). After a multistep culture protocol and induction of *Pdx1* expression, other pancreatic-specific genes such as insulin, carboxyl peptidase A, *Nkx2.2*, *Ngn3*, *NeuroD*, *P48*, *GK*, *Pax4* and *Pax6* could be detected by RT-PCR. This system demonstrated the similarity of the key role for *Pdx-1* in ES cell differentiation and the *in vivo* development of pancreatic tissues. Shiroi *et al.*[63] established an *Nkx2.2*-expressing mES cell line, Nkx-ES, that differentiated into insulin-producing cells as much as two weeks earlier in EB compared with those from the parental ES cells by both RT-PCR and dithizone (DTZ)-staining. The percentage of insulin-producing cells increased by 10-fold to 1% in Nkx-ES-derived EB cells.

To investigate the function of the *Ngn3* gene during ES cell differentiation into pancreatic cells, Treff *et al.*[64] recently generated a *Dox*-inducible Ngn3 ES cell line, Ngn3-D6, that expressed more than 200-fold Ngn3 after Dox induction. During EB formation from uninduced cells, RT-PCR analysis indicated that the expression of genes associated with endoderm (*Sox17* and *Foxa2*), and pancreatic endoderm and endocrine development (*Pdx1*, *Ngn3*) increased in these EB over time, and the sequence of the peak levels of these genes was consistent with the sequence of gene expression during *in vivo* embryonic development of the pancreas. Moreover, RT-PCR analysis after seven days of differentiation in EB culture and induction of *Ngn3* expression with Dox for another three days in adherent culture enhanced expression of insulin, glucagon and somatostatin, which could be detected during the 13 and 28 days of culture as compared with uninduced controls. These studies further supported the conclusion that ES cell differentiation recapitulates important molecular events that occur during the development of embryonic pancreas *in vivo*, and the transcription factors associated with pancreatic development increase the efficiency of pancreatic endocrine differentiation *in vitro*.

2.2. Differentiation of non-human primate ES cells into insulin-producing cells

The non-human primate is physiologically and phylogenetically similar to the human, and therefore is a clinically relevant animal model for biomedical research. Non-human primate ES cells, like those derived from mouse and human embryos, can remain undifferentiated on feeder cells and can differentiate into all three embryonic germ layers.[68] In 2004, Lester et al.[69] reported that by RT-PCR analysis, pancreatic hormone genes, insulin, glucagon and amylase could be detected after spontaneous rhesus ES cell differentiation. Pancreatic development-specific genes such *Pdx1*, *NeuroD* and *Nkx6.1* could also be detected after the differentiation. Further maturation by the addition of Exendin-4 resulted in detectable levels of c-peptide by immunostaining. A recent study by Yue et al.[70] showed that glucagon-like peptide-1 (GLP-1) could promote cynomolgus monkey ES cell maturation into insulin-producing cells. In the Yue et al. study, the differentiation method was similar to that used by Lumelsky et al.[50] After selection of nestin positive cells, 100 nM GLP-1 was added to the medium at the last stage of differentiation for 10 to 30 days. Insulin, *Glut-2*, *Pdx1*, *IAPP*, *Ngn3* and somatostatin expression was significantly stronger in the GLP-1 treatment group than that without the added factor, as measured by RT-PCR. Glucagon and PP had similar expression levels in cultures under both conditions. Western blot and ELISA analyses demonstrated that insulin protein expression was increased in the GLP-1 treatment group, and glucose-responsive insulin expression was found with c-peptide in most of the cells in the cultures by immunostaining. Transplantation of induced cells into the renal capsules and spleens of

STZ-treated mice resulted in decreased blood glucose after two days. The evidence indicated that GLP-1 induced the maturation of insulin-producing cells during differentiation. In the past, non-human primate models provided critical information in islet transplantation studies.[71,72] Using non-human primate ES cells as an *in vitro* model may help to elucidate the mechanism of pancreas development of primates, including the human.

2.3. Differentiation of human ES cells into insulin-producing cells

As early as 2001, a report was published which showed that insulin was produced by hES-derived cells.[59] In this study, after EB formation, hES cells were subjected to spontaneous differentiation in adherent cultures for 20–30 days. Insulin was detected at levels up to 150 μM/ml in the EB culture after 20–22 days, and at more than 300 μM/ml in adherent cultures after 31 days. RT-PCR analysis demonstrated the expression of insulin, glucokinase, *Glut-2*, and *Ngn3* after 30 days of differentiation, both in EB and adherent cultures. Table 3 summarizes studies attempting to differentiate hES cells into insulin-producing cells.

Selection of embryoid bodies by size was used for the differentiation of hES cells into insulin-producing cells by Xu *et al.*,[61] who reported that selection of EB greater than 70 μm improved endoderm formation and insulin production. Embryoid bodies formed by selection of hES colonies under 3D culture conditions for two weeks, were cultured in matrigel-coated plate, with spontaneous adherent differentiation for up to 12 weeks. The RT-PCR data indicated the definitive endoderm and pancreas-associated gene expression, including *Foxa2*, *Pdx1*, *Nkx6.1*, *Nkx2.2*, and *Glut2* at day 14 of EB formation (EB14). After another 2 weeks of differentiation, RT-PCR analysis showed that EB formation supported the differentiation of definitive endoderm and the development of Pdx1-positive cells. At EB14, insulin could also be detected by RT-PCR. Up to 12 weeks of differentiation after EB14, insulin was more strongly expressed as shown by immunostaining, as compared to that at day 14.

Other studies have provided further insights into human pancreatic differentiation. Islet-like cluster formation may be important for enriching insulin-producing cells and maintaining the cells for longer periods in culture without losing their potential to express insulin.[60] Still more complex differentiation methods were used to parallel developmental patterns *in vivo*.[58] A five-stage differentiation method with cocktails of growth factors was performed with hES cells to reproduce sequential differentiation to

Table 3. Human ES Cells Differentiation into Pancreatic Cells

Publication	Assady 2001 (59)	Segev 2004 (60)	Xu 2006 (61)	D'Amour 2006 (58)
hES Cell Line	H9	H9	H9, H1	CyT203
Differentiation Strategy	Spontaneous	Induced	Spontaneous	Induced
Assay				
Insulin detection by immunohistochemistry	Positive	Positive	Positive	Positive
Insulin detection in medium	Positive	Positive	N/A	Positive
Insulin secretion response to glucose level	N/A	Positive	N/A	No effect
C-peptide	N/A	Positive	N/A	Positive
Granule formation	N/A	N/A	N/A	Positive
Pancreas markers assayed by RT-PCR				
Insulin	Positive	Positive	Positive	Positive
Pdx1	Positive	Positive	Positive	Positive
Foxa2	N/A	N/A	Positive	Positive
Ngn3	Positive	Positive	Positive	Positive
Glut2	Positive	Positive	N/A	N/A
GK	Positive	Positive	N/A	N/A

N/A: Not available.

mesendoderm, definitive endoderm, primitive gut tube, posterior foregut endoderm, pancreatic endoderm and endocrine precursors, mimicking the embryonic pancreatic endocrine cell differentiation. During the whole process, several factors were added: Activin A and Wnt3a at the putative definitive the endoderm stage; FGF10 and cyclopamine at the proposed primitive gut tube stage; RA, cyclopamine and FGF10 for the posterior foregut stage; DAPT (γ-secretase inhibitor) and exendin 4 at the proposed pancreatic endoderm and endocrine precursors stage; Later, exendin 4, IGF1 and HGF were added to attempt to reproduce the hormone-expressing stage. While gene expression patterns did not indicate that the stages fully replicated *in vivo* development, several endoderm- and pancreas-specific genes were detected by RT-PCR or Western blot analysis at the conclusion of the differentiation program. The highest levels of Sox17 protein, which is a definitive endoderm marker, were observed during the stage from hES cells to definitive endoderm; Foxa2 could be seen at all the stages from the end of the definitive endoderm; Pdx1, associated with pancreatic commitment, accumulated during the posterior foregut endoderm stage and diminished at the end of the process; insulin was expressed at the last stage, when c-peptide detected by ELISA, and insulin granules were observed under electron microscopy in some cells, although these cells were not glucose-responsive.

In summary, although the field of generating β cells from ES cells is still in its infancy, gene expression patterns during differentiation of both mouse and human ES cells suggest that the *in vitro* system reflects the normal pancreatic development. Further, although no demonstration of functioning β cells has been made so far from human cells, the successful generation of insulin-producing cells is promising. In the future, embryonic stem cells have great potential of being used as a model to study the development of the pancreas, to identify detailed regulatory mechanisms, and as a possible source of insulin-producing cells for the treatment of diabetes.

References

1. Kim SK, MacDonald R. (2002) Signaling and transcriptional control of pancreatic organogenesis. *Curr Opin Genet Dev* 12: 540–547.
2. Murtaugh LC, Melton DA. (2003) Genes, signals, and lineages in pancreas development. *Annu Rev Cell Dev Biol* 19: 71–89.
3. Wild S, Roglic G, Green A, *et al.* (2004) Global prevalence of diabetes: estimates for the year 2000 and projections for 2030. *Diabetes Care* 27: 1047–1053.
4. Pipeleers D, Ling Z. (1992) Pancreatic beta cells in insulin-dependent diabetes. *Diabetes Metab Rev* 8: 209–227.
5. Butler AE, Janson J, Bonner-Weir S, *et al.* (2003) Beta-cell deficit and increased beta-cell apoptosis in humans with type 2 diabetes. *Diabetes* 52: 102–110.
6. Clark A, Wells CA, Buley ID, *et al.* (1988) Islet amyloid, increased A-cells, reduced B-cells and exocrine fibrosis: quantitative changes in the pancreas in type 2 diabetes. *Diabetes Res* 9: 151–159.
7. Hopt UT, Drognitz O. (2000) Pancreas organ transplantation. Short and long-term results in terms of diabetes control. *Langenbecks Arch Sreg* 385: 379–389.
8. Venstrom JM, McBride MA, Rother KI, *et al.* (2003) Survial after pancreas transplantation in patients with diabetes and preserved kidney function. *JAMA* 290: 2817–2823.

9. Larsen JL. (2004) Pancreas transplantation: indications and consequences. *Endocr Rev* **25:** 919–946.

10. Shapiro AM, Lakey JR, Ryan EA, *et al.* (2000) Islet transplantation in seven patients with type 1 diabetes mellitus using a glucocorticoid-free immuno-suppressive regimen. *N Engl J Med* **343:** 230–238.

11. Robertson RP. (2004) Islet transplantation as a treatment for diabetes — a work in progress. *N Engl J Med* **350:** 694–705.

12. Rother KI, Harlan DM. (2004) Challenges facing islet transplantation for the treatment of type 1 diabetes mellitus. *J Clin Invest* **114:** 877–883.

13. Ryan EA, Paty BW, Senior PA, *et al.* (2005) Five-year follow-up after clinical islet transplantation. *Diabetes* **54:** 2060–2069.

14. Bonner-Weir S. (2000) Islet growth and development in the adult. *J Mol Endocrinol* **24:** 297–302.

15. Dor Y, Brown J, Martinez OI, Melton DA. (2004) Adult pancreatic β-cells are formed by self-duplication rather than stem-cell differentiation. *Nature* **429:** 41–46.

16. Gordon Keller. (2005) Embryonic stem cell differentiation: emergence of a new era in biology and medicine. *Gene & Dev* **19:** 1129–1155.

17. Soria B, Skoudy A, Martin F. (2001) From stem cells to beta cells: new strategies in cell therapy of ciabetes mellitus. *Diabetologia* **44:** 407–415.

18. Kumar M, Melton D. (2003) Pancreas specification: a budding question. *Curr Opin Genet Dev* **13:** 401–407.

19. Magliocca JF, Odorico JS. (2006) Embryonic stem cell-based therapy for the treatment of diabetes mellitus: a work in progress. *Curr Opin Organ Transplant* **11:** 88–93.

20. Madsen OD. (2005) Stem cells and diabetes treatment. *APMIS* **113:** 858–875.

21. Slack JM. (1995) Developmental biology of the pancreas. *Development* **121:** 1569–1580.

22. Golosow N, Grobstein C. (1962) Epithelio-mesenchymal interaction in pancreatic morphogenesis. *Dev Biol* **4:** 242–255.

23. Gittes GK, Galante PE, Hannahan D, *et al.* (1996) Lineage-specific morphogenesis in the developing pancreas: role of mesenchymal factors. *Development* **122:** 439–447.

24. Banaei-Bouchareb L, Peuchmaur M, Czernichow P, Polak M. (2006) A transient microvironment loaded mainly with macrophages in the early developing human pancreas. *J Endocrinol* **188:** 467–480.

25. Cabrera O, Berman DM, Kenyon NS, *et al.* (2006) The unique cytoarchitecture of human pancreatic islets has implications for islet cell function. *Proc Natl Acad Sci USA* **103:** 2334–2339.

26. Meier JJ, Bhushan A, Butler PC. (2006) The potential for stem cell therapy in diabetes. *Pediatr Res* **59:** 65R–73R.

27. Habener JF, Kemp DM, Thomas MK. (2005) Minireview: transcriptional regulation in pancreatic development. *Endocrinology* 146: 1025–1034.
28. Jensen J. (2004) Gene regulatory factors in pancreatic development. *Dev Dyn* 229: 176–200.
29. Mfopou JK, Willems E, Leyns L, Bouwens L. (2005) Expression of regulatory genes for pancreas development during murine embryonic stem cell differentiation. *Int J Dev Biol* 49: 915–922.
30. Servitja JM, Ferrer J. (2004) Transcriptional networks controlling pancreatic development and beta cell function. *Diabetologia* 47: 597–613.
31. Huang H, Vogel SS, Liu N, *et al.* (2001) Analysis of pancreatic development in living transgenic zebrafish embryos. *Mol Cell Endocrinol* 177(1-2): 117–124.
32. Kim SK, Hebrok M, Melton DA. (1997) Pancreas development in the chick embryo. In: *Cold Spring Harbor Symposia on Quant Biol* 62: 377–383.
33. Kanai-Azuma M, Kanai Y, Gad JM, *et al.* (2002) Depletion of definitive gut endoderm in Sox17-null mutant mice, *Development* 129: 2367–2379.
34. Bort R, Martinez-Barbera JP, Beddington RS, Zaret KS. (2004) Hex homeobox gene-dependent tissue positioning is required for organogenesis of the ventral pancreas. *Development* 131: 797–806.
35. Ahlgren U, Jonsson J, Jonsson L, *et al.* (1998) β-cell-specific inactivation of the mouse *Ipf1/Pdx1* gene results in loss of the β-cell phenotype and maturity onset diabetes. *Genes Dev* 12: 1763–1768.
36. Gerrish K, Gannon M, Shih D, *et al.* (2000) Pancreatic β cell-specific transcription of the pdx-1 gene. The role of conserved upstream control regions and their hepatic nuclear factor 3β sites. *J Biol Chem* 275: 3485–3492.
37. Gittes GK, Galante PE, Hanahan D, *et al.* (1996) Lineage-specific morphogenesis in the developing pancreas: role of mesenchymal factors. *Development* 122: 439–447.
38. Sussel L, Kalamaras J, Hartigan-O'Connor DJ. (1998) Mice lacking the homeodomain transcription factor Nkx2.2 have diabetes due to arrested differentiation of pancreatic beta cells. *Development* 125: 2213–2221.
39. Ang SL, Rossant J. (1994) HNF-3β is essential for node and notochord formation in mouse development. *Cell* 78: 561–574.
40. Naya FJ, Huang H-P, Qiu Y, *et al.* (1997) Diabetes, defective pancreatic morphogenesis, and abnormal enteroendocrine differentiation in BETA2/NeuroD-deficient mice. *Genes Dev* 11: 2323–2334.
41. Pontoglio M, Sreenan S, Roe M, *et al.* (1998) Defective insulin secretion in hepatocyte nuclear factor 1α-deficient mice. *J Clin Invest* 101: 2215–2222.
42. Jacquemin P, Durviaux SM, Jensen J, *et al.* (2000) Transcription factor hepatocyte nuclear factor 6 regulates pancreatic endocrine cell differentiation and controls expression of the proendocrine gene Ngn3. *Mol Cell Biol* 20: 4445–4454.

43. Weinstein DC, Ruiz i Altaba A, Chen WS, *et al.* (1994) The winged-helix transcription factor HNF-3ß is required for notochord development in the mouse embryo. *Cell* **78**: 575–588.
44. Evans MJ, Kaufman MH. (1981) Establishment in culture of pluripotential cells from mouse embryos. *Nature* **292**: 154–156.
45. Martin GR. (1981) Isolation of a pluripotent cell line from early mouse embryos culture in medium conditioned by teratocarcinoma stem cells. *Proc Natl Acad Sci USA* **78**: 7634–7638.
46. Thomson JA, itskovitz-Eldor J, Shapiro SS, *et al.* (1998) Embryonic stem cell lines derived from human blastocysts. *Science* **282**: 1145–1147.
47. Bradley A, Evans M, Kaufman MH, Robertson E. (1984) Formation of germ-line chimaeras from embryo-derived teratocarcinoma cell lines. *Nature* **309**: 255–256.
48. Soria B, Roche E, Berna G, *et al.* (2000) Insulin-secreting cells derived from embryonic stem cells normalize glycemia in streptozotocin-induced diabetic mice. *Diabetes* **49**: 1–6.
49. Leon-Quinto T, Jones J, Skoudy BM, Soria B. (2004) *In vitro* directed differentiation of mouse embryonic stem cells into insulin-producing cells. *Diabetologia* **47**: 1442–1458.
50. Lumelsky N, Blondel O, Laeng P, *et al.* (2001) Differentiation of embryonic stem cells to insulin-secreting structures similar to pancreatic islets. *Science* **293**: 1389–1394.
51. Hori Y, Rulifson IC, Tsai BC, *et al.* (2002) Growth inhibitors promote differentiation of insulin-producing tissue from embryonic stem cells. *Proc Natl Acad Sci USA* **99**: 16105–16110.
52. Blyszczuk P, Asbranad C, Rozzo A, *et al.* (2004) Embryonic stem cells differentiate into insulin-producing cells without selection of nestin-expressing cells. *Int J Dev Biol* **48**: 1095–1104.
53. Shi Y, Hou L, Tang F, *et al.* (2005) Inducing embryonic stem cells to differentiate pancreatic beta cells by a novel three-step approach with activin and all-trans retinoic acid. *Stem Cells* **23**: 656–662.
54. Kaufman MH. (1992) *The Atlas of Mouse Development*. Academic Press, San Diego.
55. O'Rahilly R, Muller F. (1987) *Development Stages in Human Embryos*. Carnegie Institution of Washington, Washington, DC.
56. Choi D, Lee HJ, Jee S, *et al.* (2005) *In vitro* differentiation of mouse embryonic stem cells: enrichment of endodermal cells in the embroid body. *Stem Cells* **23**: 817–827.
57. Mfopou JK, Willems E, Leyns L, Bouwens L. (2005) Expression of regulatory genes for pancreas development during murine embryonic stem cell differentiation. *Int J Dev Biol* **49**: 915–922.

58. D'Amour KA, Bang AG, Eliazer S, *et al.* (2006) Production of pancreatic hormone-expressing endocrine cells from human embryonic stem cells. *Nature Biotechnol* **24**: 1392–1401.

59. Assady S, Maor G, Amit M, *et al.* (2001) Insulin production by human embryonic stem cells. *Diabetes* **50**: 1691–1697.

60. Segev H, Fishman B, Ziskind A, *et al.* (2004) Differentiation of human embryonic stem cells into insulin-producing clusters. *Stem Cells* **22**: 265–274.

61. Xu X, Kahan B, Forgianni A, *et al.* (2006) Endoderm and pancreatic islet lineage differentiation from human embryonic stem cells. *Cloning and Stem Cells* **8**: 96–107.

62. Miyazaki S, Yamato E, Miyazaki JI. (2004) Regulated expression of pdx-1 promotes *in vitro* differentiation of insulin-producing cells from embryonic stem cells. *Diabetes* **53**: 1030–1037.

63. Shiroi A, Ueda S, Ouji Y, *et al.* (2005) Differentiation of embryonic stem cells into insulin-producing cells promoted by Nkx2.2 gene transfer. *World J Gastroenterol* **11**: 4161–4166.

64. Treff NR, Vincent RK, Budde ML, *et al.* (2006) Differentiation of embryonic stem cells conditionally expressing neurogenin 3. *Stem Cells* **24**: 2529–2537.

65. Kahan BW, Jacobson LM, Hullett DA, *et al.* (2003) Pancreatic precursors and differentiated islet cell types from murine embryonic stem cells. *Diabetes* **52**: 2016–2024.

66. Sipione S, Eshpeter A, Lyon JG, *et al.* (2004) Insulin expressing cells from differentiated embryonic stem cells are not beta cells. *Diabetologia* **47**: 499–508.

67. Rajagopal J, Anderson WJ, Kume S, *et al.* (2003) Development: insulin staining of ES cell progeny from insulin uptake. *Science* **299**: 363–364.

68. Pau KY, Wolf DP. (2004) Derivation and characterization of monkey embryonic stem cells. *Reprod Biol Endocrinol* **2**: 41.

69. Lester LB, Kuo HC, Andrews L, *et al.* (2004) Directed differentiation of rhesus monkey ES cells into pancreatic cell phenotypes. *Reprod Biol Endocrinol* **2**: 42–46.

70. Yue F, Cui L, Johkura K, *et al.* (2006) Glucagon-like peptide-1 differentiation of primate embryonic stem cells into insulin-producing cells. *Tissue Eng* **12**: 2105–2116.

71. Hirshberg B, Mog S, Patterson N, *et al.* (2002) Histopathological study of intrahepatic islets transplanted in the nonhuman primate model using Edmonton protocol immunosuppression. *J Clin Endocrinol Metab* **87**: 5425–5429.

72. Hirshberg B, Montgomery S, Wysoki MG, *et al.* (2002) Pancreatic islet transplantation using the nonhuman primate (rhesus) model predicts that the portal vein is superior to the celiac artery as the islet infusion site. *Diabetes* **51**: 2135–2140.

3

Approaches to Identifying Nuclear Reprogramming Factors from Embryonic Stem Cell Biology and Somatic Cell Nuclear Cloning

Nobuaki Kikyo*

Undifferentiated cells, such as embryonic stem cells (ES cells) and vertebrate eggs, contain powerful nuclear reprogramming factors that can erase previous cell memory and induce dedifferentiation of various somatic cells. The existence of these nuclear reprogramming factors has been most clearly demonstrated by somatic cell nuclear cloning and cell fusion between ES cells and differentiated somatic cells. This chapter will discuss several examples of the reprogramming factors identified from cell fusion, somatic cell nuclear cloning, and comparisons of mRNA and chromatin structure between ES cells and differentiated cells. Further analyses of the molecular mechanisms underlying the nuclear reprogramming by undifferentiated cells will significantly contribute to the development of novel cell-based therapy in regenerative medicine.

1. Introduction

Somatic cell nuclear cloning and cell fusion between somatic cells and ES cells are two distinct procedures in which the ability of undifferentiated cells to reprogram the transcriptional patterns of differentiated cells can be observed (see Ref. 1 for a recent review). The identification of the specific cellular mechanisms underlying such nuclear reprogramming is an

*Correspondence: Stem Cell Institute, Department of Medicine, and Division of Hematology, Oncology, and Transplantation, University of Minnesota, 2001 6th St SE, 2-216 MTFR, Minneapolis, MN 55455, USA. Tel: 612-624-0498; Fax: 612-624-2436; E-mail: kikyo001@umn.edu

important undertaking that could have a tremendous impact not only on basic biology but also on regenerative medicine. More specifically, a greater understanding of nuclear reprogramming could allow for the creation of autologous tissues from patients' own somatic cells.

Somatic cell nuclear cloning has been practiced since the middle of the 20th century. The first successful vertebrate clones were created in 1952 using embryonic nuclei taken from the frog *Rana pipiens*.[2] Subsequently, the cloning of fertile adult *Xenopus* frogs from intestinal nuclei in 1966 demonstrated the acquisition of totipotency by donor nuclei during cloning.[3] The recent surge of cloning research, however, did not occur until the achievement of the first mammalian clone, Dolly the sheep, in 1997.[4] The increased cloning research after the creation of Dolly was aided by the development of mouse clones from the cumulus cells of the ovary, as this brought cloning research within the reach of standard laboratory settings.[5]

Research into cell fusion has also been in existence for several decades initiated in the 1960s by multiple investigators.[6] Since then, cell fusion has been extensively used not only in the study of nuclear reprogramming but also in the identification of tumor suppressor genes, genome mapping and the production of monoclonal antibodies. It was not until the recent explosion in stem cell biology, however, that cell fusion began to be used as a research tool for understanding nuclear dedifferentiation and pluripotence.

During the long history of nuclear cloning and cell fusion, several factors have been identified as key regulators for nuclear reprogramming. This article will discuss various ways in which cell fusion and somatic cell nuclear cloning have been applied to the study of nuclear reprogramming. In addition to cell fusion and somatic cell nuclear cloning, it will discuss recent studies of differential chromatin structure and gene expression between ES cells and somatic cells. These studies are important to address here because they provide a valuable way to understand and explore many of the events underlying the nuclear reprogramming seen during cell fusion and nuclear cloning.

2. Approaches to Identifying Nuclear Reprogramming Factors

2.1. Cell fusion

It has long been known that fusion between two types of cells suppresses the expression of tissue-specific genes in each nucleus in a bidirectional manner. This phenomenon is termed "extinction" (see Refs. 6 and 7 for reviews). An example of extinction is that fusion between rat hepatoma cells and any of a number of other different types of cells leads to the suppression of several liver-specific genes from the hepatoma cells, such as albumin, tyrosine aminotransferase and alcohol dehydrogenase. This suppression of gene expression is dosage-dependent with respect to genomic DNA. Therefore, if two hepatoma cells and a fibroblast cell are fused, liver-specific genes prevail and fibroblast-specific genes are suppressed. In the case of two fibroblasts being fused with a single hepatoma cell, the reverse is true. To identify the genetic basis of this dominant negative effect, Fournier's group prepared a series of microcell hybrids, each composed of a rat hepatoma cell and a single mouse fibroblast chromosome.[8] After screening the microcell hybrids, they found that mouse chromosome 11 contained a genetic locus necessary for suppressing expression of at least two liver-specific genes, tyrosine aminotransferase and phosphoenolpyruvate carboxykinase, from the hepatoma nucleus. They named this genetic locus tissue-specific extinguisher-1 [TSE1; Fig. 1(A)]. Utilizing cDNA subtractive hybridization, Fournier's group discovered that TSE1 codes for the regulatory subunit RIα of cAMP-dependent protein kinase (PKA).[9] This discovery is supported by the fact that a different group, working independently at around the same time as Fournier *et al.*, also identified that TSE1 codes for the RIα subunit. This second group's discovery was based on their finding that TSE1 functions through a cAMP response element.[10] Although RIα is ubiquitously expressed throughout the body, its expression level in hepatoma cells is less than 10% of that in other cells. A differential level of expression between two partner cells during fusion might explain how a housekeeping gene such as TSE1 can function as a tissue-specific extinguisher. Mouse chromosome 1 is known to contain a second extinguisher (TSE2) that has been shown to suppress expression of three other liver-specific genes (albumin, alcohol dehydrogenase and liv-10) in microcell hybrids, but the locus of this gene has not been identified.[11]

Several years after these early findings, interest in nuclear reprogramming by cell fusion was renewed and a more detailed understanding of these reprogramming events was achieved utilizing fusion between ES cells or embryonic germ cells (EG cells) and differentiated cells.[12,13] The tetraploid cells resulting from this fusion can stably maintain full chromosomal compositions for a prolonged period of time and the differentiated nuclei adopt several ES cell-like characteristics [Fig. 1(B)]. These characteristics include activation of the pluripotency gene *Oct4* accompanied by

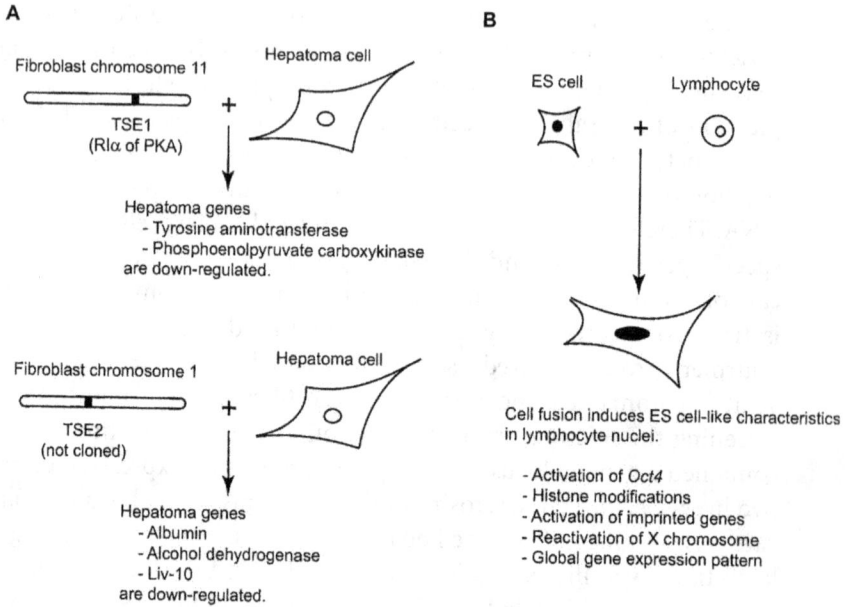

Figure 1. Nuclear reprogramming by cell fusion. (A) Gene suppression after the formation of microcell hybrids between mouse fibroblast chromosomes and rat hepatoma cells. TSE1 on mouse chromosome 11 suppresses expression of two hepatoma-specific genes, and TSE2 on chromosome 1 suppresses three hepatoma-specific genes. (B) Cell fusion between ES cells and lymphocytes causes the lymphocytes to display several ES cell-like characteristics. These characteristics include activation of *Oct4*, modification of histones, activation of imprinted genes, reactivation of the X chromosome, and a shift toward an ES cell-specific gene expression pattern.

ES cell-like histone modifications at the promoter region, DNA demethylation and reactivation of silent imprinted genes, activation of the inactive X chromosome, and reversion of the global gene expression pattern to an ES cell-like status. Consistent with the above pluripotent characteristics, the cell hybrid can contribute to all three germ layers upon injection into blastocysts (for the creation of chimeric mice) and injection into nude mice (for teratoma formation). Not surprisingly, the nuclear reprogramming activities of ES cells appear to reside in their nuclei, as demonstrated by a comparison of reprogrammability between ES cell karyoplasts and cytoplasts.[14] Clearly, ES and EG cell possess nuclear reprogramming activities that can induce differentiated cells to dedifferentiate and acquire pluripotency.

One potential problem with the cell fusion model above is that genomic DNA derived from the ES cells remains in the cell hybrids, and it is not clear whether the somatic nuclei have been sufficiently reprogrammed to sustain pluripotency independently. The pluripotency of the cell hybrids could be due to the dominance of transcriptional activities derived from the ES cell nuclei. One approach to addressing this issue is to remove ES cell-derived chromosomes after cell fusion. Although it may not be feasible to remove the entire ES cell-derived chromosome set, it is possible to selectively eliminate specific chromosomes by genetically engineering these chromosomes before fusion. A recent report used a Cre-loxP-mediated chromosome elimination cassette to successfully remove ES cell-derived chromosome 6, which harbors the self-renewal gene *Nanog*, after fusion.[15] Since the resulting cell hybrid can still sustain pluripotency after the removal of both copies of ES cell-derived chromosome 6, it is believed that the activated *Nanog* gene in the fusion partner can compensate for the loss of ES cell-derived *Nanog*. If systematically applied, this approach could be used to identify which chromosomal regions of ES cells are essential for conferring pluripotency on the fusion partner.

ES cells express a set of specific transcription factors that are essential for maintaining undifferentiated status, such as Oct4, Sox2 and Nanog. Silva *et al.* demonstrated that increased levels of *Nanog* in ES cells prior to cell fusion with neural stem cells can facilitate reprogramming of the neural stem cells.[16] Transfection of the *Nanog* gene alone, however, was insufficient to convert neural stem cells into ES-like cells. Obviously, Oct4, Sox2 and/or other unknown factors are also required to compel the fusion partner cells toward an ES cell-like status. In addition, it has been shown that Nanog, Oct4 and Sox2 bind to promoter regions of 9%, 3% and 7% of the whole genome in ES cells, respectively.[17,18] The sheer volume of interactions involving these three proteins indicates that it would be a valuable next step toward investigating whether expression patterns of these target genes in the partner nuclei in cell fusion closely follow those in normal ES cells.

Another important outcome of research into cell fusion has been the provision of a possible explanation for the phenomenon of stem cell plasticity after injection into a host. Beginning with the discovery of spontaneous fusion between ES cells and somatic cells,[19,20] nuclear reprogramming by cell fusion has been shown to take place spontaneously in the body.[21] Until these recent studies, the fact that tissue-specific stem cells expressed other tissue-specific proteins after transplantation was interpreted as a sign of unexpected plasticity in differentiation capability beyond the original lineage. It has now become tenable, however, to argue that what was originally believed to indicate stem cell plasticity is in fact a result of cell fusion at various locations in the body.

2.2. Comparison of chromatin structure and gene expression patterns between ES cells and somatic cells

2.2.1. *Differential chromatin structure*

When one is examining nuclear reprogramming as it occurs in cell fusion and nuclear cloning, various chromatin-modifying proteins are likely also involved. The gene suppressor complex, polycomb repressive complex 2 (PRC2), is one example. This complex binds to over 200 genetic loci important for development of ES cells and suppresses the genes through trimethylation of lysine 27 of histone H3 in ES cells.[22,23] Importantly, these loci cover approximately one-third of the putative target genes for each of *Oct4*, *Nanog* and *Sox2*. These findings suggest the possibility that ES cells utilize an active mechanism to suppress developmental genes until they become necessary during the differentiation process. It follows, then, that PRC2 and related proteins may be involved in genetic reprogramming during cell fusion via suppression of developmental genes in somatic partner nuclei.

Another support for the above repressive model is that a divalent domain, composed of a large genomic region containing methylated lysine 27 of histone H3 and a small region characterized by methylated lysine 4 of histone H3 (a mark of active genes), is frequently observed in developmentally regulated transcription factor genes in undifferentiated ES cells.[24] During ES cell differentiation this divalent pattern disappears, and only one of the two histone modifications remains. This observation may imply that these transcription factor genes, while repressed, are essentially "poised" to prepare for the later needs of ES cell differentiation. This conjecture is reinforced by the observation that ES cells are unusually enriched with the combination of a suppressive histone modification (trimethylated lysine 27 of histone H3) and chromatin modifications associated with active genes (early replication timing, acetylated lysine 9 of histone H3 and methylated lysine 4 of histone H3).[25] These data related to epigenetic regulation in ES cells are important for contrasting ES cells with their somatic partners since the data can be applied to the study of reprogramming.

2.2.2. *Differential gene expression*

The next approach also aims at identifying ES-cell-specific nuclear proteins that could someday be used to create ES-like cells from differentiated cells. This approach begins with the comprehensive comparison of gene expression patterns between ES cells and differentiated cells by a variety of methods, such as DNA microarray[26,27] and digital differential display[28] (Fig. 2). Once differentially expressed genes are identified and confirmed by Northern blotting or RT-PCR, these genes can be inactivated in ES cells to see if this leads to spontaneous differentiation (loss of self-renewal) and/or loss of pluripotency in the cells. This loss-of-func-

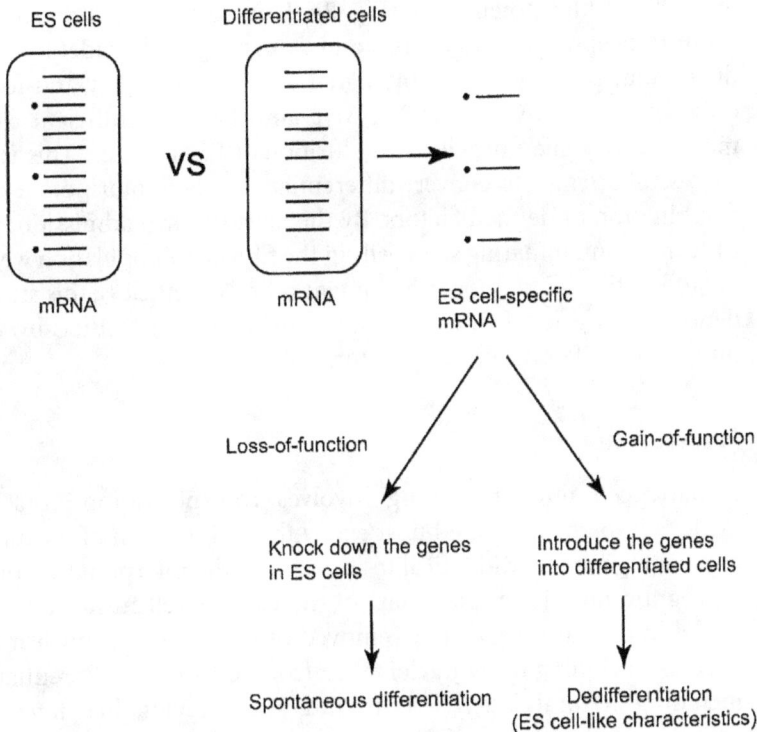

Figure 2. A strategy to identify key genes for self-renewal and pluripotency in ES cells. Comparison of the mRNA populations between ES cells and differentiated cells identifies ES cell-specific mRNAs, which are marked by •. A loss-of-function approach depletes these mRNAs in ES cells to examine the possibility of spontaneous differentiation of the cells. A gain-of-function approach introduces these genes into differentiated cells to monitor acquisition of ES cell-like characteristics in the cells.

tion approach in ES cells was exemplified by a recent report by Ivanova *et al.*[29] Here, by using DNA microarray, the authors found 901 candidate self-renewal genes that were rapidly down-regulated during ES cell differentiation induced by retinoic acid. From these candidates, a systematic short hairpin RNA (shRNA) screen of ES cells identified Esrrb and Tbx3 (both transcription factors) and Tcl1 (a cofactor for Akt1 kinase) as novel self-renewal factors. This observation will potentially expand our understanding of transcriptional regulations and signaling pathways specific to ES cells.

Another approach to the study of the functions of differentially expressed genes is to introduce a subset of these genes into differentiated cells in an attempt to identify a minimum set of genes sufficient to convert unipotent cells to pluripotent cells (Fig. 2). This gain-of-function strategy was beautifully applied by Takahashi and Yamanaka.[30] They identified 24 candidate pluripotency factors by digital differential display and found that only four factors, Oct4, Sox2, c-Myc and Klf4, are sufficient to convert mouse embryonic fibroblasts to pluripotent ES-like cells. This is the first successful attempt to convert differentiated cells to pluripotent cells by a combination of defined factors. By the authors' own admission, it is possible that contaminating stem cells in the fibroblast population were in fact dedifferentiated into more pluripotent cells. Nonetheless, this study is evidence that a gain-of-function approach is a viable method by which to directly identify pluripotency factors.

2.3. Nuclear cloning

Somatic cell nuclear cloning involves transplantation of a somatic nucleus into an enucleated host egg with the intention of creating a new organism genetically identical to the donor. The interpretation of nuclear reprogramming is simpler than in the case of cell fusion because host nuclei in eggs are completely removed or inactivated in nuclear cloning. Successful cloning using nuclei taken from terminally differentiated cells, including natural killer T cells and granulocytes, has unequivocally demonstrated that powerful nuclear reprogramming activities exist in egg cytoplasm.[31,32] These reprogramming activities can induce dedifferentiation of the donor nuclei leading to a topipotent status, sufficiently undifferentiated to support the normal development of animals albeit with quite a low efficiency (less than 3% in mouse cloning).[5] The effect of nuclear reprogramming encompasses a wide spectrum of nuclear physiology, ranging from modifications of nucleosomal and transcriptional machinery all the way to modifications of the global nuclear structure. The magnitude and scale of nuclear reprogramming is most obvious in *Xenopus* cloning, in which the donor nuclei can swell up to 100-fold in volume within one hour after injection into eggs.[33] The physiological functions of these reprogramming activities are assumed to include the maintenance of the undifferentiated state of eggs as well as the promotion of normal development during early embryogenesis. In this sense, the study of the nuclear reprogramming process occurring during cloning, especially as it pertains to transcriptional activities and chromatin regulation, is highly relevant to the study of physiological nuclear activities in eggs and early embryos.

If judged solely by the number of publications, mouse and *Xenopus* are two of the most frequently used species for cloning study. The study of mouse cloning, especially, has the advantage of already having a substantial body of knowledge concerning genetic and epigenetic gene regulations. This knowledge is important because it is vital to the investigation of the abnormal development of cloned embryos in comparison with fertilized embryos. Examples of genetic and epigenetic reprogramming in cloned mice are reactivation of the inactive X chromosome, demethylation and reactivation of imprinted genes, modifications of histone acetylation and methylation, and restoration of shortened telomere length.[34,35] Not surprisingly, these reprogramming processes are in many cases erratic in surviving embryos. For instance, DNA methylation in clones can be greater, the same as, or less than that in fertilized embryos. Telomere length is similarly variable between clones and fertilized embryos. Thus, it is quite inefficient to create a completely normal individual by thoroughly reprogramming differentiated donor nuclei; however, reprogramming can be more frequently accomplished at the cellular level. The greater efficiency of reprogramming at the cellular level is evidenced by the fact that abnormal embryos often still contain many well-differentiated cells, a remarkable finding given the complexity of cellular differentiation mechanisms.

When biochemical approaches are to be applied to identify nuclear reprogramming activities, *Xenopus* eggs become a natural first choice as a material due to the feasibility of collecting a large number of eggs. A commonly used strategy is to purify from *Xenopus* egg extract, using column chromatography, a protein factor that can bring about a specific nuclear change. Methods of making *Xenopus* egg extract are long-established, and have been extensively used for studies of the cell cycle, nuclear transport and other nuclear activities. This makes *Xenopus* egg extract readily available for the study of nuclear reprogramming. The following sections give some examples of nuclear reprogramming factors which have been identified from *Xenopus* egg extract. Note that these sections distinguish *Xenopus* oocytes (prophase of meiosis II) and eggs (metaphase of meiosis II) for accuracy since they significantly differ in terms of RNA transcription and DNA replication.

2.3.1. *Release of transcription factors from nuclei by the chromatin remodeling protein ISWI*

The discovery of the nuclear reprogramming activity of the ATPase ISWI provides an example of how proteins involved in nuclear cloning can be identified through protein purification from *Xenopus* egg extract. An important background for this work was the observation that when somatic nuclear proteins were radiolabeled prior to injection into *Xenopus* oocytes, 80% of the radioactivity disappeared from the injected nuclei within three days.[33] Conversely, when radiolabeled amino acids were injected into oocytes prior to nuclear injection, the injected nuclei strongly accumulated the radioactivity within two hours. This high level of protein exchange between injected nuclei and oocyte cytoplasm is highly likely to be a critical factor in reprogramming of injected nuclei. In order to discover the molecular mechanism behind this high level of exchange, tissue culture cell nuclei were incubated in *Xenopus* egg extract as an *in vitro* model for oocyte injection and selective release of nuclear proteins was monitored.[36] Fractionation of egg extract demonstrated that one of these proteins, the transcription factor TATA-binding protein (TBP), could be released from the nuclei by combining two of the egg fractions and ATP. Continued fractionation identified that the ATP-dependent chromatin remodeling protein ISWI was the active component in the first fraction [Fig. 3(A)]. ISWI can relax histone-DNA interactions and can increase the fluidity of chromatin structure,[37] which probably contributes to the detachment of TBP from DNA. Since TBP is essential for the transcriptional activity of all three RNA polymerases, loss of TBP from the nuclei initiates global shutoff of almost all existing transcriptional activity, probably facilitating transition to a new transcriptional pattern. Loss of TBP from injected nuclei was also reported in mouse cloning.[38]

2.3.2. *Reactivation of Oct4 by the chromatin remodeling protein Brg1*

One interesting finding about reprogramming during nuclear cloning is that *Xenopus* oocytes have the capability of reactivating *Oct4* in injected differentiated mouse cell nuclei, accompanied by DNA demethylation in the promoter region of the gene.[39,40] As mentioned earlier, *Oct4* plays an important role in the maintenance of pluripotency in undifferentiated cells and it is one of the first factors to be lost upon differentiation. In a study utilizing *Xenopus* egg extract, the ATP-dependent chromatin-remodeling protein *Brg1* was demonstrated to be crucial for the reactivation of *Oct4*[41] [Fig. 3(A)]. An *Oct4*-Brg1 connection was also supported by the fact that *Brg1*-null mice die around the implantation period,[42] the time when *in vivo* equivalents of ES cells are formed. In addition to Brg1, lack of any one of three other chromatin remodeling proteins, SNF5, SSRP1 and SNF2H, also causes peri-implantation death of mice (see Ref. 43). Due to the importance of ISWI and Brg1, the link between these other three chromatin remodeling proteins and nuclear cloning/ES cell development merits further research.

2.3.3. *Nucleolar disassembly by FRGY2a/b*

Xenopus oocytes contain more than 1,500 nucleoli that actively transcribe ribosomal RNA (rRNA) and vigorously synthesize ribosomes to prepare for the large demand of protein synthesis during early embryogenesis. When oocytes mature and become eggs (fertilization occurs at this stage), synthesis of rRNA and ribosomes stops and nucleoli become dispersed.[44] The nucleoli reappear and rRNA synthesis commences around the time of the 12th cell division after fertilization. This disappearance and reappearance of endogenous nucleoli is faithfully recapitulated by the nucleoli of donor nuclei after injection into *Xenopus* oocytes. While donor nucleoli enlarge and vigorously transcribe rRNA upon injection into oocytes, the nucleoli disappear within four hours of injection into eggs. The nucleoli of donor nuclei are later reassembled in accordance with the physiological timing of nucleolar reappearance. Similar, albeit less conspicuous, nucleolar reorganization is observed in pig, rabbit and bovine cloning (see Ref. 45).

Loss of nucleoli can be induced in tissue culture cell nuclei via incubation in egg extract. Biochemical purification identified that the Y box protein

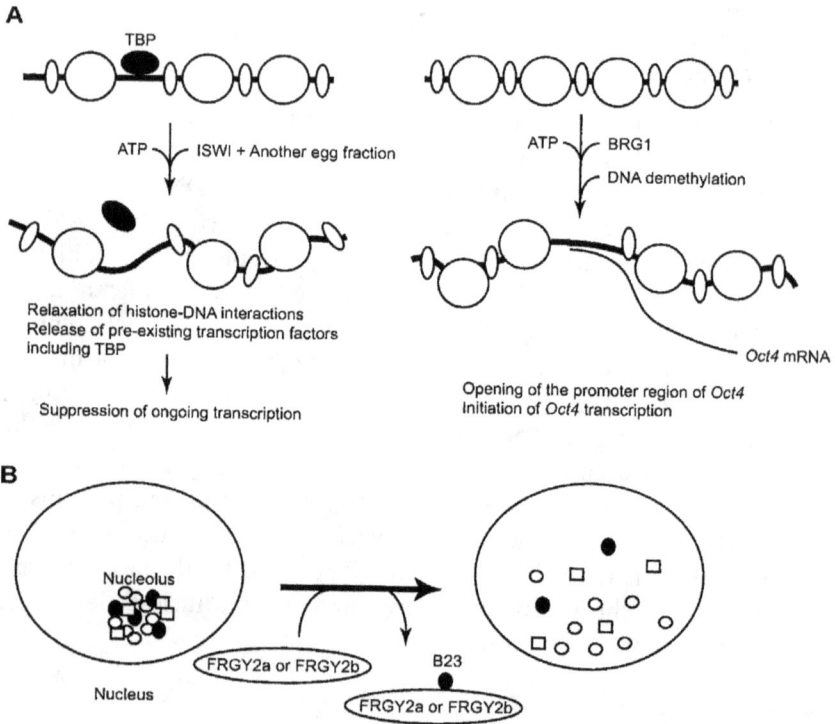

Figure 3. Nuclear reprogramming factors identified from *Xenopus* egg extract. **(A)** The chromatin remodeling ATPase ISWI relaxes nucleosomes and releases transcription factors from chromatin in the presence of a single uncharacterized egg fraction. Another chromatin remodeling ATPase, Brg1, relaxes nucleosomes and activates *Oct4* transcription. DNA demethylation in the promoter region is also involved in the *Oct4* activation. The white circles in the figure represent nucleosomes and the white ovals indicate linker histones. **(B)** Egg proteins FRGY2a and FRGY2b sequester the nucleolar protein B23, triggering nucleolar disassembly. **(C)** The histone-binding protein nucleoplasmin replaces erythrocyte-specific linker histone H1⁰ or H5 with egg-type histone variant B4. This induces decondensation of tightly packed and transcriptionally quiescent erythrocyte chromatin, leading to activation of gene transcription. **(D)** Nucleoplasmin decondenses chromatin in regular somatic cells, which is accompanied by phosphorylation and acetylation of histone H3, and the release of heterochromatin proteins, HP1β and TIF1β.

FRGY2a and its relative FRGY2b are responsible for the nucleolar disassembly in egg extract[45] [Fig. 3(B)]. FRGY2a and FRGY2b were the first proteins discovered to directly disassemble nucleoli. These two proteins share 83% identity at the amino acid level and only about 200 amino acids at the carboxy terminus of one of them are sufficient to disassemble nucleoli. It is well established that transcriptional repression of rRNA leads to

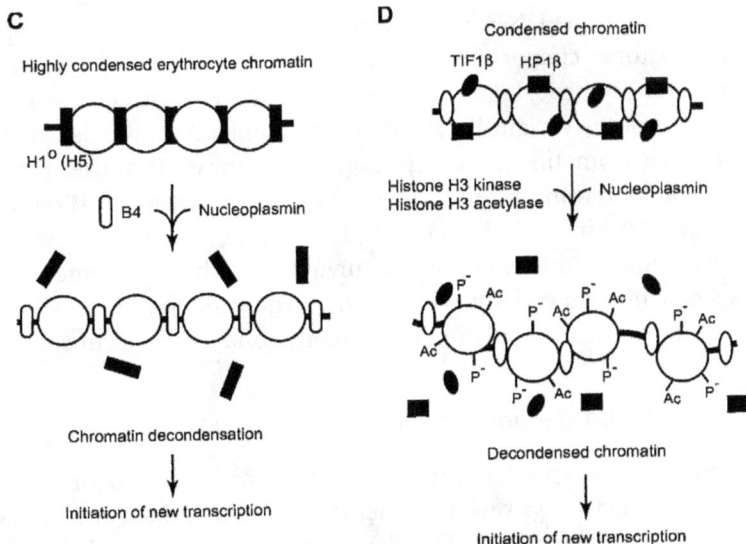

Figure 3. (*Continued*).

nucleolar disassembly but the disassembly by FRGY2a and FRGY2b preserves ongoing rRNA synthesis, uncoupling nucleolar maintenance and rRNA synthesis. Further study revealed that disassembled nucleoli can be reassembled by transferal to oocyte extract, indicating that the nucleolar disassembly by FRGY2a and FRGY2b is not a simple process of destruction. Subsequently, it was shown that FRGY2a's ability to disassemble nucleoli comes from its sequestration of the abundant nucleolar protein B23.[46] The functional implications of nucleolar disassembly in the context of nuclear reprogramming remain to be studied. However, given the function of the nucleolus as a storage area for a variety of nuclear proteins, including cell cycle regulators, tumor suppressor proteins, telomerase, signal recognition particles and tRNA,[47] it is possible that redistribution of some of the nucleolar components significantly contributes to reprogramming of injected nuclei.

2.3.4. *Replacement of linker histone by nucleoplasmin*

Xenopus erythrocyte nuclei, which contain highly condensed chromatin, are quiescent in both transcription and DNA replication. When the nuclei are injected into eggs, however, the chromatin decondenses and the nuclei start to replicate DNA. This chromatin decondensation is accomplished by the histone chaperone nucleoplasmin[48] [Fig. 3(C)]. Nucleoplasmin replaces erythrocyte-specific linker histone $H1^0$ (or H5) with egg type linker histone variant B4, making the chromatin more relaxed. As a result of this chromatin decondensation, the erythrocyte nuclei reacquire the capability to transcribe new genes. This histone exchange is reminiscent of nucleoplasmin's role in a more physiological setting. In the natural egg environment, nucleoplasmin is involved in the replacement of sperm-specific histone variants with egg-derived histones, thus creating somatic-type nucleosomal structure from highly condensed sperm chromatin.[49]

2.3.5. *Chromatin decondensation by nucleoplasmin*

Condensation-specific histone variants do not play a major role in regular somatic nuclei and yet the nuclei show chromatin decondensation when incubated in egg extract. This decondensation is also carried out by nucleoplasmin, but is based on a different mechanism. In this case, histone exchange is undetectable and histone modifications play a prominent role[50] [Fig. 3(D)]. Specifically, nucleoplasmin induces phosphorylation of several amino acids on histone H3, acetylation of lysine 4 on histone H3 and release of heterochromatin proteins HP1β and TIF1β from somatic chromatin. Chromatin decondensation by nucleoplasmin confers on the nuclei the capability to express new genes when appropriate transcription factors are provided. Importantly, chromatin decondensation by nucleoplasmin is observed only in nuclei of undifferentiated such as ES cells or embryonal carcinoma cells. Accumulating evidence suggests that these undifferentiated cells contain more relaxed chromatin than differentiated cells do.[51] Differential sensitivity to nucleoplasmin can be one manifestation of this underlying difference in overall chromatin structure between the two types of the cells. If so, detailed analysis of the mechanism by which decondensation occurs may provide new insight into the chromatin and nuclear organization of undifferentiated cells.

Conclusion

The ultimate goal of the above approaches is to identify nuclear factors sufficient to reprogram differentiated cells into pluripotent cells with the hope that these dedifferentiated cells can be redifferentiated for transplantation purposes. Looking back on the rapid progress of this field in the last few years, this goal seems to be attainable within the next 10 or 20 years. To reach this goal, however, further investigation into the basic cellular and developmental biology of stem cells is clearly needed. The main emphasis in this field is currently placed on transcription factors, chromatin proteins and signaling molecules. However, an increased examination of other avenues of stem cell research, including microRNA, translational regulation and DNA repair, will likely also prove to be prerequisites for research into nuclear reprogramming.

Note added in proof

The progress of the differentiation study by defined factors is quite rapid. Two groups have recently succeeded in inducing pluripotent cells (called induced pluripotent stem cells, iPS) from various differentiated human cells by introducing different sets of four genes. See Takahashi *et al.* (2007) *Cell* **131**: 1–12, and Yu *et al.* (2007) *Science* Nov 20; [Epub ahead of print].

Acknowledgments

I thank Sam Rayner for critical reading of the manuscript. Our work related to this manuscript has been supported by the NIH grant GM 068027.

References

1. Hochedlinger K, Jaenisch R. (2006) Nuclear reprogramming and pluripotency. *Nature* **441**: 1061–1067.
2. Briggs R, King TJ. (1952) Transplantation of living nuclei from blastula cells into enucleated frogs' eggs. *Proc Natl Acad Sci USA* **38**: 455–463.
3. Gurdon JB, Uehlinger V. (1966) "Fertile" intestine nuclei. *Nature* **210**: 1240–1241.
4. Wilmut I, Schnieke AE, McWhir J, *et al.* (1997) Viable offspring derived from fetal and adult mammalian cells [see comments] [erratum appears in *Nature* **386**(6621): 200]. *Nature* **385**: 810–813.

5. Wakayama T, Perry ACF, Zucotti M, *et al.* (1998) Full term development of mice from enucleated oocytes injected with cumulus cell nuclei. *Nature* **394:** 369–374.

6. Gourdeau H, Fournier RE. (1990) Genetic analysis of mammalian cell differentiation. *Annu Rev Cell Biol* **6:** 69–94.

7. Blau HM. (1992) Differentiation requires continuous active control. *Annu Rev Biochem* **61:** 1213–1230.

8. Killary AM, Fournier RE. (1984) A genetic analysis of extinction: transdominant loci regulate expression of liver-specific traits in hepatoma hybrid cells. *Cell* **38:** 523–534.

9. Jones KW, Shapero MH, Chevrette M, Fournier RE. (1991) Subtractive hybridization cloning of a tissue-specific extinguisher: TSE1 encodes a regulatory subunit of protein kinase A. *Cell* **66:** 861–872.

10. Boshart M, Weih F, Nichols M, Schutz G. (1991) The tissue-specific extinguisher locus TSE1 encodes a regulatory subunit of cAMP-dependent protein kinase. *Cell* **66:** 849–859.

11. Chin AC, Fournier RE. (1989) Tse-2: a trans-dominant extinguisher of albumin gene expression in hepatoma hybrid cells. *Mol Cell Biol* **9:** 3736–3743.

12. Tada M, Tada T, Lefebvre L, *et al.* (1997) Embryonic germ cells induce epigenetic reprogramming of somatic nucleus in hybrid cells. *EMBO J* **16:** 6510–6520.

13. Cowan CA, Atienza J, Melton DA, Eggan K. (2005) Nuclear reprogramming of somatic cells after fusion with human embryonic stem cells. *Science* **309:** 1369–1373.

14. Do JT, Scholer HR. (2004) Nuclei of embryonic stem cells reprogram somatic cells. *Stem Cells* **22:** 941–949.

15. Matsumura H, Otsuji T, Yasuchika K, *et al.* (2006) Targeted chromosome elimination from ES-somatic hybrid cells. *Nat Methods* **4:** 23–25.

16. Silva J, Chambers I, Pollard S, Smith A. (2006) Nanog promotes transfer of pluripotency after cell fusion. *Nature* **441:** 997–1001.

17. Boyer LA, Lee TI, Cole MF, *et al.* (2005) Core transcriptional regulatory circuitry in human embryonic stem cells. *Cell* **122:** 947–956.

18. Loh YH, Wu Z, Chew JL, *et al.* (2006) The *Oct4* and Nanog transcription network regulates pluripotency in mouse embryonic stem cells. *Nat Genet* **38:** 431–440.

19. Ying QL, Nichols J, Evans EP, Smith AG. (2002) Changing potency by spontaneous fusion. *Nature* **416:** 545–548.

20. Terada N, Hamazaki T, Oka M, *et al.* (2002) Bone marrow cells adopt the phenotype of other cells by spontaneous cell fusion. *Nature* **416:** 542–545.

21. Vassilopoulos G, Russell DW. (2003) Cell fusion: an alternative to stem cell plasticity and its therapeutic implications. *Curr Opin Genet Dev* **13:** 480–485.

22. Lee TI, Jenner RB, Boyer LA, *et al.* (2006) Control of developmental regulators by Polycomb in human embryonic stem cells. *Cell* 125: 301–313.

23. Boyer LA, Plath K, Zeitlinger J, *et al.* (2006) Polycomb complexes repress developmental regulators in murine embryonic stem cells. *Nature* 441: 349–353.

24. Bernstein BE, Mikkelsen TS, Xie X, *et al.* (2006) A bivalent chromatin structure marks key developmental genes in embryonic stem cells. *Cell* 125: 315–326.

25. Azuara V, Perry P, Sauer S, *et al.* (2006) Chromatin signatures of pluripotent cell lines. *Nat Cell Biol* 8: 532–538.

26. Ivanova NB, Dimon JT, Schaniel C, *et al.* (2002) A stem cell molecular signature. *Science* 298: 601–604.

27. Ramalho-Santos M, Yoon S, Matsuzaki Y, *et al.* (2002) "Stemness": transcriptional profiling of embryonic and adult stem cells. *Science* 298: 597–600.

28. Mitsui K, Tokuzawa Y, Itoh H, *et al.* (2003) The homeoprotein Nanog is required for maintenance of pluripotency in mouse epiblast and ES cells. *Cell* 113: 631–642.

29. Ivanova N, Dobrin R, Lu R, *et al.* (2006) Dissecting self-renewal in stem cells with RNA interference. *Nature* 442: 533–538.

30. Takahashi K, Yamanaka S. (2006) Induction of pluripotent stem cells from mouse embryonic and adult fibroblast cultures by defined factors. *Cell* 126: 663–676.

31. Inoue K, Wakao H, Ogonuki N, *et al.* (2005) Generation of cloned mice by direct nuclear transfer from natural killer T cells. *Curr Biol* 15: 1114–1118.

32. Sung LY, Gao S, Shen H, *et al.* (2006) Differentiated cells are more efficient than adult stem cells for cloning by somatic cell nuclear transfer. *Nat Genet* 38: 1323–1328.

33. Gurdon JB, Laskey RA, De Robertis EM, Partington GA. (1979) Reprogramming of transplanted nuclei in amphibia. *Int Rev Cytol Suppl* 9: 161–178.

34. Jeanisch R, Eggan K, Humpherys D, *et al.* (2002) Nuclear cloning, stem cells, and genomic reprogramming. *Cloning Stem Cells* 4: 389–396.

35. Mullins LJ, Wilmut I, Mullins JJ. (2004) Nuclear transfer in rodents. *J Physiol* 554: 1–12.

36. Kikyo N, Wade PA, Guschin D, *et al.* (2000) Active remodeling of somatic nuclei in egg cytoplasm by the nucleosomal ATPase ISWI. *Science* 289: 2360–2362.

37. Corona DF, Langst G, Clapier CR, *et al.* (1999) ISWI is an ATP-dependent nucleosome remodeling factor. *Mol Cell* 3: 239–245.

38. Kim JM, Ogura A, Nagata M, Aoki F. (2002) Analysis of the mechanism for chromatin remodeling in embryos reconstructed by somatic nuclear transfer. *Biology of Reproduction* 67: 760–766.

39. Byrne JA, Simonsson S, Western PS, Gurdon JB. (2003) Nuclei of adult mammalian somatic cells are directly reprogrammed to *Oct-4* stem cell gene expression by amphibian oocytes. *Curr Biol* 13: 1206–1213.

40. Simonsson S, Gurdon J. (2004) DNA demethylation is necessary for the epigenetic reprogramming of somatic cell nuclei. *Nat Cell Biol* 6: 984–990.

41. Hansis C, Barreto G, Maltry N, Niehrs C. (2004) Nuclear reprogramming of human somatic cells by *Xenopus* egg extract requires *Brg1*. *Curr Biol* 14: 1475–1480.

42. Bultman S, Gebuhr T, Yee D, *et al.* (2000) A *Brg1* null mutation in the mouse reveals functional differences among mammalian SWI/SNF complexes. *Mol Cell* 6: 1287–1295.

43. de la Serna IL, Ohkawa Y, Imbalzano AN. (2006) Chromatin remodelling in mammalian differentiation: lessons from ATP-dependent remodellers. *Nat Rev Genet* 7: 461–473.

44. Gurdon JB. (1965) Cytoplasmic regulation of RNA synthesis and nucleolus formation in developing embryos of *Xenopus laevis*. *J Mol Biol* 12: 27–35.

45. Gonda K, Fowler J, Katoku-Kikyo N, *et al.* (2003) Reversible disassembly of somatic nucleoli by the germ cell proteins FRGY2a and FRGY2b. *Nat Cell Biol* 5: 205–210.

46. Gonda K, Wudel J, Nelson D, *et al.* (2006) Requirement of the protein B23 for nucleolar disassembly induced by the FRGY2a family proteins. *J Biol Chem* 281: 8153–8160.

47. Olson MO, Hingorani K, Szebeni A. (2002) Conventional and nonconventional roles of the nucleolus. *Int Rev Cytol* 219: 199–266.

48. Dimitrov S, Wolffe AP. (1996) Remodeling somatic nuclei in *Xenopus* laevis egg extracts: molecular mechanisms for the selective release of histones H1 and H1^0 from chromatin and the acquisition of transcriptional competence. *EMBO J* 15: 5897–5906.

49. Philpott A, Leno GH, Laskey RA. (1991) Sperm decondensation in *Xenopus* egg cytoplasm is mediated by nucleoplasmin. *Cell* 65: 569–578.

50. Tamada H, Van Thuan N, Reed P, *et al.* (2006) Chromatin decondensation and nuclear reprogramming by nucleoplasmin. *Mol Cell Biol* 26: 1259–1271.

51. Meshorer E, Misteli T. (2006) Chromatin in pluripotent embryonic stem cells and differentiation. *Nat Rev Mol Cell Biol* 7: 540–546.

4

Cell Cycle Control of Stem Cell Fate

Lucas Nacusi and Robert Sheaff*

1. Introduction

Regenerative medicine is being transformed from hypothetical Valhalla to viable reality with the identification, isolation, and characterization of distinct stem cell populations. A better understanding of molecular events controlling their proliferation is now required so stem cells can be expanded for therapeutic purposes.

The logical imperatives of cell division entail increasing mass, accurately replicating DNA once and only once, and properly segregating genetic material between daughter cells during cellular division. The overall process encompassing these events is called the cell cycle. In order to protect genomic integrity and ensure proper execution of the proliferative program, DNA replication in the S phase is bracketed by Gap phases 1 and 2 to separate it from cell division in the M phase. Work in experimental systems like yeast, invertebrates, and mammalian tissue culture cells identified evolutionarily conserved components controlling proliferation (termed cell cycle machinery), suggesting that common regulatory mechanisms are employed in diverse cell types. In multicellular organisms, the individual cells must subordinate this proliferative potential to the greater good, so division is tightly regulated to prevent inappropriate cellular expansion. Thus, for most cell types proliferation is but a brief interlude on the way to becoming a differentiated, nonreplicating cell of specialized function. Stem cells are unique in that their proliferative potential must be maintained because they are ultimately responsible for generating all cells in the fully developed organism, as well as replacing lost or injured cells in the adult.

*Correspondence: Department of Chemistry and Biochemistry, University of Tulsa, Tulsa, OK, USA.
E-mail: robert-sheaff@utulsa.edu

The importance of understanding how stem cells maintain and control the cell cycle at the molecular level is highlighted by the emerging field of regenerative medicine, which hopes to generate sufficient numbers of these remarkably malleable cells to treat human diseases and repair damaged tissues. This article describes evolutionarily conserved mechanisms thought to control duplication of eukaryotic cells, and investigates their applicability to understanding stem cell expansion. Similarities and differences between non-stem-cell and stem cell proliferation are discussed, with particular emphasis on recent results from transgenic mice that question fundamental assumptions about the role of cell cycle machinery. These surprising observations suggest that unique mechanisms might regulate stem cell proliferation, a possibility with important implications for their effective application in regenerative medicine.

2. Nature of the Stem Cell

Discussing stem cell proliferation requires a workable definition of this distinct cell type, which is most dramatically illustrated during the development of a multicellular organism.[1,2] As Virchow stated in 1885, "*Omnis cellula e cellula*" ("All cells come from cells"); so the two initial cells of the fertilized egg must be capable of self-renewal and differentiation into various cell lineages required to form the adult. Stem cell populations are also present in the fully developed organism, where they help maintain homeostasis by replacing cells lost to normal turnover or injury. How stem cells coordinate self-renewal with differentiation remains one of the great mysteries of stem cell biology.

2.1. Self-renewal and differentiation

Stem cells function as a cell "factory" for increasing cell number and diversity during embryonic development, organismal growth, and tissue regeneration.[1,2] During human development total cell number must increase from two to trillions, while at the same time generating approximately 220 different cell types composing the tissues and organs.[2] Embryonic stem cells (ES cells) accomplish this remarkable feat by undergoing asymmetric divisions in which one daughter commits to a differentiation pathway while the other remains a stem cell (Fig. 1). Adult stem cells have a similar

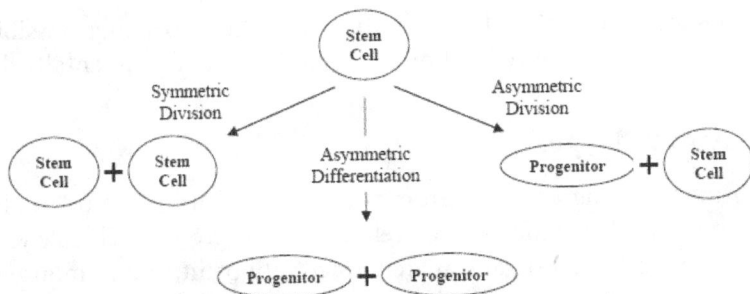

Figure 1. Stem cell proliferative outcomes. A dividing stem cell can undergo symmetric division to generate two identical stem cells, symmetric differentiation to generate two progenitor cells committed to a differentiation pathway(s), and asymmetric division to produce a copy of the original stem cell (self-renewal) and one progenitor cell committed to differentiation.

capability, although differentiation outcomes are generally more restricted and determined by the tissue in which they reside. For instance, stem cells located in the bone marrow generate eight separate cell lineages making up cellular components of blood.[1] For an adult mouse this means replacing $\sim 2.4 \times 10^8$ red blood cells and 4×10^6 nonlymphoid peripheral blood cells every day, not including cells lost due to injury or apoptosis (e.g. blood loss, infection, and/or ingestion of cytotoxic chemicals).[3] As a consequence of this remarkable regenerative capacity, in two years a mouse generates blood cell components equal to 60% of its body weight, while adult humans synthesize their body weight in only seven years.[4] The stem cell properties of self-renewal and differentiation can be functionally analyzed by transplantation and rescue of mice that have been lethally irradiated to ablate host stem cell populations. Self-renewal properties can be further demonstrated by serial transfer of transplanted stem cells into secondary recipients.[5]

Like a cellular sword of Damocles, stem cells must maintain their self-renewing ability while at the same time preventing inappropriate expansion that would threaten organism survival. This balancing act is accomplished by transiently exiting the cell cycle for extended periods (quiescence) rather than irreversibly withdrawing (senescence). Throughout its lifespan the stem cell must remain vigilant about maintaining genomic integrity, because any mistakes will be perpetuated through cell lineages and possibly passed to future generations. Thus, the unique demands of stem cell proliferation raise the intriguing possibility that their proliferative control mechanisms may be fundamentally different.

2.2. Stem cell types

Athena sprang complete from the head of Zeus, but most multicellular organisms begin life as a two-cell fertilized egg. Stem cells able to generate a complete adult organism are termed totipotent, and in mouse embryogenesis, totipotency is retained up until the eight-cell stage of the morula. At this point the blastocyst is formed, in which outer trophoblast cells surround an inner cell mass (ICM) of undifferentiated, pluripotent stem cells that will eventually develop into the essential embryonic cell types — ectoderm, mesoderm, and endoderm — as well as primordial germ cells giving rise to male and female gametes (Fig. 2).[2] Pluripotency was demonstrated by transplanting cultured mouse embryonic stem cells (mES cells) into blastocysts and showing that they contribute to all cell lineages of the resulting chimeric mouse.[2]

In the completed organism, distinct stem cell populations are present in different tissues and exhibit identifiable individual characteristics. These adult stem cells are described as multipotent and help maintain homeostasis by replacing cells lost to normal turnover or injury. They can be distinguished from each other based on the identity of the tissue from which they are isolated, their differentiation ability, the stage of development at which they are present, gene expression patterns, etc.[1] Adult stem cells also exhibit shared properties such as absence of differentiation

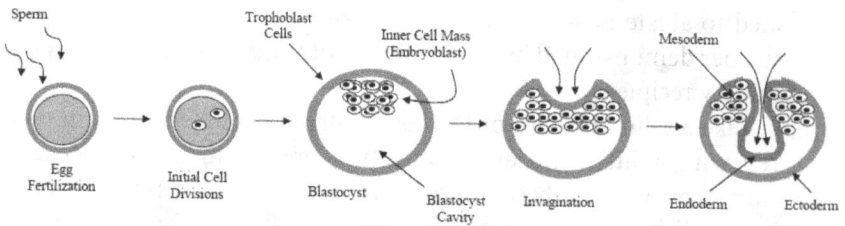

Figure 2. Origin of embryonic cell types. Egg fertilization is followed by initial division of totipotent stem cells to generate the eight-cell morula. These cells then differentiate to form the blastocyst, which is composed of trophoblast cells encompassing the inner cell mass. Blastular invagination forms the three different germ layers — ectoderm, endoderm, and mesoderm. Coordinated growth of different cell types in these layers generates the embryo.

markers, a criterion often used to define and enrich them for therapeutic purposes. Stem cells isolated from specific tissues can be quite Proteus-like under different environmental or *in vitro* culturing conditions, generating differentiated cells from other cell lineages. A hematopoietic stem cell (HSC), for example, can transdifferentiate into neural tissue and vice versa.[1]

Regardless of identity, all eukaryotic cells must complete the same basic tasks in order to replicate and divide. Since stem cells give rise to all other cell types, proliferative control mechanisms are likely to be conserved at some level. It is therefore instructive to discuss the wealth of information on duplication of nonstem or somatic cells, and then determine its applicability to the ancestral parent.

3. Mechanisms of Proliferative Control

A cell can be considered the fundamental unit of complex organisms because it is the smallest entity capable of copying itself. All cells are made up of the same basic components, and so regardless of type must accomplish the same basic tasks during duplication: increase mass, accurately replicate DNA, and properly segregate genetic material during division. While the decision to proliferate is determined in large part by extracellular signaling, internal mechanisms regulate cell cycle progression and ensure accuracy of the process.[6]

3.1. The cell cycle

Historically cell division has been envisioned as generating two identical daughter cells (e.g. yeast and mammalian tissue culture cells).[7] A dividing stem cell has other options, however, which may require additional and/or different mechanisms to ensure desired outcomes. Addressing this possibility first requires discussing how an idealized, nonstem somatic cell is thought to replicate and divide.

Cellular division is the universal process by which all cells duplicate their DNA and separate it into newly formed daughter cells.[7] In eukaryotes DNA replication and cell division occur sequentially and are coordinated with cellular growth in a highly ordered sequence of events. Separation of duplicated chromosomes does not begin until all DNA has been fully replicated,

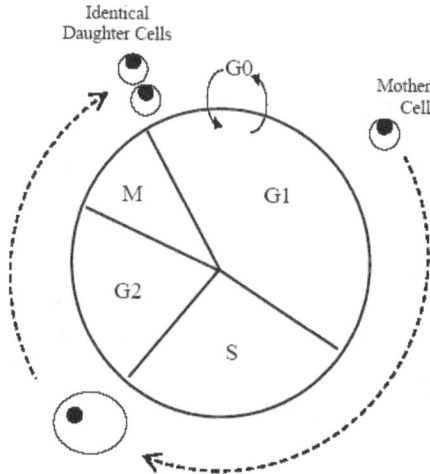

Figure 3. Eukaryotic cell cycle. In order to proliferate a cell must enter the cell cycle and proceed in a temporal and regulated manner through G1 (Gap1), S (DNA replication), G2 (Gap2), and M (mitosis or cell division) phases. If proliferative conditions are not satisfactory (e.g. low growth factors) or the cell adopts an alternative fate, it will exit the cycle and enter a quiescent state termed G0.

and DNA replication is not reinitiated until mitosis is completed. In most cell types, the S and M phases are separated by two gap phases such that a complete cycle can be represented by the sequence G1/S/G2/M (Fig. 3). This organization ensures that each daughter receives an identical copy of the hereditary material and so avoids genomic instability through succeeding generations. If conditions are not appropriate for division, the cell can exit the proliferative cycle and transiently enter a quiescent G0 phase.[8]

3.2. Cell cycle machinery

The cell cycle field developed from physical observation of cell growth and division.[7] Biochemical studies of amphibian egg extracts and genetic studies of yeast led to the discovery of molecules controlling cell cycle progression.[9–12] A "maturation-promoting factor" (MPF) purified from frog oocytes allowed egg extracts to pseudocycle between the S and M phases.[9] Characterization of yeast mutants led to identification of cdc28, a cell division control protein required for cell cycle transitions.[10,11] These disparate findings converged with the realization that MPF and cdc28 are related protein kinases. They were subsequently found to be associated with cyclin subunits, proteins originally identified in sea urchin eggs that were synthesized and destroyed at specific points in the cell cycle.[12]

The activity of the cyclin-dependent kinase (CDK) oscillates throughout the cell cycle due to periodic synthesis and destruction of its cyclin binding partner, activating/inhibiting phosphorylations, and tight binding to cyclin-dependent kinase inhibitors (CKIs) (Fig. 4).[13,14] The expression of specific cyclins at defined points is thought to play an important role in maintaining the temporal order of events.[15] Cyclin binding alone is insufficient to activate CDK; the kinase must also be phosphorylated by the CDK-activating kinase (CAK). Together these modifications cause a conformational change in the T loop region so substrates can bind the catalytic pocket, and reorient the bound ATP so its phosphate can be transferred.[13] In contrast, phosphorylation of CDK residues T14 and Y15 by Wee1/Myt-1 inhibit cyclin–CDK activity, which can be reversed by the CDC25 phosphatase.[13,16] CDK activity can also be negatively regulated by small tight binding inhibitors that are divided into two families based on sequence homology and affinity for CDKs: (a) The INK family (p15^{INK4b}, p16^{INK4a}, p18^{INK4c}, and p19^{INK4d}), which preferentially bind and inhibit CDK4 and CDK6 to prevent their association with D type cyclins; and (b) The CIP/KIP family (p21^{CIP1}, p27^{KIP1}, and p57^{KIP2}), which inhibit a broad spectrum of cyclin-CDK complexes by tightly binding both components.[6]

3.3. G1 progression

For somatic cells the G1 phase is thought to be the key regulatory period for deciding to continue the proliferative cycle or withdraw and adopt an alternative fate (Fig. 4). This decision is influenced by reception and interpretation of signals from other cells, nutrient availability, and presence or absence of mitogens (growth factors, hormones, cytokines, etc.), in conjunction with an internal evaluation of the cell's readiness and ability to successfully carry out all the operations necessary for faithful duplication.[6] CDK activation is considered essential for cell cycle progression and completion, and is linked to extracellular events in the following manner. During early G1, mitogens initiate signaling pathways that generate functional cyclin D–CDK4/6, which phosphorylates pocket proteins (retinoblastoma protein [pRb] and related family members p107 and p130) that are inhibiting E2F-Dp1/2 transcription (Fig. 5).[17] E2F then transcribes a broad range of genes required for cell cycle progression and DNA synthesis.[18] Cyclin D–CDK4/6 is viewed as partially rate-limiting for

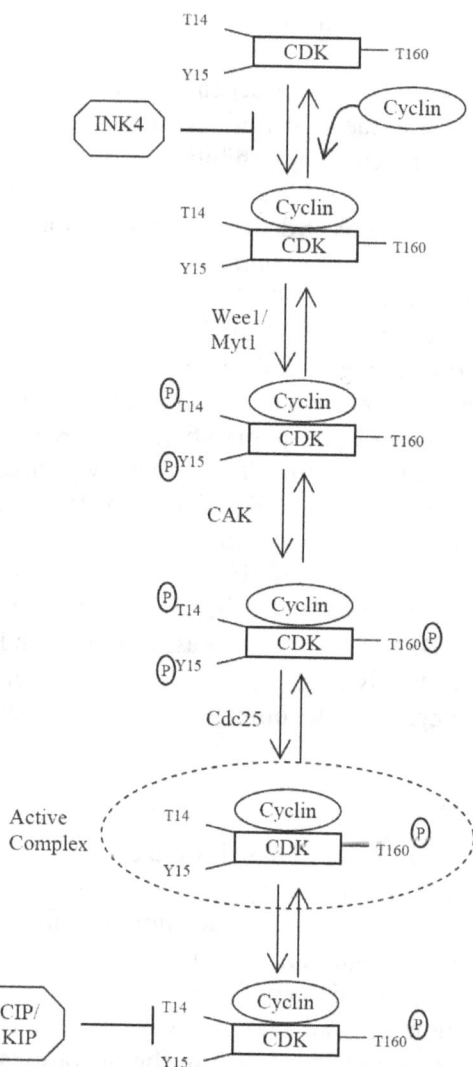

Figure 4. Mechanism of CDK regulation. CDK activation is a multistep process requiring cyclin binding and CDK phosphorylation by CDK-activating kinase (CAK) at T160 (or equivalent). Together these events permit binding of the protein substrate and induce conformational change of the CDK active site to reorient the ATP so phosphate transfer can occur. In contrast, phosphorylation of CDK residues T14 and/or Y15 by Wee1/Myt1 is inhibitory, so their removal by Cdc25 leads to cyclin–CDK activation. CDK activation and activity can also be blocked by small, tight-binding inhibitory proteins called cyclin-dependent kinase inhibitors. Members of the INK4 family block association of cyclin D family members with CDK4/6, while members of the CIP/KIP family are more general inhibitors targeting the cyclin–CDK complex.

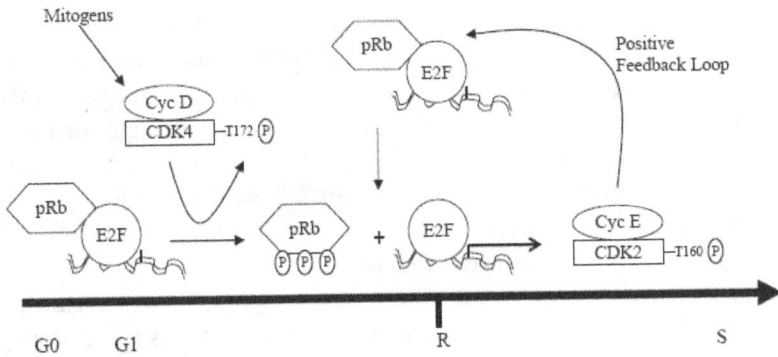

Figure 5. G1 progression and cell fate. Mitogen stimulation leads to formation of active cyclin D–CDK4/6 complexes, which phosphorylate pRb to release E2F and initiate transcription of genes involved in growth and cell cycle progression. One of these is cyclin E, which associates with CDK2 and also phosphorylates pRb in a positive feedback loop that may signify commitment to completion of the cell cycle (the Restriction point).

G1 progression because overexpressing cyclin D1 shortens G1 and can overcome various forms of G1 arrest.[6] Furthermore, inhibiting cyclin D–CDK4/6 activity via microinjected cyclin D1 antibodies, cyclin D1 antisense, or p16[INK4a] inhibitory peptides all cause G1 arrest.[6]

One of the main transcriptional targets of E2F-Dp1/2 is cyclin E, which when synthesized forms a complex with CDK2. Cyclin E–CDK2 phosphorylates additional pRb and activates more E2F-Dp1/2, resulting in a burst of cyclin E synthesis via a positive feedback loop.[6] This reaction may contribute to progression through the Restriction point, the switch to mitogen independence, and commitment to completing the proliferative cycle (Fig. 5).[19] Cyclin E–CDK2 also drives G1 progression and prepares for future events by targeting its own inhibitor p27[KIP1] for proteasomal degradation, phosphorylating DNA replication machinery, and initiating duplication of the centrosome (microtubule organizing center for chromosome segregation).[6] Once these tasks are complete, cyclin E–CDK2 self-destructs by phosphorylating the cyclin and targeting it for proteasomal degradation.[6] Like cyclin D, cyclin E is considered essential and partially rate-limiting for G1 progression because its overexpression shortens the G1 phase, and DNA replication and cell cycle progression are both blocked by dominant negative CDK2, microinjection of CDK2 antibodies, CDK2 antisense, and CDK2 inhibitors.[19]

As a consequence of cyclin E–CDK2 and E2F activity, cyclin A is synthesized and associates with CDK2 to monitor DNA replication and prepare for the M phase. As soon as DNA is completely replicated once and only once, the cell proceeds to the G2/M phase, where cyclin B–CDK1 ensures that chromosomes are properly segregated and coordinates division into two identical daughter cells (Fig. 3).[8] Throughout the cycle, an additional layer of molecular checkpoints monitor progress, scan for and correct errors, and ensure orderly transitions.[7] After dividing, new cells are in G1 and must decide whether to initiate a new proliferative cycle or withdraw and adopt an alternative fate.[13]

This well-supported model is generally accepted as an accurate description of how cells regulate their proliferative cycle. Its basic premises hold from yeast to human cells, and so it seems a foregone conclusion that the same cycle machinery should regulate stem cell proliferation in a like manner. As discussed below, however, fundamental differences in purposes and potential outcomes may necessitate unique mechanisms and/or strategies of proliferative control for stem cells.

4. The Stem Cell Proliferative Cycle

Understanding control of stem cell proliferation is essential for achieving their full potential in regenerative medicine. Stem cells are unique in that self-renewal must be coordinated with differentiation in order to fulfill the needs of the organism (Fig. 1). Their extensive proliferative capabilities require continual monitoring and tight regulation to avoid inappropriate cellular expansion associated with the tumorigenic phenotype. Finally, the status of stem cells as repositories of genetic information that will be passed to future generations means that protecting genomic integrity is paramount. These unique characteristics raise the possibility that stem cell proliferative control mechanisms are different from those in nonstem cells.

4.1. Deciding to divide

Manipulating stem cell proliferation to bypass limitations on expansion is essential for obtaining sufficient quantities for regenerative therapies, but it remains a Herculean task because multiple factors affect the decision to divide and often vary depending on the stem cell type. During early development, embryonic stem cells (ES cells) must replicate rapidly and repeatedly to generate the large number of cells required by the growing organism.[2]

This extensive proliferation is likely driven by intrinsic mechanisms since extracellular signaling pathways are either nonexistent or still developing. Consistent with this idea, ES cells retain their proliferative robustness when cultured *in vitro*.[20] During development, ES cells must still communicate to regulate cell number and direct production of properly sized tissues and organs; they accomplish this in part by using gap junctions that allow direct cytoplasmic exchange of small molecules between adjacent cells. As intercellular signaling pathways are established, gap junctions become less important.[2] In *C. elegans*, for instance, blocking differentiation of germ-line stem cells and maintaining proliferative capacity requires activation of the Notch-related receptor GLP-1.[5] This receptor is activated by a neighboring cell that displays its membrane-bound ligand LAG-2.

The ability to isolate and maintain ES cell lines in culture allowed further identification and analysis of signals promoting proliferation.[20] These lines were derived from the inner cell mass of blastocyctes (Fig. 2), and cultured on a feeder layer of mouse embryo fibroblasts (MEFs) or using conditioned media. The ability of MEFs to support these lines implied that they provide factor(s) necessary for self-renewal and/or blocking differentiation. Leukemia inhibitory factor (LIF) was identified as a soluble glycoprotein of the interleukin-6 (IL-6) family of cytokines that prevents differentiation.[2] LIF acts through membrane-bound gp130 signaling to induce activation of the signal transducer and activator of transcription-3 (STAT3) (Fig. 6). Varying LIF concentration changed the probability of differentiation but did not affect the growth rate, indicating that appropriate levels and types of cytokine stimulation are required to maintain self-renewal. The complete absence of IL-6 family members, removal of MEFs, or STAT3 inactivation promotes spontaneous differentiation of ESCs *in vitro*.[2]

Undifferentiated mouse embryonic stem cells (mES cells) are characterized by expression of specific cell surface antigens, membrane-bound receptors, and alkaline phosphatase/telomerase activity. They also express the epiblast/germ cell restricted transcription factor Oct-3/4, which is essential for development and maintenance of pluripotency in the ICM.[2] A major function of Oct3/4 may be to repress expression of differentiation genes.[21] Its complete absence induces formation of trophectoderm and loss of pluripotency, while modest increases in its levels (< twofold) causes differentiation into endoderm and mesoderm.[2] These different outcomes emphasize the importance of precise transcriptional regulation

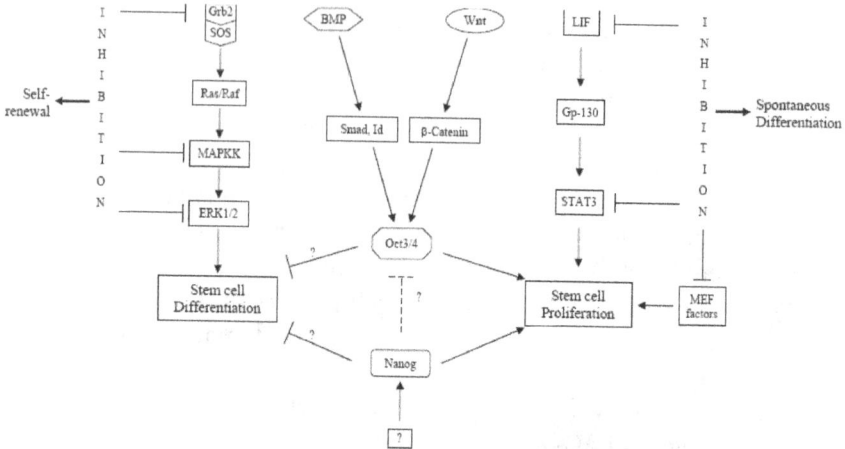

Figure 6. Signaling pathways determining stem cell fate. Depicted are some of the well-known signal transduction pathways implicated in stem-cell-fate decisions. The choice between self-renewal and differentiation involves multiple regulatory inputs that are only partially understood, many of which converge on transcription factors like Oct3/4 and Nanog. The duration and extent of their activation are crucial for determining the outcome, and they can operate by inhibiting differentiation signals (BMP pathway) and by promoting proliferative signals (Wnt pathway). Poorly understood regulatory interactions between Nanog and Oct3/4 also appear to influence stem-cell-fate decisions.

in controlling cell fate determination. Nanog is another recently identified homeodomain protein that is a key regulator of ES cell pluripotency.[2] It is restricted to and required in epiblast cells, where its forced expression causes constitutive self-renewal in the absence of LIF. Nanog may act by blocking the differentiation-inducing effect of Oct3/4 (Fig. 6).[2] STAT3 and Oct3/4 may interact to coordinate transcriptional regulation of genes involved in cell fate decisions.[22]

LIF alone does not inhibit differentiation of human embryonic stem cells (hES cells) — unlike the mouse version — revealing surprising species-dependent differences in stem cells from the same origin.[2] They must be maintained on a layer of MEFs or other feeder cells derived from human tissues. This unexpected difference emphasizes the perils of generalizing about molecular mechanisms controlling stem cell fate, and led to the search for additional required factors. LIF can sustain self-renewal and differentiation capabilities of hES cells in the presence of bone morphogenic protein (BMP), which functions by inducing expression of inhibitor-of-differentiation (Id) genes via activation of the Smad signaling pathway (Fig. 6).[2] Lineage-specific transcription factors causing differentiation are blocked by Id proteins, thereby allowing the self-renewal response to LIF/STAT3. Consistent with this idea, forced Id gene expression mediates release from BMP and serum dependence so that LIF alone maintains pluripotency.[2]

This Daedalian labyrinth of interacting signaling pathways is further complicated by the observation that other well-known signal transduction pathways have also been implicated in control of embryonic stem cell fate, oftentimes with consequences different than those observed in nonstem cells. Activation of the Ras/MAPK (mitogen-activating protein kinase) pathway is commonly associated with somatic cell proliferation, yet in the case of stem cells it promotes differentiation (Fig. 6).[2] Inhibiting Ras/MAPK blocks the commitment to specific lineages and helps maintain ES cell self-renewal, suggesting that Ras/MAPK influences Nanog and Oct3/4 activity.[2] Additional pathways like Wnt signaling also help maintain pluripotency by modulating Nanog and Oct3/4 expression.[1] Ultimately, ES cell self-renewal depends on a balance among diverse signaling inputs and direct interactions with neighboring cells. These pathways likely affect proliferative control mechanisms in ways that may be stem-cell-type and even species-specific, but the precise interactions remain to be identified and characterized.

Adult stem cells operate under much different constraints than their embryonic counterparts. They reside in a defined niche and proliferate only rarely to replace cells lost to normal turnover or injury. In addition to their infrequent replication under normal conditions, adult stem cells are known for long cell cycle transit times that may contribute to their survival, self-renewal, and differentiation.[5] Adhesion molecules like integrins play an important role in this process by restraining stem cells within their microenvironment and helping to maintain their proliferative potential.[23] This quiescence was originally thought to reflect complete cell cycle arrest, and under normal conditions to describe the majority of the stem cell population. In blood, for example, this clonal-succession model posited that although a large supply of hematopoietic stem cells (HSCs) exist, only a small percentage are proliferating and hence responsible for mature blood cells at any one time.[24] The majority were reserved in a quiescent state until needed or the proliferative capacity of active stem cells was exhausted. BrdU labeling experiments revealed that HSCs typically exhibit a cell cycle transit time of three days, but it can be decreased to <12 hours in the presence of hematopoietic damaging agents like cyclophosphamide.[3] At any one time only ~5% of mouse HSCs were in S/G2/M and another 20% in G1, with ~ 8% entering the cell cycle each day.[3] Nevertheless, by 6 days approximately half the HSCs incorporated BrdU, and by 30 days >90% were labeled. These experiments indicate that although ~75% of HSCs are quiescent at any one time, all are eventually recruited into the cycle and divide on average once every 57 days.[3] Consistent with this interpretation, when chimeric mice were generated that contain HSCs from different genotypes, many contributed to hematopoiesis simultaneously.[3]

Recent work suggests that distinct osteoblastic and vascular niches work together in the bone marrow to regulate the HSC proliferative decision (Fig. 7).[25] A subset of osteoblasts produce various signals (e.g. Ang-1) that inhibit proliferation and maintain HSCs in a quiescent state. Cell cycle re-entry requires relocating from the osteoblastic to the vascular niche, which promotes proliferation due to its nutrient-rich environment, higher oxygen levels, and elevated growth factors. HSC recruitment involves the growth factor FGF-4 and chemokines such as SDF-1, which activates Rho GTPases Rac1/2 to regulate HSC mobilization and homing. Activation of signal transduction pathways (e.g. PI3K, PKC, p38-MAPK) also appears necessary, with the well-known transcription factor c-myc being a likely downstream target. Mitogenic stimulation up-regulates

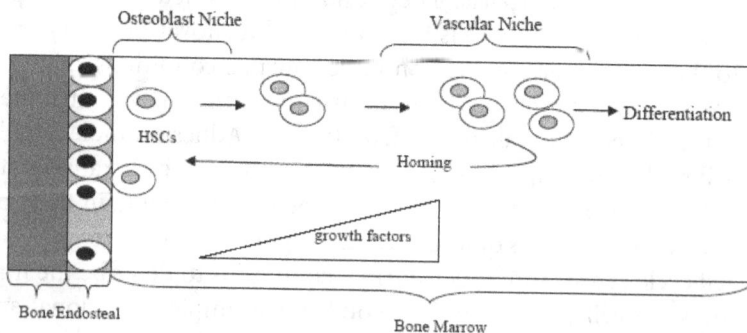

Figure 7. Bone marrow niches. Proposed model for how distinct microenvironments might control the fate of hematopoietic stem cells (HSCs). They are maintained in the osteoblast niche in a quiescent state due to signals from osteoblasts in the endosteal region that inhibit HSC proliferation and promote expression of adherent proteins limiting motility. To replace lost or damaged cells, extracellular signals promote relocation of HSCs from the osteoblastic to the vascular niche, where they can proliferate and differentiate to be released into the circulatory system. In some instances HSCs in the vascular niche return to the osteoblast microenvironment in a process called homing.

c-myc and causes HSC release from the osteoblastic niche, while antipro-liferative signals like TGF-β cause c-myc down-regulation. Artificially eliminating c-myc in the bone marrow leads to accumulated HSCs over-expressing adhesion molecules like N-cadherin and integrin, with subse-quent failure to generate the progenitor cells required for differentiation.[25] C-myc protein appears to be differentially segregated during HSC divi-sion, with the daughter receiving low levels retained in the osteoblastic niche in a quiescent state due to up-regulated N-cadherin and integrin. High c-myc levels in the other daughter down-regulate N-cadherin and integrin, causing this cell to depart the osteoblastic niche for additional proliferation and differentiation. Once mature blood cells have been gen-erated in the vascular niche, they are released into the circulatory system to provide space for further cell expansion.

4.2. Proliferative outcomes

A dividing somatic cell produces two identical daughters, but a stem cell must select among additional options to generate the required number of differentiated progeny while at the same time perpetuating itself. It can accomplish this task employing a variety of strategies (Fig. 1).[5] The stem cell population can be expanded by symmetric division to generate two identical daughters. Asymmetric division maintains the stem cell pool and creates a committed cell destined for an alternative fate. Symmetric differ-entiation depletes the stem cell pool by producing two differentiated prog-enies. These options provide the exquisite control necessary for properly responding to requests for cell expansion and/or replacement, and can be accomplished in several ways: (1) specific genes can be disproportionately localized prior to division and then differentially segregated between daughter cells; (2) an intrinsic circadian mechanism can "count" the num-ber of symmetric divisions prior to differentiation; (3) external signals and/or the microenvironment can influence division in response to chang-ing environmental conditions. In *Drosophila* germ cells, for instance, the decision is made based in part on whether or not the mitotic spindle is ori-ented perpendicular or parallel to the niche in which they are located.[5]

Establishing a specific cell lineage requires exiting the uncommitted state and entering a developmental pathway. Experiments on *C. elegans* indicate that these events are independently controlled, requiring loss of a zinc finger protein called PIE-1 which normally represses differentiation genes.[5] During duplication of blastomere stem cells, PIE-1 is asymmetrically

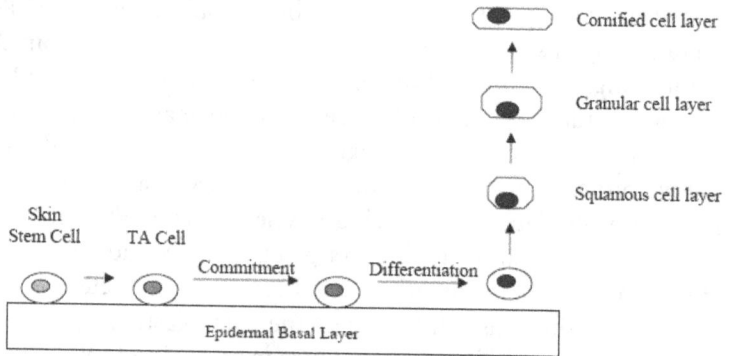

Figure 8. Skin cell regeneration. Epidermal stem cells associated with the basal layer asymmetrically divide to generate a transient amplifying (TA) cell that is committed to divide and further differentiate into the various skin cell lineages.

distributed so that its absence induces commitment in conjunction with positive factors like SKN-1 and local cell-to-cell interactions.[5] In mammals the deciding factors may be different, with differentiation in some cases being a default pathway activated after removal of the stem cell from its microenvironment.

Formation of epidermal tissue provides a classic example of coordinated stem cell renewal and differentiation (Fig. 8).[23] Differentiated surface cells are continually being lost due to damage (e.g. UV exposure) and so must be replaced to maintain skin integrity. The epidermal stem cell therefore divides asymmetrically to recreate a stem cell and produce a rapidly dividing transient amplifying (TA) cell that is committed to generating a specific lineage. TA cells transit the cycle significantly faster than their ancestors in order to amplify the tissue, but exhibit age and tissue effects on proliferation capacity that are absent in stem cells.[26] Human epidermis is renewed approximately every 14 days, and because each TA cell divides 3–6 times before further differentiating, the self-renewing capacity of primitive stem cells must be tremendous.[23]

The cell cycle transit times of embryonic and adult stem cells vary widely, depending on the developmental stage and environmental conditions. In general, mES cells proceed through the cycle much more rapidly than their adult brethren, consistent with the need to rapidly expand the cell number during growth and development.[20] They can also replicate through many generations while maintaining the ability to differentiate. Once new daughter cells commit to a specific lineage, their cell cycle rate and that of any resulting progeny slow, progressively becoming more similar to nonstem somatic cells. Human ES cells have also been cloned, propagated, and cultured from blastocysts and aborted fetuses. In some respects they are similar to mES cells in that they express Oct3/4, exhibit telomerase activity, and promote formation of teratomas in immunodeficient mice that contain cells of ectodermal, endodermal, and mesodermal origin.[2] They can also be cultured *in vitro* and maintain their proliferative potential for long periods, but in contrast to mES cells, they form mainly cystic embroid bodies, display proteoglycans, and exhibit different subtypes of stage-specific antigens.[2] Human ES cells also have a significantly longer average population doubling time of 30–35 hours. These differences are likely controlled at the molecular level, and may reflect subtle yet significant variations in proliferative control mechanisms between species.

4.3. Molecular control of cell cycle progression

The Restriction point in the G1 phase is the critical decision-making period for somatic cells (Fig. 5). Most cell growth (i.e. increase in mass) occurs during this period, and they typically spend up to 70% of the total transit time in this stage. The length of the overall proliferative cycle is therefore determined mainly by the rate of G1 progression.[6] On a molecular level this means that the G1 cell cycle machinery plays a major role in both the overall proliferative rate and cell fate determination. The central regulatory event defining this checkpoint is Rb phosphorylation and inactivation, which releases E2F so it can activate a transcriptional program driving the increase in cell mass and progression through the proliferative cycle (Fig. 5).[17] Consistent with this idea, the overall proliferation rate of somatic cells can generally be accelerated by overexpressing G1 cyclins.[6]

The cycle of pluripotent ES cells differs in fundamental ways from that of the idealized somatic cell described above.[2] ES cells have a significantly shortened G1 and spend more than half their time in the S phase, which explains their rapid cell cycle transit time of ~10–12 hours and implies a reduced contribution of G1 regulatory events. In support of this interpretation, their proliferation is relatively unaffected by serum starvation, contact inhibition, or anchorage dependence, all of which inhibit cycling of nontransformed somatic cells.[2] In some ways ES cell proliferation resembles that of transformed cells, although unlike cancer cells they rigorously maintain a normal diploid karyotype throughout extensive rounds of proliferation.

The ability to isolate and culture mES cells provided an ideal opportunity to analyze proliferative control mechanisms at the molecular level.[20] Cell cycle machinery, whose amounts usually vary dramatically (e.g. cyclins), was constitutively expressed at high levels in ES cells, while levels of inhibitors like p21[CIP1] and p27[KIP1] were generally lower. Constitutive expression of positive proliferative factors correlated with elevated mRNA due to its enhanced stability and acetylation of histone H3 at promoter regions. As expected, CDKs were active throughout the cycle, with modest oscillation only apparent for cyclin B-CDC2.[20] The retinoblastoma protein pRb was therefore constitutively phosphorylated and not associated with E2F, resulting in continual transcription of its target genes. Similar differences were observed in regulation of DNA replication machinery forming and firing replication origins (Fig. 9).[20] In nonstem cells these events require low G1 CDK activity to allow assembly of prereplication complexes (pre-RCs) at origins. Activated CDKs then initiate the bidirectional replication fork by phosphorylating pre-RC components. In ES cells, however, levels of the loading factor Cdc6 — which facilitates formation of the pre-RC on DNA during the late M to early G1 phase — were substantially elevated, and it was not degraded by the proteasome in early G1 (as in somatic cells).[20] The DNA binding proteins MCM2 and MCM4 were also much more abundant. Despite these differences, cell-cycle-dependent phosphorylation and chromatin association/dissociation of replication factors was still regulated and similar to that observed in somatic cells. Furthermore, conditional knockout of the Cdc7 kinase that helps fire replication origins also caused ES cells to arrest in the S phase.[20] These observations suggest that while many of the same components are involved in proliferative control of stem and nonstem cells, their regulation and even activity may differ in fundamental ways.

The proliferative control mechanisms of a daughter cell committed to differentiation undergo significant changes. The pRb-E2F regulatory network is established by decreasing CDK activity and enforcing periodic expression of cell cycle machinery, resulting in temporal extension of G1 and its control by cyclin–CDK regulation.[27] Transcriptional mechanisms are responsible for decreasing levels of cell cycle and DNA replication machinery, just as they played the opposite role during ES cell self-renewal. Together these events establish G1 control and lead to substantial lengthening of cell cycle transit time. In *Drosophila* development this transition is initiated after the 16th mitotic event.[27]

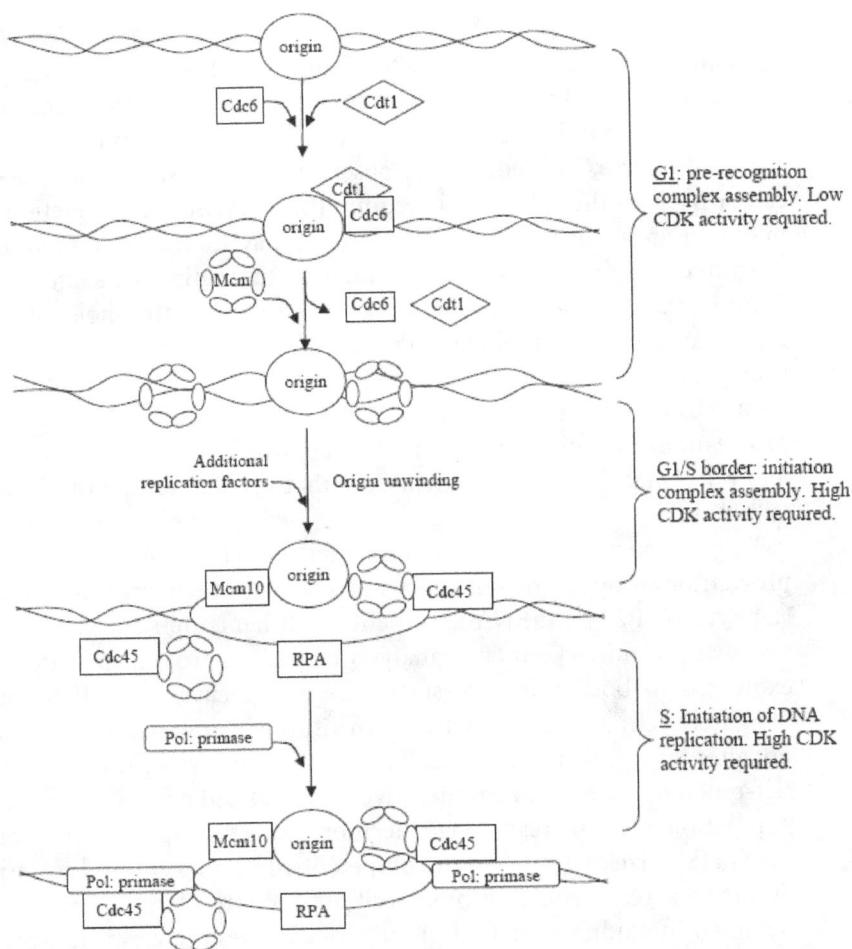

Figure 9. Initiation of DNA replication. During the G1 phase the prereplication complex assembles at the origin of DNA replication. At the G1/S transition high cyclin–CDK activity leads to recruitment of additional replication components and origin unwinding. DNA polymerases are then able to access the single strands and begin replication at the beginning of the S phase, which also requires high cyclin–CDK activity. See text for details.

The CIP/KIP CDK inhibitors deserve special mention because they help determine the pool size of stem cells (p21^{CIP1}) and the progenitor populations (p27^{KIP1}) derived from them.[28,29] In hematopoiesis, p21^{CIP1} maintains stem cells in a quiescent state; its mRNA levels are high in nondividing human HSCs and reduced upon commitment to progenitor populations.[28]

Artificially reducing p21^{CIP1} leads to increased HSC proliferation under normal homeostatic conditions. Nevertheless, p21−/− mice undergoing bone marrow transplant or hematopoietic injury exhibited higher mortality because required replacement cells were not generated and active stem cells were exhausted.[28] These observations suggest that limiting stem cell entry into the cell cycle is required to prevent their depletion and maintain an adequate injury response. It may therefore be possible to uncouple proliferation from differentiation to manipulate stem cells for therapeutic purposes.[29] Overcoming the inhibitory threshold blocking cell cycle re-entry, combined with exposure to various combinations of proproliferative cytokines, could potentially provide the large number of stem cells required by regenerative medicine. Such advances are essential in situations like injury to neuronal tissue, which responds poorly to cell loss yet contains neural precursor cells that proliferate extensively when cultured.[24]

P27^{KIP1} also plays an important role in stem cell biology, but it inhibits progenitor repopulation efficiency rather than influencing uncommitted stem cells.[29] P27^{KIP1} mRNA is present in all hematopoietic cell populations, and a flow-cytometric analysis of p27^{KIP1} protein indicated it is expressed in both primitive stem cells and their more differentiated daughters.[30] In a hematopoietic reconstitution assay, lack of p27^{KIP1} did not alter the number of HSCs, cell cycling parameters, or the capacity for self-renewal; instead, the proliferative potential and pool size of the HSC population were increased. Furthermore, in a transplant competition assay, HSCs lacking p27^{KIP1} outcompeted wild type cells and were predominantly responsible for blood cell production.[30] Similar results were obtained in studies of neural progenitors, suggesting a general role for p27^{KIP1} as an intrinsic timer controlling the number of allowable proliferation cycles during progenitor expansion.[24] Intriguingly, recent results suggest that microRNAs may contribute to regulation of p27^{KIP1} levels.[31] *Drosophila* germline cells require these small, biologically active RNAs to bypass normal G1/S checkpoint controls and ensure continuous cell division.[31] Mutating a key component in their production called Dicer arrested cells in G1 due to upregulated p27^{KIP1}, suggesting that microRNAs help maintain self-renewal in part by keeping p27^{KIP1} levels low. By artificially decreasing p27^{KIP1} it may be possible to greatly increase the progenitor pool without significantly affecting the overall number of pluripotent stem cells. Such an approach would be quite useful for enhancing the efficiency of therapies relying on a small number of available stem cells or attempting to expand a genetically modified subpopulation.[24]

Other cell cycle inhibitors also play important and distinct roles in control of stem cell proliferation.[24] The INK4 family member p18^{INK4c} targets CDK4/6 and is a strong inhibitor of stem cell self-renewal. Mice lacking p18^{INK4c} undergo increased expansion resulting in a larger stem cell pool.[24] Transplanted p18–/– HSCs outcompete their wild type counterparts and can be serially transplanted beyond the capacity of normal cells. This effect arises through an increase in the number of primitive HSCs rather than progenitors, as was the case with p27^{KIP1}. Although p18^{INK4c} and p21^{CIP1} both influence HSC kinetics, their absence nevertheless causes distinct phenotypes. HSCs lacking p21^{CIP1} are prematurely depleted while those without p18^{INK4c} are able to efficiently self-renew during long engraftment. Thus, it appears that the absence of p18^{INK4c} causes HSCs to favor the self-renewal pathway.[24]

The applicability of current cell cycle models to ES cell proliferation is questionable given their constitutive expression of cell cycle machinery, elevated CDK activity, perpetual pRb phosphorylation, and continual E2F-mediated transcription. In somatic cells, excess CDK activity shortens G1 and accelerates S phase entry, so the higher levels in ES cells may contribute to the truncated G1. It is still unclear how the temporal order of cell cycle progression and initiation of DNA replication is controlled under these conditions. CDK2 could be weakly regulated, the modest changes in cyclin B–CDC2 might be responsible, or a specific subset of CDK involved in origin firing might be cell-cycle-regulated.[20] A more intriguing possibility is that cyclin–CDKs do not control S phase initiation in ES cells, a radical idea that nevertheless must be considered given the surprising consequences of ablating cell cycle regulators in mice.

5. Re-Evaluating the Role of Cyclin–CDKs in Stem Cell Proliferation

The fate of nonstem cells is clearly determined in G1 at the Restriction point.[6] Although regulated differently in stem cells, cyclins and CDKs are expressed at high levels and exhibit enhanced activity.[20] Cell cycle theory predicts that ablating G1 cyclins or CDKs in the mouse should block cell proliferation and prevent embryogenesis. Surprisingly, however, in most cases these mice are actually viable or develop to late stages of gestation, indicating that cell expansion and embryo formation still occur in the absence of what are generally considered "essential" cell cycle components.[32,33] Understanding these foundation-rattling results is essential for developing effective methods of expanding stem cell populations *in vitro* for use in regenerative medicine.

CDK4/CDK6 and the D-type cyclins have long been considered essential for initiating G1 progression in response to mitogenic signaling (Fig. 5).[33] Mice lacking CDK4 or CDK6 alone are still viable, probably due to compensatory activity of the remaining family member.[34] While CDK4/6 double knockouts die due to fatal anemia, their late-stage development indicates that cell proliferation and organ development still take place.[34] Mouse embryonic fibroblasts (MEFs) lacking CDK4/6 can even be immortalized and cultured *in vitro*.[34] These cells respond to mitogenic signaling and display relatively normal cell cycle kinetics, despite lower pRb phosphorylation and cyclin E expression. Similar results were obtained with deletion of individual D-type cyclins; the mice were viable but exhibited tissue-specific phenotypes, again suggesting partial compensation by remaining family members.[32,33] Remarkably, however, the triple cyclin D1-3 knockout mice survived until mid/late gestation, when they died from heart defects.[35] The development of most tissues in the absence of D-type cyclins demonstrated that they are not required for expansion of most mammalian cell types. Consistent with this interpretation, MEFs lacking D-type cyclins still proliferated in culture and responded to the presence or absence of mitogenic signals.[35] These observations suggest that the pRb-E2F control module is not controlling G1 progression during mouse development, and that initiation of DNA synthesis is not absolutely dependent on sequential activation of cyclin–CDKs.

In further support of this idea, CDK2−/− mice are also viable and survive up to two years.[36,37] MEFs derived from these animals proliferate normally — albeit with a modest delay in S phase entry — indicating that CDK2 is not essential for division of most cell types.[37] Similar results were obtained using a conditional CDK allele in MEFs, an important control ruling out the possibility that mice lacking CDK2 activate compensatory mechanisms during development.[36] Either a different CDK (e.g. CDC2) is responsible or other mechanisms are operational.[37] The importance of CDK2 was also investigated by ablating its cyclin E binding partners. Mice lacking cyclin E1 or E2 alone are viable and develop normally, but surprisingly the double cyclin E1/2 knockout causes mid-gestational embryonic lethality due to placental defects.[38] Because mice lacking CDK2 are normal, this lethality indicates that cyclin E carries out CDK2-independent functions that are required for viability. MEFs lacking E-type cyclins can be cultured and proliferate in the presence of mitogens, but cannot re-enter the cell cycle from quiescence because MCM proteins fail to bind DNA origins (Fig. 9).[32,38] These observations indicate that while the mechanics of DNA replication are maintained in stem cells, their mechanism of control is likely different.

Further evidence that stem cell proliferative control mechanisms are unique comes from experiments ablating negative regulators of G1 progression. Mice lacking pRb die at about day 14.5 of gestation, but not due to expected defects in G1 control; instead, there is excess proliferation of extra embryonic trophoblast cells that disrupts required placenta architecture.[39–41] This surprising explanation was demonstrated by successfully rescuing pRb−/− embryos with a wild-type placenta, which went to term and died shortly after birth.[42] While pRb is not required during embryogenesis, the E2F1-3 triple knockout is lethal and blocks MEF proliferation.[43] Thus, it appears that the central role of E2F transcriptional activity is maintained in cell proliferation, but other mechanisms are controlling its function in cellular expansion during early development (see Ref. 33 for a more detailed exploration of possible candidates).

Mice lacking the CDK inhibitor p27^{KIP1} develop relatively normally, but are 33% larger and exhibit female sterility.[44] P27−/− MEFs do not display excess CDK activity or accelerated progression through G1, and they still maintain responsiveness to mitogenic and antimitogenic signaling. Conversely, p27^{KIP1} elimination is not required for cell proliferation and development because mice expressing a p27T187A mutant are viable and develop normally.[45] MEFs from these animals are also healthy and display typical cell cycle parameters. Thus, overcoming a p27^{KIP1} inhibitory threshold is not a prerequisite for proliferation. These observations support the idea that cell cycle control of stem cells does not rely on regulating CDK activity. As discussed earlier, however, p27^{KIP1} is clearly involved in controlling the progenitor cell number, raising the possibility that it does so by regulating a target other than CDKs.[29]

If mechanisms controlling stem cell proliferation differ from those of their progeny, then upstream signaling pathways might have different downstream targets (or alternative effects on the same targets). Consistent with this idea, stem cells sometimes exhibit unique responses to signal transduction. Quiescent somatic cells in culture require the presence of extracellular proliferative signals like growth factors to trigger, via tyrosine kinase receptors, intracellular signaling pathways that activate classical cell cycle machinery (e.g. the Ras/MAPK activation of cyclin D–CDK4/6 described earlier) (Fig. 6). In stem cells, however, activation of Ras/MAPK does not affect cyclin D synthesis or induce cell cycle proliferation, but instead leads to stem cell differentiation.[46,47] Some signaling pathways appear to have similar effects in both stem and nonstem cells. The phosphatidylinositol-3-kinase (PI3K) pathway regulates proliferation of somatic cells by upregulating cyclin synthesis and promoting CKI relocation/turnover.[47,48] In stem cells the PI3K pathway also induces proliferation and prevents differentiation, but the mechanism remains uncertain.[49] We previously described a novel role for p27[KIP1] binding the SH2–SH3 adaptor protein Grb2 and inhibiting activation of Ras/MAPK.[50,51] P27[KIP1] targeting Grb2 requires phosphorylating p27[KIP1] on serine 10, which we recently showed can be mediated by Akt.[52] Thus, the PI3K/Akt pathway could prevent differentiation and promote proliferation in part by directing p27[KIP1] to inhibit Ras/MAPK (Fig. 10). This novel connection could

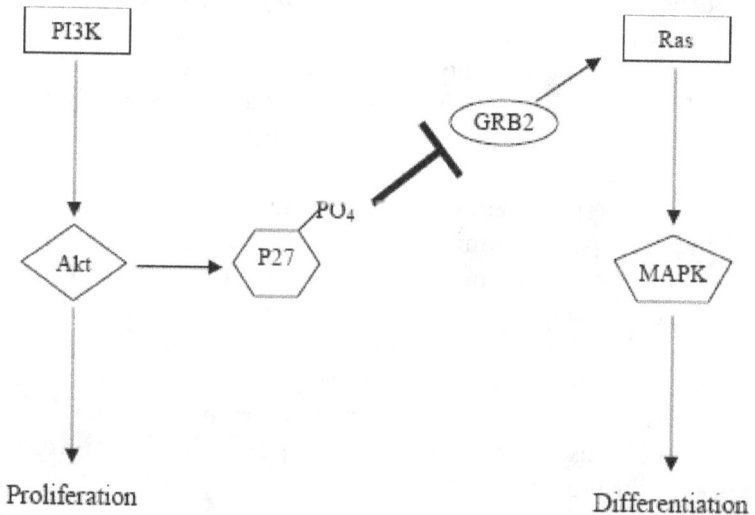

Figure 10. Role for p27[KIP1] in stem-cell-fate decisions via linkage of signal transduction pathways. Proposed model to explain how p27[KIP1] might regulate the progenitor cell number independently of its CDK inhibitor function. The PI3K/Akt pathway promotes stem cell growth and proliferation, while the Ras/MAPK pathway induces differentiation. P27[KIP1] could link these two pathways because its phosphorylation by Akt could initiate the recently described function of p27[KIP1] as an inhibitor of Ras activation (see text for details).

help explain p27^{KIP1} control of progenitor pool size independently of its role in CDK regulation.[28,29]

The greatly abbreviated G1 phase observed in stem cells, combined with the absence and/or expendability of classical G1 machinery, implies the decision to proliferate and cell cycle progression are controlled mainly by internal rather than external signaling. One rationale for this organization is that nonstem cells could be designed to remain in quiescence until the proper extracellular signals are received. In contrast, stem cells may favor proliferation unless they receive external information (e.g. from the niche) telling them to desist and/or differentiate. In other words, stem cells have less need for G1 decision-making machinery because they are permanently committed to the proliferative cycle due to the necessity of self-renewing. The bigger concern is deciding whether to divide symmetrically or asymmetrically, and so extracellular signaling may be directed toward influencing this outcome. It is tempting to speculate that absence of G1 control mechanisms may actually be required for flexibility in making this decision, an arrangement that would fit with recent ideas questioning the rigid hierarchical nature of stem cell proliferation and differentiation.[53] In this view stem cells can adopt alternative fates depending on the cell cycle position, i.e. the cell cycle controls fate decisions by influencing whether a primitive cell remains a stem cell or differentiates.[53] In support of this idea, engraftment and progenitor phenotypes of hematopoietic stem cells vary inversely with the cell cycle position, suggesting that stimuli received at distinct cell cycle positions influence cell fate decisions.[53] This hypothesis is particularly intriguing in light of the long cell cycle transit times observed in some stem cell populations, which might provide a window of opportunity for deciding upon and implementing the desired outcome.[5] In the hematopoietic system there is some evidence that the fate of stem cell progeny is restricted by environmental signals acting shortly after the S phase, with progression to the next S phase restoring pluripotency.[53] These phenotypic shifts in identity correlate with chromatin remodeling that may alter the cell surface phenotype and thus the stem cell response to its environment.[53]

6. Implications for Regenerative Medicine

Regenerative medicine seeks to replace cells lost to injury or disease utilizing a cornucopia of expandable stem cell populations that can differentiate into desired cell lineages. Realizing this goal requires sufficient quantities of multipotent human stem cells with high proliferative capacity, but that tightly control replicative capacity to maintain homeostasis and prevent exhaustion *in vivo*. This control presents a significant challenge to *in vitro* expansion of stem cells for therapeutic purposes. While current efforts have focused on identifying and mimicking the extracellular signals perceived as required for stem cell expansion, a clearer understanding of how cell cycle progression is internally regulated may provide new targets whose manipulation more effectively accomplishes this goal.

References

1. Cai J, Weiss ML, Rao MS. (2004) In search of stemness. *Exp Hematol* **32:** 585–598.
2. Wobus AM, Boheler KR. (2005) Embryonic stem cells: prospects for developmental biology and cell therapy. *Physiol Rev* **85:** 635–678.
3. Cheshier SH, Morrison SJ, Liao X, Weissman IL. (1999) *In vivo* proliferation and cell cycle kinetics of long-term self-renewing hematopoietic stem cells. *Proc Natl Acad Sci USA* **96:** 3120–3125.
4. MacKey MC. (2001) Cell kinetic status of haematopoietic stem cells. *Cell Prolif* **34:** 71–83.
5. Morrison SJ, Shah NM, Anderson DJ. (1997) Regulatory mechanisms in stem cell biology. *Cell* **88:** 287–298.
6. Sherr CJ, Roberts JM. (1999) CDK inhibitors: positive and negative regulators of G1 phase progression. *Genes Dev* **13:** 1501–1512.
7. Nurse P. (2000) A long twentieth century of the cell cycle and beyond. *Cell* **100:** 71–78.
8. Nurse P, Masui Y, Hartwell L. (1998) Understanding the cell cycle. *Nat Med* **4:** 1103–1106.
9. Masui Y, Markert CL. (1971) Cytoplasmic control of nuclear behavior during meiotic maturation of frog oocytes. *J Exp Zool* **177:** 129–145.
10. Hartwell LH. (1978) Cell division from a genetic perspective. *J Cell Biol* **77:** 627–637.
11. Nurse P, Bissett Y. (1981) Gene required in G1 for commitment to cell cycle and in G2 for control of mitosis in fission yeast. *Nature* **292:** 558–560.
12. Evans T, Rosenthal ET, Youngblom J, *et al.* (1983) Cyclin: a protein specified by maternal mRNA in sea urchin eggs that is destroyed at each cleavage division. *Cell* **33:** 389–396.
13. Morgan DO. (1997) Cyclin-dependent kinases: engines, clocks, and microprocessors. *Annu Rev Cell Dev Biol* **13:** 261–291.

14. Peter M. (1997) The regulation of cyclin-dependent kinase inhibitors (CKIs). *Prog Cell Cycle Res* **3**: 99–108.
15. Murray AW. (2004) Recycling the cell cycle: cyclins revisited. *Cell* **116**: 221–234.
16. Draetta G, Eckstein J. (1997) Cdc25 protein phosphatases in cell proliferation. *Biochim Biophys Acta* **1332**: M53–63.
17. Bartek J, Bartkova J, Lukas J. (1996) The retinoblastoma protein pathway and the restriction point. *Curr Opin Cell Biol* **8**: 805–814.
18. Stevaux O, Dyson NJ. (2002) A revised picture of the E2F transcriptional network and RB function. *Curr Opin Cell Biol* **14**: 684–691.
19. Woo RA, Poon RY. (2003) Cyclin-dependent kinases and S phase control in mammalian cells. *Cell Cycle* **2**: 316–324.
20. Fujii-Yamamoto H, Kim JM, Arai K, Masai H. (2005) Cell cycle and developmental regulations of replication factors in mouse embryonic stem cells. *J Biol Chem* **280**: 12976–12987.
21. Pesce M, Wang X, Wolgemuth DJ, Sholer H. (1998) Differential expression of the Oct-4 transcription factor during mouse germ cell differentiation. *Mech Dev* **71**: 89–98.
22. Niwa H, Miyazaki J, Smith AG. (2000) Quantitative expression of Oct-3/4 defines differentiation, dedifferentiation or self-renewal of ES cells. *Nat Genet* **24**: 372–376.
23. Fuchs E, Segre JA. (2000) Stem cells: a new lease on life. *Cell* **100**: 143–155.
24. Cheng T. (2004) Cell cycle inhibitors in normal and tumor stem cells. *Oncogene* **23**: 7256–7266.
25. Yin T, Li L. (2006) The stem cell niches in bone. *J Clin Invest* **116**: 1195–1201.
26. Dunnwald M, Chinnathambi S, Alexandrunas D, Bickenback JR. (2003) Mouse epidermal stem cells proceed through the cell cycle. *J Cell Physiol* **195**: 194–201.
27. White J, Stead E, Faast R, *et al.* (2005) Developmental activation of the Rb-E2F pathway and establishment of cell cycle-regulated cyclin-dependent kinase activity during embryonic stem cell differentiation. *Mol Biol Cell* **16**: 2018–2027.
28. Cheng T, Rodrigues N, Shen H, *et al.* (2000) Hematopoietic stem cell quiescence maintained by p21cip1/waf1. *Science* **287**: 1804–1808.
29. Cheng T, Rodrigues N, Dombkowski D, *et al.* (2000) Stem cell repopulation efficiency but not pool size is governed by p27(kip1). *Nat Med* **6**: 1235–1240.
30. Tamir A, Petrocelli T, Stetler K, *et al.* (2000) Stem cell factor inhibits erythroid differentiation by modulating the activity of G1-cyclin-dependent kinase complexes: a role for p27 in erythroid differentiation coupled G1 arrest. *Cell Growth Differ* **11**: 269–277.
31. Hatfield SD, Shcherbata HR, Fischer KA, *et al.* (2005) Stem cell division is regulated by the microRNA pathway. *Nature* **435**: 974–978.

32. Sherr CJ, Roberts JM. (2004) Living with or without cyclins and cyclin-dependent kinases. *Genes Dev* **18:** 2699–2711.

33. Moeller SJ, Sheaff RJ. (2006) G1 phase: components, conundrums, context. *Results Probl Cell Differ* **42:** 1–29.

34. Malumbres M, Sotillo R, Santamaria D, *et al.* (2004) Mammalian cells cycle without the D-type cyclin-dependent kinases Cdk4 and Cdk6. *Cell* **118:** 493–504.

35. Kozar K, Ciemerych MA, Rebel VI, *et al.* (2004) Mouse development and cell proliferation in the absence of D-cyclins. *Cell* **118:** 477–491.

36. Ortega S, Prieto I, Odajima J, *et al.* (2003) Cyclin-dependent kinase 2 is essential for meiosis but not for mitotic cell division in mice. *Nat Genet* **35:** 25–31.

37. Berthet C, Aleem E, Coppola V, *et al.* (2003) Cdk2 knockout mice are viable. *Curr Biol* **13:** 1775–1785.

38. Geng Y, Yu Q, Sicinska E, *et al.* (2003) Cyclin E ablation in the mouse. *Cell* **114:** 431–443.

39. Clarke AR, Maandag ER, van Roon M, *et al.* (1992) Requirement for a functional Rb-1 gene in murine development. *Nature* **359:** 328–330.

40. Jacks T, Fazeli A, Schmitt EM, *et al.* (1992) Effects of an Rb mutation in the mouse. *Nature* **359:** 295–300.

41. Lee EY, Chang CY, Hu N, *et al.* (1992) Mice deficient for Rb are nonviable and show defects in neurogenesis and haematopoiesis. *Nature* **359:** 288–294.

42. Wu L, de Bruin A, Saavedra HI, *et al.* (2003) Extra-embryonic function of Rb is essential for embryonic development and viability. *Nature* **421:** 942–947.

43. Wu L, Timmers C, Maiti B, *et al.* (2001) The E2F1-3 transcription factors are essential for cellular proliferation. *Nature* **414:** 457–462.

44. Kiyokawa H, Koff A. (1998) Roles of cyclin-dependent kinase inhibitors: lessons from knockout mice. *Curr Top Microbiol Immuno* **227:** 105–120.

45. Malek NP, Sundberg H, McGrew S, *et al.* (2001) A mouse knock-in model exposes sequential proteolytic pathways that regulate p27Kip1 in G1 and S phase. *Nature* **413:** 323–327.

46. Burdon T, Stracey C, Chambers I, *et al.* (1999) Suppression of SHP-2 and ERK signalling promotes self-renewal of mouse embryonic stem cells. *Dev Biol* **210:** 30–43.

47. Jirmanova L, Afanassieff M, Gobert-Gosse S, *et al.* (2002) Differential contributions of ERK and PI3-kinase to the regulation of cyclin D1 expression and to the control of the G1/S transition in mouse embryonic stem cells. *Oncogene* **21:** 5515–5528.

48. Liang J, Slingerland JM. (2003) Multiple roles of the PI3K/PKB (Akt) pathway in cell cycle progression. *Cell Cycle* **2:** 339–345.

49. Takahashi K, Murakami M, Yamanaka S. (2005) Role of the phosphoinositide 3-kinase pathway in mouse embryonic stem (ES) cells. *Biochem Soc Trans* **33**: 1522–1525.

50. Cheng AM, Saxton TM, Sakai R, *et al.* (1998) Mammalian Grb2 regulates multiple steps in embryonic development and malignant transformation. *Cell* **95**: 793–803.

51. Moeller SJ, Head ED, Sheaff RJ. (2003) p27Kip1 inhibition of GRB2-SOS formation can regulate Ras activation. *Mol Cell Biol,* **23**: 3735–3752.

52. Nacusi LP, Sheaff RJ. (2006) Akt1 sequentially phosphorylates p27kip1 within a conserved but non-canonical region. *Cell Div* **1**: 11.

53. Quesenberry PJ, Colvin GA, Lambert JF. (2002) The chiaroscuro stem cell: a unified stem cell theory. *Blood* **100**: 4266–4271.

ADULT STEM CELLS

5

Multipotent Adult Progenitor Cells

Yuehua Jiang*, April Breyer and Catherine Verfaillie

1. Introduction

Stem cells are uncommitted cells with the potential to differentiate into cells of a specific tissue. The embryonic stem (ES) cell is the prototypical stem cell with unlimited self-renewal and differentiation capacity. It can give rise to all of the somatic and germ line cells of the fully developed organism. Adult stem cells have been identified in most tissues, and include hematopoietic stem cells (HSCs), neural stem cells (NSCs), hepatic oval cells, epidermal stem cells, mesenchymal stem cells (MSCs), etc. Hematopoietic stem cells are among the most extensively characterized and have long been used therapeutically. The more recent discovery of mesenchymal stem cells offers abundant and very promising resources for repair of bone, cartilage, and certain other tissues. Compared with ES cells, tissue-specific adult stem cells were believed to have less self-renewal ability and only differentiate into a limited number of cell types. We have isolated a population of primitive cells from normal human, rodent, and swine bone marrow that have, at the single cell level, pluripotent differentiation and extensive proliferation potential, and we have named them multipotent adult progenitor cells (MAPCs).[1–3]

1.1. Culture conditions

MAPCs are derived from mesenchymal stem cell culture after 25–35 population doublings. Cells are cultured on plates coated with fibronectin and media containing a low dose of fetal bovine serum (FBS), epidermal growth factor (EGF), and platelet-derived growth factor (PDGF).

*Corresponding: Stem Cell Institute and Department of Medicine, University of Minnesota, 2001 6th St SE, Minneapolis, MN 55455, USA. Tel: (612) 626-4217; E-mail: jian0028@ umn.edu

Leukemia inhibitor factor (LIF) is needed in mouse and rat MAPC culture. Recently we modified cell culture media to contain additional lipids, and started to culture rodent MAPCs at low (5%) O_2 level.[4,5] Under low O_2 level culture conditions, we obtained a rodent MAPC line with a higher Oct-4 level (see "Characteristics of MAPCs" below).

Isolation and maintenance of undifferentiated state MAPCs are technically demanding. Successful isolation and culture of MAPCs are affected by many factors, including the age of the animal at the time of isolation, the bone marrow (BM) collection procedure, the culture cell density, trypsinization time, the CO_2 concentration in the incubator, the pH of the media, the lot of the fetal bovine serum used in the media, and the type of plastic which cells are cultured on.

Generally, the younger the age of the animal, the better the chance of getting an MAPC line. In rodent MAPC isolation, we found that it is important to include the ends of the bones and to flush the bone marrow vigorously in order to get the cells best able to begin MAPC lines, as it has been shown that the cells present in the growth plates and endosteum have the most potential to form MAPC lines and can be cultured for longer than the cells from the middle of the bone.

Cell density is a key factor that plays a role in maintaining the undifferentiated state of MAPCs. MAPCs from different species have different optimal maintenance density, ranging from 100 to 400 cells/cm². When cultured at higher density, MAPCs tend to lose pluripotent potential. Successful maintenance of MAPCs requires not only overall low cell density, but also even distribution throughout culture plates in order to minimize cell–cell contact. Trypsinization may affect the culture of MAPCs; hence MAPCs should only be trypsinized for a short period of time (less than 30 seconds), and the plates should be tapped gently. This length of time is long enough for MAPCs to detach, but is not sufficient for removing any differentiating cells that may be present, as they are more adherent. This short term exposure to trypsin also minimizes its toxic effect on the cells. After the cells are detached, MAPC media or serum is added to deactivate the trypsin.

Screening of serum lots is needed to optimize MAPC culture. Some serum lots do not support MAPC growth. Serum can affect parameters such as cell growth speed, level of Oct-4 expression, amount of cytogenetic abnormality, differentiation capability, and cell morphology. The implications of factors such as faster growth can be more cytogenetic abnormalities, differentiation, or loss of differentiation potential. Therefore, testing should be done to ensure that the serum lot chosen maintains properly all of these aspects.

As MAPCs die under alkaline conditions, special attention should be paid to media pH and incubator CO_2 concentration. It is best to keep CO_2 concentrations at 5.5–6% and allow media to equilibrate in the incubator for at least 30 minutes before use.

1.2. MAPCs derived from other tissues

Besides bone marrow, MAPCs can be isolated from mouse brain and muscle.[6] Gupta *et al.* cultured cells with multipotent adult progenitor characteristics from rat kidneys and termed the cells multipotent renal progenitor cells (MRPCs).[7] The cells from different tissues have similar morphology, phenotype, and *in vitro* differentiation ability. Microarray data showed that mouse MAPCs from bone marrow, brain, and muscle have a highly similar expressed gene profile.[6]

1.3. Characteristics of MAPCs

MAPCs are small cells. The size of a human MAPC is between 10 and 15 μm, a rodent MAPC is between 9 and 12 μm, and a swine MAPC is less than 10 μm in diameter. They have a spindle or triangular shape except while dividing, at which time they become round. MAPCs have large nuclei and scant cytoplasm.

MAPCs show robust proliferation capability. When maintained under proper conditions, human MAPCs have been expanded for 50–80 population doublings, mouse and rat MAPCs for 80–150 population doublings, and swine MAPCs for more than 100 population doublings. Karyotype analysis indicated that this expansion was not associated with the presence of obvious cytogenetic abnormalities. Cytogenetic analysis has rarely shown abnormalities in human MAPCs. However, as tetraploidy and aneuploidy can be seen once rodent MAPCs have been expanded for more than 60–70 population doublings, monthly karyotyping is needed to ensure that cells are cytogenetically normal.

MAPCs have active telomerase. Telomere lengths were compared at different population doublings in human, rodent, and swine MAPCs. It has been shown that the long telomeres of MAPCs do not shorten after extensive culture. Also, MAPCs derived from older donors (or animals) have the same telomere length as those from the younger donors.

MAPCs express Oct-3/4, which is critical in maintaining embryonic stem cells in undifferentiated status and is down-regulated when ES cells undergo somatic cell commitment and differentiation. In rodent MAPC culture, the Oct-4 levels can vary, depending on the conditions under which cells are cultured. In a low O_2 environment, it is possible to obtain MAPC clones with Oct-4 expression levels of 10–60% of the ES cell level, and Rex-1 expression at 5–30% of ES levels. However, even at low O_2 levels, Nanog is not expressed in MAPCs, where it is expressed in ES cells.

Flow cytometry is used to assess the MAPC phenotype. No matter which species MAPCs are derived from, there are CD34, CD45, MHC class I, MHC class II, CD44, and other markers of more differentiated cells at negative or very low positive levels. Mouse MAPCs express low levels of stage-specific embryonic antigen (SSEA)-1. Interestingly, high Oct-4 mouse MAPCs cultured under low O_2 conditions are positive for c-Kit. High Oct-4 rat MAPCs are CD31-positive.

1.4. *In vitro* differentiation

MAPCs have pluripotent differentiation potential. It has been shown that at high density, they can differentiate not only into mesenchymal lineage cells but also endothelium, neuroectoderm, and endoderm.

1.4.1. *Mesenchymal lineage differentiation*

Like mesenchymal stem cells, MAPCs can be easily differentiated into cells of mesenchymal lineage.[1,8,9] Induced with ascorbic acid, dexamethasone (Dex), and β-glycerophosphate, MAPCs differentiate into osteoblasts (osteopontin, bone sialoprotein, osteonectin positive and with calcium deposition). Transforming growth factor beta 1 (TGFβ-1) coaxes MAPCs into chondroblasts (type II collagen positive), and horse serum or insulin is used for the differentiation of adipocytes (oil red staining positive).

MAPCs can be differentiated into myocytes by using 5-azacytizine.[10] After differentiation, MyoD was first detected, and mature muscle proteins like fast-twitch myosin and sarcomeric actin showed up in the late differentiation stage. By using a more physiological condition [media containing 5% FBS, vascular endothelial growth factor (VEGF), basic fibroblast growth factor (bFGF), and insulin-like growth factor (IGF) 1], MAPCs were induced to form multinucleated tubes. These tubes stained positive for α-actinin and SK-myosin, which indicated the presence of striated myofibers in the cytoplasm.

1.4.2. *Neuroectodermal differentiation*

To induce differentiation of rodent MAPCs to the neural lineage, MAPCs were cultured in a serum-free medium with bFGF for 7 days, followed by 7 days with FGF-8 plus sonic hedgehog (SHH), then brain-derived neurotrophic factor (BDNF) in an N2 medium for an additional 7 days.[11] Differentiation of MAPCs into neuron-like cells was correlated with the activation of genes involved in neural development, and sequential expression of early and late neural and glial markers was seen. RT-PCR data showed that mRNA levels of Otx2, Otx1, Pax2, Pax5, and nestin increased during the early differentiation stage. As differentiation progressed, cells acquired neuronal morphology. On day 5, MAPC-derived cells expressed nestin, a neuroprogenitor specific marker, but more mature neural and glial proteins were not yet seen. By day 7, some cells stained positive for Nurr1, a midbrain specific dopaminergic transcription factor. By day 14, some cells were NF-200 positive, and some expressed glial fibrillary acidic protein (GFAP) or myelin basic protein (MBP). After 21 days of culture, MAPCs differentiated into cells with even more mature neuronal morphology. Among the neuron-like cells, some expressed markers of dopaminergic neurons, including dopadecarboxylase (DDC) and dopamine, some of serotonergic neurons, such as thyrotropin-releasing hormone (TrH), and some were gamma amino butyric acid (GABA)–positive.

It has been shown that neuron-like cells are polarized, as Tau is expressed only in the neural axon and microtubule-associated protein 2 (MAP-2) in dendrites. Even though MAPC-derived neuron-like cells expressed neurotransmitters at this stage, they did not exhibit neural electrophysiological characteristics. In order to further differentiate MAPCs into functional neurons, differentiated cells were cocultured with fetal astrocytes on cover slips. Patch clamp experiments demonstrated that coculture of step 3 cells (cells treated with all cytokines through BDNF) with fetal brain astrocytes for 8 days resulted in functional voltage-gated sodium channels and synaptic potentials, which is characteristic of mature neurons.

1.4.3. *Hepatocyte differentiation*

When cultured on matrigel with FGF4 and hepatocyte growth factor (HGF), MAPCs can differentiate into cells with morphological, phenotypic, and functional characteristics of hepatocytes.[12] During differentiation, cells first expressed hepatocyte nuclear factor-3β (HNF-3β), GATA4, cytokeratin 19 (CK19), transthyretin, and α-fetoprotein by day 7, and then expressed CK18, hepatocyte nuclear factor (HNF)-4, and HNF-1α on days 14–28. By day 21, some cells were binucleated, and most cells stained positive for albumin and CK18. These cells also nearly acquired full functional properties of hepatocytes: secretion of urea and albumin, presence of phenobarbital-inducible cytochrome p450, uptake of LDL, and storage of glycogen.

Snykers *et al.*[13] differentiated MAPCs into hepatocytes using the method of sequential addition of cytokines. Sequential growth factor signaling more closely mimics the natural developmental environment of hepatocytes during embryogenesis than the original method. Compared to the original cytokine cocktail method, sequential addition of liver-specific factors (FGF-4, HGF, insulin-transferrin-sodium-selenite (ITS), and dexamethasone) in a time-dependent order significantly improved the differentiation efficacy.

1.4.4. *Endothelial differentiation*

Treated with VEGF-B, MAPCs can be differentiated into endothelial cells.[14] During differentiation of MAPCs to endothelia, angioblast-like cells (CD34$^+$, VE-cadherin$^+$, Flk1$^+$) first appeared. These cells subsequently expressed more mature endothelial markers such as von Willebrand factor (vWF) and platelet endothelium cell adhesion molecule (PECAM). Functional studies indicated that MAPCs can differentiate into mature endothelial cells, which can release vWF upon histamine stimulation, and can form vascular tubes. Compared to AC133$^+$ endothelial progenitor cells (hAC 133$^+$), human MAPCs are more versatile in endothelial cell differentiation.

Aranguren *et al.*[15] demonstrated that human MAPCs can differentiate into both arterial and venous endothelial cells, while hAC 133$^+$ cells only adopt a venous and microvascular endothelial cell (EC) phenotype. VEGF165 treatment of human MAPCs induced significant levels of the arterial markers Hey-2, Dll-4, EphrinB1, and EphrinB2, as well as the venous marker EphB4. The resultant hMAPC-ECs were functional, as demonstrated by acetylated-LDL uptake and vascular tube formation. Arterial EC differentiation of human MAPCs is attenuated by blocking SHH or the Notch pathway and is boosted by simultaneous Notch and PTC activation with delta-like 4 (Dll-4), a Notch ligand, and SHH.

1.4.5. *Smooth muscle differentiation*

TGFβ-1 alone or combined with PDGF-BB induces smooth muscle cells (SMCs).[16] During differentiation, expression of SMC contractile apparatus-specific genes resembles what has been observed during vascular development *in vivo*. Smooth muscle (SM) 22α and α-SMA transcripts and proteins are induced by day 2 and day 4 respectively, followed by calponin-1 and SM-MHC. The expression of the transcription factor myocardin is also increased significantly from day 0 to day 6. When MAPC-derived SMCs were passed, all of their expression levels were increased further.

Several functional studies demonstrated that SMCs derived from MAPCs display functional attributes similar to those of neonatal SMCs, as they have functional L-type calcium channels, capability to remodel fibrin when entrapped, and increased responses after 3-week cyclic distention.

1.4.6. *Pancreatic β-cell differentiation*

Previously we tried to differentiate MAPCs into pancreatic β-cells. Although early pancreatic and endocrine pancreas transcription factors could be induced, insulin-secreting pancreatic β-cells could not be obtained. However, when we improved culture conditions and maintained MAPCs at 5% O_2 concentration, we derived MAPC clones with a high Oct-4 level. MAPCs with a high Oct-4 level showed more robust potential for differentiation into cells of several mesoderm and endoderm lineages *in vitro*. When high Oct-4 MAPCs were treated with activin, bone morphogenetic protein (BMP) 4, anti-sonic-hedgehog antibody (Ab), EGF, exendin, β-cellulin, and growth and differentiation factor (GDF) 11, they sequentially expressed Hnf-3β, Hnf-1, Pdx-1, Ngn-3, Neuro-D1, Pax-4, and Nkx-6.1, as well as insulin-1, insulin-2, glucagon, and somatostatin at mRNA level in a timely order. This indicates that the MAPCs first committed to endocrine pancreas endoderm and subsequently insulin$^+$ β-cells. Toward the end of differentiation, cell clusters form. Immunofluorescence data showed that 10–20% of cells in the cluster expressed Pdx-1, 5–10% glucagon, and 1–2% insulin. Additional study showed that c-peptide is secreted in response to glucose *in vitro*. This verifies that insulin found in differentiated β-cells is not exogenous, and that MAPCs might be useful in generating β-cells to treat type I diabetes (unpublished data).

1.5. Blastocyst injection

Ten to twelve Rosa 26 MAPCs (containing the Neo/β-gal transgene) were microinjected into each blastocyst, or to show that a single cell is capable of chimaerism, one cell was injected.[2,20] Animals born from microinjected blastocysts had a normal size and appearance. Chimaerism was determined by levels of Neo/β-gal in tail clippings using Q-PCR. Chimaerism following blastocyst injection occurred from 10–12 MAPCs as well as from a single MAPC. Around 30% of mice derived from blastocysts microinjected with a single MAPC were chimaeric, ranging between 0.1% and 45%. MAPC contribution was confirmed by x-gal staining and staining with anti-β-gal fluorescein isothiocyanate (FITC) antibody. In the brain, skeletal muscle, cardiac muscle, liver, intestine, lung, kidney, spleen, marrow, blood, and skin, costaining was seen for β-gal and tissue-specific markers, which indicates that MAPCs differentiated into cells of the organ in which they were incorporated. Thus, a single MAPC can contribute to and differentiate into cells of the three germ layers.

1.6. *In vivo* postnatal engraftment and *in vivo* differentiation

1.6.1. *Hematopoietic cell differentiation*

Previously we reported that when *in vitro* cultured MAPCs from Rosa 26 mice were intravenously transplanted into noninjured NOD-SCID mice, engraftment was detected in multiple tissues by Q-PCR.[2] Double-staining for β-gal and tissue-specific markers was used to assure the differentiation of MAPCs into cells of a specific tissue. Costaining for β-gal and tissue-specific markers confirmed *in vivo* differentiation of the donor MAPCs in response to local cues to hematopoietic cells, and epithelia of the lung, liver, and intestine. Skeletal muscle, myocardium, and brain did not have MAPC engraftment. No teratomas were found.

Tolar *et al.*[17,18] reported that host factors may affect the biodistribution and persistence of reporter gene (Luciferase/DsRed)-labeled MAPCs after injection. Host irradiation and NK depletion may help more widespread biodistribution and persistence of MAPCs, as MHC-class I negative MAPCs are targets of NK cells.

In a previous study, hematopoietic contribution of donor MAPCs was low. Since then, several changes have been made to improve the MAPC transplantation. MAPCs isolated under low O_2 conditions were used for infusion. The Oct-4 level of these MAPCs was much higher than that of previous MAPCs, almost equivalent to the ES cell level. Also, NOD-SCID mice were sublethally irradiated and treated with an anti-NK cell antibody to deplete NK cells. With these modifications, the new MAPC injection studies show that MAPCs are capable of reconstituting the hematopoietic system and generating functional lymphoid cells when coinjected with host bone marrow cells.[19] MAPCs robustly contributed to multilineage engraftment. Donor-derived hematopoietic cells with markers for B- or T-lymphocytes and myeloid cells were detected. Donor-derived dendritic cells, NK cells, and megakaryocytes were also detected in the spleen and peripheral blood of long term engrafted mice. Donor-MAPC-derived hematopoietic stem cells were present and capable of serial multilineage hematopoietic reconstitution.

1.6.2. *Endothelium differentiation*

Intravenously injected human MAPCs contribute to vasculature formation in tumors.[14] Twelve weeks after human MAPCs were infused intravenously into non-irradiated NOD-SCID mice, some of the mice were sacrificed. A host cell thymic lymphoma was found in one mouse. Approximately 12% of the vasculature costained with an antihuman $\beta 2$-microglobulin–FITC Ab and an anti–vWF-Cy3 Ab, but not antimouse CD31, which indicated that those endothelia were derived from human MAPCs. Among the NOD-SCID mice in which ROSA MAPCs were injected, one animal developed a lymphoma of recipient B-lymphocyte origin. Endothelial cells in the tumor were partially of donor origin.

MAPC-derived endothelial cells also contributed to tumor angiogenesis as well as wound-healing angiogenesis.[15] In the study, murine Lewis lung carcinoma spheroids were transplanted near the shoulder blades of NOD-SCID mice. It was shown that the mice given human-MAPC-derived endothelial cells had tumors with significantly higher vascular masses compared to the tumors in control mice. Human-MAPC-derived endothelial cells also played a part in tumor neoangiogenesis *in vivo*, as seen by immunofluorescent staining. To see if human-MAPC-derived endothelial cells contributed to wound-healing angiogenesis, researchers looked at the areas of the ear that had been clipped to tag the mouse. They found that neoangiogenesis in these wounds partially came from the MAPC-derived endothelial cells.

1.7. Brain transplantation

A rat stroke model was used to test MAPC engraftment in and therapeutic effect on the brain.[21] Cortical brain ischemia was created through permanent ligation of the right middle cerebral artery (MCA) distal to the striatal branch. One week after brain ischemia, human MAPCs were injected into three sites in the cortex surrounding the infarction. Cyclosporin A was used as an immunosuppressant. Two and six weeks after transplantation, a limb placement test *and* a tactile stimulation test were used to assess neurological deficits, and immunohistochemistry was done to trace the injected human MAPCs.

Compared with two control groups, results showed that human MAPC injection significantly ameliorated ischemia-induced neurological deficits. It was shown that transplanted human MAPCs migrated along the corpus callosum. Immunohistology showed that some transplanted MAPCs were immunopositive for the astrocyte marker GFAP, the oligodendrocyte marker GalC, and the neuronal markers tubulin-III, neuro-filament-200, neurofilament-160, hNSE, and hNF-70. However, as the morphology of these cells was immature (rounded in shape with few fiber processes) six weeks after transplantation, functional improvement may be attributed to trophic effects of human MAPCs instead of new neuron integration into the brain.

Rat MAPCs were also injected into the hippocampus in neonatal rats with a hypoxic–ischemic (HI) brain injury.[22] Similarly, transplated MAPCs survived in the hippocampal region, and the coordinated motor functions were significantly improved two weeks after MAPC transplantation.

1.8. Muscle transplantation

A study was done to test whether MAPCs incorporate and differentiate into skeletal muscles *in vivo*.[10] Cardiotoxin was injected into tibiaris anterior (TA) muscles to induce muscle regeneration. Twenty-four hours later, enhanced green fluorescent protein (EGFP) transduced undifferentiated or predifferentiated MAPCs were injected into TA muscles of irradiated NOD-SCID mice. When undifferentiated MAPCs were injected, a few EGFP-positive cells existed in the scar tissue, but they did not express myocyte markers. However, when MAPCs were predifferentiated toward myocytes for three days before injection, the engraftment was greatly enhanced. Numerous EGFP-positive cells were seen in the periphery of regenerating muscles. Those cells expressed myocyte markers (α-actinin and SK-myosin) and could fuse and form mature myotubes. This demonstrated that *in vivo* myocyte differentiation from MAPCs needs not only local cues from host muscles but also driving signals to push the conversion of undifferentiated MAPCs.

1.9. Mechanisms underlying pluripotency of MAPCs

There are several possible explanations for pluripotency of MAPCs. Understanding the mechanisms of pluripotency of MAPCs is important to the field of stem cell biology and will provide clues for the use of MAPCs in tissue repopulation and regeneration.

First, pluripotency of MAPCs could be attributed to multiple different somatic stem cells. However, retroviral marking studies demonstrated that differentiation of MAPCs into cells of neuroectoderm, mesoderm, and endoderm lineages occurs at the single cell level. The blastocyst studies indicated that a single MAPC contributes to the formation of most somatic tissues. This data does not support the idea that MAPCs are a mixture of multiple stem cells.

There are quite a few pluripotent adult stem cells reported, namely MIAMI cells, hBMSCs, USSCs, etc.[23-25] One common characteristic of these cells is that they are all cultured cells. This raises the possibility that cells may undergo dedifferentiation/transdifferentiation during the long term cell culture. As MAPCs were derived from mesenchymal stem cell culture after 25–35 population doublings, it is possible that MAPCs are dedifferentiated MSCs. It has been shown in numerous studies that after allogeneic bone marrow transplantation, MSCs, like stromal cells, are of host origin. However, data from syngeneic bone marrow transplants into lethally irradiated C57BL6 mice indicated that MAPCs are of donor origin.[26] This study used irradiation to eliminate the cells in bone marrow from which MAPCs are derived, then showed that bone marrow transplantation transfers these cells from the donor into the recipient.

The different origins of MAPCs and MSCs revealed that *in-vitro-* cultured MAPCs are physiologically different from both stromal cells and MSCs. This suggests that despite the fact they all reside in the BM adherent cell component, they could be ontologically unrelated. Although MAPCs still may be the product of dedifferentiation, we can rule out the possibility that they are dedifferentiated cells from MSCs.

The other possibility is cell fusion, which has been reported both *in vitro* and *in vivo*.[27–30] *In vitro* coculture of genetically marked ES cells with genetically marked NSCs or BM cells has resulted in cell fusion. The fused cells were tetraploid cells that contained both the genetic information of the neural or BM cells and the pluripotency of the ES cells. *In vivo*, it has been shown that hepatocytes derived from transplanted HSCs in FAH$^{-/-}$ mice were actually derived from cell fusion. It was also shown that fusion could occur in skeletal muscle, cardiac muscle, and Purkinje cells. Recent studies of muscle regeneration suggest that fusion does not occur directly between HSCs and muscle fibers. Instead, a descendant of HSCs, such as myelomonocytic precursors, may fuse with muscle fibers after muscle injury.[31] In liver regeneration of FAH$^{-/-}$ mice, a similar mechanism might exist.[32,33] Circulating myelomonocytic cells may fuse with hepatic cells, triggering the liver nodule's proliferation from these tetraploid cells.

The pluripotency of MAPCs *in vitro* is definitely not the result of cell fusion. Gene insertion indicated that MAPCs were derived from a single cell. Also, MAPCs are diploid, not tetraploid. They have never been cocultured with embryonic stem cells or other tissue-specific cells. Although the possibility cannot be ruled out that *in vivo* cell fusion causes multiple tissue engraftment of MAPCs in uninjured mice, the chance is slim as the phenomenon of cell fusion both *in vitro* and *in vivo* is extremely rare (1/100,000 cells). This suggests that the rate of cell fusion should be too low to account for the multiple tissue engraftment of MAPCs.

MAPCs have characteristics similar to those of embryonic stem cells, namely proliferation without senescence, the need for LIF in mouse MAPC culture, single cell differentiation to all three germ layers, and contribution to all somatic cell types when injected in a blastocyst. MAPCs could be considered to be equivalent to embryonic stem cells, which have persisted in adult tissues. If this is true, it is important to find their location, original phenotypes, and their reactions when injury occurs *in vivo*.

2. MAPC Application

No matter what their origin is or what the mechanism underlying their pluripotency is, MAPCs could be a new potent and abundant source of pluripotent adult stem cells. They hold a great deal of promise for developing medical therapy for many diseases, including diabetes, Parkinson's disease, hemophilia, and other tissue degenerative diseases.

As MAPCs allow autologous transplantation, they might overcome a substantial obstacle to the success of cell transplantation — the immune response from the recipient. Furthermore, since MAPCs are easily genetically manipulated, they could be good candidates for *ex vivo* gene correction of autologous stem cells. However, while they hold immense therapeutic potential, there is a long way to go before reaching applicable cell therapy. A large number of studies are still required to fully characterize MAPCs.

First and foremost is the safety problem. Derivation of MAPCs requires long term culture. Although karyotyping revealed no obvious cytogenetic abnormalities in MAPCs, it is unavoidable that cells in culture eventually accumulate genetic mutations. Whether MAPC transplantation will lead to malignancy and long term effects will need to be clarified. For the treatment of various diseases, we also need to test when and where to inject MAPCs and what kind of cells (undifferentiated vs lineage-committed vs terminally differentiated cells) are going to be used. Furthermore, as MAPC culture needs stringent conditions, a reliable culture system needs to be built to provide steady, high quality MAPCs. Generally, the application of MAPCs for human disease will require much more knowledge about their biological properties.

References

1. Reyes M, Lund T, Lenvik T, *et al.* (2001) Purification and *ex vivo* expansion of postnatal human marrow mesodermal progenitor cells. *Blood* **98**: 2615–2625.
2. Jiang Y, Jahagirdar B, Reyes M, *et al.* (2002) Pluripotent nature of adult marrow derived mesenchymal stem cells. *Nature* **418**: 41–49.
3. Zeng L, Rahrmann E, Hu Q, *et al.* (2006) Multipotent adult progenitor cells from swine bone marrow. *Stem Cells* **24**: 2355–2366.
4. Breyer A, Estharabadi N, Oki M, *et al.* (2006) Multipotent adult progenitor cell isolation and culture procedures. *Exp Hematol* **34**: 1596–1601.

5. Jiang Y, Breyer A, Lien L, *et al.* (2004) Culture of multipotent adult progenitor cells (MAPCs). *Blood (ASH Annual Meeting Abstracts)* 104: 2329.
6. Jiang Y, Vaessen B, Lenvik T, *et al.* (2002) Multipotent progenitor cells can be isolated from post-natal murine bone marrow, muscle, and brain. *Exp Hematol* 30: 896–904.
7. Gupta S, Verfaillie CM, Chmielewski D, *et al.* (2006) Isolation and characterization of kidney-derived stem cells. *J Am Soc Nephrol* 17: 3028–3040.
8. Qi H, Aguiar DJ, Williams SM, *et al.* (2003) Identification of genes responsible for osteoblast differentiation from human mesodermal progenitor cells. *Proc Natl Acad Sci USA* 100: 3305–3310.
9. Maes C, Coenegrachts L, Sotkcmans I, *et al.* (2006) *J Clin Invest* 116: 1230–1242.
10. Muguruma Y, Reyes M, Nakamura Y, *et al.* (2003) *In vivo* and *in vitro* differentiation of myocytes from human bone marrow-derived multipotent progenitor cells. *Exp Hematol* 31: 1323–1330.
11. Jiang Y, Henderson D, Blackstadt M, *et al.* (2003) Neuroectodermal differentiation from mouse multipotent adult progenitor cells. *Proc Natl Acad Sci USA* 100 Suppl 1: 11854–11860.
12. Schwartz RE, Reyes M, Koodie L, *et al.* (2002) Multipotent adult progenitor cells from bone marrow differentiate into functional hepatocyte-like cells. *J Clin Invest* 109: 1291–1302.
13. Snykers S, Vanhaecke T, Papeleu P, *et al.* (2006) Sequential exposure to cytokines reflecting embryogenesis: the key for *in vitro* differentiation of adult bone marrow stem cells into functional hepatocyte-like cells. *Toxicol Sci* 94: 330–341.
14. Reyes M, Dudek A, Jahagirdar B, *et al.* (2002) Origin of endothelial progenitors in human post-natal bone marrow. *J Clin Invest* 109: 337–346.
15. Aranguren XL, Luttun A, Clavel C, *et al.* (2006) *In vitro* and *in vivo* arterial differentiation of human multipotent adult progenitor cells. *Blood* 109: 2634–2642.
16. Ross JJ, Hong Z, Willenbring B, *et al.* (2006) Cytokine-induced differentiation of multipotent adult progenitor cells into functional smooth muscle cells. *J Clin Invest* 116: 3139–3149.
17. Tolar J, Osborn M, Bell S, *et al.* (2005) Real-time *in vivo* imaging of stem cells following transgenesis by transposition. *Mol Ther* 12: 42–48.
18. Tolar J, O'Shaughnessy MJ, Panoskaltsis-Mortari A, *et al.* (2006) Host factors that impact the biodistribution and persistence of multipotent adult progenitor cells. *Blood* 107: 4182–4188.
19. Serafini M, Dylla SJ, Oki M, *et al.* (2007) Hematopoietic reconstitution by multipotent adult progenitor cells: precursors to long-term hematopoietic stem cells. *J Exp Med* 204: 129–139.

20. Keene CD, Ortiz-Gonzalez XR, Jiang Y, *et al.* (2003) Neural differentiation and incorporation of bone marrow-derived multipotent adult progenitor cells after single cell transplantation into blastocyst stage mouse embryos. *Cell Transplant* **12**: 201–213.
21. Zhao LR, Duan WM, Reyes M, *et al.* (2002) Human bone marrow stem cells exhibit neural phenotypes and ameliorate neurological deficits after grafting into the ischemic brain of rats. *Exp Neurol* **174**: 11–20.
22. Yashuhara T, Matsukawa N, Yu G, *et al.* (2006) Transplantation of cryopreserved human bone marrow-derived multipotent adult progenitor cells for neonatal hypoxic–ischemic injury: targeting the hippocampus. *Rev Neurosci* **17**: 215–225.
23. D'Ippolito G, Diabira S, Howard GA, *et al.* (2004) Marrow-isolated adult multilineage inducible (MIAMI) cells, a unique population of postnatal young and old human cells with extensive expansion and differentiation potential. *JCS* **117**: 2971–2981.
24. Yoon Y, Wecker A, Heyd L, *et al.* (2005) Clonally expanded novel multipotent stem cells from human bone marrow regenerate myocardium after myocardial infarction. *J Clin Invest* **115**: 326–338.
25. Kögler G, Sensken S, Airey JA, *et al.* (2004) A new human somatic stem cell from placental cord blood with intrinsic pluripotent differentiation potential. *J Exp Med* **200**: 123–135.
26. Reyes M, Li S, Foraker J, *et al.* (2005) Donor origin of multipotent adult progenitor cells in radiation chimeras. *Blood* **106**: 3646–3649.
27. Ying QL, Nichols J, Evans EP, Smith AG. (2002) Changing potency by spontaneous fusion. *Nature* **416**: 545–548.
28. Terada N, Hamazaki T, Oka M, *et al.* (2002) Bone marrow cells adopt phenotypes of other cells by spontaneous cell fusion. *Nature* **416**: 542–545.
29. Lagasse E, Connors H, Al-Dhalimy M, *et al.* (2000) Purified hematopoietic stem cells can differentiate into hepatocytes *in vivo*. *Nat Med* **6**: 1229–1234.
30. Grompe M. (2003) The role of bone marrow stem cells in liver regeneration. *Semin Liver Dis* **23**: 363–372.
31. Doyonnas R, LaBarge MA, Sacco A, *et al.* (2004) Hematopoietic contribution to skeletal muscle regeneration by myelomonocytic precursors. *Proc Natl Acad Sci USA* **101**: 13507–13512.
32. Willenbring H, Bailey AS, Foster M, *et al.* (2004) Myelomonocytic cells are sufficient for therapeutic cell fusion in liver. *Nat Med* **10**: 744–748.
33. Camargo FD, Finegold M, Goodell MA. (2004) Hematopoietic myelomonocytic cells are the major source of hepatocyte fusion partners. *J Clin Invest* **113**: 1266–1270.

6

Ocular Surface Epithelial Stem Cells

De-Quan Li, Stephen C. Pflugfelder and Andrew J. W. Huang*

1. Introduction

Stem cell research has revolutionized the fields of reproductive and regenerative biology. While its application in clinical medicine is expanding, the understanding of human stem cells has been challenged by their restricted availability and limited characterization. Significant strides have been made in the past two decades regarding the ocular surface epithelial stem cells. This article will review the current knowledge of adult stem cells of ocular surface epithelia, and discuss the clinical implications of stem cells in ocular surface rehabilitation for pathological conditions.

2. Adult Stem Cells and Niches

As discussed elsewhere in this book, stem cells are known to have the capacity for self-renewal and the ability to generate differentiated progeny or multiple cell lineages. While recent studies indicate that embryonic stem cells have pluripotent potentials, organ-specific stem cells are known to reside in many adult tissues, such as bone marrow, liver, brain, intestine, skin and cornea.[1-3] By nature, true stem cells such as embryonic stem cells can turn into any type of cells, whereas progenitor cells such as adult stem cells are committed to becoming specific cell types of a particular tissue. Adult stem cells, somatic stem cells, or organ-specific stem cells are small subpopulations of progenitor cells with unique characteristics in

*Correspondence: Department of Ophthalmology and Visual Sciences, Washington University, 660 S. Euclid Ave, Campus Box 8096, St. Louis, MO 63110, USA. Tel: 314-362-0622, Fax: 314-362-0627. E-mail: huanga@vision.wustl.edu.

morphology, phenotype, and growth potential. Currently accepted criteria for adult stem cells include (1) quiescent slow cycling or long cell cycle time during homeostasis *in vivo*, (2) undifferentiated small cells with primitive cytoplasm, (3) high proliferative and/or pluripotent potential after wounding or placement in culture, (4) capacity for self-renewal through the life span, and (5) ability to regenerate functional tissues.[1–5]

Adult stem cells may generate differentiated progenies through asymmetric divisions. The asymmetric cell division gives rise to one daughter stem cell (for self-renewal) and one nonstem cell committed to differentiating into specific phenotypes (for tissue regeneration). Hence, progenitor cells are not all adult stem cells. Consequently, there are adult stem cells and related "transient amplifying cells (TACs)" or committed precursor cells or progenitor cells. In turn, these committed cells, with more rapid yet limited proliferation, less self-renewal and only limited differentiation, can increase the number of differentiated cells produced by a single adult stem cell. Therefore, these TACs, also known as transit amplifying cells or transit cells, are committed progenitors linking the adult stem cells with their terminally differentiated progenies.[2,6]

Adult stem cells typically reside in specialized and often more protected locations known as niches,[1,6,7] which constitute three-dimensional microenvironments containing stem cells, neighboring differentiated cells, mesenchymal cells and extracellular matrix. Niches provide sheltering environments that sequester stem cells from differentiative stimuli, apoptotic stimuli and/or other stimuli that would otherwise challenge the stem cell reserves. The niche may also safeguard against excessive stem cell production that could eventually lead to unrestricted growth such as cancer. Thus, a niche is a specific location in a tissue where adult stem cells can reside infinitely, produce progeny cells while self-renewing, and substitute damaged cells to maintain the structural and functional integrity of the tissue. Similar stem cell biology has been observed in ocular surface epithelia.

3. Ocular Surface Epithelial Stem Cells

The ocular surface consists of the cornea, the conjunctiva and a junctional zone, which is known as the limbus, at the corneoscleral junction. A healthy and transparent cornea is essential as a refractive interface for the vision and as a protective barrier for the intraocular milieu. Histologically, the ocular surface epithelium comprises three distinct phenotypes, i.e. corneal, limbal and conjunctival epithelia. It has been demonstrated by Wei *et al.* that corneal and conjunctival epithelial cells belong to two separate and distinct lineages.[8] The stem cells of corneal and conjunctival epithelia are small, quiescent and undifferentiated subpopulations of the ocular surface epithelial cells. Corneal epithelial stem cells are responsible for maintenance of the corneal epithelial homeostasis and wound healing, whereas the conjunctiva epithelia are generally renewed by conjunctival stem cells.

The corneal epithelium includes approximately one to three outermost layers of flattened cells called squames, two to three subjacent layers of suprabasal or wing cells, and a single layer of basal columnar cells. Numerous evidences have indicated that corneal epithelial stem cells reside in the basal epithelial layer of the limbus, a straddling zone 1.5–2 mm in width between the peripheral cornea and bulbar conjunctiva.[4,9–11] Although it was recognized as early as 1971 that corneal epithelial cells are repopulated by centripetal migration of the cells from the limbus,[12] it was not until 1986 that Schermer *et al.*[13] proposed that limbal basal epithelial cells, devoid of differentiation markers, may represent corneal epithelial stem cells and are responsible for the regeneration of the corneal epithelium.

Other supporting evidence for this limbal location of corneal epithelial stem cells includes the fact that limbal basal cells (1) contain quiescent, slow-cycling cells identified as the "label-retaining cells" following pulse-chase labeling of cells with [3H]-thymidine or bromodeoxyuridine (BrdU);[14] (2) lack the corneal epithelial differentiation-associated cytokeratin pair K3[13] and K12;[15] and (3) exhibit a much higher proliferative potential than the central corneal epithelium in culture.[14,16,17] Furthermore, abnormal corneal epithelial wound healing occurs when the limbal epithelium is partially or completely damaged in experimental models and human patients,[18,19] and transplantation of limbal cells can be used to reconstitute the entire corneal epithelium in patients with a defective limbus due to injury or inflammation.[10,20] Collectively, these data succinctly indicate that limbal basal epithelia exhibit the full complement of well-defined properties for keratinocyte (epithelial) stem cells and represent corneal epithelial stem cells. As a result, based on their location, limbal basal epithelia are often referred to as corneal epithelial stem cells.

By contrast, the identity and physical location of conjunctival stem cells have been much less established and remain somewhat controversial. Unlike the nonsecretory and stratified corneal epithelium, the secretory conjunctival epithelium consists of both stratified keratinocytes and mucin-secreting goblet cells. These two types of cells have been shown to arise from a common conjunctival progenitor cell. Evidence suggests that conjunctival stem cells or progenitors of goblet cells appear to aggregate in the conjunctival fornix (cul-de-sac),[21,22] although epithelial cells in other conjunctival regions, such as the mucocutaneous junctions at the lid margin, may also display characteristics of conjunctival stem cells.[23]

Despite the convincing evidence regarding the distinct lineage of corneal and conjunctival epithelial stem cells, it remains unclear if the corneal stem cells can differentiate into conjunctival cells or vice versa.

4. Candidate Markers of Corneal Epithelial Stem Cells

Limbal basal epithelia consist of at least three functionally different cell types, i.e. corneal stem cells, transient amplifying cells and terminally differentiated cells. The limbal stem cells are only a small subpopulation, estimated to be less than 1% of those basal cells,[24,25] or even as few as 100 cells per limbus to be true stem cells.[11,26] Although putative corneal epithelial stem cells characterized by various purported stem cell markers have been known to exist in the basal limbus for more than two decades,[15] to date none of these markers by itself can unequivocally confirm the identity of limbal stem cells.[11,24,25,27]

The proposed epithelial stem cell markers can be categorized into three major groups: (1) nuclear proteins such as transcription factor p63; (2) cell membrane or transmembrane proteins, including integrins (integrin $\beta1$, $\alpha6$ and $\alpha9$), receptors [epidermal growth factor receptor (EGFR), transferrin receptor CD71] and drug resistance transporters (ABCG2); and (3) cytoplasmic proteins such as cytokeratins (K19), nestin and α-enolase. In addition, a variety of differentiation markers have also been used to distinguish undifferentiated stem cells from more differentiated cells. These include cytokeratins (K3 and K12), involucrin, E-cadherin (an intercellular adhesive molecule) and connexin 43 (a gap junction protein).

The nuclear transcription factor p63, a member of the p53 family, is highly expressed in the basal epithelial cells of many human tissues, and the truncated dominant-negative ΔNp63 isoform is the predominant species in these cells.[28] It has been reported that p63 knockout mice lack stratified epithelia and contain clusters of terminally differentiated keratinocytes on the exposed dermis,[29] and that the expression of p63 is associated with proliferative potential in human keratinocytes.[29,30] ABCG2 (ATP-binding cassette, subfamily G, member 2), a member of the ATP-binding cassette transporters and formally known as breast cancer resistance protein 1 (BCRP1), has been used as a molecular determinant for the side population enriched in hematopoietic stem cells.[31] It has also been proposed as a universal marker for stem cells, including those for corneal epithelial cells.[24,32]

Integrins are cell surface adhesion molecules which consist of noncovalently associated heterodimeric transmembrane receptors with an α and a β subunit binding to extracellular matrix proteins. They are the major components responsible for stable adhesions to the basement membrane in hemiadherent junctions. Integrin $\alpha 9$ has been localized at the basal cells of the epidermis, conjunctiva and corneolimbus after birth, and at early adulthood in murine ocular surface.[33] Integrin $\beta 1$ has previously been proposed as a putative stem cell marker for epidermal keratinocytes. The integrin $\beta 1$-enriched human epidermal basal cells from both keratinocyte cultures and foreskin biopsies have been shown to have a higher colony-forming efficiency than the unfractionated cells.[34,35] In addition, the slow-cycling, BrdU-label retaining cells in the adult murine cornea express high levels of integrin $\beta 1$ and $\beta 4$ but only little integrin $\alpha 9$.[36] It has been reported that limbal basal epithelial cells express higher levels of EGFR than the limbal suprabasal cells, and the more mature and differentiated cells express the lowest levels of EGFR.[37,38]

As a member of the cytokeratin family of intermediate filaments, K19 has been suggested as a marker for the epidermal stem cells in skin hair follicles. K19 is expressed in the hair follicle and absent in the interfollicular epidermis at hairy sites. K19 is also associated with the slow-cycling [3H]-thymidine-label-retaining cells.[39] A cytoplasmic glycolytic enzyme, α-enolase, has also been proposed to be a candidate marker for corneal epithelial stem cells. It is preferentially expressed by the limbal basal cells, as well as the basal cells of other stratified epithelia.[11,40]

Intercellular communication plays an important role in regulating cellular development and differentiation. Communication via the gap junctions is mediated by connexins, a family of related amphipathic polypeptides. These intercellular communicating channels allow direct passive diffusion of low molecular weight solutes between neighboring cells.[41] E-cadherin is a transmembrane Ca^{2+}-dependent homophilic adhesion receptor that plays an important role in cell–cell adhesion. Cell to cell contact mediated by E-cadherin results in cell activation and an increase of signaling molecules involved in cell differentiation and survival.[42] Since connexin 43 and E-cadherin are primarily expressed by differentiated epithelial cells, the absence of these intercellular communication molecules in limbal basal cells may be an inherent feature of stem cells, implying the need for stem cells to maintain their unique nature with restricted intercellular communications.[41,42] N-cadherin is another member of the classical cadherin family and has been demonstrated to be expressed by hematopoietic stem cells. Recently, it has been demonstrated to be exclusively expressed by putative limbal stem cells.[43] K3 and K12, a cytokertain pair, have been well established as cornea-specific differentiation markers.[13,15,44] They are absent in the limbal basal epithelia. Involucrin, a structural protein of the cornified cell envelope,[45] is another marker for epithelial differentiation.[46]

Using these markers, the basal epithelium of the human limbus has been characterized as small primitive cells with three expressing patterns:[47] (1) p63, ABCG2 and integrin α9 are exclusively expressed in a subset of basal cells; (2) most basal cells express higher levels of integrin β1, EGFR, K19 and α-enolase; and (3) differentiated markers such as nestin, E-cadherin, connexin 43, involucrin, K3 and K12 are absent in basal cells. Most recently, Li *et al.* also found that neurotrophic factors, namely nerve growth factor (NGF) and glial cell-derived neurotrophic factor (GDNF), and their corresponding receptors, TrkA and GFRα-1, are exclusively expressed by a subpopulation of limbal basal epithelial cells (Fig. 1). These molecules may serve as exclusive markers to characterize the stem cell phenotypes of limbal basal cells, similar to those markers in pattern 1.

Taken together, the candidate stem cell markers expressed by the corneal and limbal epithelia are summarized in Table 1. These unique expression patterns may facilitate the identification of putative limbal stem cells.[43,47] The presence of p63 in limbal basal cells suggests a higher proliferative potential in these cells, while the presence of ABCG2 may be a feature that protects limbal stem cells from damage by drugs and toxins.

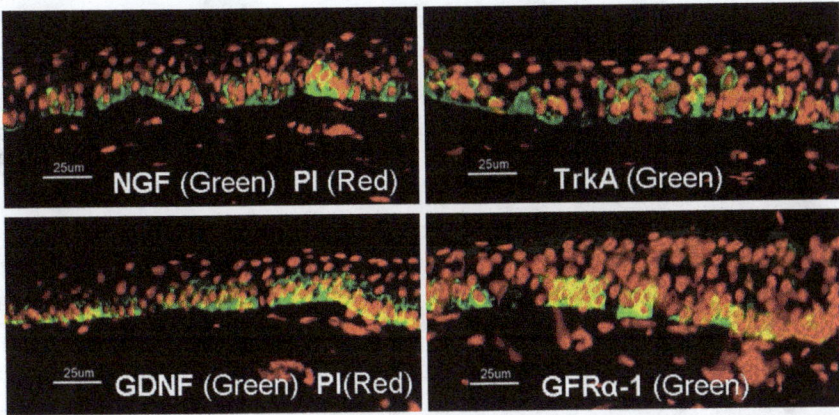

Figure 1. Immunofluorescent stainings show that NGF, GDNF, TrkA and GFRα-1 (green) were exclusively expressed in a subset of the basal cells of the human limbal epithelium, with cell nuclei being counterstained by propidium iodide (PI) (red).

Table 1. Expression Patterns of Candidate Stem Cell Markers

Pattern	Markers	Limbal Epithelium		Corneal Epithelium	
		Basal	Suprabasal	Basal	Suprabasal
I	p63	+++	−	−	−
	ABCG2	+++	−	−	−
	Integrin α9	+++	−	−	−
	N-cadherin	+++	−	−	−
	NGF	+++	−	−	−
	TrkA	+++	−	−	−
	GDNF	+++	−	−	−
	GFRα-1	+++	−	−	−
II	Integrin β1	+++	+	+++	++
	EGFR	+++	+	+++	++
	K19	+++	+	+++	+++
	α-enolase	+++	+	++	+
III	Nestin	−	+++	+	+++
	E-cadherin	−	+++	+	+++
	Connexin 43	−	+++	+	+++
	Involucrin	−	+++	+	+++
	K3	−	+++	+++	+++

The grading was based on immunofluorescent staining: −, undetectable; +, weakly positive; ++, moderately positive; +++, strongly positive.

Co-localization of NGF and GDNF with their receptors in the limbal basal epithelium may implicate their roles as critical survival factors for the stem cells. The presence of high levels of EGFR may enable these basal cells to undergo rapid cell division during development and following wounding in response to growth factors. With more abundant integrins ($\beta1$ and $\alpha9$) and the lack of connexin 43 and E-cadherin, limbal stem cells may adhere strongly to their underlying extracellular matrix with better stability in their niche and more resistance to shear forces. These characteristics may also allow individual stem cells to be mobilized more rapidly and exit from their niche during self-renewal. Furthermore, the lack of expression of K3, K12 and involucrin indicates that the limbal basal cells are relatively undifferentiated.

5. Enrichment of Corneal Epithelial Stem Cells

Although it is well accepted that corneal epithelial stem cells are derived from the basal limbus, isolation of these stem cells from the limbus remains challenging. This may in part be due to the lack of a bona fide stem cell marker, thereby impeding the identification and further characterization of corneal stem cells. Based on adult stem cell criteria and the putative limbal stem cell phenotype, Li *et al.* have attempted to enrich human limbal stem cells through various novel approaches, including cell sizing, differential adherence to extracellular matrix (collagen type IV), and cell sorting for the side population to separate cells expressing ABCG2 or connexin 43 cell surface markers. Using these strategies, they have successfully isolated five clonogenic cell populations from limbal epithelia and their cultures.[32,48–50] As summarized in Table 2, these five clonogenic populations isolated by different methods all display some characteristics of adult stem cells. They are relatively undifferentiated with the presence of stem cell-associated markers (ABCG2, p63 or integrin $\beta1$) and the absence of differentiation-associated markers (K3, K12, involucrin or connexin 43). They have high proliferative potential, namely greater clonal forming efficiency (2–4-fold) and growth capacity in culture. Lastly, they have the capacity for self-renewal, containing quiescent, slow-cycling BrdU label-retaining cells in culture. Although these isolated populations showed varying degrees of enrichment for certain stem cell properties, they were far from being pure and accounted for as much as 10% of the

Table 2. Properties of Isolated Clonogenic Populations Enriched with Corneal Epithelial Stem Cells

Properties	Small Cells	RAC	SP	ABCG2+	Cx 43dim	ALL
Reference	48	49	32	32	50	
Cell markers	Size	Adhesion	Hoechst	ABCG2	Connexin 43	None
Population %	11.0 ± 4.5	10.4 ± 1.5	2.9 ± 1.1	2.3 ± 1.8	9.3 ± 1.7	100
BrdU-LRC %	11.6 ± 1.5	16.0 ± 2.5			13.5 ± 3.6	3.2 ± 0.6
Positive cells%						
ABCG2 +	38.1 ± 1.6				26.7 ± 10.5	10.2 ± 5.5
p63 +	37.3 ± 5.2	47.5 ± 8.6			63.3 ± 13.7	19.6 ± 4.5
Integrin β1 +		57.1 ± 9.8			85.4 ± 15.2	26.6 ± 3.1
Connexin 43 +					5.1 ± 3.0	59.3 ± 15.2
Involucrin +	1.9 ± 0.2	1.8 ± 0.6			15.2 ± 5.9	41.8
Keratin 3 +	5.2 ± 0.5				12.5 ± 5.3	40.9 ± 5.0
mRNA levels						
ABCG2	+++	+++	++	++	+++	+
ΔNp63	+++	+++	+++	+++	+++	++
Integrin β1		++++			+++	++
Connexin 43					+	++
Involucrin	+	−			+	++
Keratin 3	+				−	++
Keratin 12	+	−				++
GAPDH	++++	++++	++++	++++	++++	++++
CFE on day 6 (%)	5.4 ± 1.0	3.8 ± 0.4	4.8 ± 1.6	5.7 ± 0.5	8.7 ± 1.2	2.0 ± 0.5
Confluent days	12–14	10–14	14	14	12–14	18–24

RAC, rapidly adherent cells; SP, side population; Cx, connnexin; ALL, unfractioned whole limbal epithelial cells.

total cells, a percentage much higher than the estimated number of putative stem cells. It remains promising that by combining multiple candidate markers, including positive and negative markers, the corneal epithelial stem cells can be eventually isolated and characterized in the future.[25,51]

6. Corneal Epithelial Stem Cell Niche

A niche is a microenvironment essential for the homeostasis of stem cells. It is maintained by factors intrinsic to the stem cells as well as other exogenous factors in the environment. The exogenous factors include many molecules produced by nearby fibroblasts, vascular endothelial cells and

Limbal Palisades of Vogt

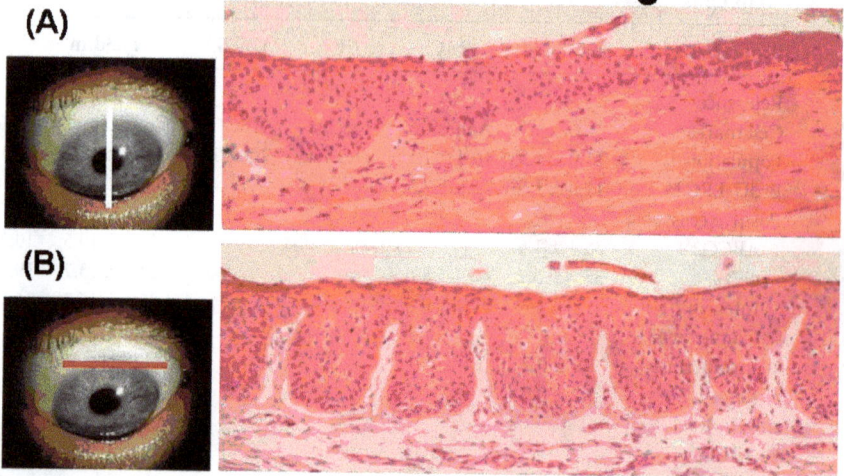

(A)

(B)

Figure 2. Architecture of the limbal palisades of Vogt (Haematoxylin–Eosin stain). (A) A radial section through the central cornea showing about 7–10 layers of stratified limbal epithelia. (B) A tangential cross-section through the superior limbus showing the papilla-like epithelial columns and interspersed connective tissue in the palisades of Vogt (reproduced from Ref. 47 with permission).

perivascular smooth muscle cells, as well as molecules released from those blood vessels. The niche for corneal epithelial stem cells is likely to be located in the limbal palisades of Vogt, a unique microenvironment sequestering the quiescent corneal epithelial progenitor cells.[11,47,51] Anatomically, the palisades of Vogt are composed of papilla-like columns, with smaller and densely packed basal cells, larger but less densely packed cells within the columns and flattened squamous cells in the apical layers (Fig. 2[47]). Underneath the basement membrane, the limbal stroma is heavily innervated[52] and vascularized.[53] Blood vessels, nerves and connective tissue intersperse between the epithelial columns. The column architecture of the limbus offers several advantages for the stem cells. It provides a greater surface area to house larger quantities of stem cells that can be mobilized for self-renewal. The undulating architecture also provides stem cells with various layers of overlying epithelial cells, which can provide better protection of the basal layer and allow more efficient wound healing response at different corneal epithelial depths. The melanin pigments are primarily concentrated in the limbal papillae and may shield the stem cells from untoward UV damage. With stromal matrices and rich neurovascular networks being uniquely inserted between limbal papilla-like columns, the basal layer of limbal epithelia can readily access critical nutrients and other supportive factors. In addition, factors released by the adjacent conjunctival cells and the transient amplifying cells may also play a role in regulating the proliferation of corneal epithelial stem cells.[9,11,54]

Based on their roles in epithelial–mesenchymal interaction, three types of cytokines in the human ocular surface have been proposed.[55] Type I cytokines are released from epithelial cells (transforming growth factor α, interleukin-1β, platelet-derived growth factor BB) to modulate fibroblasts. Type II cytokines [insulin-like growth factor 1, transforming growth factor (TGF) β1 and TGF-β2, basic fibroblast growth factor and leukemia inhibitory factor] can modulate both epithelial cells and fibroblasts. Type III cytokines (keratinocyte growth factor and hepatocyte growth factor) are produced by fibroblasts to modulate epithelial cells. Intricate interactions may exist between epithelial cells and fibroblasts in the cornea and limbus to modulate cellular proliferation and differentiation.[55-58] Limbal fibroblasts have been noted to produce a different repertoire of growth factors from the keratocytes in the central cornea.[55,58-61] The renewal of limbal stem cells may be preferentially stimulated by keratinocyte growth factor produced by limbal fibroblasts,[60] and their survival may be preferentially controlled by antiapoptotic factors derived from limbal fibroblasts (as well as mouse 3T3 cells).[62]

Compositional heterogeneity of the extracellular matrix has been shown to exist between cornea and limbus.[27,51] Other than laminin-1 and laminin-5, the limbal basement membrane expresses laminin α2β2 chains, while the corneal basement membrane does not. Moreover, α1, α2 and α5 chains of type IV collagen are present in the limbal basement membrane, while α3 and α5 chains are in the corneal counterpart.[63,64] These basement membrane components might help determine the distribution of stem cells in the limbal niche.

Functionally, the limbal stem cell niche can be regarded as having two distinct compartments: a quiescent compartment and a proliferative compartment.[11] In homeostatic states, the balance is shifted toward maintaining the corneal epithelial stem cells in their undifferentiated state, in which cell survival is favored over cell division and cells remain quiescent. Survival of the corneal epithelial stem cells is most likely regulated via mesenchymal cells in the stroma, which produce survival factors to keep the stem cells undifferentiated. Wounding or other pathological processes may cause the limbal stem cells to proliferate.[65] During epithelial migration after small corneal epithelial wounds, there is no immediate loss of stem cells, suggesting that the stem cell niche remains unperturbed during re-epithelialization. When the epithelial migration is complete, the niche functions to enhance proliferation by signaling the stem cells to divide until the corneal epithelial layers are brought back to a normal thickness. During large wounding or repeated epithelial trauma, the niche is forced to maintain active proliferation for an extended duration. Such a proliferative pressure may damage the limbal niche and result in corneal stem cell deficiency or dysfunction.

7. Conjunctival Epithelial Stem Cells

The nonkeratinized conjunctival epithelium extends from the corneolimbus to the lid margin, where there is a gradual transition to the keratinized, stratified squamous epithelium of skin. The conjunctival epithelium along with mucin-secreting goblet cells not only contributes to the stability of the tear film, but also serves as a mechanical and immunological barrier for the ocular surface.

As discussed above, conjunctival stem cells have not been as well investigated as corneal stem cells.[66] Most importantly, current evidence indicates that these two epithelia are of two different lineages arising from distinct stem cell populations.[8] Under similar culture conditions, conjunctival and corneal epithelial cells displayed different cytokeratin profiles, with K4 and K13 being expressed by the conjunctival epithelial cells, and K3 and K12 by the corneal epithelial cells.[8,21] When corneal, limbal and conjunctival epithelial cell suspensions were injected subcutaneously into athymic mice, the cysts derived from limbal and corneal epithelial cells retained the features of the normal corneal epithelium with a stratified squamous epithelium without goblet cells, while cysts derived from conjunctival epithelial cells displayed a normal conjunctival morphology with the presence of numerous goblet cells.[8,22]

At present, the two conjunctival cell types, namely keratinocytes (conjunctival epithelium) and goblet cells, are believed to arise from a common bipotent progenitor.[22] It was found that mixed populations of keratinocytes and goblet cells in the above-mentioned cysts were derived ultimately from a cultured single conjunctival epithelial cell. Similarly, Pellegrini et al.[17] have found goblet cells in cultures of transient amplifying cells derived from a single conjunctival clone. They have also noted that the differentiation to goblet cells occurs at fairly specific time points in the life span of these transient amplifying cells, suggesting that the decision for the conjunctival keratinocytes to differentiate into goblet cells is dependent upon an intrinsic "cell doubling clock." As such, conjunctival epithelial stem cells are likely to be at least bipotent, and they can undergo intrinsically divergent pathways leading to keratinocytes and goblet cells, respectively.

In several rodent models, the conjunctival fornix has been shown to be enriched with conjunctival epithelial stem cells. As slow cycling is a well-established feature of stem cells, the forniceal conjunctival epithelium of mice displayed a significantly higher percentage of label-retaining (slow-cycling) cells, as compared with the bulbar and palpebral conjunctiva.[67] Following stimulation with a tumor promoter, the forniceal basal cells underwent a significantly greater and more sustained proliferative response.[68] Taken together, a stable population of slow-cycling progenitor cells with a greater proliferative reserve exists in the conjunctival fornix that can undergo cell divisions in response to various perturbations of the ocular surface.

Other *in vitro* evidence also supports the fornix being the location of conjunctival epithelial stem cells in rabbits. Cells from the forniceal conjunctiva were shown to have greater growth potential in culture with a higher colony forming efficiency and a greater number of serial subcultures, whereas cells from the bulbar and palpebral regions only generated small colonies and could not be subcultured more than once.[21]

Although the conjunctival fornix contains the greatest density of conjunctival stem cells, pockets of stem cells may also exist throughout the conjunctival epithelium. Other investigators have analyzed the proliferative capacities of various conjunctival regions *in vitro* and found that stem cells may be uniformly distributed over the bulbar and fornical conjunctiva.[17] The stem cells scattered in the bulbar and palpebral conjunctiva may give rise to the dispersed pockets of goblet cells seen within these regions. Conversely, the higher density of goblet cells noted in the fornix[67,69] may be due to the bipotent conjunctival stem cells being more concentrated in this region.

The molecular markers for conjunctival epithelial stem cells are even less established as corneal stem cells. The basal cells of the human conjunctival epithelium expressed exclusive markers such as K5 and K14 cytokeratins, α-enolase and TGF-β receptor II, as well as higher levels of TGF-β receptor I, EGFR, integrins $\alpha 3$, $\alpha 6$, αv and $\beta 4$, when compared with the expression of these markers by the conjunctival suprabasal epithelial cells. Li *et al.* have recently observed that neurotrophic factors, such as NGF and GDNF, and their corresponding receptors, TrkA and GFRα-1, were exclusively localized in the basal layer of the conjunctival epithelium, while these markers were undetectable in the suprabasal layers of the conjunctival epithelium.

Clustering of conjunctival epithelial stem cells in the fornix implicates this region as a niche for the conjunctival stem cells. The conjunctival fornix is located within the upper and lower conjunctival recesses created by the inflection of conjunctival tissues conjoining the undersurface of the eyelid and globe. This location is farther away from the external environment than other regions of the ocular surface and provides an ideal haven for protecting the conjunctival stem cells from untoward insults. The subjacent stroma of the fornix comprises a network of flexible collagen and elastic fibers, which guards the epithelial cells against shearing and mechanical forces. The fornix is also the most richly vascularized and innervated region of the conjunctiva, which enables the stem cells to draw necessary nutrients readily and to respond promptly to various stimuli through cytokine-mediated or neuronal mechanisms. All these features are important for the fornix to function as a niche for conjunctival stem cells.

8. Limbal Stem Cell Deficiency

Based upon the concept of ocular surface epithelial stem cells, a group of potentially blinding ocular surface diseases, collectively known as limbal stem cell deficiency, has evolved. These conditions, in which the limbal basal epithelium and/or limbal stroma are damaged, display common features of defective stem cells with conjunctivalization of the cornea and recurrent epithelial breakdown. The etiology of limbal stem cell deficiency or dysfunction is diverse and it can be caused by a variety of hereditary or acquired disorders.[19,20,70,71] Inherited disorders include aniridia,[72] familial dysautonomia, ectodermal dysplasia and keratitis associated with multiple endocrine deficiencies, in which limbal stem cells may be congenitally absent or dysfunctional. Acquired conditions that may result in limbal stem cell deficiency include Stevens–Johnson syndrome, chemical injuries, ocular cicatricial pemphigoid (or mucous membrane pemphigoid), contact lens-induced keratopathy, multiple surgeries or cryotherapies to the limbal region, neurotrophic keratopathy and peripheral ulcerative keratitis with limbal damage.[10,73]

Clinically, limbal stem cell deficiency is often associated with persistent or recurrent epithelial defects, ulceration, corneal vascularization, chronic inflammation, scarring and conjunctivalization (conjunctival epithelial ingrowth) of the cornea, with resultant loss of the clear demarcation between the corneal and the conjunctival epithelium at the limbal region.[10,73] Chronic instability of the abnormal epithelium and related ulceration may lead to progressive corneal melting of the cornea with the risk of perforation. Because some of the clinical features of limbal deficiency can also be found in other ocular surface disorders, the pathognomonic feature for limbal deficiency is the existence of conjunctival epithelial phenotypes on the corneal surface.[10,20] The conjunctival phenotype with goblet cells on the corneal surface is pathognomonic and can be confirmed by impression cytology.[73] Other clinical signs of limbal deficiency include corneal haziness, leukocytic stromal infiltration and increased corneal permeability to topical fluorescein. Greater permeability of the ingrown conjunctival epithelium permits influx of inflammatory cells or fluorescein dye into corneal tissue more readily than the normal corneal epithelium. Limbal deficiency may be localized (with partial or sectorial limbal damage) or complete (with 360° limbal damage). In partial limbal stem cell deficiency, some sectors of the limbal and corneal epithelium remain normal, and conjunctival ingrowth with abnormal whorl or vortex epitheliopathy is present only at the regions devoid of healthy limbal barriers.

9. Surgical Therapy for Corneal Epithelial Stem Cell Deficiency

Accurate diagnosis of limbal deficiency is crucial for ocular surface rehabilitation, as the patients are poor candidates for conventional penetrating keratoplasty (corneal transplantation) due to their poor epithelial reserves at the limbus and the increased risks of immunological rejection. Over the past decade, the therapeutic potential of restoring limbal stem cells for treatment of ocular surface disorders associated with limbal deficiency has been well recognized. Restoration of the limbal barriers and stem cells by various modalities of limbal stem cell transplantation is often required prior to penetrating keratoplasty.[10,20]

Table 3. Surgical Modalities for Ocular Surface Stem Cell Deficiency

I. Epithelial Stem Cell Transplantation

Tissue Type	Conjunctival Transplantation	Limbal Transplantation		
Donor origin		Conjunctival limbal transplantation	Keratolimbal transplantation (+/– AMT)	*Ex vivo* stem cell expansion (+/– AMT)
Autograft	CAU	CLAU	N/A	EVELAU
Cadeveric allograft	c-CAL	c-CLAL	KLAL	c-EVELAL
Living-related allograft	lr-CAL	lr-CLAL	N/A	lr-EVELAL

II. Substrate Transplantation   **AMT**

CAU, conjunctival autograft; c-CAL, cadaveric conjunctival allograft; lr-CAL, living-related conjunctival allograft; CLAU, conjunctival limbal autograft; c-CLAL, cadaveric conjunctival limbal allograft; lr-CLAL, living-related conjunctival limbal allograft; KLAL, keratolimbal allograft; EVELAU, *ex vivo* expanded limbal autograft; c-EVELAL, cadaveric *ex vivo* expanded limbal allograft; lr-EVELAL, living-related *ex vivo* expanded limbal allograft; AMT, amniotic membrane transplantation.

The surgical procedures to restore limbal stem cell deficiency are usually classified by the type of the tissue to be transplanted, including conjunctival transplantation, limbal transplantation and amniotic membrane transplantation (for stromal replacement). Based on the source of the donor, conjunctival or limbal transplantation can be sub-classified as auto- or allografts. Table 3 shows a classification scheme for these procedures.

9.1. Conjunctival transplantation

According to the source of the donor tissue, conjunctival transplantation can be either a conjunctival autograft (CAU) or a conjunctival allograft (CAL). The procedure was first described in 1977 by Thoft,[74] who was considered the pioneer of modern ocular surface reconstruction. A conjunctival autograft utilizes a patch of conjunctival tissue obtained from a healthy area of the same eye or the fellow eye of the patient to replace the area with damaged ocular surface epithelium in the peripheral cornea. An epithelial front spreads onto the corneal surface from the edges of each graft during the re-epithelialization process. It can successfully restore a minimally vascularized or scarred ocular surface.

The concept of using the conjunctival epithelium to restore the corneal surface was based on Thoft's hypothesis of "conjunctival transdifferentiation," which surmised that the conjunctival epithelium could transform into a cornea-like epithelium.[75-77] Even though this procedure has been shown to restore the ocular surface adequately, it remains controversial whether the transplanted conjunctival epithelium can transform into a normal corneal epithelium. Although the conjunctival epithelium can morphologically transform into tissue that is indistinguishable from the corneal epithelium in certain cases, its full transdifferentiation to a corneal phenotype has not been substantiated.[78-80] Progressive conjunctivalization following conjunctival transplantation for corneas with limbal deficiency has been reported.[81] Despite its limitations, conjunctival transplantation remains a viable modality for management of pterygium and for treatment of partial limbal stem cell deficiency.[82,83]

When autologous conjunctival or donor limbal tissues are not readily available, conjunctival tissue obtained from a cadaver (c-CAL) or from a living blood relative (lr-CAL) can be used. Conjunctival allografts can be utilized for ocular surface conditions with bilateral involvement, such as Stevens–Johnson syndrome, ocular cicatricial pemphigoid, bilateral chemical burns, and in patients lacking healthy conjunctival tissue in fellow eyes due to prior trauma or glaucoma surgery. Kwitko *et al.* utilized first degree living relatives as donors for conjunctival allografts with satisfactory stabilization of the corneal surface.[84] They observed rejection episodes in patients who had a complete mismatch of HLA alleles between the donor and the recipient. Therefore, preoperative selection of a closely HLA-matched donor is crucial prior to conjunctival allografts, especially when a future corneal transplantation is being contemplated for visual rehabilitation.

9.2. Limbal transplantation

Limbal transplantation procedures can be classified into auto- or allografts based on the source of the donor tissue. Because corneal epithelial stem cells are sequestered in the basal limbus, they are better delivered in more malleable surrounding carrier tissue to avoid damage. Depending on the surrounding tissue to be used as a carrier, limbal transplantation can be classified as conjunctival limbal transplantation or keratolimbal transplantation. Conjunctival limbal donor tissue can be obtained from the fellow eye for a conjunctival limbal autograft (CLAU), thereby avoiding immunological rejection. In contrast, a conjunctival limbal allografts can be obtained from either a cadaver (cadaveric conjunctival limbal allograft, c-CLAL) or a living relative (living-related conjunctival limbal allograft, Ir-CLAL). A keratolimbal allograft (KLAL) is a procedure in which allogeneic limbal stem cells are transplanted with the peripheral cornea as a carrier. Regardless of the type of carrier tissue, all limbal allografts have a significant risk of graft rejection.

9.2.1. Conjunctival Limbal Autograft (CLAU)

A CLAU can be harvested from the healthy fellow eye using the perilimbal conjunctivae as a carrier for the limbal stem cells. The donor tissue provides a new source of corneal epithelial stem cells. CLAUs are indicated for patients with unilateral limbal deficiency of various etiologies. A history of prior surgeries involving the limbus or long term contact lens wear[85] precludes the use of fellow eyes as limbal donors. A potential risk of limbal autograft transplantation is the development of iatrogenic limbal stem cell deficiency in the donor eye. Animal studies have suggested that full-thickness excision of the sectorial limbus, similar to the surgery performed in a donor eye for limbal autografting, can eventually compromise corneal epithelial healing in the presence of a subsequently induced epithelial defect.[86] Nonetheless, development of clinical limbal deficiency has yet to be reported in fellow or living-relative donor eyes. Because of these potential concerns, excision of less than four clock hours of the limbal circumference and accompanying conjunctiva as donor tissues has been recommended. This donor tissue can be obtained by making a circumferential thin dissection of the conjunctiva starting 3 mm posterior to the limbus to its limbal insertion. A lamellar dissection of the limbus approximately 150 μm thick and 1 mm into the peripheral cornea is then performed to obtain the donor limbal epithelium. Since the donor tissue is autologous in CLAUs, there is no concern about immunological rejection of the donor tissue.

9.2.2. Cadaveric Conjunctival Limbal Allograft (c-CLAL)

A c-CLAL uses alternative donor tissues for ocular surface reconstruction. This procedure is indicated for patients with bilateral limbal damage and no autologous tissue or available living-related donors. It was described as homotransplantation of limbal stem cells in 1994 by Pfister, who performed the transplantation with conjunctiva and limbus from a cadaveric corneoscleral rim in which the central donor cornea had been used for another patient.[87]

When using the donor corneoscleral rim for a c-CLAL, the rim needs to be thinned, thereby yielding a thin doughnut-shaped conjunctival limbal graft and avoiding adding too much tissue thickness to the peripheral cornea of the recipient. Alternatively, lamellar dissection of the entire donor corneoscleral button using a lamellar dissecting blade or viscoelastic agents can be performed, followed by trephination of the central donor cornea to obtain a thin donor ring.

While cadaveric allografts can circumvent the potential risk of developing limbal deficiency in the fellow donor eye after aurografts or in the donor eyes of living relatives, they predispose the recipient eye to a greater risk of allograft rejection. The availability of viable conjunctiva is also limited on cadaveric donor tissue. Postoperatively, conjunctivolimbal injection and significant chemosis are often noted in conjunctival limbal allografts. Conjunctival vascular engorgement and hemorrhage are also observed in recipients and may represent early signs of limbal allograft rejection or reactive hyperemia from vascular anastomosis of large conjunctival vessels (Fig. 3).[88] Topical steroids and systemic immunosuppression are usually required to reduce the risks of limbal allograft rejection and related

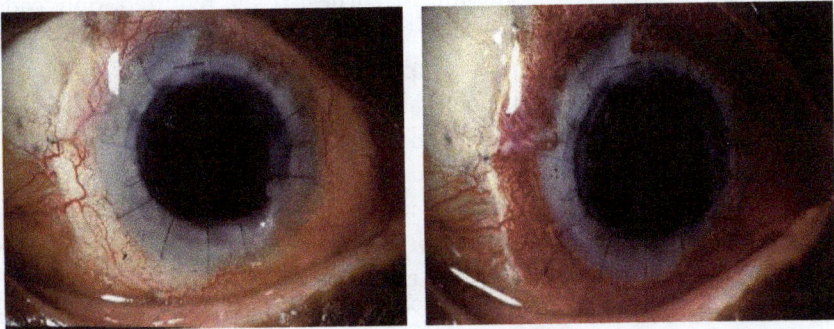

Figure 3. Keratolimbal allograft rejection. *Left*: A patient received a keratolimbal allograft, followed by a penetrating keratoplasty for chemical burns. *Right*: After the patient self-discontinued the oral cyclosporine, marked vascular engorgement of perilimbal vessels was noted. The vascular congestion improved after the systemic immunotherapy was resumed.

complications. With proper immunosuppression after conjunctival limbal allografts, a stable ocular surface and subsequent successful penetrating keratoplasty can be achieved.[87]

9.2.3. *Living-related Conjunctival Limbal Allograft (lr-CLAL)*

The lr-CLAL has been used to further reduce the risk of rejection associated with the use of unrelated cadaveric donor tissue. The indications for lr-CLALs are similar to those for c-CLALs or keratolimbal allografts (KLALs) when the fellow, or less involved, eye is deemed unsuitable for donating limbal stem cells. The living-related tissue has the advantage of providing better histocompatibility. In lr-CLALs, rapid transfer of the freshly harvested graft from the donor to the recipient may prevent the unexpected stem cell death associated with preservation and storage of cadaveric donor tissue.[88] However, due to the risk of developing limbal deficiency in the donor eye, it has been recommended that the amount of tissue to be obtained from the living donor should be limited to two clock hours of limbal tissue at the superior and inferior limbus of each eye.[88] Regardless of the degree of immune histocompatibility, moderate systemic immunosuppression is still recommended to prevent limbal allograft rejection.

9.2.4. *Keratolimbal Allograft (KLAL)*

Because of the larger amount of corneal tissue required, only cadaveric donor tissue is suitable for KLALs. KLALs are indicated for most ocular surface disorders requiring restoration of the limbus. Several different surgical techniques for keratolimbal allotransplantation have been proposed. They include the corneoscleral ring,[89] the corneoscleral crescent[90] and homologous penetrating central limbokeratoplasty.[91,92]

Initial experience on KLALs indicated a moderate (50–75%) but satisfactory success in restoring ocular surface integrity.[88,92] Later reviews revealed a long term survival rate of less than 30% in KLALs performed for severe ocular surface disorders.[93,94] Many of the patients who received limbal allografts also required penetrating keratoplasty (PKP) for visual rehabilitation. Patients who underwent a KLAL alone seemed to fare better than those who underwent a KLAL combined with PKP. In patients who received a KLAL and PKP simultaneously, corneal graft rejection was a major complication. Within this subset of patients, rejection episodes were more frequently encountered in patients with chemical or thermal injury (69%) than in patients with inflammatory conjunctivitis such as Stevens–Johnson syndrome or ocular pemphigoid (27%).[89]

Possible explanations for this observation include increased antigenic loads from the central corneal graft and peripheral keratolimbal rim. The combined procedure may have also resulted in a greater wound healing response and increased postoperative inflammation. Primary failure of the limbal stem cell transplant may lead to further conjunctivalization and vascularization of the cornea with subsequent increase of immune sensitization and eventual rejection of the corneal graft. Consequently, some investigators have recommended PKP to be performed no earlier than 3–6 months after a KLAL.[90,93] The major disadvantage of a KLAL is the high risk of graft rejection. Intensive immunosuppression by systemic corticosteroids and cyclosporine has been reported to improve the survival of limbal allografts.[95]

9.3. *Ex vivo* limbal stem cell expansion

To generate sufficient limbal stem cells and minimize the risk of developing subsequent limbal deficiency in the donor eye, *ex vivo* expansion of corneal epithelial stem cells by culturing limbal explants on an amniotic membrane or other carrier (i.e. collagen shield or hydrophilic soft contact lens) has been investigated. A much smaller amount of limbal tissue is needed for tissue culture from the donor eye than is required for conjunctival limbal auto- or allografts, thereby minimizing potential complications in the donor eye.

Transplanting cultured corneal epithelial cells was described in the mid-1980s.[96–98] However, cells derived from the central corneal epithelium consisted primarily of terminally differentiated cells that could not be subcultivated more than twice after senescence.[99] Transplanting cultured limbal stem cells was subsequently achieved by McCulley *et al.*[100,101] Following transplantation, the cultured cells formed hemidesmosomal adhesion to the underlying stroma within 24 hours.[101] Transplantation of cultured autologous limbal stem cells with or without carrier amniotic membrane has been reported to manage a group of ocular surface disorders with limbal deficiencies successfully.[99,102,103] Minimal stem cell loss in the donor eye can be achieved by harvesting only 1–3.1 mm^2 of limbal tissue from the donor eye. However, variable success rates have been observed in transplanting limbal allografts cultivated on amniotic membrane.[104,105] A variety of biological and immunological factors may have contributed to the variable success rates in these reported series.[104,105]

As discussed, limbal stem cells for *ex vivo* cultivation may be harvested from autologous or allogeneic donor sites.[103-105] The antigenicity of limbal tissue from allogeneic sources remains high because HLA Class I antigens are expressed on the limbal epithelium and HLA Class II antigen-positive Langerhans cells may be present. Elimination of certain antigenic elements, such as Langerhans cells, may occur during the culture process and provide a theoretical advantage of *ex vivo* stem cell expansion. Postoperative immunosuppression is recommended for recipients of allogeneic epithelium expanded *ex vivo*. The long term success of this procedure remains to be determined.[103-105]

9.4. Amniotic membrane transplantation

Amniotic membrane is the innermost layer of the placenta and consists of three layers: epithelium, basement membrane and avascular stroma. Since the first report of using amniotic membrane for conjunctival reconstruction by de Rotth,[106] studies have demonstrated the successful use of amniotic membrane for ocular surface reconstruction.[107]

In addition to being used as a tissue carrier for the cultivation of limbal stem cells, the amniotic basement membrane may serve as a matrix replacement for diseased conjunctival and corneal epithelial basement membrane.[108] It may support the epithelial proliferation and facilitate the maintenance of appropriate epithelial phenotypes.[109] In patients with partial limbal deficiency with superficial corneal surface disease, amniotic membrane transplantation alone is often sufficient to restore the corneal surface by reducing inflammation or corneal vascularization and providing a conducive substrate for the remaining stem cells to proliferate.[110] For complete limbal deficiency requiring allogeneic limbal transplant, adjunctive amniotic membrane transplant to replace the damaged corneal stroma may suppress perilimbal inflammation and enhance the survival of limbal allografts.[110]

Besides being used as an epithelial substrate, amniotic membrane may provide a favorable niche for limbal stem cells. Soluble factors derived from amniotic membrane likely play active roles in its supportive functions. Amniotic membrane can suppress the production of TGF-β1, a cytokine responsible for fibroblast activation in wound healing, and may limit scar formation on the ocular surface.[111] In addition, amniotic membrane can promote the nongoblet cell differentiation of cultured rabbit conjunctival epithelial cells.[112] Other studies have suggested that removal of the amniotic epithelium to expose its basement membrane can increase Connexin 43 and K3 cytokeratin expression in actively dividing limbal cells cultured on amniotic membrane.[113] It remains to be determined if stem cells can be activated into transient amplifying cells (TACs) following environmental cues such as exposure to the basement membrane of amniotic grafts.

9.5. Management of limbal stem cell transplants

Prior to surgical rehabilitation of eyes with limbal deficiency, the ocular surface should be adequately managed. These eyes are often associated with tear film instability due to damage of the neural reflex or lacrimal apparatus. Aqueous tear deficiency can be improved by topical lubricants or punctal occlusion. However, ocular surface inflammation should also be controlled with topical and/or oral anti-inflammatory agents if needed. Mechanical irritations of the ocular surface from lid pathologies such as trichiasis or entropion should be properly managed to ensure the survival of newly transplanted limbal tissues.

Topical and systemic immunosuppressive agents are usually necessary for limbal allografts, regardless of the donor origin. Commonly used agents include topical and systemic corticosteroids, mycophenylate mofetil and cyclosporin or FK506. The systemic cyclosporin and mycophenylate are often continued for an extended duration to prevent the rejection of limbal allografts or corneal transplants. Patients should be monitored for toxicity associated with these medications. Because of their anti-inflammatory and other supportive properties, autologous serum or plasma drops can be used to treat severe dry eye and to support the nourishment of transplanted limbal tissues.

Summary

(1) There is convincing evidence for the existence of corneal stem cells and their niche in the basal limbus.
(2) Currently, there is no universal stem cell marker for ocular surface epithelial stem cells.

(3) Newer strategies may facilitate the isolation and characterization of limbal stem cells.

(4) Surgical modalities based on the concept of limbal stem cells have been successful in restoring limbal deficiency and related ocular surface morbidities.

(5) Limbal allografting is the most commonly used procedure for ocular surface reconstruction; however, its long term success is limited by immunological rejection.

(6) Immunosuppression is required for limbal allografts.

(7) *Ex vivo* expansion of corneal epithelial stem cells shows promise for generating sufficient stem cells for treating limbal deficiency.

References

1. Watt FM, Hogan BL. (2000) Out of Eden: stem cells and their niches. *Science* 287: 1427–1430.

2. Slack JM. (2000) Stem cells in epithelial tissues. *Science* 287: 1431–1433.

3. Blau HM, Brazelton TR, Weimann JM. (2001) The evolving concept of a stem cell: entity or function? *Cell* 105: 829–841.

4. Lavker RM, Sun TT. (2000) Epidermal stem cell: properties, markers, and location. *Proc Natl Acad Sci USA* 97: 13473–13475.

5. Cotsarelis G, Kaur P, Dhouailly D, *et al.* (1999) Epithelial stem cells in the skin: definition, markers, localization and functions. *Exp Dermatol* 8: 80–88.

6. Diaz-Flores L, Jr, Madrid F, Gutierrez R, *et al.* (2006) Adult stem and transit-amplifying cell location. *Histol Histopathol* 21: 995–1027.

7. Moore KA, Lemischka IR. (2006) Stem cells and their niches. *Science* 311: 1880–1885.

8. Wei ZG, Sun TT, Lavker RM. (1996) Rabbit conjunctival and corneal epithelial cells belong to two separate lineages. *Invest Opthalmol Vis Sci* 37: 523–533.

9. Dua HS, Azuara-Blance A. (2000) Limbal stem cells of the corneal epithelium. *Surv Ophthalmol* 44: 415–425.

10. Tseng SC. (1989) Concept and application of limbal stem cells. *Eye* 3: 141–157.

11. Stepp MA, Zieske JD. (2005) The corneal epithelial stem cell niche. *Ocul Surf* 3: 15–26.

12. Davanger M, Evensen A. (1971) Role of the pericorneal papillary structure in renewal of corneal epithelium. *Nature* 229: 560–561.

13. Schermer A, Galvin S, Sun TT. (1986) Differentiation-related expression of a major 64K corneal keratin *in vivo* and in culture suggests limbal location of corneal epithelial stem cells. *J Cell Biol* 103: 49–62.

14. Cotsarelis G, Cheng SZ, Dong G, *et al.* (1989) Existence of slow-cycling limbal epithelial basal cells that can be preferentially stimulated to proliferate: implications on epithelial stem cells. *Cell* **57**: 201–209.

15. Kurpakus MA, Maniaci MT, Esco M. (1994) Expression of keratins K12, K4 and K14 during development of ocular surface epithelium. *Curr Eye Res* **13**: 805–814.

16. Ebato B, Friend J, Thoft RA. (1998) Comparison of limbal and peripheral human corneal epithelium in tissue culture. *Invest Ophthalmol Vis Sci* **29**: 1533–1537.

17. Pellegrini G, Golisano O, Paterna P, *et al.* (1999) Location and clonal analysis of stem cells and their differentiated progeny in the human ocular surface. *J Cell Biol* **145**: 769–782.

18. Chen JJ, Tseng SC. (1991) Abnormal corneal epithelial wound healing in partial-thickness removal of limbal epithelium. *Invest Ophthalmol Vis Sci* **32**: 2219–2233.

19. Huang AJ, Tseng SC. (1991) Corneal epithelial wound healing in the absence of limbal epithelium. *Invest Ophthalmol Vis Sci* **32**: 96–105.

20. Dua HS, Saini JS, Azuara-Blanco A, Gupta P. (2000) Limbal stem cell deficiency: concept, aetiology, clinical presentation, diagnosis and management. *Indian J Ophthalmol* **48**: 83–92.

21. Wei ZG, Wu RL, Lavker RM, Sun TT. (1993) *In vitro* growth and differentiation of rabbit bulbar, fornix, and palpebral conjunctival epithlia; implication on conjunctival epithelial transdifferentiation and stem cells. *Invest Ophthalmol Vis Sci* **34**: 1814–1828.

22. Wei ZG, Lin T, Sun TT, Lavker RM. (1997) Clonal analysis of the *in vivo* differentiation potential of keratinocytes. *Invest Ophthalmol Vis Sci* **38**: 753–761.

23. Wirtschafter JD, Ketcham JM, Weinstock RJ, *et al.* (1999) Mucocutaneous junction as the major source of replacement palpebral conjunctival epithelial cells. *Invest Ophthalmol Vis Sci* **40**: 3138–3146.

24. Budak MT, Alpdogan OS, Zhou M, *et al.* (2005) Ocular surface epithelia contain ABCG2-dependent side population cells exhibiting features associated with stem cells. *J Cell Sci* **118**: 1715–1724.

25. Pajoohesh-Ganji A, Stepp MA. (2005) In search of markers for the stem cells of the corneal epithelium. *Biol Cell* **97**: 265–276.

26. Collinson JM, Morris L, Reid AI, *et al.* (2002) Clonal analysis of patterns of growth, stem cell activity, and cell movement during the development and maintenance of the murine corneal epithelium. *Dev Dyn* **224**: 432–440.

27. Schlotzer-Schrehardt U, Kruse FE. (2005) Identification and characterization of limbal stem cells. *Exp Eye Res* **81**: 247–264.

28. Yang A, Schweitzer R, Sun D, *et al.* (1999) p63 is essential for regenerative proliferation in limb, craniofacial and epithelial development. *Nature* **398**: 714–718.

29. Parsa R, Yang A, McKeon F, Green H. (1999) Association of p63 with proliferative potential in normal and neoplastic human keratinocytes. *J Invest Dermatol* **113**: 1099–1105.

30. Pellegrini G, Dellambra E, Golisano O, *et al.* (2001) p63 identifies keratinocyte stem cells. *Proc Natl Acad Sci USA* **98**: 3156–3161.

31. Zhou S, Schuetz JD, Bunting KD, *et al.* (2001) The ABC transporter Bcrp1/ABCG2 is expressed in a wide variety of stem cells and is a molecular determinant of the side-population phenotype. *Nat Med* **7**: 1028–1034.

32. de Paiva CS, Chen Z, Corrales RM, *et al.* (2005) ABCG2 transporter identifies a population of clonogenic human limbal epithelial cells. *Stem Cells* **23**: 63–73.

33. Stepp MA, Zhu L, Sheppard D, Cranfill RL. (1995) Localized distribution of alpha 9 integrin in the cornea and changes in expression during corneal epithelial cell differentiation. *J Histochem Cytochem* **43**: 353–362.

34. Jones PH, Watt FM. (1993) Separation of human epidermal stem cells from transit amplifying cells on the basis of differences in integrin function and expression. *Cell* **73**: 713–724.

35. Watt FM. (1998) Epidermal stem cells: markers, patterning and the control of stem cell fate. *Philos Trans R Soc Lond B Biol Sci* **353**: 831–837.

36. Pajoohesh-Ganji A, Pal-Ghosh S, Simmens SJ, Stepp MA. (2006) Integrins in slow-cycling corneal epithelial cells at the limbus in the mouse. *Stem Cells* **24**: 1075–1086.

37. Zieske JD, Wasson M. (1993) Regional variation in distribution of EGF receptor in developing and adult corneal epithelium. *J Cell Sci* **106**: 145–152.

38. Liu Z, Carvajal M, Carraway CA, *et al.* (2001) Expression of the receptor tyrosine kinases, epidermal growth factor receptor, ErbB2, and ErbB3, in human ocular surface epithelia. *Cornea* **20**: 81–85.

39. Michel M, Torok N, Godbout MJ, *et al.* (1996) Keratin 19 as a biochemical marker of skin stem cells *in vivo* and *in vitro*: keratin 19 expressing cells are differentially localized in function of anatomic sites, and their number varies with donor age and culture stage. *J Cell Sci* **109**: 1017–1028.

40. Zieske JD, Bukusoglu G, Yankauckas MA. (1992) Characterization of a potential marker of corneal epithelial stem cells. *Invest Ophthalmol Vis Sci* **33**: 143–152.

41. Matic M, Petrov IN, Chen S, *et al.* (1997) Stem cells of the corneal epithelium lack connexins and metabolite transfer capacity. *Differentiation* **61**: 251–260.

42. Scott RAH, Lauweryns B, Snead DMJ, *et al.* (1997) E-cadherin distribution and epithelial basement membrane characteristics of the normal human conjunctiva and cornea. *Eye* **11**: 607–612.

43. Hayashi R, Yamato M, Sugiyama H, *et al.* (2007) N-cadherin is expressed by putative stem/progenitor cells and melanocytes in the human limbal epithelial stem cell niche. *Stem Cells* **25**: 289–296.

44. Liu CY, Zhu G, Converse R, *et al.* (1994) Characterization and chromosomal localization of the cornea-specific murine keratin gene *Krt1.12. J Biol Chem* **260**: 24627–24636.

45. Tong L, Corrales RM, Chen Z, *et al.* (2006) Expression and regulation of cornified envelope proteins in human corneal epithelium. *Invest Ophthalmol Vis Sci* **47**: 1938–1946.

46. Banks-Schlegel S, Green H. (1981) Involucrin synthesis and tissue assembly by keratinocytes in natural and cultured human epithelia. *J Cell Biol* **90**: 732–737.

47. Chen Z, de Paiva CS, Luo L, *et al.* (2004) Characterization of putative stem cell phenotype in human limbal epithelia. *Stem Cells* **22**: 355–366.

48. de Paiva CS, Pflugfelder SC, Li DQ. (2006) Cell size correlates with phenotype and proliferative capacity in human corneal epithelial cells. *Stem Cells* **24**: 368–375.

49. Li DQ, Chen Z, Song XJ, *et al.* (2005) Partial enrichment of a population of human limbal epithelial cells with putative stem cell properties based on collagen type IV adhesiveness. *Exp Eye Res* **80**: 581–590.

50. Chen Z, Evans WH, Pflugfelder SC, Li DQ. (2006) Gap junction protein connexin 43 serves as a negative marker for a stem cell-containing population of human limbal epithelial cells. *Stem Cells* **24**: 1265–1273.

51. Li W, Hayashida Y, Chen YT, Tseng SC. (2007) Niche regulation of corneal epithelial stem cells at the limbus. *Cell Res* **17**: 26–36.

52. Lawrenson JG, Ruskell GL. (1991) The structure of corpuscular nerve endings in the limbal conjuctiva of the human eye. *J Anat* **177**: 75–84.

53. Goldberg MF, Bron AJ. (1982) Limbal palisades of Vogt. *Trans Am Ophthalmol Soc* **80**: 155–171.

54. Zieske JD. (1994) Perpetuation of stem cells in the eye. *Eye* **8**: 163–169.

55. Li DQ, Tseng SC. (1995) Three patterns of cytokine expression potentially involved in epithelial–fibroblast interactions of human ocular surface. *J Cell Physiol* **163**: 61–79.

56. Wilson SE, He YG, Weng J, *et al.* (1994) Effect of epidermal growth factor, hepatocyte growth factor, and keratinocyte growth factor, on proliferation, motility and differentiation of human corneal epithelial cells. *Exp Eye Res* **59**: 665–678.

57. Wilson SE, Schultz GS, Chegini N, *et al.* (1994) Epidermal growth factor, transforming growth factor alpha, transforming growth factor beta, acidic fibroblast growth factor, basic fibroblast growth factor, and interleukin-1 proteins in the cornea. *Exp Eye Res* **59**: 63–71.

58. Li DQ, Lee SB, Tseng SC. (1999) Differential expression and regulation of TGF-β1, TGF-β2, TGF-β3, TGF-βRI, TGF-βRII and TGF-βRIII in cultured human corneal, limbal, and conjunctival fibroblasts. *Curr Eye Res* **19**: 154–161.

59. Li DQ, Tseng SCG. (1996) Differential regulation of cytokine and receptor transcript expression in human corneal and limbal fibroblasts by EGF, TGF-α, PDGF-BB and IL-1β. *Invest Ophthalmol Vis Sci* **37**: 2068–2080.

60. Li DQ, Tseng SCG. (1997) Differential regulation of keratinocyte growth factor and hepatocyte growth factor/scatter factor by different cytokines in cultured human corneal and limbal fibroblasts. *J Cell Physiol* **172**: 361–372.

61. Tseng SC, Li DQ, Ma X. (1999) Suppression of transforming growth factor-beta isoforms, TGF-beta receptor type II, and myofibroblast differentiation in cultured human corneal and limbal fibroblasts by amniotic membrane matrix. *J Cell Physiol* **179**: 325–335.

62. Tseng SC, Kruse FE, Merritt J, Li DQ. (1996) Comparison between serum-free and fibroblast-cocultured single-cell clonal culture systems: evidence showing that epithelial anti-apoptotic activity is present in 3T3 fibroblast-conditioned media. *Curr Eye Res* **15**: 973–984.

63. Ljubimov AV, Burgeson RE, Butkowski RJ, *et al.* (1995) Human corneal basement membrane heterogeneity: topographical differences in the expression of type IV collagen and laminin isoforms. *Lab Invest* **72**: 461–473.

64. Tuori A, Uusitalo H, Burgeson RE, *et al.* (1996) The immunohistochemical composition of the human corneal basement membrane. *Cornea* **15**: 286–294.

65. Pal-Ghosh S, Pajoohesh-Ganji A, Brown M, Stepp MA. (2004) A mouse model for the study of recurrent corneal epithelial erosions: alpha9beta1 integrin implicated in progression of the disease. *Invest Ophthalmol Vis Sci* **45**: 1775–1788.

66. Ang LP, Tan DTH, Beuerman RW, Lavker RM. (2004) Ocular surface epithelial stem cells: implications for ocular surface homeostasis. In: Pflugfelder SC, Beuerman RW, Stern ME (eds.), *Dry Eye and Ocular Surface Disorders*, Marcel Dekker, New York, pp. 225–246.

67. Wei ZG, Cotsarelis G, Sun TT, Lavker RM. (1995) Label-retaining cells are preferentially located in fornical epithelium: implication on conjunctival epithelial homeostasis. *Invest Ophthalmol Vis Sci* **36**: 236–246.

68. Lavker RM, Wei ZG, Sun TT. (1998) Phorbol ester preferentially stimulates mouse fornical conjunctival and limbal epithelial cells to proliferate *in vivo*. *Invest Ophthalmol Vis Sci* **39**: 301–307.

69. Huang AJ, Tseng SC, Kenyon KR. (1988) Morphogenesis of rat conjunctival goblet cells. *Invest Ophthalmol Vis Sci* **29**: 969–975.

70. Miyashita H, Shimmura S, Kobayashi H, *et al.* (2006) Collagen-immobilized poly (vinyl alcohol) as an artificial cornea scaffold that supports a stratified corneal epithelium. *J Biomed Mater Res B Appl Biomater* **76**: 56–63.

71. Lavker RM, Tseng SC, Sun TT. (2004) Corneal epithelial stem cells at the limbus: looking at some old problems from a new angle. *Exp Eye Res* **78**: 433–446.

72. Nishida K, Kinoshita S, Ohashi Y, *et al.* (1995) Ocular surface abnormalities in aniridia. *Am J Ophthalmol* **120**: 368–375.
73. Puangsricharern V, Tseng SCG. (1995) Cytologic evidence of corneal diseases with limbal stem cell deficiency. *Ophthalmology* **102**: 1476–1485.
74. Thoft RA. (1977) Conjunctival transplantation. *Arch Ophthalmol* **95**: 1425–1427.
75. Huang AJ, Watson BD, Hernandez E, Tseng SC. (1988) Photothrombosis of corneal neovascularization by intravenous rose bengal and argon laser irradiation. *Arch Ophthalmol* **106**: 680–685.
76. Kinoshita S, Friend J, Thoft RA. (1983) Biphasic cell proliferation in transdifferentiation of conjunctival to corneal epithelium in rabbits. *Invest Ophthalmol Vis Sci* **24**: 1008–1014.
77. Shapiro MS, Friend J, Thoft RA. (1981) Corneal re-epithelialization from the conjunctiva. *Invest Ophthalmol Vis Sci* **21**: 135–142.
78. Harris TM, Berry ER, Pakurar AS, Sheppard LB. (1985) Biochemical transformation of bulbar conjunctiva into corneal epithelium: an electrophoretic analysis. *Exp Eye Res* **41**: 597–605.
79. Kinoshita S, Kiorpes TC, Friend J, Thoft RA. (1982) Limbal epithelium in ocular surface wound healing. *Invest Ophthalmol Vis Sci* **24**: 577–581.
80. Chen WY, Mui MM, Kao WW, *et al.* (1994) Conjunctival epithelial cells do not trans-differentiate in organotypic cultures: expression of K12 keratin is restricted to corneal epithelium. *Curr Eye Res* **13**: 765–778.
81. Tsai RJ, Sun TT, Tseng SC. (1990) Comparison of limbal and conjunctival autograft transplantation in corneal surface reconstruction in rabbits. *Ophthalmology* **97**: 446–455.
82. Gomez-Marquez J. (1931) New operative procedure for pterygium. *Arch de Oftal Hispano-Am* **31**: 87.
83. Kenyon, Wagoner MD, Hettinger ME. (1985) Conjunctival autograft transplantation for advanced and recurrent pterygium. *Ophthalmology* **92**: 1461–1470.
84. Kwitko S, Marinho D, Barcaro S, *et al.* (1995) Allograft conjunctival transplantation for bilateral ocular surface disorders. *Ophthalmology* **102**: 1020–1025.
85. Jenkins C, Tuft S, Liu C, Buckley R. (1993) Limbal transplantation in the management of chronic contact-lens-associated epitheliopathy. *Eye* **7**: 629–633.
86. Chen JJ, Tseng SC. (1990) Corneal epithelial wound healing in partial limbal deficiency. *Invest Ophthalmol Vis Sci* **31**: 1301–1314.
87. Pfister RR. (1994) Corneal stem cell disease: concepts, categorization, and treatment by auto- and homotransplantation of limbal stem cells. *CLAO J* **20**: 64–72.
88. Daya SM, Ilari L. (2001) Living-related conjunctival limbal allograft for the treatment of stem cell deficiency. *Ophthalmology* **108**: 126–133.

89. Tsubota K, Satake Y, Kaido M, *et al.* (1999) Treatment of severe ocular-surface disorders with corneal epithelial stem-cell transplantation. *N Engl J Med* **340:** 1697–1703.

90. Croasdale CR, Schwartz GS, Malling JV, Holland EJ. (1999) Keratolimbal allograft: recommendation for tissue procurement and preparation by eye banks, and standard surgical technique. *Cornea* **18:** 52–58.

91. Sundmacher R, Reinhard T. (1996) Central corneolimbal transplantation under systemic cyclosporine A cover for severe stem cell deficiencies. *Graefe's Arch Clin Exp Ophthalmol* **234:** 122–125.

92. Reinhard T, Sundmacher R, Spelsberg H, Althaus C. (1999) Homologous penetrating central limbo-keratoplasty (HPCLK) in bilateral limbal stem cell deficiency. *Acta Ophthalmol Scand* **77:** 663–667.

93. Solomon A, Ellies P, Anderson DF, *et al.* (2002) Long-term outcome of keratolimbal allograft with or without penetrating keratoplasty for total limbal stem cell deficiency. *Ophthalmology* **109:** 1159–1166.

94. Ilari L, Daya SM. (2002) Long-term outcomes of keratolimbal allograft for the treatment of severe ocular surface disorders. *Ophthalmology* **109:** 1278–1284.

95. Tsubota K, Toda I, Saito H, *et al.* (1995) Reconstruction of the corneal epithelium by limbal allograft transplantation for severe ocular surface disorders. *Ophthalmology* **102:** 1486–1495.

96. Friend J, Kinoshita S, Thoft RA, Eliason JA. (1982) Corneal epithelial cell cultures on stroma carriers. *Invest Ophthalmol Vis Sci* **23:** 41–49.

97. Gipson IK, Friend J, Spurr JJ. (1985) Transplant of corneal epithelium to rabbit corneal wound *in vivo*. *Invest Ophthalmol Vis Sci* **26:** 425–433.

98. Geggel HS, Friend J, Thoft RA. (1985) Collagen gel for ocular surface. *Invest Ophthalmol Vis Sci* **26:** 901–905.

99. Pellegrini G, Traverso CE, Franzi AT, *et al.* (1997) Long-term restoration of damaged corneal surface with autologous cultivated corneal epithelium. *Lancet* **349:** 990–993.

100. He YG, McCulley JP. (1991) Growing human corneal epithelium on collagen shield and subsequent transfer to denuded corneal *in vitro*. *Curr Eye Res* **10:** 851–863.

101. He YG, Alizadeh H, Kinoshita K, McCulley JP. (1999) Experimental transplantation of cultured human limbal and amniotic epithelial cells onto the corneal surface. *Cornea* **18:** 570–579.

102. Torfi H, Schwab IR, Isseroff RR. (1996) Transplantation of cultured autologous limbal stem cells for ocular surface diseases. *In Vitro* **32:** 47A.

103. Tsai RJ, Li LM, Chen JK. (2000) Reconstruction of damaged corneas by transplantation of autologous limbal epithelial cells. *N Eng J Med* **343:** 86–93.

104. Shimazaki J, Aiba M, Goto E, *et al.* (2002) Transplantation of human limbal epithelium cultivated on amniotic membrane for the treatment of severe ocular surface disorders. *Ophthalmology* **109:** 1285–1290.

105. Schwab IR, Reyes M, Isseroff RR. (2000) Successful transplantation of bio-engineered tissue replacements in patients with ocular surface disease. *Cornea* **19**: 421–426.
106. de Rotth A. (1940) Plastic repair of conjunctival defects with fetal membrane. *Arch Ophthalmol* **23**: 522–525.
107. Kim JC, Tseng SC. (1995) Transplantation of preserved human amniotic membrane for surface reconstruction in severely damaged rabbit corneas. *Cornea* **14**: 473–484.
108. Fukuda K, Chikama T, Nakamura M, Nishida T. (1999) Differential distribution of subchains of the basement membrane components type IV collagen and laminine among the amniotic membrane, cornea, and conjunctiva. *Cornea* **18**: 73–79.
109. Grueterich M, Espana EM, Touhami A, *et al.* (2002) Phenotypic study of a case with successfully transplantation of *ex vivo* expanded human limbal epithelium for unilateral total limbal stem cell deficiency. *Ophthalmology* **109**: 1547–1552.
110. Tseng SC, Prabhasawat P, Barton K, *et al.* (1998) Amniotic membrane transplantation with or without limbal allogrants for corneal surface reconstruction in patients with limbal stem cell deficiency. *Arch Ophthalmol* **116**: 431–441.
111. Ma X, Li D, Tseng SC. (1997) Cytokine expression by human limbal and corneal fibroblast is modulated by amniotic membrane. *Invest Ophthalmol Vis Sci* **38**: S512.
112. Meller D, Tseng SC. (1999) Conjunctival epithelial cell differentiation on amniotic membrane. *Invest Ophthalmol Vis Sci* **40**: 878–886.
113. Grueterich M, Espana K, Tseng SC. (2002) Connexin 43 expression and proliferation of human limbal epithelium on intact and denuded amniotic membrane. *Invest Ophthalmol Vis Sci* **43**: 63–71.

STEM CELLS FOR MUSCLE AND HEART

7

Stem Cells in Skeletal Muscle Regeneration

Atsushi Asakura*

Skeletal muscle is a highly regenerative tissue in the adult vertebrate body. Adult muscle contains muscle satellite cells (myogenic stem cells) which give rise to daughter myogenic precursor cells (myoblasts) in adult skeletal muscle, where they function in postnatal tissue growth and regeneration. Satellite cells are normally mitotically quiescent. Following injury, they initiate proliferation to produce their progeny of myoblasts to mediate the regeneration of muscle. Adult skeletal muscle also contains a novel stem cell population purified as a side population (SP), which actively excludes Hoechst 33342 dye. Muscle SP cells that express the stem cell marker Sca-1 possess the ability to differentiate into myogenic precursor cells, as well as satellite cells following transplantation. Recent work demonstrates that the heart also contains a novel stem cell-like cell population purified as SP/Sca-1$^+$ cells. Thus, these novel cardiac stem/progenitor cells that can participate in heart regeneration appear to have similar characteristics to muscle SP cells. In this article, I outline recent findings regarding stem cell populations in skeletal muscle and their involvement in muscle regeneration. In addition, I discuss cardiac stem/progenitor cells and their similarity to muscle SP cells.

1. Introduction

The study of skeletal muscle differentiation has been intensive since the 1960s because of readily discerned cell-biological criteria for differentiation

*Correspondence: Stem Cell Institute, University of Minnesota Medical School, McGuire Translational Research Facility, 2001 6th St SE, Mail Code 2873, Minneapolis, MN, 55455, USA. Tel: 612-624-7108; Fax: 612-624-2436; E-mail: asakura@umn.edu

as a multinucleated syncitium and the availability of many biochemical markers, such as sarcomeric proteins, including actin and myosin. In addition, the muscle cell lines successfully isolated from skeletal muscle were essential for early studies that elucidated mechanisms of cellular and molecular myogenesis. Consequently, the study of myogenesis has provided a powerful biological system for investigating the molecular regulation of the developmental program that controls the genesis, growth, migration, and differentiation of specific cell lineages as well as stem cells.[1–3]

Our knowledge of the molecular mechanisms for myogenesis was dramatically accelerated about 20 years ago, following the discovery of the MyoD family of transcription factors, termed the myogenic regulatory factors (MRFs). The MRFs comprise a group of muscle-specific bHLH (basic helix-loop-helix) transcription factors, consisting of MyoD, Myf5, myogenin, and MRF4.[4] They can initiate transcription of muscle-specific genes through direct binding to E-boxes, a consensus DNA motif, existing in the regulatory regions of the muscle-specific genes. Expression of the MRFs in various cell types *in vitro* and *in vivo* is able to initiate a muscle differentiation program. In addition, the restricted muscle-specific expression of the MRFs, together with their ability to dominantly induce a myogenic program, led to the suggestion that the MyoD family are the master regulators of the muscle developmental program. Furthermore, gene targeting in mice allowed a genetic dissection of these regulatory pathways and clearly defined the roles played by the MyoD family of transcription factors in myogenesis during embryogenesis as well as postnatal muscle development and regeneration. These MRFs, together with other transcription factors such as Pax3 and Pax7, a paired-type homeodomain protein, also regulate differentiation and self-renewal properties of muscle stem cells.[2,3]

During vertebrate development, the mesodermal progenitors of skeletal muscle originate from precursor cells that arise from the somites and prechordal mesoderm. The specification of muscle progenitors from the somites is regulated by signals from the neural tube, notochord, and other surrounding tissues that provide both positive and negative cues. Satellite cells are believed to represent myogenic stem cells of postnatal skeletal muscle. They reside close to skeletal muscle fibers beneath the basal lamina. They appear in muscle fibers during late embryogenesis and they originate exclusively from the somites for body and limb muscle. Satellite cells in adult muscle are mitotically quiescent, but initiate proliferation when activated by growth factors induced by weight bearing or other trauma such as injury to mediate the postnatal growth and regeneration of muscle. After multiple rounds of cell division, satellite cells undergo terminal differentiation to form multinucleated myotubes.[1–3] Adult skeletal muscle also contains a novel stem cell population purified as a side population (SP), which actively excludes Hoechst 33342 dye. These cells possess the ability to differentiate into myogenic precursor cells, as well as satellite cells following transplantation.[5]

The purpose of this article is to provide the reader with an introduction and an in-depth review of the molecular mechanisms that regulate muscle stem cell development and differentiation. The topics discussed include muscle stem cells, the regulation of adult muscle regeneration, the therapeutic potential of muscle stem cells for muscular dystrophy, and a comparison of cardiac stem/progenitor cells and muscle stem cells.

2. The Embryonic Origin of Skeletal Muscles

2.1. The terminology of muscle cells

Defined terms are used for specific muscle cell types during myogenesis.[3,6] During vertebrate development, Pax3- and/or Pax7-positive *muscle precursor (progenitor) cells* derived from the mesoderm become committed myogenic cells called myoblasts which following determination, initiate expression of the MRFs, such as MyoD and/or Myf5. Proliferating *myoblasts* exit their cell cycle and differentiate into postmitotic *myocytes* that express muscle-specific genes such as myogenin, the myosin heavy chain (MHC), and muscle creatine kinase (MCK). These mononucleated myocytes fuse with each other to form multinucleated *myotubes* that initiate construction of sarcomeric structures and eventually become mature *muscle fibers* that initiate muscle contraction. *Satellite cells*, which are stem cells for postnatal muscle, histologically reside close to skeletal muscle fibers beneath the basal lamina. They originate from the central region of dermomyotome, which is a cell source for the trunk and limb muscle. In adult muscle, satellite cells are mitotically quiescent but are activated (initiate proliferation) in response to stress to produce proliferating daughter *myogenic precursor cells* (*MPCs*) that mediate the postnatal growth and regeneration of muscle. This is a novel stem cell population purified as side population cells (*SP cells*), which actively exclude Hoechst 33342 dye. These cells also possess the ability to differentiate into myogenic precursor cells, as well as satellite cells in adult muscle.[5]

2.2. Embryonic muscle cells originate from mesodermal precursors (Fig. 1)

During vertebrate development, the skeletal muscle of the trunk and limbs originates from the somites, which are epithelial spheres of paraxial mesoderm.[2,3] Craniofacial muscles originate from cephalic paraxial mesoderm (somitomeres), prechordal mesoderm, and occipital somites.[7] Trunk and limb myogenesis is relatively well-understood compared to craniofacial muscle. During early embryogenesis, murine embryos form 65 pairs of somites. Specification of somitic lineages occurs following the exposure of somitic cells to patterning signals, including the sonic hedgehog (Shh), Wnt proteins, bone morphogenetic proteins (BMPs) and Noggin derived from the surrounding tissues, such as the neural tube, the notochord, and the lateral plate.[4,8] The ventral part of the somites gives rise to the sclerotome, which eventually forms the cartilage of the vertebral column and ribs. The dorsal part of the somites maintains an epithelial structure, called the dermomyotome, which expresses Pax3 and Pax7.[2,3] Cells in the dorsomedial region of the dermomyotome gradually extend laterally beneath the entire width of the dermomyotome. These cells longitudinally elongate beneath the dorsomedial part of the dermomyotome to form MRFs-positive committed myogenic cells called myoblasts. Proliferating myoblasts eventually begin to express myogenin and become the first terminally differentiated myocytes in a structure called the myotome. The compartment beneath the dorsomedial portion of the dermomyotome is called the epaxial myotome and gives rise to the epaxial muscles of the deep back. The developmental process of the ventrolateral region of the dermomyotome is the mirror image of the dorsomedial part. Ventrolateral region-derived cells begin to longitudinally elongate beneath the ventrolateral part of the dermomyotome to form the ventral myotome. The compartment residing beneath the ventrolateral part of the dermomyotome is called the nonmigratory hypaxial myotome and gives rise to lateral skeletal muscles in the trunk, such as intercostal muscle and bodywall muscle. Pax3-positive epithelial cells in the ventrolateral part of the

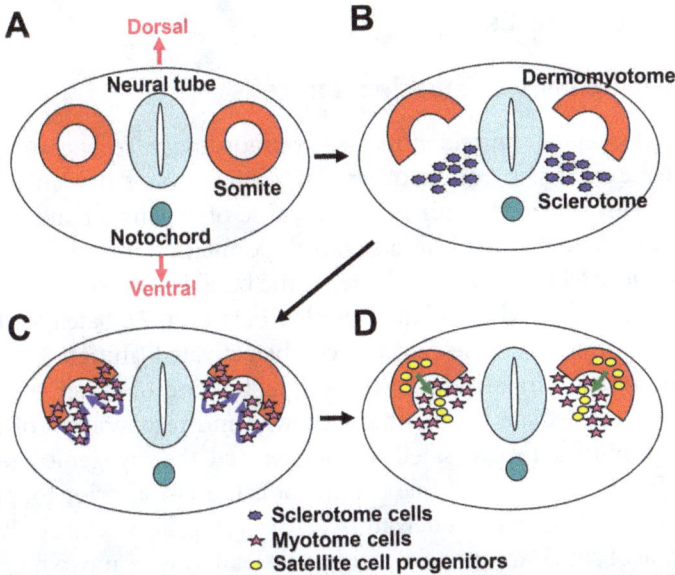

Figure 1. Schematic representation of the mouse embryo at thoracic level somites.[2] By 8.0–8.5 dpc in the mouse embryo, the ventral part of the newly formed somites undergoes de-epithelialization to form mesenchymal cells which give rise to the sclerotome **(A, B)**. By 8.5 dpc, the dorsal part of the somites still maintains a proliferating epithelial structure called the dermomyotome **(B)**. By 8.5–9.5 dpc, cells in the dorsomedial region of the dermomy-otome (dorsomedial lip) withdraw from the cell cycle and migrate beneath the dorsome-dial part of the dermomyotome to form the dorsal myotome (epaxial myotome), which later gives rise to deep back muscles **(C)**. Cells in the ventrolateral region of the der-momyotome (ventrolateral lip or hypaxial somite bud) withdraw from the cell cycle and migrate beneath to the ventral part of the dermomyotome to form the ventral myotome (hypaxial myotome), which later gives rise to intercostal and body-wall muscles **(C)**. The Pax3/Pax7 positive cells are derived from the central region of the dermomyotome **(D)**. These cells do not express any MRFs and they continuously proliferate until they initiate to express MRFs to undergo muscle terminal differentiation in the myotome. These cell pop-ulations contribute to the late (fetal) stage of myogenesis and satellite cells as postnatal muscle stem cells.

dermomyotome delaminate and ventrally migrate out from the der-momyotome. Such migratory muscle precursor cells enter the ventral region of trunk and four limb buds, and initiate expression of MRFs with continuous proliferation. Migratory muscle precursors eventually give rise to the so-called migratory hypaxial muscles: pectoralis, abdominal and diaphragm muscles of the trunk, and all limb muscles.

3. Muscle Stem Cells

3.1. Satellite cells as muscle stem cells

Adult skeletal muscle possesses extraordinary regeneration capability. After exercise or muscle injury, large numbers of new muscle fibers are normally formed within a week because of expansion and differentiation of muscle satellite cells as a stem cell population for muscle regeneration. Muscle satellite cells reside beneath the basal lamina of adult skeletal muscle juxtaposed with skeletal muscle fibers (Fig. 2). Satellite cells are normally mitotically quiescent but can be activated (initiate proliferation) in response to stress induced by weight bearing or other trauma such as injury to mediate the postnatal growth and regeneration of muscle. The progeny of activated satellite cells, termed the myogenic precursor cells (MPCs), undergo multiple rounds of cell division prior to their terminal differentiation to form new muscle fibers (Fig. 2). Satellite cells account for 2–5% of sublaminar nuclei in adult skeletal muscle at two months in mice. The number of quiescent satellite cells in adult muscle remains relatively constant over multiple cycles of degeneration and regeneration, indicating an inherent capacity for self-renewal of satellite cells, an important characteristic of stem cells.[1–3,9,10]

3.2. Molecular mechanisms in satellite cell activation and differentiation

The MRF expression program during satellite cell activation, proliferation, and differentiation is analogous to the program manifested during the

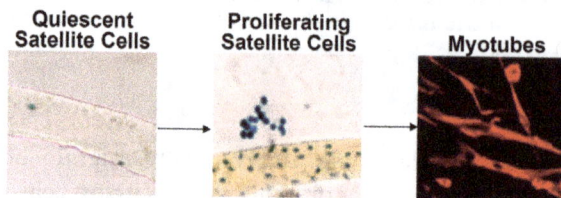

| Quiescent Satellite Cells | Proliferating Satellite Cells | Myotubes |

Figure 2. Muscle satellite cells are stem cells for muscle regeneration. Single muscle fibers can be isolated from adult mouse muscle. Myf5–nlacZ is expressed in quiescent satellite cells in fresh (day 0) single muscle fibers as well as proliferating satellite cells on the dish and myonuclei in the muscle fiber (day 3) derived from Myf5–nlacZ mice.[13] Following expansion of satellite cells, these cells exit the cell cycle and fuse with each other to form myosin heavy chain positive (red) multinucleated myotubes.

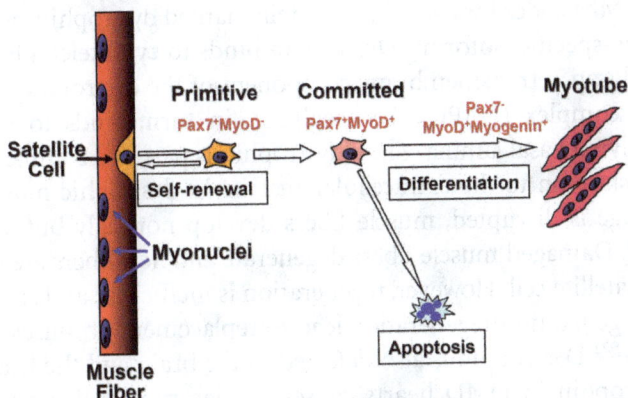

Figure 3. Schematic model for muscle satellite cell differentiation. Mitotically quiescent satellite cells (Pax7$^+$MyoD$^-$) which reside between the basal lamina and sarcolemma of muscle fibers are activated (initiate cell division) and initiate expression of MyoD during muscle regeneration. The activated satellite cells give rise to the muscle precursor cells (Pax7$^+$MyoD$^+$), which eventually down-regulate Pax7 expression. These Pax7$^-$ satellite cells exit their cell cycle and then initiate myogenin expression to differentiate into myotubes (Pax7$^-$MyoD$^+$Myogenin$^+$). Alternatively, these Pax7$^-$ satellite cells undergo apoptosis. Some proportions of satellite cells down-regulate MyoD expression and go back to the quiescent satellite cells as a self-renewal population (Pax7$^+$MyoD$^-$).

embryonic myogenesis. Quiescent satellite cells express no detectable MRFs but do express c-met, a receptor tyrosine kinase, M-cadherin, a cell adhesion molecule,[11,12] CD34, a cell surface marker,[13] CXCR4, a chemokine receptor,[14] integrin α7, a cell adhesion molecule,[15] nestin, an intermediate filament,[16] Pax7,[17] Pax3,[18] and Myf5–nlacZ[13] in transgenic mice in which the nuclear lacZ gene is knocked into the Myf5 locus.[19] During *in vitro* single muscle fiber culture and *in vivo* muscle regeneration in which satellite cells are activated to enter the cell cycle, either MyoD or Myf5 is rapidly up-regulated concomitantly with the entrance of the cell cycle, followed soon after by coexpression of MyoD and Myf5 (Fig. 3). Following proliferation, MPCs exit their cell cycle and initiate expression of myogenin and MRF4 to start their terminal differentiation program.[2,3,10,11] Therefore, MyoD and Myf5 may play an early role in the satellite cell proliferation and differentiation program.

Duchenne muscular dystrophy (DMD) is an X-linked progressive disorder affecting 1/3,500 males.[20,21] Clinical features manifest at 2–3 years of age. Patients become wheelchair-bound before the age of 13 and often die from cardiac arrest or respiratory insufficiency. The dystrophin gene

(2.4 Mb) codes for a 427 kDa protein, named dystrophin, as well as several tissue-specific isoforms. Dystrophin binds to cytoskeletal F-actin and dystroglycan, a transmembrane component of the dystrophin-associated protein complex (DAPC). Dystroglycan, in turn, binds to merosin in the overlying basal lamina. Thus, dystrophin is part of a complex that links the cytoskeleton to the extracellular matrix. In dystrophic muscle, where this linkage is disrupted, muscle fibers develop normally but are easily damaged. Damaged muscle fibers degenerate and new fibers are recruited from the satellite cell. However, regeneration is inefficient, and successive rounds of degeneration/regeneration lead to replacement of muscle by connective tissue.[22] Dystrophin is also detected in the brain and the heart. Absence of dystrophin in DMD hearts causes cardiac muscle degeneration and the gradual development of dilated cardiomyopathy, which is a major lethality for DMD. The mdx mouse, an animal model for DMD, lacks dystrophin due to a mutation. Skeletal muscle necrosis occurs during the weaning stage, with visible muscle weakness. Many cellular and biochemical features resemble those characteristic of the early myopathic phase of DMD.[21]

Mice lacking the MyoD gene display profound deficits in satellite cell function. MyoD$^{-/-}$ mice interbred with mdx mice exhibit increased penetrance of the mdx phenotype characterized by muscle atrophy and increased myopathy leading to premature death.[23] In spite of the presence of morphologically normal satellite cells in MyoD$^{-/-}$ mice, skeletal muscle from such mice displays a significantly reduced capacity for regeneration following injury. While proliferating MyoD$^{-/-}$ myoblasts express a normal level of c met, expression of M-cadherin and desmin, markers for myoblasts, are significantly reduced.[24] In contrast, Myf5 expression is highly up-regulated in these cells, suggesting the compensation mechanism of Myf5. Under conditions that normally induce differentiation of wild type myoblasts, MyoD$^{-/-}$ myoblasts continue to proliferate and exhibit delayed terminal differentiation[12,23,25] and perturbed neuromuscular junctions of the diaphragm.[26] Supporting this, MyoD$^{-/-}$ myoblasts showed delayed induction of myogenin and MRF4. The Grounds group transplanted MyoD$^{-/-}$ muscle into mice and found increased migration of MyoD$^{-/-}$ myoblasts in host muscle compared to *wild type* control.[27] In addition, even though the terminal differentiation is delayed, MyoD$^{-/-}$ myoblasts eventually contribute to muscle fiber formation in host muscle.[25,28] Therefore, MyoD$^{-/-}$ myoblasts display characteristics that are more primitive than those of wild type myoblasts and may represent an intermediate stage between a stem cell and a myogenic precursor cell.[24,29]

3.3. Self-renewal of satellite cells regulated by Pax3 and Pax7 (Fig. 3)

The majority of quiescent satellite cells express Pax7, and Pax3 expression is also detected in a minor population.[17,18,30] Upon activation, Pax-positive satellite cells initiate expression of Myf5 and/or MyoD. During differentiation of satellite cells, Pax genes are down-regulated and myogenin expression is up-regulated to induce muscle-specific genes.[3,31–33] The molecular hierarchy of MyoD and myogenin is well understood: myogenin gene expression is directly regulated by MyoD and MEF2 through direct binding to their DNA-binding sites located in the regulatory region of myogenin.[4,34,35] Some proportion of satellite cells down-regulates MyoD expression to remain in an undifferentiated state and stay as quiescent cells, termed reserve cells.[36] These quiescent cells continuously express Pax genes and reconstitute the satellite cell compartment as self-renewal *in vivo*. Forced expression of Pax7 suppresses myogenic differentiation and progression of the cell cycle.[31] Taken together, these data suggest that expression of Pax genes in satellite cells is important for self-renewal, whereas MyoD, downstream of Pax genes, is required for activation, proliferation and subsequent myogenic differentiation through myogenin induction.[3]

Gene knockout mouse experiments demonstrate the essential role of the Pax7 gene in satellite cell maintenance and/or survival.[3,17,30,33,37] At birth, the number of satellite cells in Pax7$^{-/-}$ mice is nearly normal. However, satellite cells are progressively lost within two weeks because of cell death, suggesting that Pax7 is essential for antiapoptotic roles in satellite cells. Consistent with the cell survival function of Pax7, the phenotype of Pax3 mutant mice is partially rescued by crossing them with p53 knockout mice. p53 is a tumor suppressor gene and is involved in cell cycle arrest, DNA repair and progression of apoptosis. The double mutant mice display reduced neural crest cell death seen in Pax3 mutant mice through suppression of apoptosis by the loss of p53 function.[38]

3.4. Developmental origin of satellite cells

A population of cells that express Pax3 and Pax7 has been identified in the central part of the myotome during chicken and mouse embryogenesis (Fig. 1).[2,3,39–41] These Pax3/Pax7 positive cells are derived from the central region of the dermomyotome in which Pax3/Pax7 genes are already expressed. These cells do not express any MRFs and they continuously proliferate until they start to express MRFs to undergo muscle terminal differentiation in the myotome. These cell populations contribute to the late (fetus) stage of myogenesis during embryogenesis. In addition, these cells move to where satellite cells reside in muscle fibers during late embryogenesis (embryonic day 17 for mice). Therefore, the satellite cells that emerge during late embryogenesis account for the majority of muscle progenitors. Pax3:Pax7 double mutant mice display a remarkable muscle loss during the fetal development stage. In the muscle, there are no myogenic progenitors because the upstream cells for myogenic progenitors cannot activate MRF gene expression, and thus undergo apoptosis or differentiate into non-myogenic cells. These results strongly suggest that Pax3 or Pax7 genes are required for fetal myogenesis as well as development of the satellite cell population.[2,3]

Lower vertebrates such as amphibians maintain remarkable regeneration capacities.[2] During tail-bud regeneration in *Xenopus*, the Pax7-positive population reconstitutes muscle fibers.[42] Similarly, the Pax7-positive population is also required for muscle regeneration during limb regeneration in the salamander.[43] Taken together, these data highlight essential roles of the Pax gene in muscle regeneration including survival of myogenic progenitor cells in vertebrates. In contrast to vertebrates in which Pax3 and Pax7 genes play essential roles in neural tube formation, neural crest cell migration and myogenesis, ascidians which are phylogenically very close to vertebrates possess only one Pax3/7 gene (HrPax-37). In addition, ascidians express this gene in the central nervous system and neural crest cells but not in muscle during embryogenesis.[44] Therefore, one hypothesis is that during vertebrate evolution, animals acquired a new expression niche for Pax3/7 genes in myogenic lineages to evolve into more complicated muscle systems, such as those that support migratory myogenic cells for limb muscle and muscle regeneration by satellite cells.[45]

3.5. Multipotential differentiation capability of satellite cells

During vertebrate embryogenesis, mesodermal progenitors give rise to distinct cell lineages, including skeletal myocytes, osteocytes, chondrocytes and adipocytes, in response to different signals derived from the surrounding niche.[46] The existence of multipotential mesodermal progenitor cells in the embryo has been extensively studied by using the multipotential

C3H10T1/2 cell line derived from the embryonic mesoderm. 10T1/2 cells can differentiate into three mesodermal cell lineages — myocytes, adipocytes and chondrocytes — following treatment with 5-azacytidine, a demethylation reagent.[47] Treatment with BMP can also induce osteogenic, chondrogenic and adipogenic differentiation of 10T1/2 cells.[48] In addition, multipotential mesenchymal stem cells (MSCs) isolated from bone marrow can differentiate into skeletal myocytes, adipocytes, osteocytes and chondrocytes following treatment with various inducers as well as after *in vivo* transplantation.[49,50] Therefore, these results strongly suggest that there are common mesenchymal progenitor cells which give rise to distinct mesenchymal progenies during embryogenesis.

Recently *in vitro* experiments have demonstrated that satellite cells, which are believed to be already committed to the muscle lineage, can differentiate into adipocytes or osteocytes by treatment with adipogenic inducers such as thiazolidinedione/fatty acids, or BMPs, respectively (Fig. 2).[51–53] Satellite cells are readily isolated from adult mouse skeletal muscle by enzymatic digestion. Under normal culture conditions such as high serum with basic fibroblast growth factor (bFGF), myoblasts can proliferate while expressing myoblast markers such as MyoD, Myf5, desmin, M-cadherin and Pax7, indicating that satellite cells are committed myogenic stem cells. Under low serum conditions, these myoblasts exit their cell cycle and fuse to each other to form multinucleated myotubes. These observations clearly indicate that the default differentiation fates of satellite cells are myogenic. Therefore, treatment with adipogenic or osteogenic inducers may transform satellite cells to a dedifferentiation status as a prerequisite, in which satellite cells are suppressed for their myogenic properties and have acquired multidifferentiation potential prior to initiation of nonmuscle differentiation (osteocytes and adipocytes). Alternatively, nonmuscle differentiation of satellite cells by inducers may occur without thorough dedifferentiation processes. To clarify these two possibilities, expression of master transcription factors for adipogenesis (PPARγ2) and osteogenesis (RUNX2/ALM3) was examined. Interestingly, proliferating satellite cells do not only express MyoD but also coexpress PPARγ2 and RUNX2 (Ref. 52 and Asakura, unpublished data). Therefore, satellites are multipotential stem cells that possess intrinsic differentiation properties for myogenesis, adipogenesis and osteogenesis by expressing three master transcription factors. However, it is clear that the default differentiation pathway must be myogenic.

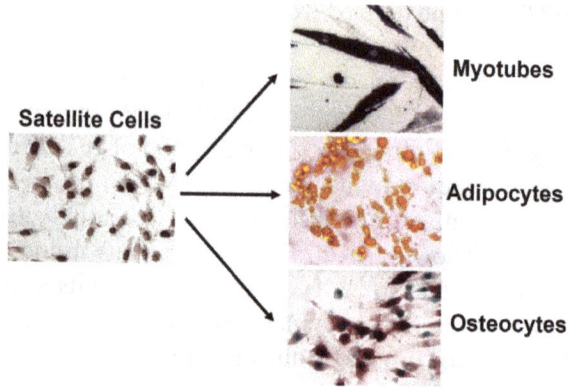

Figure 4. Muscle satellite cells are multipotential stem cells.[51] After low serum conditions, adipogenic inducers (MDI-I) or treatment with BMP4, Pax7-positive satellite cells form myosin heavy chain positive myotubes, or Oil Red-O-positive adipocytes, or alkaline phosphatase-positive osteocytes, respectively.

Therefore, nonmyogenic differentiation fates must be suppressed in myoblasts by unknown mechanisms (Fig. 4).

3.6. Micro-RNAs regulate muscle differentiation

Micro-RNAs (miRNAs) are an abundant class of noncoding small 21- to 24-nucleotide (nt) regulatory RNAs which control many gene expressions at the posttranscriptional level.[54] miRNAs have a specific secondary stem-loop hairpin structure within their primary transcripts. They bind to 3′-untranslated regions (3′-UTR) of the target genes with homologies to the miRNA's sequences, and then either suppress translation from the mRNA or promote degradation of the mRNA. Until now, hundreds of miRNAs have been identified in mammalian and shown to play essential roles in many developmental systems, including cell proliferation, differentiation, apoptosis, cancer cells and stem cells, through suppressing target gene expression. Some miRNAs are transcribed in a tissue-specific manner, indicating that tissue-specific miRNAs may be involved in tissue-specific gene repression.[55] For example, miR-1 (miR-1-1 and miR-1-2), miR-133 (miR-133a-1 and miR-133a-2) and miR-206 are specifically expressed during myoblast–myotube transition (sequences between miR-1-1 and miR-1-2 or miR-133a-1 and miR-133a-2 are highly conserved but these genes are located on different loci). Chromatin immunoprecipitation

(ChIP) assay demonstrates that MRFs such as MyoD and myogenin can bind to E-boxes located upstream of these miRNA genes as well as miR-100, miR-191, miR-138-2 and miR-22, indicating the direct transcriptional regulation by MRFs during muscle differentiation.[55,56] Target genes for these miRNAs have been identified and shown to be involved in muscle differentiation: miR-1 promotes myogenic differentiation by targeting histone deacetylase 4 (HDAC4), a transcriptional repressor of muscle gene expression.[57] One of the targets of HDAC4 is an MEF2. Therefore, miR-1 may promote myogenic differentiation through MEF2 activity. By contrast, miR-133 enhances myoblast proliferation by repressing the serum responsible factor (SRF). Multiple target genes of miR-206 have been identified and shown to control myogenic differentiation. They include DNA polymerase $\alpha1$, a factor for DNA replication, B-ind1, a proliferation factor interacting with RAC, connexion 43, a gap junction factor involved in formation of neuromuscular junction, follistatin-like 1, a BMP4 family, and utrophoin, a dystrophin-related protein.[58–60] These gene expressions indeed do need to be suppressed during myogenic differentiation. MyoD is a potent transcription activator and regulates myogenic differentiation as a master regulator. However, very little is known about the ability of MyoD to suppress gene expression during myogenic differentiation. Since some miRNAs including miR-206 expression are regulated by MyoD during myogenic differentiation, in addition to activating muscle-specific genes, MyoD may utilize these miRNAs including miR-206 to suppress gene expression for myogenic differentiation.[58] Taken together, these evidence highlights a novel mechanism by which miRNAs regulate many different gene expressions during myoblast proliferation and differentiation. Therefore, in the near future, the mechanisms by which miRNAs regulate the self-renewal, proliferation and differentiation processes of muscle satellite cells will be elucidated.

3.7. Muscle side population (SP) cells (Fig. 5)

Goodell *et al.* first demonstrated that hematopoietic stem cells (HSCs) in bone marrow from many different species can be isolated by fluorescence-activated cell sorting (FACS) of side population (SP) cells.[61] The SP cells exclude Hoechst 33342 dye through the activity at the cell surface of multidrug resistance (MDR) pomp proteins such as ABCG2/BCRP1.[62] Gussoni *et al.* then demonstrated that bone marrow-derived SP cells have the capacity to contribute to regenerating muscle fibers following

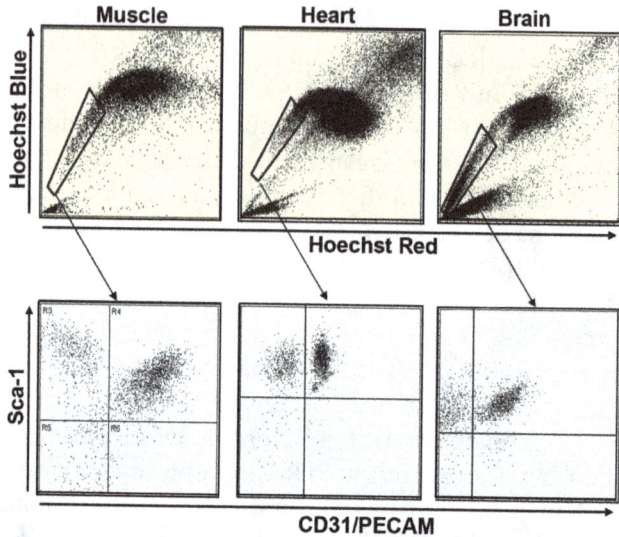

Figure 5. SP cells in adult mouse muscle, heart and brain. FACS fractionation of SP cells from muscle, heart and brain was employed for the isolation of adult stem cells by Hoechst dye exclusion.[66] These three tissue-derived SP cells contain hematopoietic marker CD45[+] and CD45[-] cells. CD45[-]SP cells were further fractionated by Sca-1 (a stem cell marker) and CD31/PECAM (an endothelial cell marker) antibodies. These three tissue-derived CD45[-] SP fractions mainly contain Sca-1[+]CD31[+] and Sca-1[+]CD31[-] cells.

intravenous injection into lethally irradiated mice.[63] The same group also showed that adult skeletal muscle contains SP cells which have the ability to contribute to the entire hematopoietic repertoire and to regenerate muscle fibers following intravenous injection into lethally irradiated mice.[63] These data strongly suggest the presence of myogenic progenitor cells in both SP fractions. Muscle SP cells express stem marker Sca-1 but not any satellite cell markers including Myf5-nlacZ or Pax7, and do not readily undergo myogenic differentiation *in vitro*. By contrast, satellite cells are positive for Pax7 and Myf5-nlacZ but normally negative for Sca-1, and readily undergo myotube formation.[64] In addition, satellite cell populations are not copurified during FACS/Hoechst purification. These results clearly indicate that muscle SP cells are a cell population distinct from satellite cells. Consistent with these data, mutant mice lacking the Pax7 gene display significant reduction of satellite cells, yet their muscle contains cells with a normal proportion of SP cells.[17] An important question was whether or not muscle SP cells include myogenic progenitor cells that contribute to muscle regeneration and can give rise to satellite cells in the muscle environment. Cultures of freshly isolated muscle SP cells alone fail to differentiate into myogenic cells, indicating that the default differentiation fate of muscle SP cells is not myogenic.[64] However, cell transplantation experiments have clearly demonstrated that the muscle SP cells contain cells that contribute to regenerating muscle fibers.[63–65] In addition, muscle SP cells have the potential to give rise to satellite cells in regenerating muscle following transplantation. Therefore, the novel myogenic progenitor cells fractionated into SP undergo myogenic differentiation in response to the muscle niche. Myogenic differentiation of muscle SP cells was observed to occur *in vitro* when cocultured with primary myoblasts.[64] Taken together, the inability of muscle SP cells to undergo myogenic differentiation except in the presence of myoblasts suggests that the process is regulated via mechanisms that involve muscle cell-mediated niche interactions (Fig. 6). Interestingly, many other adult tissues, including heart and brain, contain some proportion of SP cells (Fig. 5),[66–71] some of which

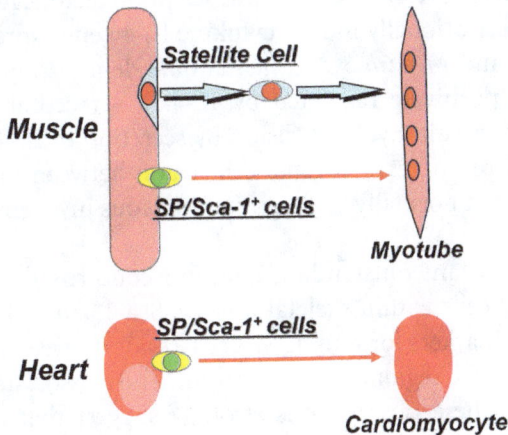

Figure 6. Model for the relationship between satellite cells, muscle stem cells and cardiac stem/progenitor cells. Muscle SP cells or Sca-1+ cells located outside of adult muscle fibers and closely associated with vasculatures have the capacity to participate in muscle regeneration and to give rise to satellite cells, together with satellite cells residing beneath the basal lamina of muscle fibers. In the adult heart, cardiac stem/progenitor cells (SP cells or Sca-1+ cells) can be isolated as a similar population of muscle SP cells or Sca-1+ cells. These cardiac stem/progenitor cells have the potential to differentiate into cardiomyocytes. However, since the heart does not possess powerful stem cells like the satellite cells for skeletal muscle, the regeneration capacity of the heart is still very poor compared to skeletal muscle regeneration.

also express Sca-1,[67,70–72] indicating that tissue-specific stem/progenitor cells might be commonly divided into SP fractions.

Similar muscle niche-mediated myogenic induction was observed for neural stem cells (NSCs). NSCs are able to undergo myogenic differentiation, either when cocultured with myoblasts or intramuscularly injected.[73] In addition, bone marrow-derived cells also undergo myogenic differentiation when intramuscularly or intravenously injected.[63,74] In this case, a prior myelomonocytic precursor cell differentiation is a prerequisite for differentiation into myogenic cells and/or for fusion with regenerating muscle fibers, in response to the muscle niche. Therefore, these observations suggest that induction of myogenic differentiation and/or myogenic fusion for muscle SP cells and other stem cells is mediated by the muscle niche including cell–cell adhesion, secreted factor(s) and/or the extracellular matrix. For example, NFAT/IL-4 signaling induces muscle fiber hypertrophy through promotion of muscle fusion.[75] Muscle hypertrophy induced by calcineurin, a calmodulin-dependent calcium-activated protein phosphatase, is mediated by transnuclear location of dephosphorylated forms of NFAT c2 and c3 transcription factors. Active NFAT proteins then transcriptionally induce cytokine IL-4 gene expression. Gene knockout mice and *in vitro* experimental data demonstrate that myogenic cell fusion is positively regulated by IL-4/IL-4 receptor signaling and that myotubes recruit myoblast fusion by secretion of IL-4. Indeed, treatment with IL-4 promotes myogenic cell fusion between myoblasts and MSCs which do not normally undergo spontaneous myogenic differentiation on their own.[76]

An important outstanding question concerns the *in situ* location of muscle SP cells within skeletal muscle. Sca-1+ and ABCG2+ cells, both of which are markers for muscle SP cells, reside between muscle fibers and in the interstitial region, and are prominently associated with blood vessels.[64,77–79] These observations strongly suggest that muscle SP cells are located as vascularly associated cells in skeletal muscle. Recent data have also confirmed the close relationship between satellite cells and vasculatures *in vivo*.[80] In adult muscle sections, almost all satellite cells are located at the juxtaposition of vasculatures. Coculture with endothelial cells increases satellite cell growth mediated by endothelial cell-secreted factors, including vascular endothelial growth factor (VEGF), platelet-derived growth factor (PDGF-BB), hepatic growth factor (HGF), insulin-like growth factor-I (IGF-I) and bFGF. In addition, *in vitro* angiogenesis also increases by coculture with satellite cells. These results suggest that vascular cells including endothelial cells are important for proliferation of satellite cells and thus for muscle regeneration. Consistent with these observations, coculture with muscle SP cells containing vascularly associated cells increases satellite cell proliferation and suppresses myogenic differentiation mediated by BMP4.[81] Therefore, while muscle SP cells have potential to differentiate into myogenic cells, it is also possible that they play novel roles in myogenesis as niche cells by promoting expansion, suppressing terminal differentiation, and possibly the regulating self-renewal of the satellite cells.

3.8. Other myogenic progenitor cells in muscle

Several groups have isolated a multipotential stem cell-like population from postnatal muscle that can differentiate into several cell types, including muscle cells *in vitro* and/or *in vivo*. Torrente *et al.* reported a myogenic progenitor population, termed muscle-derived stem cells, isolated from neonatal muscle using a preplating technique in which cells that are slow to attach to the tissue culture plate can be purified through multiple passages.[82] Muscle-derived stem cells express both Sca-1 and CD34 but not desmin. When these cells were intra-arterially injected, cells first adhered to the endothelium in the host muscle tissue, and then migrated into the muscle to contribute to regenerating muscle fibers. The Huard group used a similar preplating technique to isolate a multipotential stem cell-like population from neonatal muscle, also called muscle-derived stem cells (MDSCs).[83] Their MDSCs also include Sca-1+ and CD34+ cells and have a multipotential differentiation capacity to give rise to myocytes, glial cells, endothelial cells and even hematopoietic cells following transplantation. Tamaki *et al.* isolated myogenic–endothelial progenitor cells in a CD34+/CD45− fraction from adult skeletal muscle using FACS.[78] These cells express Sca-1 but do not express any endothelial cell or myogenic markers such as CD31, Flk-1, MyoD or Myf5. Clonal culture experiments demonstrate that these cells have the potential to differentiate into three cell lineages: adipocytes, endothelial cells and myocytes. These cells reside in the interstitial spaces of skeletal muscle and differentiate into endothelial cells and muscle fibers following intramuscular transplantation. They also express a muscle SP cell-marker, BCRP1/ABCG2, and some of these cell populations (if not all) are indeed fractionated with the SP cells. Interestingly, Sca-1+/CD45− SP cells isolated from adult skeletal muscle also have the potential to give rise to adipocytes, osteocytes and myocytes (data not shown).

The Verfaillie group identified multipotential stem cells, termed multipotential adult progenitor cells (MAPCs), from adult bone marrow. MAPCs express Sca-1, Flk-1 and CD13 but do not express any hematopoietic or endothelial markers such as CD45 or CD31.[84] Clonal MAPCs differentiate into a wide variety of cell lineages including skeletal muscle and the entire hematopoietic repertory *in vivo* after transplantation. The same group identified an MAPC-like cell population in adult skeletal muscle and brain, which has the potential to differentiate into endothelium, neurons, glia and hepatocytes[85] *in vitro*.

During development, the vascularly associated myogenic progenitors, termed mesoangioblasts in the fetus by Cossu and colleagues, persist in close association with the vasculature as adult stem cells in skeletal muscle. These cells have the potential to differentiate into skeletal muscle, cardiac muscle, smooth muscle, osteocytes and endothelial cells *in vitro* and *in vivo*.[86,87] Mesoangioblasts contribute efficiently to regenerating muscle in muscular dystrophy model (dy/dy) mice following intra-arterial transplantation.[88] Recently, the same group isolated a mesoangioblast-like cell population from canine neonatal and human fetus muscle. Then, they demonstrated that canine mesoangioblasts contribute efficiently to muscle fiber formation in muscular dystrophy model dogs following intra-arterial transplantation.[89] Some dogs which had received their own mesoangioblasts expressing a minidystrophin gene or mesoangioblasts derived from other healthy donor dogs displayed dystrophin expression in their muscle and improved muscle function. These results indicate that mesoangioblasts are a potentially useful cell type for therapeutic cell transplantation for muscular dystrophy. In addition, arterial injection is an efficient stem cell delivery method. The same group also identified that mesoangioblasts isolated from muscle are indeed pericytes.[90] Blood vessels are composed of two interacting cell types, endothelial cells and perivascular cells, referred to as pericytes, vascular smooth muscle cells, or mural cells, which develop on the surface of the vascular tubes.[91] The main function of pericytes is to give rise to vascular smooth muscle and to communicate with endothelial cells. However, recent data demonstrate that pericytes are multipotential progenitor cells that have the potential to undergo osteogenic, chondrogenic, adipogenic and myogenic differentiation after treatment with specific inducers or transplantation.[92] Interestingly, recent results demonstrate that smooth muscle cells in the dorsal aorta and myotome cells are derived from a common progenitor during embryogenesis.[93] Since pericytes exist in all adult tissues, stem cell transplantation of the multipotential pericytes isolated from adult tissues could be a potential therapeutic method in the near future. Interestingly, the muscle SP population also contains a pericyte-like population that possesses mesenchymal differentiation capacities: muscle SP cells can differentiate into myogenic cells, osteocytes, adipocytes and smooth muscle following proper inductions (Asakura, unpublished observation). Taken together, one possibility is that individually isolated multipotential progenitor cells from muscle may be of similar cell origin to the pericytes. Furthermore, since the main function of pericytes is to give rise to vascular smooth muscle and to communicate with endothelial cells, it is also possible that pericytes regulate satellite cell proliferation, differentiation and self-renewal processes by acting as juxtaposed niche cells during muscle development and regeneration.

3.9. Hematopoietic potential cells in muscle

As described above, Gussoni *et al.* demonstrated that muscle SP cells have the potential to give rise to entire hematopoietic cells following intravenous injection into lethally irradiated mice,[63] indicating that the muscle SP fraction contains muscle hematopoietic potential cells (HPCs). Jackson *et al.* and other groups reported that cultured muscle-derived cells exhibit the capacity to reconstitute the entire hematopoietic repertoire following intravenous injection into lethally irradiated mice, suggesting that muscle HPCs can be cultured and expanded *ex vivo* while maintaining their hematopoietic stem cell activity.[94-96] In addition, *in vitro* assays demonstrate that muscle contains a remarkably high level of hematopoietic progenitor activities that form multiple types of hematopoietic colonies.[64,66,97-100] These muscle HPCs are also enriched in the muscle SP fraction, as they are in bone marrow-derived SP cells.[64,66] These observations raised an interesting question about the origin of these muscle HPCs within muscle. However, the answer was relatively simple since only CD45+ (a hematopoietic lineage marker) muscle-derived cells exhibit the capacity to give rise to hematopoietic cells *in vitro* and reconstitute the entire hematopoietic repertoire following intravenous injection into lethally irradiated mice.[97,99-101] In addition, bone marrow transplantation experiments demonstrate that muscle HPCs are indeed of bone-marrow origin and that circulating HPCs originating from bone marrow may reside within skeletal muscle. The reason why muscle contains HPCs that possess such a remarkable capability for hematopoietic differentiation potential remains an interesting question. However, it was reported that not only muscle but also many other adult tissues, such as brain, heart, lung, liver, spleen, kidney and small intestine, contain different proportions of CD45[+] hematopoietic progenitors that are enriched in the SP fraction (Fig. 5).[66] Several experiments ruled out the possibility that these adult tissue-derived HPCs are simply contaminating peripheral blood cells within the tissue preparation.[66,97] Therefore, these HPCs are normal residents in many adult tissues. However, hematopoietic reconstitution of irradiated mice has not yet been reported for cells isolated from these adult tissues, except for adult skeletal muscle and the liver, which have been shown to contain HSCs.[102,103] Interestingly, bone marrow, skeletal muscle, the spleen and the liver appear to contain more primitive multipotential myeloid progenitors than the other tissues, including brain, heart, lung, kidney and small intestine, which appear to contain more committed myeloid progenitors.[66,103] This implies that there is something unique about the skeletal muscle niche that allows it to support the survival and maintenance of such primitive progenitors.

Transplantation of bone marrow-derived cells including SP cells and HSCs has also resulted in contribution to regenerating muscle fibers.[15,63,74,104–106] In addition, several reports clearly show that bone marrow-derived cells have the potential to give rise to a satellite cell compartment following transplantation.[105] As described above, myogenic differentiation and/or fusion of bone marrow-derived cells occurs through a prior myelomonocytic precursor cell differentiation in response to the muscle niche.[107,108] Supporting this phenomenon, the muscle-derived CD45[+]SP cells exhibit the potential to become myogenic cells following coculture with primary myoblasts.[64] Furthermore, muscle-derived CD45[+] cells contribute to regenerating muscle fibers following intramuscular injection.[98] Finally, muscle-derived Sca-1[+]CD45[+] cells undergo myogenic differentiation by forced expression of Pax7 or activation of the canonical Wnt signaling pathway.[109,110] Taken together, these observations suggest that muscle HPCs and bone marrow-derived cells including HSCs possess similar biological characteristics. Indeed, bone marrow transplantation experiments suggested that muscle HPCs may originate in the bone marrow.[97] However, myogenic differentiation of HSCs and bone marrow cells is very poor, which may hinder the therapeutic use of bone marrow cells for muscular dystrophy.[111–114]

4. Cardiac Stem/Progenitor Cells

Cardiomyocytes in the adult heart are believed to be terminally differentiated cells, and hypertrophy, not regeneration or replacement of damaged cardiomyocytes, is the major form of cardiomyocyte growth in the heart. However, recent experimental results have provided evidence that cardiomyocyte replication occurs under the physiological and pathological conditions of the adult heart, suggesting the possibility of myocardium regeneration mediated through cardiomyocyte proliferation or putative cardiac stem/progenitor cells that are capable of generating new cardiomyocytes.[115,116] The origins of these dividing cardiomyocytes remain unclear. There are two possible origins of these cells. First, these cells might be a small population of replicating cardiomyoblasts — committed cardiomyocyte precursor cells. Second, the cycling cardiomyocytes might be derived from uncommitted stem cell-like population cells that expand and differentiate into progeny of cardiomyocytes in response to proper stimulation. Recent work has demonstrated the existence of stem cell-like populations in the adult heart. Anversa *et al.* have reported stem cell-related antigen-positive cells in the interstitial region of the adult heart. They used three surface markers: c-kit, the receptor for the stem cell factor; MDR1, a p-glycoprotein involved in drug exclusion; and Sca-1, a stem cell antigen.[115] As described before, many other adult tissues, including heart, have been shown to contain SP cells (Fig. 5).[66–71] These results suggest that the stem cell population in muscle, such as Sca-1$^+$ and Hoechst dye low/negative, may commonly reside in many adult tissues.[5] Such stem cells might be responsible for tissue repair after damage. It would be important to explore whether some tissue-derived SP cells, such as skeletal muscle or heart SP cells, have the capacity to form cardiomyocytes and/or vascular cells in the transplanted heart. Importantly, heart SP cells can differentiate into cardiomyocytes *in vitro* and following transplantation into the heart (Fig. 6).[67,71,117–119] Recent work has also demonstrated the existence of stem cell-like populations in the adult heart. Beltrami *et al.* have reported a c-kit$^+$ stem cell-like population in the interstitial region of the adult heart.[120] The heart-derived c-kit$^+$ cells have the potential to differentiate into cardiomyocytes *in vitro* and *in vivo* following transplantation into the infarcted heart. In addition, transplanted c-kit$^+$ cells could partially repair the myocardial contractile function of the damaged heart. Furthermore, adult heart-derived Sca-1$^+$ cardiomyocyte progenitor cells localized in the interstitial region of the heart have the potential to give rise to cardiomyocytes *in vitro* and *in vivo*.[72,121] More recent work has identified the embryonic cardiomyocyte progenitors. LIM-homeodomain transcription factor Islet-1 (Isl1) marks a multipotential progenitor population which makes a substantial contribution to the embryonic heart through differentiation into cardiomyocytes, smooth muscle and endothelial cells.[122,123] The Orkin group has also identified a similar cell population from the embryonic heart[124]: these progenitor cells express cardiac specific homeodomain transcription factor Nkx2.5 and a stem cell marker c-kit and have the potential to differentiate into cardiomyocytes and smooth muscle cells. Taken together, recent intensive work successfully identified endogenous cardiac stem/progenitor cells that play important roles in heart development and regeneration. These cardiac stem/progenitor cells residing in the heart might be more effective than the other exogenous stem cells in rebuilding functional heart structure. Therefore, therapeutic cardiac stem/progenitor cell transplantation or recruitment and expansion of cardiac stem cells would promote a pool of young replicating cardiomyocytes and develop a novel application for cardiovascular diseases.

Despite the successful finding of endogenous cardiac stem/progenitor cells which have been shown to contribute to cardiomyocytes and other cell types *in vitro* and in the damaged heart, the regeneration capacity of the heart is still very poor compared to that of the skeletal muscle, in which regeneration is almost complete within a few weeks (Fig. 6). In the skeletal muscle system, satellite cells are potent myogenic stem cells and play prime and essential roles in muscle regeneration, and other cell types, such as muscle SP cells, MDSCs and mesoangioblasts, may play minor roles in recreating muscle fibers. Clearly, the heart does not possess powerful stem cells like the satellite cells for skeletal muscle. Since the marker expression and differentiation fate of cardiac stem/progenitor cells are similar to those of muscle SP cells, MDSCs and mesoangioblasts, one possibility is that cardiac stem/progenitor cells in the heart may be corresponding cells to muscle SP cells, MDSCs and mesoangioblasts found in skeletal muscle. Therefore, such cell populations may exist in many adult tissues and contribute to tissue regeneration after damage.

5. Future Directions

Many molecular mechanisms must be involved in the development, activation, proliferation, differentiation and self-renewal processes of satellite cells. Some of these include secreted factor(s) such as bFGF and IL-4, an extracellular matrix and a muscle niche including cell–cell interactions such as with endothelial cells or pericytes. The molecular mechanism regulating these processes for satellite cells will be elucidated by combining information from *in vitro* culture systems and from different animal systems, including genetic analysis afforded by gene knockout and knockdown experiments. In the near future, the finding of molecular and cell-based signaling cascades for myogenesis and muscle stem cells would not only facilitate our understanding of the mechanisms of muscle development and regeneration, but also lead us to understanding other stem cell and regeneration systems, such as cardiac stem cells and heart regeneration. Finally, our outcomes would develop novel applications of stem cell transplantation to treat many degenerative diseases, including muscular dystrophy.

Acknowledgments

I thank Mayank Verma for critical reading of the manuscript. I am supported by a Research Grant from the Muscular Dystrophy Association (MDA) and the Korean Institute of Science and Technology (KIST).

References

1. Wagers AJ, Conboy IM. (2005) Cellular and molecular signatures of muscle regeneration: current concepts and controversies in adult myogenesis. *Cell* 122: 659–667.
2. Buckingham M. (2006) Myogenic progenitor cells and skeletal myogenesis in vertebrates. *Curr Opin Genet Dev* 16: 525–532.
3. Buckingham M, Bajard L, Daubas P, *et al.* (2006) Myogenic progenitor cells in the mouse embryo are marked by the expression of Pax3/7 genes that regulate their survival and myogenic potential. *Anat Embryol* (Berl) 211 **Suppl** 1: 51–56.
4. Tapscott SJ. (2005) The circuitry of a master switch: Myod and the regulation of skeletal muscle gene transcription. *Development* 132: 2685–2695.
5. Asakura A. (2003) Stem cells in adult skeletal muscle. *Trends Cardiovasc Med* 13: 123–128.
6. Stockdale FE. (1992) Myogenic cell lineages. *Dev Biol* 154: 284–298.
7. Noden DM, Francis-West P. (2006) The differentiation and morphogenesis of craniofacial muscles. *Dev Dyn* 235: 1194–1218.

8. Cossu G, Borello U. (1999) Wnt signaling and the activation of myogenesis in mammals. *Embo J* **18**: 6867–6872.
9. Seale P, Asakura A, Rudnicki MA. (2001) The potential of muscle stem cells. *Dev Cell* **1**: 333–342.
10. Collins CA. (2006) Satellite cell self-renewal. *Curr Opin Pharmacol* **6**: 301–306.
11. Cornelison DD, Wold BJ. (1997) Single-cell analysis of regulatory gene expression in quiescent and activated mouse skeletal muscle satellite cells. *Dev Biol* **191**: 270–283.
12. Cornelison DD, Olwin BB, Rudnicki MA, Wold BJ. (2000) MyoD$^{(-/-)}$ satellite cells in single-fiber culture are differentiation defective and MRF4 deficient. *Dev Biol* **224**: 122–137.
13. Beauchamp JR, Heslop L, Yu DS, *et al.* (2000) Expression of CD34 and Myf5 defines the majority of quiescent adult skeletal muscle satellite cells. *J Cell Biol* **151**: 1221–1234.
14. Ratajczak MZ, Majka M, Kucia M, *et al.* (2003) Expression of functional CXCR4 by muscle satellite cells and secretion of SDF-1 by muscle-derived fibroblasts is associated with the presence of both muscle progenitors in bone marrow and hematopoietic stem/progenitor cells in muscles. *Stem Cells* **21**: 363–371.
15. LaBarge MA, Blau HM. (2002) Biological progression from adult bone marrow to mononucleate muscle stem cell to multinucleate muscle fiber in response to injury. *Cell* **111**: 589–601.
16. Day K, Shefer G, Richardson JB, *et al.* (2007) Nestin-GFP reporter expression defines the quiescent state of skeletal muscle satellite cells. *Dev Biol* **304**: 246–259.
17. Seale P, Sabourin LA, Girgis-Gabardo A, *et al.* (2000b) Pax7 is required for the specification of myogenic satellite cells. *Cell* **102**: 777–786.
18. Montarras D, Morgan J, Collins C, *et al.* (2005) Direct isolation of satellite cells for skeletal muscle regeneration. *Science* **309**: 2064–2067.
19. Tajbakhsh S, Bober E, Babinet C, *et al.* (1996) Gene targeting the myf-5 locus with nlacZ reveals expression of this myogenic factor in mature skeletal muscle fibres as well as early embryonic muscle. *Dev Dyn* **206**: 291–300.
20. Hartigan-O'Connor D, Chamberlain, JS. (1999) Progress toward gene therapy of Duchenne muscular dystrophy. *Semin Neurol* **19**: 323–332.
21. Kunkel LM, Bachrach E, Bennett RR, *et al.* (2006) Diagnosis and cell-based therapy for Duchenne muscular dystrophy in humans, mice, and zebrafish. *J Hum Genet* **51**: 397–406.
22. Bell CD, Conen PE. (1968) Histopathological changes in Duchenne muscular dystrophy. *J Neurol Sci* **7**: 529–544.
23. Megeney LA, Kablar B, Garrett K, *et al.* (1996) MyoD is required for myogenic stem cell function in adult skeletal muscle. *Genes Dev* **10**: 1173–1183.

24. Sabourin LA, Girgis-Gabardo A, Seale P, *et al.* (1999) Reduced differentiation potential of primary MyoD$^{-/-}$ myogenic cells derived from adult skeletal muscle. *J Cell Biol* **144:** 631–643.

25. White JD, Scaffidi A, Davies M, *et al.* (2000) Myotube formation is delayed but not prevented in MyoD-deficient skeletal muscle: studies in regenerating whole muscle grafts of adult mice. *J Histochem Cytochem* **48:** 1531–1544.

26. Staib JL, Swoap SJ, Powers SK. (2002) Diaphragm contractile dysfunction in MyoD gene-inactivated mice. *Am J Physiol Regul Integr Comp Physiol* **283:** R583–590.

27. Smythe GM, Grounds MD. (2001) Absence of MyoD increases donor myoblast migration into host muscle. *Exp Cell Res* **267:** 267–274.

28. Huijbregts J, White JD, Grounds MD. (2001) The absence of MyoD in regenerating skeletal muscle affects the expression pattern of basement membrane, interstitial matrix and integrin molecules that is consistent with delayed myotube formation. *Acta Histochem* **103:** 379–396.

29. Seale P, Rudnicki MA. (2000a) A new look at the origin, function, and "stem-cell" status of muscle satellite cells. *Dev Biol* **218:** 115–124.

30. Oustanina S, Hause G, Braun T. (2004) Pax7 directs postnatal renewal and propagation of myogenic satellite cells but not their specification. *Embo J* **23:** 3430–3439.

31. Olguin HC, Olwin BB. (2004) Pax-7 up-regulation inhibits myogenesis and cell cycle progression in satellite cells: a potential mechanism for self-renewal. *Dev Biol* **275:** 375–388.

32. Relaix F, Montarras D, Zaffran S, *et al.* (2006) Pax3 and Pax7 have distinct and overlapping functions in adult muscle progenitor cells. *J Cell Biol* **172:** 91–102.

33. Zammit PS, Relaix F, Nagata Y, *et al.* (2006) Pax7 and myogenic progression in skeletal muscle satellite cells. *J Cell Sci* **119:** 1824–1832.

34. Gerber AN, Klesert TR, Bergstrom DA, Tapscott SJ. (1997) Two domains of MyoD mediate transcriptional activation of genes in repressive chromatin: a mechanism for lineage determination in myogenesis. *Genes Dev* **11:** 436–450.

35. Bergstrom DA, Penn BH, Strand A, *et al.* (2002) Promoter-specific regulation of MyoD binding and signal transduction cooperate to pattern gene expression. *Mol Cell* **9:** 587–600.

36. Yoshida N, Yoshida S, Koishi K, *et al.* (1998) Cell heterogeneity upon myogenic differentiation: down-regulation of MyoD and Myf-5 generates "reserve cells." *J Cell Sci* **111**(Pt 6): 769–779.

37. Kuang S, Charge SB, Seale P, *et al.* (2006) Distinct roles for Pax7 and Pax3 in adult regenerative myogenesis. *J Cell Biol* **172:** 103–113.

38. Pani L, Horal M, Loeken MR. (2002) Rescue of neural tube defects in Pax-3-deficient embryos by p53 loss of function: implications for Pax-3-dependent development and tumorigenesis. *Genes Dev* **16:** 676–680.

39. Relaix F, Rocancourt D, Mansouri A, Buckingham M. (2005) A Pax3/Pax7-dependent population of skeletal muscle progenitor cells. *Nature* **435**: 948–953.

40. Kassar-Duchossoy L, Giacone E, Gayraud-Morel B, *et al.* (2005) Pax3/Pax7 mark a novel population of primitive myogenic cells during development. *Genes Dev* **19**: 1426–1431.

41. Gros J, Manceau M, Thome V, Marcelle C. (2005) A common somitic origin for embryonic muscle progenitors and satellite cells. *Nature* **435**: 954–958.

42. Chen Y, Lin G, Slack JM. (2006) Control of muscle regeneration in the *Xenopus* tadpole tail by Pax7. *Development* **133**: 2303–2313.

43. Morrison JI, Loof S, He P, Simon A. (2006) Salamander limb regeneration involves the activation of a multipotent skeletal muscle satellite cell population. *J Cell Biol* **172**: 433–440.

44. Wada H, Holland PW, Sato S, *et al.* (1997) Neural tube is partially dorsalized by overexpression of HrPax-37: the ascidian homologue of Pax-3 and Pax-7. *Dev Biol* **187**: 240–252.

45. Holland LZ, Laudet V, Schubert M. (2004) The chordate amphioxus: an emerging model organism for developmental biology. *Cell Mol Life Sci* **61**: 2290–2308.

46. Brand-Saberi B, Wilting J, Ebensperger C, Christ B. (1996) The formation of somite compartments in the avian embryo. *Int J Dev Biol* **40**: 411–420.

47. Taylor SM, Jones PA. (1979) Multiple new phenotypes induced in 10T1/2 and 3T3 cells treated with 5-azacytidine. *Cell* **17**: 771–779.

48. Katagiri T, Yamaguchi A, Ikeda T, *et al.* (1990) The non-osteogenic mouse pluripotent cell line, C3H10T1/2, is induced to differentiate into osteoblastic cells by recombinant human bone morphogenetic protein-2. *Biochem Biophys Res Commun* **172**: 295–299.

49. Pittenger MF, Mackay AM, Beck SC, *et al.* (1999) Multilineage potential of adult human mesenchymal stem cells. *Science* **284**: 143–147.

50. Liechty KW, MacKenzie TC, Shaaban AF, *et al.* (2000) Human mesenchymal stem cells engraft and demonstrate site-specific differentiation after *in utero* transplantation in sheep. *Nat Med* **6**: 1282–1286.

51. Asakura A, Komaki M, Rudnicki M. (2001) Muscle satellite cells are multipotential stem cells that exhibit myogenic, osteogenic, and adipogenic differentiation. *Differentiation* **68**: 245–253.

52. Wada MR, Inagawa-Ogashiwa M, Shimizu S, *et al.* (2002) Generation of different fates from multipotent muscle stem cells. *Development* **129**: 2987–2995.

53. Shefer G, Wleklinski-Lee M, Yablonka-Reuveni Z. (2004) Skeletal muscle satellite cells can spontaneously enter an alternative mesenchymal pathway. *J Cell Sci* **117**: 5393–5404.

54. Michalak P. (2006) RNA world — the dark matter of evolutionary genomics. *J Evol Biol* **19**: 1768–1774.

55. Rao PK, Kumar RM, Farkhondeh M, *et al.* (2006) Myogenic factors that regulate expression of muscle-specific microRNAs. *Proc Natl Acad Sci USA* **103**: 8721–8726.

56. Cao Y, Kumar RM, Penn BH, *et al.* (2006) Global and gene-specific analyses show distinct roles for Myod and Myog at a common set of promoters. *Embo J* **25**: 502–511.

57. Chen JF, Mandel EM, Thomson JM, *et al.* (2006) The role of microRNA-1 and microRNA-133 in skeletal muscle proliferation and differentiation. *Nat Genet* **38**: 228–233.

58. Rosenberg MI, Georges SA, Asawachaicharn A, *et al.* (2006) MyoD inhibits Fstl1 and Utrn expression by inducing transcription of miR-206. *J Cell Biol* **175**: 77–85.

59. Kim HK, Lee YS, Sivaprasad U, *et al.* (2006) Muscle-specific microRNA miR-206 promotes muscle differentiation. *J Cell Biol* **174**: 677–687.

60. Anderson C, Catoe H, Werner R. (2006) MIR-206 regulates connexin43 expression during skeletal muscle development. *Nucleic Acids Res* **34**: 5863–5871.

61. Goodell MA, Brose K, Paradis G, *et al.* (1996) Isolation and functional properties of murine hematopoietic stem cells that are replicating *in vivo. J Exp Med* **183**: 1797–1806.

62. Zhou S, Schuetz JD, Bunting KD, *et al.* (2001) The ABC transporter Bcrp1/ABCG2 is expressed in a wide variety of stem cells and is a molecular determinant of the side-population phenotype. *Nat Med* **7**: 1028–1034.

63. Gussoni E, Soneoka Y, Strickland CD, *et al.* (1999) Dystrophin expression in the mdx mouse restored by stem cell transplantation. *Nature* **401**: 390–394.

64. Asakura A, Seale P, Girgis-Gabardo A, Rudnicki MA. (2002a) Myogenic specification of side population cells in skeletal muscle. *J Cell Biol* **159**: 123–134.

65. Bachrach E, Li S, Perez AL, *et al.* (2004) Systemic delivery of human microdystrophin to regenerating mouse dystrophic muscle by muscle progenitor cells. *Proc Natl Acad Sci USA* **101**: 3581–3586.

66. Asakura A, Rudnicki MA. (2002b) Side population cells from diverse adult tissues are capable of *in vitro* hematopoietic differentiation. *Exp Hematol* **30**: 1339–1345.

67. Hierlihy AM, Seale P, Lobe CG, *et al.* (2002) The post-natal heart contains a myocardial stem cell population. *FEBS Lett* **530**: 239–243.

68. Summer R, Kotton DN, Sun X, *et al.* (2003) Side population cells and Bcrp1 expression in lung. *Am J Physiol Lung Cell Mol Physiol* **285**: 97–104.

69. Alvi AJ, Clayton H, Joshi C, *et al.* (2003) Functional and molecular characterisation of mammary side population cells. *Breast Cancer Res* **5**: R1–8.

70. Montanaro F, Liadaki K, Volinski J, *et al.* (2003) Skeletal muscle engraftment potential of adult mouse skin side population cells. *Proc Natl Acad Sci USA* **100**: 9336–9341.

71. Martin CM, Meeson AP, Robertson SM, *et al.* (2004) Persistent expression of the ATP-binding cassette transporter, Abcg2, identifies cardiac SP cells in the developing and adult heart. *Dev Biol* **265**: 262–275.

72. Oh H, Bradfute SB, Gallardo TD, *et al.* (2003) Cardiac progenitor cells from adult myocardium: homing, differentiation, and fusion after infarction. *Proc Natl Acad Sci USA* **100**: 12313–12318.

73. Galli R, Borello U, Gritti A, *et al.* (2000) Skeletal myogenic potential of human and mouse neural stem cells. *Nat Neurosci* **3**: 986–991.

74. Ferrari G, Cusella-De Angelis G, Coletta M, *et al.* (1998) Muscle regeneration by bone marrow-derived myogenic progenitors. *Science* **279**: 1528–1530.

75. Horsley V, Jansen KM, Mills ST, Pavlath GK. (2003) IL-4 acts as a myoblast recruitment factor during mammalian muscle growth. *Cell* **113**: 483–494.

76. Schulze M, Belema-Bedada F, Technau A, Braun T. (2005) Mesenchymal stem cells are recruited to striated muscle by NFAT/IL-4-mediated cell fusion. *Genes Dev* **19**: 1787–1798.

77. Zammit P, Beauchamp J. (2001) The skeletal muscle satellite cell: stem cell or son of stem cell? *Differentiation* **68**: 193–204.

78. Tamaki T, Akatsuka A, Ando K, *et al.* (2002) Identification of myogenic-endothelial progenitor cells in the interstitial spaces of skeletal muscle. *J Cell Biol* **157**: 571–577.

79. Uezumi A, Ojima K, Fukada S, *et al.* (2006) Functional heterogeneity of side population cells in skeletal muscle. *Biochem Biophys Res Commun* **341**: 864–873.

80. Christov C, Chretien F, Abou-Khalil R, *et al.* (2007) Muscle satellite cells and endothelial cells: close neighbors and privileged partners. *Mol Biol Cell* **18**: 1397–1409.

81. Frank NY, Kho AT, Schatton T, *et al.* (2006) Regulation of myogenic progenitor proliferation in human fetal skeletal muscle by BMP4 and its antagonist Gremlin. *J Cell Biol* **175**: 99–110.

82. Torrente Y, Tremblay JP, Pisati F, *et al.* (2001) Intraarterial injection of muscle-derived CD34$^{(+)}$Sca-1$^{(+)}$ stem cells restores dystrophin in mdx mice. *J Cell Biol* **152**: 335–348.

83. Qu-Petersen Z, Deasy B, Jankowski R, *et al.* (2002) Identification of a novel population of muscle stem cells in mice: potential for muscle regeneration. *J Cell Biol* **157**: 851–864.

84. Jiang Y, Jahagirdar BN, Reinhardt RL, *et al.* (2002a) Pluripotency of mesenchymal stem cells derived from adult marrow. *Nature* **418**: 41–49.

85. Jiang Y, Vaessen B, Lenvik T, *et al.* (2002b) Multipotent progenitor cells can be isolated from postnatal murine bone marrow, muscle, and brain. *Exp Hematol* **30**: 896–904.

86. De Angelis L, Berghella L, Coletta M, *et al.* (1999) Skeletal myogenic progenitors originating from embryonic dorsal aorta coexpress endothelial and myogenic markers and contribute to postnatal muscle growth and regeneration. *J Cell Biol* **147**: 869–878.

87. Minasi MG, Riminucci M, De Angelis L, *et al.* (2002) The meso-angioblast: a multipotent, self-renewing cell that originates from the dorsal aorta and differentiates into most mesodermal tissues. *Development* **129**: 2773–2783.

88. Sampaolesi M, Torrente Y, Innocenzi A, *et al.* (2003) Cell therapy of alpha-sarcoglycan null dystrophic mice through intra-arterial delivery of mesoangioblasts. *Science* **301**: 487–492.

89. Sampaolesi M, Blot S, D'Antona G, *et al.* (2006) Mesoangioblast stem cells ameliorate muscle function in dystrophic dogs. *Nature* **444**: 574–579.

90. Dellavalle A, Sampaolesi M, Tonlorenzi R, *et al.* (2007) Pericytes of human skeletal muscle are myogenic precursors distinct from satellite cells. *Nat Cell Biol* **9**: 255–267.

91. Bergers G, Song S. (2005) The role of pericytes in blood-vessel formation and maintenance. *Neuro-oncol* **7**: 452–464.

92. Brachvogel B, Moch H, Pausch F, *et al.* (2005) Perivascular cells expressing annexin A5 define a novel mesenchymal stem cell-like population with the capacity to differentiate into multiple mesenchymal lineages. *Development* **132**: 2657–2668.

93. Esner M, Meilhac SM, Relaix F, *et al.* (2006) Smooth muscle of the dorsal aorta shares a common clonal origin with skeletal muscle of the myotome. *Development* **133**: 737–749.

94. Jackson KA, Mi T, Goodell MA. (1999) Hematopoietic potential of stem cells isolated from murine skeletal muscle. *Proc Natl Acad Sci USA* **96**: 14482–14486.

95. Howell JC, Yoder MC, Srour EF. (2002) Hematopoietic potential of murine skeletal muscle-derived CD45$^{(-)}$Sca-1$^{(+)}$c-kit$^{(-)}$ cells. *Exp Hematol* **30**: 915–924.

96. Dell'Agnola C, Rabascio C, Mancuso P, *et al.* (2002) *In vitro* and *in vivo* hematopoietic potential of human stem cells residing in muscle tissue. *Exp Hematol* **30**: 905–914.

97. Kawada H, Ogawa M. (2001) Bone marrow origin of hematopoietic progenitors and stem cells in murine muscle. *Blood* **98**: 2008–2013.

98. McKinney-Freeman SL, Jackson KA, Camargo FD, *et al.* (2002) Muscle-derived hematopoietic stem cells are hematopoietic in origin. *Proc Natl Acad Sci USA* **99**: 1341–1346.

99. Issarachai S, Priestley GV, Nakamoto B, Papayannopoulou T. (2002) Cells with hemopoietic potential residing in muscle are itinerant bone marrow-derived cells. *Exp Hematol* **30**: 366–373.

100. Geiger H, True JM, Grimes B, *et al.* (2002) Analysis of the hematopoietic potential of muscle-derived cells in mice. *Blood* 100: 721–723.

101. McKinney-Freeman SL, Majka SM, Jackson KA, *et al.* (2003) Altered phenotype and reduced function of muscle-derived hematopoietic stem cells. *Exp Hematol* 31: 806–814.

102. Taniguchi H, Toyoshima T, Fukao K, Nakauchi H. (1996) Presence of hematopoietic stem cells in the adult liver. *Nat Med* 2: 198–203.

103. Uchida N, Leung FY, Eaves CJ. (2002) Liver and marrow of adult mdr-1a/1b$^{(-/-)}$ mice show normal generation, function, and multi-tissue trafficking of primitive hematopoietic cells. *Exp Hematol* 30: 862–869.

104. Bittner RE, Schofer C, Weipoltshammer K, *et al.* (1999) Recruitment of bone-marrow-derived cells by skeletal and cardiac muscle in adult dystrophic mdx mice. *Anat Embryol* (Berl) 199: 391–396.

105. Fukada S, Miyagoe-Suzuki Y, Tsukihara H, *et al.* (2002) Muscle regeneration by reconstitution with bone marrow or fetal liver cells from green fluorescent protein-gene transgenic mice. *J Cell Sci* 115: 1285–1293.

106. Corti S, Strazzer S, Del Bo R, *et al.* (2002) A subpopulation of murine bone marrow cells fully differentiates along the myogenic pathway and participates in muscle repair in the mdx dystrophic mouse. *Exp Cell Res* 277: 74–85.

107. Corbel SY, Lee A, Yi L, Duenas J, *et al.* (2003) Contribution of hematopoietic stem cells to skeletal muscle. *Nat Med* 9: 1528–1532.

108. Doyonnas R, LaBarge MA, Sacco A, *et al.* (2004) Hematopoietic contribution to skeletal muscle regeneration by myelomonocytic precursors. *Proc Natl Acad Sci USA* 101: 13507–13512.

109. Seale P, Ishibashi J, Scime A, Rudnicki MA. (2004) Pax7 is necessary and sufficient for the myogenic specification of CD45$^+$:Sca1$^+$ stem cells from injured muscle. *PLoS Biol* 2004 May; 2(5): E130. Epub 2004 May 11 2: 664–672.

110. Polesskaya A, Seale P, Rudnicki MA. (2003) Wnt signaling induces the myogenic specification of resident CD45$^+$ adult stem cells during muscle regeneration [comment]. *Cell* 113: 841–852.

111. Ferrari G, Stornaiuolo A, Mavilio, F. (2001) Failure to correct murine muscular dystrophy. *Nature* 411: 1014–1015.

112. Lapidos KA, Chen YE, Earley JU, *et al.* (2004) Transplanted hematopoietic stem cells demonstrate impaired sarcoglycan expression after engraftment into cardiac and skeletal muscle. *J Clin Invest* 114: 1577–1585.

113. Dell'Agnola C, Wang Z, Storb R, *et al.* (2004) Hematopoietic stem cell transplantation does not restore dystrophin expression in Duchenne muscular dystrophy dogs. *Blood* 104: 4311–4318.

114. Cossu G. (2004) Fusion of bone marrow-derived stem cells with striated muscle may not be sufficient to activate muscle genes. *J Clin Invest* 114: 1540–1543.

115. Anversa P, Nadal-Ginard B. (2002) Myocyte renewal and ventricular remodelling. *Nature* **415**: 240–243.
116. Urbanek K, Quaini F, Tasca G, *et al.* (2003) From the cover: intense myocyte formation from cardiac stem cells in human cardiac hypertrophy. *Proc Natl Acad Sci USA* **100**: 10440–10445.
117. Pfister O, Mouquet F, Jain M, *et al.* (2005) CD31⁻ but not CD31⁺ cardiac side population cells exhibit functional cardiomyogenic differentiation. *Circ Res* **97**: 52–61.
118. Tomita Y, Matsumura K, Wakamatsu Y, *et al.* (2005) Cardiac neural crest cells contribute to the dormant multipotent stem cell in the mammalian heart. *J Cell Biol* **170**: 1135–1146.
119. Oyama T, Nagai T, Wada H, *et al.* (2007) Cardiac side population cells have a potential to migrate and differentiate into cardiomyocytes *in vitro* and *in vivo*. *J Cell Biol* **176**: 329–341.
120. Beltrami AP, Barlucchi L, Torella D, *et al.* (2003) Adult cardiac stem cells are multipotent and support myocardial regeneration. *Cell* **114**: 763–776.
121. Matsuura K, Nagai T, Nishigaki N, *et al.* (2004) Adult cardiac Sca-1-positive cells differentiate into beating cardiomyocytes. *J Biol Chem* **279**: 11384–11391.
122. Laugwitz KL, Moretti A, Lam J, *et al.* (2005) Postnatal isl1⁺ cardioblasts enter fully differentiated cardiomyocyte lineages. *Nature* **433**: 647–653.
123. Moretti A, Caron L, Nakano A, *et al.* (2006) Multipotent embryonic isl1⁺ progenitor cells lead to cardiac, smooth muscle, and endothelial cell diversification. *Cell* **127**: 1151–1165.
124. Wu SM, Fujiwara Y, Cibulsky SM, *et al.* (2006) Developmental origin of a bipotential myocardial and smooth muscle cell precursor in the mammalian heart. *Cell* **127**: 1137–1150.

8

Myogenic Precursor Cells in the Extraocular Muscles

Kristen M. Kallestad and Linda K. McLoon*

1. Introduction: Extraocular Muscle Properties Distinct from Limb Skeletal Muscle

The extraocular muscles (EOMs) are the muscles that move the eye in the orbit. They consist of a total of 12 muscles, 6 within each orbit: superior, inferior, lateral, and medial rectus muscles and superior and inferior oblique muscles. The EOMs control complex eye movements in order to maintain binocular vision and eye alignment. These movements are exquisitely controlled, and abnormalities in them result in a lack of binocular alignment called strabismus. The EOMs have the fastest contractile properties and the smallest motor units when compared to the other skeletal muscles in the body. They display many properties that are different from those of limb skeletal muscles. Physiologically, they differ in calcium handling abilities[1,2] and in their antioxidant enzyme activity patterns[3] compared to limb skeletal muscles. One of the long-standing enigmas of the EOMs is their differential involvement or sparing in muscle diseases. The EOMs are preferentially involved in diseases like oculopharyngeal dystrophy, Graves' ophthalmopathy, and ocular myasthenia gravis, yet they are preferentially spared in Duchenne and Becker muscular dystrophy,[4,5] as well as sarcoglycan deficiency (limb-girdle muscular dystrophy), congenital muscular dystrophy, and merosin-deficient muscular dystrophy.[6,7] Despite the lack of dystrophin in Duchenne muscular dystrophy, normal ocular motility and normal saccadic velocities are maintained throughout

*Correspondence: Department of Ophthalmology, University of Minnesota, Room 374 LRB, 2001 6th Street SE, Minneapolis, MN 55455, USA. Tel: 612-626-0777; E-mail: mcloo001@umn.edu.

life.[5] The explanation for this sparing in these muscular dystrophies is unknown; however, it suggests that the EOMs have significant adaptive properties that allow them to maintain muscle mass, force and motility despite these genetic problems.

The EOMs have a complex anatomical structure, and based on myofiber cross-sectional area, they have two anatomically distinct layers — an orbital and a global layer. These layers have different myosin heavy chain (MyHC) isoform expression patterns. Additional complexity in MyHC expression patterns is demonstrated by the observation that there are significant differences in the overall pattern in the middle region of the muscle in comparison with the tendon ends[8,9] and even within the middle and ends of individual myofibers.[10] This complexity is compounded in adult EOMs by the presence of MyHC isoforms not normally expressed in adult limb skeletal muscle, including immature MyHC isoforms[11] and an EOM-specific myosin.[12] In fact, the eye muscles continue to express a large number of properties that are normally completely down-regulated in adult limb skeletal muscle and only expressed in limb skeletal muscle during development or regeneration. These include such properties as multiply-innervated myofibers,[13] retention of the immature isoform of the acetylcholine receptor,[14] unique patterns of calcium regulation,[2] expression of N-CAM,[15] and up-regulation of a number of myogenic regulatory factors and their receptors.[16–18] These properties again suggest that the sparing of the EOMs in various forms of muscular dystrophy may be related to the retention of pathways known to be active in muscle development, repair and regeneration.

2. Myogenic Precursor Cells in Skeletal Muscle

Adult skeletal muscles are composed of large multinucleated myofibers, and the myonuclei within them have long been known to be postmitotic.[19] The ability to regenerate or repair themselves after injury or disease resides in a population of positionally defined cells called satellite cells.[20] These cells sit between the sarcolemma and the basal lamina, where they reside in a quiescent state in normal adult skeletal muscle. When a muscle is diseased or injured, these satellite cells become activated,[21,22] divide, and either fuse together to form new myofibers or fuse into existing fibers to aid in their repair. In the last few years, it has become increasingly evident that this regenerative cell population is heterogeneous, both in the proteins they express[23] and in their properties during muscle injury, repair, and regeneration.[24]

Initially, the myogenic precursor cells were divided into those that are quiescent — and therefore not active in the cell cycle — and those that are activated. Distinct markers can be used to define these populations both *in vitro* and *in vivo*,[25,26] and while there is some dissension in the literature, several generally accepted markers for activated satellite cells include hepatocyte growth factor[21] as well as some of the myogenic signaling factors, including MyoD,[24,27] myogenin,[24] and myf5.[25] A number of markers have been linked specifically to the identification of quiescent satellite cells, and these include Pax7,[28] CD34,[25] cMet,[21,23] notch,[29] and m-cadherin.[25] From these studies, it is clear that there are a number of markers available for identifying "quiescent cells in the satellite cell position." It is unclear, however, whether they all identify the same cell population. For example, using a preplate isolation procedure, myogenic precursor cells were separated based on variable adhesion characteristics;[30] one population developed into muscle *in vitro* while the other population was able to differentiate into multiple tissue types.[31] In another study, isolated mononucleated cells placed *in vitro* demonstrated that two populations were present, one proliferating rapidly and the other at markedly slower rates.[32] In addition, the rapidly proliferating cells, both *in vitro* and *in vivo*, fused primarily with differentiated myotubes or myofibers, while the slowly proliferative cells fused mainly with each other. Using methods for following myogenic precursor cell proliferation, the skeletal muscles of growing rats again were shown to contain two proliferative compartments, one rapidly dividing and the other a more "slow-cycling" population hypothesized to be "reserve cells."[33] Clonal analysis demonstrated that clones derived from individual satellite cells show diverse behavior when induced to differentiate.[34–37] Based on these studies, it would appear that these populations contain cells with significantly different potentials for proliferation and differentiation. Further work needs to be done to distinguish the role of the local cellular environment; however, we will discuss data to support the hypothesis that the myogenic precursor cells can be demonstrated to have an array of intrinsic characteristics when they are removed from mature muscle tissue.

3. Myogenic Precursor Cells in Extraocular Muscle

The anatomic and physiologic differences between mature skeletal muscle myofibers in the EOMs compared with limb muscles, as outlined previously, suggested that there might be intrinsic differences in their populations of myogenic precursor cells as well. Morphometric analysis of myogenic precursor cell numbers shows that when compared to limb skeletal muscles, the EOMs contain significantly more myogenic precursor cells. The average myofiber from limb muscles averages 0–20 myogenic precursor cells per myofiber,[25,38,39] while the EOMs average 60–100 myogenic precursor cells per myofiber.[40]

One significant difference between these myogenic precursor-cell populations in normal adult limb muscles compared to normal adult EOMs is the presence of activated satellite cells in normal EOM. Examination of adult rabbit, mouse, monkey and human normal adult EOMs demonstrated the presence of activated satellite cells, as visualized with both MyoD and HGF expression.[41–44] When the frequency of these activated satellite cells is quantified, it appears that the percentage varies somewhat from rabbit to human, but approximately 1–5% of the myofibers in any given cross-section had associated with them an activated satellite cell (Tables 1 and 2).

Table 1. Percent of Quiescent Satellite Cells in EOMs and Leg Muscles

Quiescent Satellite Cell Marker	Percent in Rabbit EOMs (Based on Myofiber Number)		Percent in Human EOMs (Based on Myofiber Number)		Percent in Rabbit Leg Muscles (Based on Myofiber Number)
	Orbital	Global	Orbital	Global	
Pax7	14.7%	12%	8%	10.2%	2.4%
cMet	17%	11%	NA	NA	1.9%

Table 2. Percent of Activated Satellite Cells in EOMs and Leg Muscles

Activated Satellite Cell Marker	Percent in Rabbit EOMs (Based on Myofiber Number)		Percent in Human EOMs (Based on Myofiber Number)		Percent in Rabbit Leg Muscles (Based on Myofiber Number)
	Orbital	Global	Orbital	Global	
MyoD	5.7%	3.2%	2.8%	2.5%	0.0%
HGF	5.0%	4.5%	2.6%	1.5%	0.0%
myogenin	2.1%	1.7%	3.5%	0.8%	0.0%

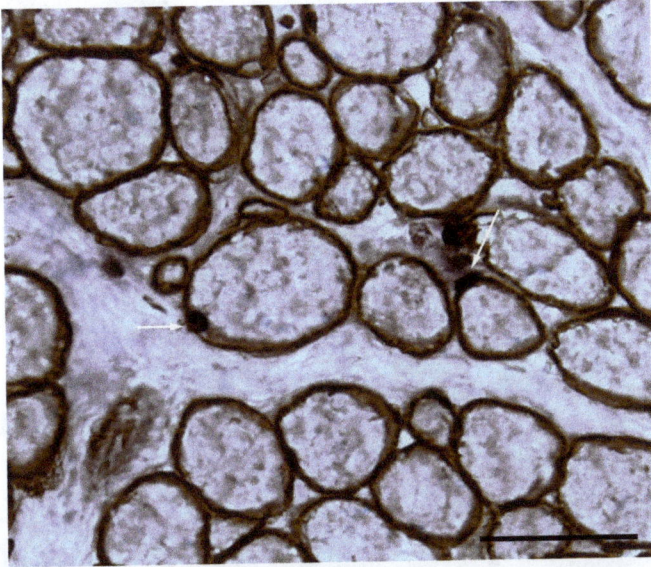

Figure 1. Photomicrograph from a normal adult rabbit superior rectus muscle with a brdU-positive myonucleus within the dystrophin-positive sarcolemma (horizontal arrow) and a brdU-positive cell in the satellite cell position (slanted arrow). The bar is 20 microns.

The presence of a significant number of activated satellite cells suggested that ongoing myofiber remodeling may be occurring throughout adult life in the EOMs. This was confirmed using bromodeoxyuridine (BrdU) labeling methods. BrdU is a thymidine analog, and when injected into animals becomes incorporated into replicating DNA in the place of thymidine. The labeled progeny can be visualized subsequently by immunohistochemical localization of BrdU-positive nuclei. Using this strategy, we have demonstrated that the EOMs continuously add myonuclei to existing myofibers.[41,42,44] These new myonuclei are added in random positions along the length of individual myofibers, and on average, minimally 12 new myonuclei are added to normal adult EOM myofibers every day (Fig. 1). The ongoing nature of this process is demonstrated by administration of BrdU over multiple days or weeks. When the muscles from rabbits were treated daily with BrdU for 7 days and examined 21 days later for evidence of BrdU-positive myonuclei, approximately 5% of the myofibers in cross-section had labeled nuclei.[41] The presence of ongoing myonuclear apoptosis was demonstrated in the adult EOMs,[44] suggesting that there is a continuous process of myonuclear loss and myonuclear

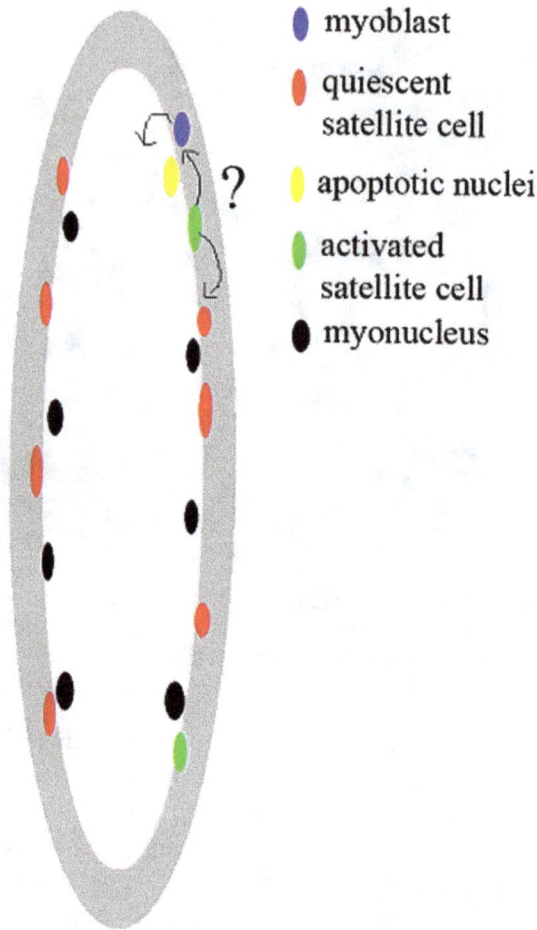

Figure 2. Model of continuous myofiber remodeling. A myonucleus becomes apoptotic (yellow). This process signals to an overlying satellite cell to become activated (green). One daughter cell (blue) will fuse into the existing myofiber, while the other daughter cell will stay in the satellite cell positive and become quiescent (red).

addition (Fig. 2). Our current working model suggests that some localized factor causes an individual myonucleus to apoptose. This results in a signal to a nearby satellite cell, causing its activation and division. One daughter cell is then thought to fuse into the fiber, while the other daughter cell is hypothesized to remain in the satellite cell position and ultimately become quiescent. This hypothesis is supported by our observation of short segments of individual adult EOM myofibers that are positive for activated caspase (McLoon *et al.*, 2004).

The EOMs are not alone in this phenomenon, and this may be a property of craniofacial muscles generally. Thus far, the laryngeal muscles also were shown to undergo continuous myofiber remodeling at approximately the same rate as the EOMs.[45] Additionally, a recent study demonstrated that the laryngeal muscles are spared in the mdx mouse model of muscular dystrophy.[46] As has been suggested a number of years ago, it would appear that the craniofacial muscles, including the masseter, the EOMs, the laryngeal muscles and others, are distinct allotypes compared to limb skeletal muscles.[47,48] This appears to be reflected in the properties of their myogenic precursor cells as well. Using a regenerating muscle paradigm, the myogenic precursor cells derived from the jaw and limb muscles had intrinsically different properties, which they maintained when transplanted to new locations.[48] This study would support the concept that the myogenic precursor cells in craniofacial muscles are intrinsically and constitutively different from those in limb skeletal muscles.

4. Constitutive Differences in the Myogenic Precursor Cells in EOM and Limb Skeletal Muscle

Our working hypothesis is that there are subpopulations of myogenic precursor cells found within the EOMs that are constitutively and functionally different from those in limb skeletal muscles. There is a basis for this viewpoint even when one compares myogenic precursor cells from slow and fast limb skeletal muscles. Certainly innervation, including both the nerve itself and the pattern of stimulation, is known to play a crucial role in the determination of many of the properties of adult muscle, most specifically their myosin heavy and light chain isoform composition.[49,50] However, cross-innervation does not completely respecify the myosin isoform types that are expressed. More relevant to the current discussion, however, is the demonstration that despite significant alterations in myosin isoform content after cross-innervation or low frequency stimulation, the satellite cells themselves retain their intrinsic original behaviors when isolated from the manipulated muscles and grown *in vitro*.[51] In fact, during regeneration in the absence of innervation, slow and fast muscles express their original patterns of myosin expression in the face of either slow or fast patterns of direct stimulation.[52,53] This suggests that the myogenic precursor cells within these muscles retain their intrinsic properties, and that these intrinsic properties play a dominant role in determination of the muscle phenotype.[52,53]

Studies have shown that leg muscles atrophy when functionally denervated by injection of botulinum toxin;[54,55] in contrast, there is little to no change seen in muscle fiber size after functional denervation of the EOMs.[56] If the number of activated satellite cells is quantified in the EOMs after functional denervation by an injection of botulinum toxin, significant increases in the number occur.[57] The satellite cells in leg muscles show only an abortive response. In addition, there is a significant up-regulation in the process of myofiber remodeling, as evidenced by increased numbers of BrdU-positive myonuclei within the first two weeks after a single botulinum toxin injection.[57] After complete unilateral section of the recurrent laryngeal nerve in rabbits, the intrinsic laryngeal muscles react with a significant increase in activated satellite cells and myofiber remodeling.[58] Even six months after denervation, these muscles retain elevated numbers of activated satellite cells within them.[58] The long term retention of regenerative ability in the larynx after long term denervation, compared to the lack of successful regeneration of denervated, atrophied limb skeletal muscle, is well known.[59] Even reinnervation by the phrenic nerve can restore the intrinsic laryngeal muscles to full function.[60] The constitutive nature of the differences between the myogenic precursor cells of extraocular and limb muscles is further supported by recent a gene profiling study, which demonstrated significant differences.[61] The EOMs expressed 121 unique genes and 53 genes that were differentially regulated. These included a number known to be involved in myogenic determination, such as BMP4 and IGF-binding protein 3. Previous gene profiling studies comparing the EOMs and leg muscles, rather than just myogenic precursor cells, also showed significant up-regulation in the adult EOMs of genes that are specifically involved in myogenesis compared to their down-regulation in limb muscle.[17,62]

5. Enriched Populations of Myogenic Precursor Cells in the EOMs

Flow cytometry is a powerful tool for studying whole populations of mononucleated cells. Similar to immunohistochemistry in its ability to allow visualization of cells expressing particular antigens, flow cytometry is useful for rapid analysis of large numbers of cells which can be sorted

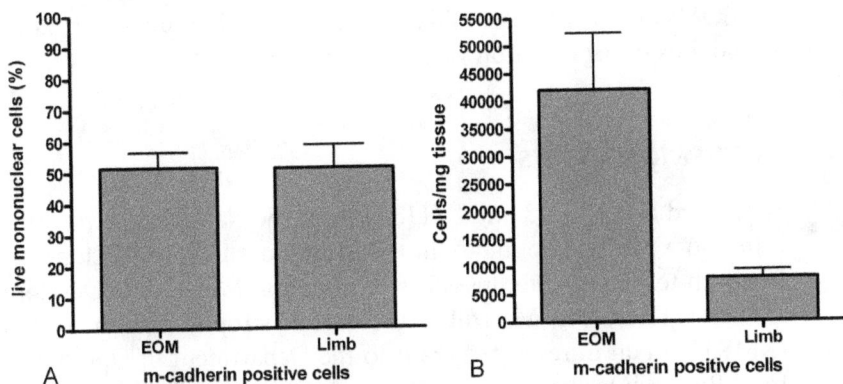

Figure 3. Flow cytometric analysis of m-cadherin-positive live mononuclear cells from mouse extraocular muscle and limb muscle. **(A)** Percent of m-cadherin-positive cells out of all live cells. **(B)** Number of m-cadherin-positive cells per mg wet weight of tissue.

based on antigen expression. Our recent studies have focused on using this approach to define the populations of myogenic precursor cells in adult EOMs and compare these populations with those in limb muscles. Myogenic precursor cell numbers are significantly elevated in the EOMs compared with limb skeletal muscles, with five- to eightfold more satellite cells per mg tissue weight.[63,64] While the total number of mononuclear cells within the EOMs is increased, the percentage of cells expressing the cell markers thus far examined, for example the myogenic markers Pax-7 and M-cadherin, is similar in the EOMs and tibialis anterior (Fig. 3).[40,64] In addition, the EOM myogenic precursor cells seem to be more resistant to apoptosis than cells from the limb.[65] A method that has become standard for separating populations of mononuclear cells from a variety of tissues is the Hoechst dye exclusion assay.[66] This assay has been used specifically with muscle tissue to separate cell populations with different myogenic characteristics.[67] It is also known that Hoechst dye is toxic to cells.[68] We examined whether the Hoechst dye had differential toxicity in mononuclear cells derived from the EOMs and leg muscles, based on the premise that normal adult EOMs have up-regulated levels of activated caspase 3[44] and respond to injury with increased proliferation and hypertrophy.[56,69] Hoechst dye was used to sort myogenic precursor cells from the EOMs and limb muscles. These studies revealed that the dye resulted in significantly more mononuclear cell death in the cells derived from limb muscles when compared to those from the EOMs.[65] Greater quantity and possible resistance to cell death make the myogenic precursor cell population from the EOMs an attractive candidate for myoblast therapy investigations, as autologous transplantation would be possible.

Acknowledgments

Supported by EY15313 and EY11375 from the National Eye Institute, the Minnesota Medical Foundation, the Minnesota Lions and Lionesses, the Research to Prevent Blindness Lew Wasserman Mid-Career Development Award (LKM), the Marzolf Muscular Dystrophy Center Fellowship (KMK), and an unrestricted grant to the Department of Ophthalmology from Research to Prevent Blindness, Inc.

References

1. Kjellgren D, Ryan M, Ohlendieck K, *et al.* (2003) Sarcoplasmic reticulum Ca2+ ATPases (SERCA-1 and -2) in human extraocular muscles. *Invest Ophthalmol Vis Sci* **44:** 5057–5062.
2. Andrade FH, McMullen CA, Rumbaut RE. (2005) Mitochondria are fast Ca2+ sinks in rat extraocular muscles: a novel regulatory influence on contractile function and metabolism. *Invest Ophthalmol Vis Sci* **46:** 4541–4547.
3. Ragusa RJ, Chow CK, St Clair DK, Porter JD. (1996) Extraocular, limb and diaphragm muscle group-specific antioxidant enzyme activity patterns in control and mdx mice. *J Neurol Sci* **139:** 180–186.
4. Karpati G, Carpenter S. (1986) Small-caliber skeletal muscle fibers do not suffer deleterious consequences of dystrophic gene expression. *Am J Med Genet* **25:** 653–658.
5. Kaminski HJ, Al-Hakim M, Leigh RJ, *et al.* (1992) Extraocular muscles are spared in advanced Duchenne dystrophy. *Ann Neurol* **32:** 586–588.
6. Pachter BR, Davidowitz J, Breinin GM. (1973) A light and EM study in serial sections of dystrophic extraocular muscle fibers. *Invest Ophthalmol* **12:** 917–923.
7. Porter JD, Merriam AP, Hack AA, *et al.* (2000) Extraocular muscle is spared despite the absence of an intact sarcoglycan complex in sarcoglycan-deficient mice. *Neuromuscul Disord* **11:** 197–207.
8. McLoon LK, Rios L, Wirtschafter JD. (1999) Complex three-dimensional patterns of myosin isoform expression: differences between and within specific extraocular muscles. *J Muscle Res Cell Motil* **20:** 771–783.
9. Rubinstein NA, Hoh JF. (2000) The distribution of myosin heavy chain isoforms among rat extraocular muscle fiber types. *Invest Ophthalmol Vis Sci* **41:** 3391–3398.

10. Jacoby J, Ko K, Weiss C, Rushbrook JI. (1990) Systematic variation in myosin expression along extraocular muscle fibres of the adult rat. *J Muscle Res Cell Motil* **11**: 25–40.

11. Wieczorek DF, Periasamy M, Butler-Browne GS, *et al.* (1985) Co-expression of multiple myosin heavy chain genes, in addition to a tissue-specific one, in extraocular muscles. *J Cell Biol* **101**: 618–629.

12. Sartore S, Mascarello F, Rowlerson A, *et al.* (1987) Fibre types in extraocular muscles: a new myosin isoform in the fast fibres. *J Muscle Res Cell Motil* **8**: 161–172.

13. Pachter BR, Davidowitz J, Breinin GM. (1976) Light and electron microscopic serial analysis of mouse extraocular muscles: morphology, innervation and topographical organization of component fiber populations. *Tissue Cell* **8**: 547–560.

14. Horton RM, Manfredi AA, Conti-Tronconi BM. (1993) The embryonic gamma subunit of the nicotinic acetylcholine receptor is expressed in adult extraocular muscle. *Neurology* **43**: 983–986.

15. McLoon LK, Wirtschafter JD. (1996) NCAM is expressed in mature extraocular muscles. *Invest Ophthalmol Vis Sci* **37**: 318–327.

16. McLoon LK, Christiansen SP. (2003) Increasing extraocular muscle strength with insulin-like growth factor II. *Invest Ophthalmol Vis Sci* **44**: 3866–3872.

17. Fischer MD, Gorospe JR, Felder E, *et al.* (2002) Expression profiling reveals metabolic and structural components of extraocular muscles. *Physiol Genomics* **9**: 71–84.

18. Anderson BC, Christiansen SP, Grandt S, *et al.* (2006) Increased extraocular muscle strength with direct injection of insulin-like growth factor-I. *Invest Ophthalmol Vis Sci* **47**: 2461–2467.

19. Moss FP, Leblond CP. (1970) Nature of dividing nuclei in skeletal muscles of growing rats. *J Cell Biol* **44**: 459–462.

20. Mauro A. (1961) Satellite cells of skeletal muscle fibers. *J Biophys Biochem Cytol* **9**: 493–495.

21. Allen RE, Sheehan SM, Taylor RG, *et al.* (1995) Hepatocyte growth factor activates quiescent skeletal muscle satellite cells *in vitro*. *J Cellul Physiol* **165**: 307–312.

22. Anderson JE. (2000) A role for nitric oxide in muscle repair: nitric oxide-mediated activation of muscle satellite cells. *Mol Biol Cell* **11**: 1859–1874.

23. Cornelison DD, Wold BJ. (1997) Single-cell analysis of regulatory gene expression in quiescent and activated mouse skeletal muscle satellite cells. *Dev Biol* **191**: 270–283.

24. Grounds MD, Garrett K, Lai MC, *et al.* (1992) Identification of skeletal muscle precursor cells *in vivo* by use of MyoD1 and myogenin probes. *Cell Tissue Res* **267**: 99–104.

25. Beauchamp JR, Heslop L, Yu DS, *et al.* (2000) Expression of CD34 and Myf5 defines the majority of quiescent adult skeletal muscle satellite cells. *J Cell Biol* **151**: 1221–1234.

26. Olguin HC, Olwin BB. (2004) Pax-7 up-regulation inhibits myogenesis and cell cycle progression in satellite cells: a potential mechanism for self-renewal. *Dev Biol* **275**: 375–388.

27. Montarras D, Chelly J, Bober E, *et al.* (1991) Developmental patterns in the expression of Myf5, MyoD, myogenin and Mrf4 during myogenesis. *New Biol* **3**: 592–600.

28. Seale P, Sabourin LA, Girgis-Gabardo A, Rudnicki MA. (2000) Pax7 is required for the specification of myogenic satellite cells. *Cell* **102**: 777–786.

29. Conboy IM, Rando TA. (2002) The regulation of Notch signaling controls satellite cell activation and cell fate determination in postnatal myogenesis. *Dev Cell* **3**: 397–409.

30. Qu Z, Balkir L, van Deutekom J, Robbins PD, *et al.* (1998) Development of approaches to improve cell survival in myoblast transfer therapy. *J Cell Biol* **142**: 1257–1267.

31. Jankowski RJ, Deasy BM, Cao B, *et al.* (2002) The role of CD34 expression and cellular fusion in the regeneration capacity of myogenic progenitor cells. *J Cell Sci* **115**: 4361–4374.

32. Rouger K, Brault M, Daval N, *et al.* (2004) Muscle satellite heterogeneity: *in vitro* and *in vivo* evidences for populations that fuse differently. *Cell Tissue Res* **317**: 319–326.

33. Schultz E. (1996) Satellite cell proliferative compartments in growing skeletal muscles. *Devel Biol* **175**: 84–94.

34. Schultz E, Lipton BH. (1982) Skeletal muscle satellite cells: changes in proliferation potential as a function of age. *Mech Ageing Dev* **20**: 377–383.

35. Baraffino A, Hamman M, Bernheim L, *et al.* (1996) Identification of self-renewing myoblasts in the progeny of single human muscle satellite cells. *Differentiation* **60**: 47–57.

36. Molnar G, Ho ML, Schroedl NA. (1996) Evidence for multiple satellite cell populations and a non-myogenic cell type that is regulated differentially in regenerating and growing skeletal muscle. *Tissue Cell* **28**: 547–556.

37. Yoshida N, Yoshida S, Koishi K, *et al.* (1998) Cell heterogeneity upon myogenic differentiation: down-regulation of MyoD and Myf5 generates "reserve cells." *J Cell Sci* **111**: 769–779.

38. Wozniak AC, Pilipowicz O, Yablonka-Reuveni Z, *et al.* (2003) C-met expression and mechanical activation of satellite cells on cultured muscle fibers. *J Histochem Cytochem* **51**: 1437–2445.

39. Kadi F, Charifi N, Denic C, Lexell J. (2004) Satellite cells and myonuclei in young and elderly women and men. *Muscle Nerve* **29**: 120–127.

40. McLoon LK, Thorstenson KM, Solomon A, Lewis MP. (2007) Myogenic precursor cells in craniofacial muscles. *Oral Sci* **13**: 134–140.

41. McLoon LK, Wirtschafter JD. (2002) Continuous myonuclear addition to single extraocular myofibers of uninjured adult rabbits. *Muscle Nerve* **25**: 348–358.

42. McLoon LK, Wirtschafter J. (2002) Activated satellite cells are present in uninjured extraocular muscles of mature mice. *Trans Am Ophthalmol Soc* **100**: 119–124.

43. McLoon LK, Wirtschafter JD. (2003) Activated satellite cells in extraocular muscles of normal, adult monkeys and humans. *Invest Ophthalmol Vis Sci* **44**: 1927–1932.

44. McLoon LK, Rowe J, Wirtschafter JD, McCormick KM. (2004) Continuous myofiber remodeling in uninjured extraocular myofibers: myonuclear turnover and evidence for apoptosis. *Muscle Nerve* **29**: 707–715.

45. Goding GS, Al-Sharif K, McLoon LK. (2005) Myonuclear addition to uninjured laryngeal myofibers in adult rabbits. *Ann Otol Rhinol Laryngol* **114**: 552–557.

46. Marques MJ, Ferretti R, Vomero VU, *et al.* (2007) Intrinsic laryngeal muscles are spared from myonecrosis in the mdx mouse model of Duchenne muscular dystrophy. *Muscle Nerve* **35**: 349–353.

47. Hoh JF, Hughes S. (1988) Myogenic and neurogenic regulation of myosin gene expression in cat jaw-closing muscles regenerating in fast and slow limb muscle beds. *J Muscle Res Cell Motil* **9**: 59–72.

48. Hoh FJ, Hughes S. (1991) Basal lamina and superfast myosin expression in regenerating cat jaw muscle. *Muscle Nerve* **14**: 398–406.

49. Gauthier GF, Burke RE, Lowey S, Hobbs AW. (1983) Myosin isozymes in normal and cross-reinnervated cat skeletal muscle fibers. *J Cell Biol* **97**: 756–771.

50. Thomas PE, Ranatunga KW. (1993) Factors affecting muscle fiber transformation in cross-reinnervated muscle. *Muscle Nerve* **16**: 193–199.

51. Barjot C, Rouanet P, Vigneron P, *et al.* (1998) Transformation of slow- or fast-twitch rabbit muscles after cross-innervation or low frequency stimulation does not alter the *in vitro* properties of their satellite cells. *J Muscle Res Cell Motil* **19**: 25–32.

52. Kalhovde JM, Jerkovic R, Sefland I, *et al.* (2005) Fast and slow muscle fibres in hindlimb muscles of adult rats regenerate from intrinsically different satellite cells. *J Physiol* **562**: 847–857.

53. Patterson MR, Stephenson GMM, Stephenson DG. (2006) Denervation produces different single fiber phenotypes in fast- and slow-twitch hindlimb muscles of the rat. *Am J Physiol Cell Physiol* **291**: C518–C528.

54. Duchen LW. (1970) Changes in motor innervation and cholinesterase localization induced by botulinum toxin in skeletal muscle of the mouse: differences between fast and slow muscles. *J Neurol Neurosurg Psychiat* **33**: 40–54.

55. Hassan SM, Jennekens FGI, Veldman H. (1995) Botulinum toxin-induced myopathy in the rat. *Brain* **118**: 533–545.

56. Spencer RF, McNeer KW. (1987) Botulinum toxin paralysis of adult monkey extraocular muscle. Structural alterations in orbital, singly innervated muscle fibers. *Arch Ophthalmol* **105**: 1703–1711.

57. Ugalde I, Christiansen SP, McLoon LK. (2005) Botulinum toxin treatment of extraocular muscles in rabbits results in increased myofiber remodeling. *Invest Ophthalmol Vis Sci* **46**: 4114–4120.

58. Shinners MJ, Goding GS, McLoon LK. (2006) Effect of recurrent laryngeal nerve section on the laryngeal muscles of adult rabbits. *Otolaryngol Head Neck Surg* **134**: 413–418.

59. Tucker HM. (1978) Human laryngeal reinnervation: long-term experience with the nerve-muscle pedicle technique. *Laryngoscope* **88**: 598–604.

60. Kingham PJ, Birchall MA, Burt R, *et al.* (2005) Reinnervation of laryngeal muscles: a study of changes in myosin heavy chain expression. *Muscle Nerve* **32**: 761–766.

61. Porter JD, Israel S, Gong B, *et al.* (2006) Distinctive morphological and gene/protein expression signatures during myogenesis in novel cell lines from extraocular and hindlimb muscle. *Physiol Genomics* **24**: 264–275.

62. Porter JD, Khanna S, Kaminski HJ, *et al.* (2001) Extraocular muscle is defined by a fundamentally distinct gene expression profile. *Proc Natl Acad Sci USA* **98**: 12062–12067.

63. McLoon LK, Thorstenson K, Asakura A. (2005) Extraocular muscles contain more satellite cells and more multipotent precursor cells than limb muscle. *Invest Ophthalmol Vis Sci ARVO Abstr* **46**: 4679.

64. Thorstenson K, McLoon LK. (2006) Extraocular muscle myogenic precursor cells appear to be more resistant to cell death than cells of limb muscles. *Invest Ophthalmol Vis Sci ARVO Abstr* **47**: 5399.

65. Thorstenson KM. (2006) Myogenic precursor cells from extraocular muscles are more resistant to cell death and may contain more primitive progenitors than limb muscle. *HHMI Meeting of Medical Fellows Abstr* **23**: 23

66. Goodell MA, Brose K, Paradis G, *et al.* (1996) Isolation and functional properties of murine hematopoietic stem cells that are replicating *in vivo. J Exp Med* **183**: 1797–1806.

67. Asakura A, Seale P, Girgis-Gabardo A, Rudnicki MA. (2002) Myogenic specification of side population cells in skeletal muscle. *J Cell Biol* **159**: 123–134.

68. Mohorko N, Kregar-Velikonja N, Repovs G, *et al.* (2005) An *in vitro* study of Hoechst 33342 redistribution and its effects on cell viability. *Hum Exp Toxicol* **24**: 573–580.

69. Christiansen SP, McLoon LK. (2006) The effect of resection on satellite cell activity in extraocular muscle. *Invest Ophthalmol Vis Sci* **47**: 605–613.

9

Treating the Continuum of Coronary Heart Disease with Progenitor-Cell-based Repair: The University of Minnesota Experience

Doris A. Taylor*, Jonathan D. McCue and Andrey G. Zenovich

1. Introduction

Results of the recent trials in patients with acute myocardial infarction (AMI) have underscored the remarkable advancements in pharmacological and interventional approaches to revascularizing infarct-related arteries and restoring blood flow to the acutely damaged myocardium.[1,2] Specifically, at one year post-AMI, the occurrence of a composite endpoint of death, reinfarction, and New York Heart Association (NYHA) class IV heart failure (HF) has repeatedly been observed below 10%.[1,2] These estimates parallel the observed reduction in mortality related to coronary heart disease (CHD) in the most recent epidemiological observations.[3] However, both clinical trials and public health data show an alarmingly high prevalence of HF-related events. In fact, at least 30% of patients exhibit HF symptomatology at one year post-AMI, and this number grows to over 50% by year 7.[3,4] Unacceptably high rates of HF-related morbidity and mortality, together with the growing number of people over 65 years of age, have created an unmet need for development of therapies to halt and prevent CHD.

Several approaches have been developed in the past two decades in an attempt to combat the consequences of acute myocardial injury and thereby lessen the prevalence of post-AMI HF, and to improve patients' quality of life. Drug therapy with beta-blockers, ACE inhibitors, and other

*Correspondence: Director, Center for Cardiovascular Repair, 312 Church Street SE, 7-105A Nils Hasselmo Hall, Minneapolis, MN 55455, USA. Tel: 612-626-1416; Fax: 612-1121; E-mail: dataylor@umn.edu

agents has exhibited measurable positive effect on survival and HF-related hospitalizations, as well as on quality of life.[5,6] However, a lack of a uniform translation of the benefits observed in trials into wide clinical settings together with inadequate patient compliance have precluded seeing those effects on a population-wide scale.[7] Drug-eluting stents, once thought to be a panacea and a "fix-all" tool for atherosclerotic CHD, have recently drawn attention because of an increased prevalence of in-stent thrombosis in long-term clinical follow-up studies.[8] Clinical trials of angiogenic growth factors, for the most part, have fallen short of expectations.[9] Therefore, it is not at all surprising that once cell therapy hinted at a possibility of reducing left ventricular (LV) remodeling studies with bone marrow mononuclear cells (BMNCs), skeletal myoblasts (SKMBs), and other cell types were initiated in a variety of disease contexts (Fig. 1), and the field virtually exploded.

The excitement and hope of cell therapy is easy to understand. After injury, cells in the heart are lost due to necrosis and apoptosis, and transplantation of new cells can theoretically fix that. Deploying a stent and/or

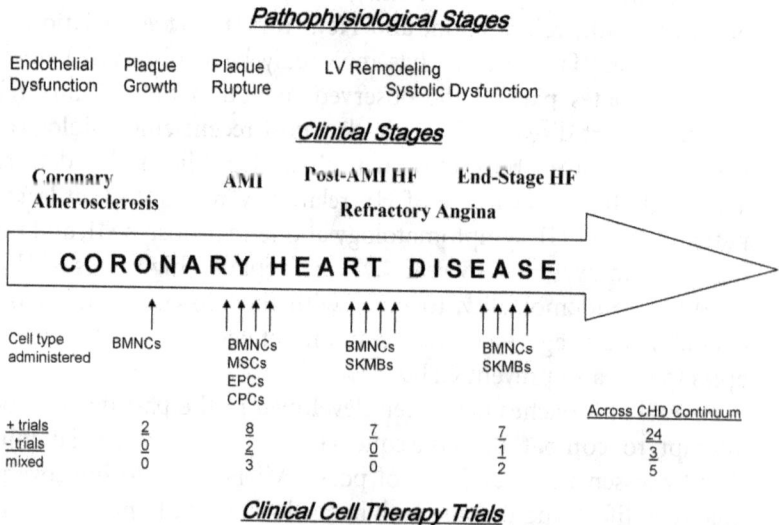

Pathophysiological Stages

Endothelial Dysfunction	Plaque Growth	Plaque Rupture	LV Remodeling Systolic Dysfunction

Clinical Stages

Coronary Atherosclerosis		AMI	Post-AMI HF	End-Stage HF
			Refractory Angina	

CORONARY HEART DISEASE →

Cell type administered	BMNCs	BMNCs MSCs EPCs CPCs	BMNCs SKMBs	BMNCs SKMBs	
					Across CHD Continuum
+ trials	2	8	7	7	24
− trials	0	2	0	1	3
mixed	0	3	0	2	5

Clinical Cell Therapy Trials

Figure 1. Representative diagram of clinical cell therapy trials across the continuum of coronary heart disease to date. Abbreviations: AMI, acute myocardial infarction; BMNCs, bone marrow mononuclear cells; CPCs, cardiac progenitor cells; CHD, coronary heart disease; EPCs, endothelial progenitor cells (defined as AC133+/CD34+); HF, heart failure; LV, left ventricular; MSCs, mesenchymal stem cells; SKMBs, skeletal myoblasts.

administering reperfusion therapy is geared toward lessening the ischemic damage and preserving existing cells. Beta-blockers, ACE inhibitors, and several other drugs target deleterious neurohormonal mechanisms, which lead to progressive LV remodeling and decline of myocardial contractility. Neither of these approaches replace dead cells or induce efficient repair of the damaged regions of the heart.

Full myocardial regeneration with cells is not coming to the office of a cardiologist near you just yet. Although major bench research efforts are ongoing at this time, we are years away from regenerating the heart or building new hearts from cells. However, cell therapy to "repair" the injured myocardium and prevent subsequent functional deterioration is becoming available. Recent data from a multicenter REPAIR-AMI trial showed that BMNCs administered intracoronarily on an average of four days following AMI improved recovery of LV function, prevented LV remodeling, and reduced occurrence of a combined clinical end-point of death, reinfarction, and revascularization at follow-up.[10] It appears that cell-based therapy is capable of shifting a balance between ischemic injury and endogenous repair toward repair and that produces clinically relevant results, at least in the phase II studies. With BMNCs, SKMBs, adipose-derived cells, cardiac progenitor populations and, potentially, umbilical cord-derived progenitors or even multipotent adult stem cells, we now have a selection of various cell types that could allow us to treat the entire continuum of CHD — from initiation of atherosclerosis to end-stage HF.

In this review, we provide an overview of several contributions that our laboratory has made to the field of cell therapy and offer some of our perspectives on how to proceed with second- and third-generation clinical trials, as the field leaves its infancy and enters adulthood.

2. Cell Therapy Can Reverse Left Ventricular Remodeling in Heart Failure

In 1998, we reported the first successful cell-based repair of the heart and restoration of function after transplantation of SKMBs into the injured portion of the rabbit's myocardium where a cryolesion had been induced.[11] Regenerative properties of SKMBs, such as expansion and formation of neofibers after injury, translated into improved regional systolic LV function and diastolic pressure–strain relationship. The success of transplantation depended on engraftment but not on strict differentiation of SKMBs into mature cardiomyocytes. In fact, myogenin-positive mononucleated myocytes connected by intercalated discs were found in the center of the infarcted region, and more primitive myogenin-negative cells were found in the peri-infarct zone.[11,12] The comparison of SKMBs and fibroblasts (FBs) helped resolve whether specific properties of cells would result in systolic or diastolic improvement. It turned out that cellular cardiomyoplasty with FBs improved diastolic performance to a similar degree as SKMBs did, but systolic function declined after transplantation of FBs.[13] Of note, SKMBs transplanted into the center of the infarcted myocardium were electrically isolated from the rest of the tissue, the contractility of the injured segment improved. This result suggested that the SKMB-treated segments responded to contraction by stretching, which is consistent with the observation of presence of myogenin-positive cells. The potential of treating both systolic and diastolic dysfunction with SKMBs, their capacities for myogenesis, and resistance to ischemia made it an attractive adjunct in treatment of HF.

Menasche *et al.* initiated the first observational clinical trial in 2000, transplanting SKMBs in addition to coronary artery bypass grafting (CABG).[14] Since then, several clinical trials have been published, including a phase II trial that was reported at the Scientific Sessions of the American Heart Association in November of 2006. To date, over 200 HF patients have been treated with SKMBs (see Ref. 15 for an extensive review of the trials). Several conclusions from these trials merit a discussion.

First, the improvements in HF symptomatology assessed by NYHA functional class are uniform throughout the trials. Patients with advanced heart failure (mean NYHA functional class of II–III) show a reduction of approximately one functional class, which is clinically significant. These positive data are corroborated by some functional outcomes related to regional myocardial function, myocardial perfusion, and viability — all showing significant increases with SKMB treatment. These improvements vary, however, most likely based on the degree of myocardial injury and a correspondent systolic impairment at the baseline. When examining the data from all clinical trials to date, one observation stands out immediately. Specifically, patients with LV ejection fractions (LVEF), a clinical measure of overall myocardial contractility, below 30% (i.e. largest scar) demonstrated less functional improvement than those patients with LVEF of 35–40%, where improvements were larger and more sustainable. When LVEF reflected a relatively mild myocardial damage (>40%), the benefit of SKMBs was similar to that in patients with severely depressed function, suggesting that there may need to be a sufficient degree of ischemia for exogenous cells to demonstrate their beneficial effects. Similar conclusions have recently been derived from the REPAIR-AMI study, which examined BMNCs given post-AMI.[10] Inversely, too much ischemia may lessen the magnitude of a positive functional outcome. Therefore, the degree of ischemia as well as the amount of scar tissue may determine the long-term success of cell therapy in HF. Of note, the number of cells administered differed in every single trial, confounding interpretation of these trends.

Secondly, SKMBs have been administered not only in patients with varying degrees of ischemia, but also in several different environments, and that could have influenced the outcomes. For instance, in three out of seven trials, the transplantation of cells was performed in addition to CABG. In the rest of the cases, administration was performed without CABG. It is possible that cells could be at a survival advantage when restoration of blood flow is performed at the time of cell transfer. However, it is also possible that alterations of the tissue environment associated with varying degrees of atherosclerosis (e.g. oxidative stress), or type 2 diabetes (e.g. hyperglycemia, "peaks and valleys" of insulin concentration), or other tissue conditions may influence engraftment signaling or even emit apoptotic stimuli that decrease immediate survival of cells. Exploration of these ideas could ideally be performed via a systematic comparison across all trials to date in the context of a registry, the creation of which we have recently proposed.[15,16]

Thirdly, arrhythmia post-SKMB treatment seems to have tamed the enthusiasm of its potential clinical application. Specifically, ventricular tachycardia alone and combined with junctional rhythm were registered in postprocedural monitoring in several patients. Taking into consideration the early data showing lack of electrical integration with the native tissue and lack of connexin-43 overexpression,[17,18] it is easy to focus on the SKMBs themselves as the origin of those arrhythmic events. Despite the appropriateness of these concerns, the data should be interpreted in the context of systolic HF, where arrhythmogenesis is an inherent part of the disease process. In fact, recently published MADIT-II and SCD-HeFT trials performed in a patient population, which would be the target for SKMB transplantation, showed survival benefits of prophylactic implantation of an internal cardioverter–defibrillator.[19,20] Segregation of responsibility for arrhythmias between the cells and the process of HF seems nearly impossible at the present time. However, overexpression of connexin-43, and N-cadherin (and possibly other molecules governing electrical coupling, as well as prophylactic administration of admiodarone[21] along with optimal pharmacological treatment of HF in those who have received SKMBs, may all be salutary to the fate of these cells as a potential therapy in HF patients.

Translation of results from basic science research into the first-generation clinical trials generated the next set of questions that need to be answered before additional clinical trials can take place. This issue is not specific to SKMBs. Development of a new therapy often rests on reiteration between bench and bedside. These cycles often refine the product itself, and/or suggest a target patient population where it could undergo additional testing. At this time, cell therapy with SKMBs is exactly at that stage. The main questions that need to be answered include (but are not limited to) the impact of cell connectivity to safety and efficacy, i.e. whether administration of cells directly into the center of the infarcted myocardium (versus the periphery of the injured zone) is responsible for arrhythmogenesis. Another important issue is the extent of the environmental influences on survival and engraftment of SKMBs. We also need to conduct comparisons of delivery techniques (intramyocardial injections versus intracoronary administration versus intravenous administration) to determine if a particular route of administration can offer superior engraftment and functional benefits without compromising safety. Answers to these questions will define how the next generation of clinical trials is to be conducted to yield maximal efficacy benefits and minimal safety concerns.

Testing whether the location of placement of cells bears any importance for the outcome has two separate important questions: (1) whether placing them close to each other offers any advantages over allowing a larger distance between injections in terms of functional benefits, and (2) whether transplantation directly into the center of the scar or into its periphery might be directly related to the degree of contractile recovery. In addition, whether a central or peri-infarct zone placement of cells results in less arrhythmia (or none at all) is a clinically important issue, since a minimal variance in catheter placement may be responsible for a major difference in patient safety and ultimately the success of the therapy itself when cell-based protocols are being administered clinically.

To address the first question preclinically, Ott *et al.*[22] injected 10^7 SKMBs as 15 microdepots (10 μL each) or 3 macrodepots (50 μL each)

Table 1. Functional Outcomes of Cellular Cardiomyoplasty Using Macrodepot and Microdepot Injection Techniques

Functional Parameter	Control Group	Macrodepot Group	Microdepot Group
LV Ejection Fraction, %	39.1 ± 6.4	53.7 ± 11.9*	70.7 ± 2.0[†,‡]
LV End-Diastolic Diameter	8.0 ± 0.9	7.9 ± 0.5	6.5 ± 0.7

*$p < 0.05$, microdepot group versus control group.
[†]$p < 0.05$, microdepot group versus control group.
[‡]$p < 0.05$ microdepot group versus macrodepot group.
See Ref. 22 for additional data and methodological details.
Abbreviation: LV, left ventricular.

in the center and periphery of rat myocardial tissue 7 days after the left anterior descending artery (LAD) had been ligated. Rats in the control group received injections of culture medium. At 6 weeks post-procedure, significant improvements were seen in both macro- and microdepot groups versus control (Table 1). However, the mean ejection fraction in the microdepot group was significantly higher than in the group that received injections in the macrodepot fashion. On histological examination, neoangiogenesis was almost twice as high in the microdepot group (23.5 ± 6.6 capillaries per high-power field, macrodepot group, versus 42.4 ± 14.5 capillaries per high-power field, microdepot group, $p < 0.02$). Consistent with these observations were histological findings of a larger degree of myotube formation and a lesser amount of tissue necrosis in the microdepot group.

Apparently, injection pressure, when applied in close proximity, benefits transplanted cells in the following two ways. First, a reduced depot diameter together with an increased number of depots allows for better engraftment because blood flow and oxygen can reach larger quantities of cells versus a macrodepot, where the areas of depots are larger, and therefore the potential for apoptosis of transplanted cells would be larger. Secondly, a closer proximity of cells allows for augmented paracrine interactions throughout the engraftment and differentiation processes. Another very distinct possibility is that with smaller volumes of injectates and a higher number of tissue injections, there is stimulation of angiogenesis secondary to needle puncture via an inflammatory response, which is supported by the histological data. On the other hand, injecting large volumes most likely causes increased hypoxia at the center of the depot and tissue edema, neither of which results in favorable conditions for tissue repair.

The answer to the question of whether transplantation of cells directly into the scar periphery or the center has recently been sought out by our group[49] (McCue JD *et al.*, manuscript, in press). Fe-labeled SKMBs (or saline) were injected into the center of the scar or into the periphery of the injured zone in 34 New Zealand rabbits. After producing myocardial injury by ligating the middle of the LAD, magnetic resonance imaging identified wall motion abnormalities and allowed us to map the center of the injured segment in each rabbit. These maps of the injured segment guided the transplantation of cells. Follow-up measurements of LV function were performed at approximately 1 month post-procedure, and arrhythmias were assessed by Holter-monitoring both pre- and post-cell transplantation up to 14 days after the procedure. There was a significantly worse LV remodeling in the centrally-treated group versus the animals where cells had been transplanted into the peri-infarct zone [Fig. 2(a)]. Holter monitoring results were consistent with the LV function data; centrally-treated animals showed significantly higher frequencies of pacing-induced ventricular ectopy, including tachycardia (VT) and fibrillation [Fig. 2(b)]. Occurrences of unstimulated premature ventricular contractions and VT at all time points outside of the 48 hour-long postoperative window were also higher in the centrally-treated animals (data not shown). These data speak to the injection location possibly being a key variable in the outcome of the procedure in terms of the safety and efficacy of exogenous cell transplantation. The famous real estate slogan "Location, location, location" might find its applicability in the field of cell therapy.

In clinical trials thus far, determination of the exact placement of cells preprocedurally has not been uniformly performed. However, targeted placement of cells (SKMBs or other types) could easily occur under a direct surgical approach, or via a NOGA-mapping system, or with a robotically assisted minimally invasive procedure utilizing a DaVinci system (Intuitive Surgical®, Sunnyvale, California). Going forward, a minimally invasive approach to cell transplantation may be an alternative for HF patients with significant comorbidities that make conventional surgical techniques a less attractive alternative due to comorbidities (prior neurological history, concomitant lung disease, inability to lie down due to musculoskeletal disorders, etc.).

After induction of HF by microembolization, Ott *et al.*[23] transplanted a combination of $2.9 \times 10^8 \pm 5.9 \times 10^7$ autologous skeletal myoblasts and

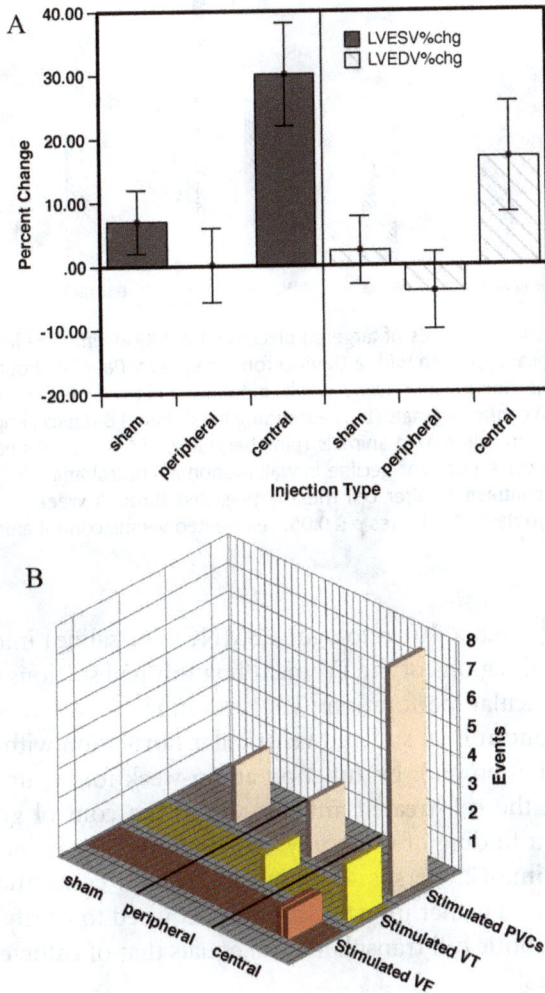

Figure 2. Functional and electrical outcomes of SKMB transplantation into the center of the scar and into the periphery of the scar. **Panel A** reflects functional outcomes of cellular cardiomyoplasty into the two locations. Both LVESV (dark bars) and LVEDV (light descending bars) increased following SKMB placement into the center of the scar reflecting LV remodeling. These effects were absent following cell transplantation into the periphery of the scar, indicating anti-remodeling effects of cell therapy ($p \leq 0.05$, central versus peripheral location). Changes seen in sham-operated animals followed the trends observed in rabbits that received SKMBs into the center of the scar. **Panel B** depicts electrical outcomes (number of observed events) of SKMB placement into the two locations in the animal myocardium. The frequency of PVCs, VT and VF events was higher in the centrally treated animals. Abbreviations: LVEDV, left ventricular end-diastolic volume; LVESV, left ventricular end-systolic volume; PVC, premature ventricular contraction; VF, ventricular fibrillation; VT, ventricular tachycardia.

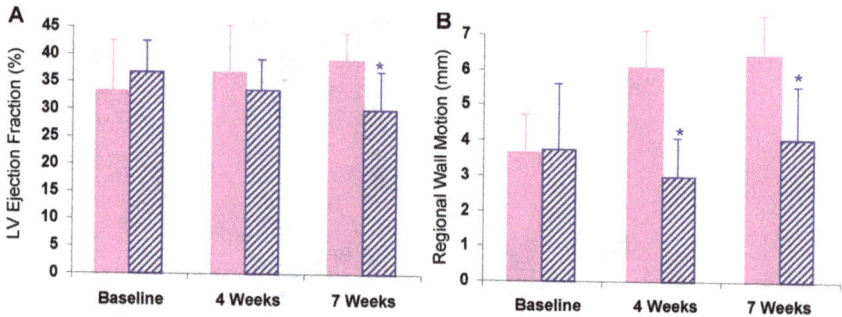

Figure 3. Functional outcomes of targeted placement of SKMBs into the left ventricular wall by a direct surgical approach with a DaVinci robotic system. Panel A shows improvement in the mean LV ejection fraction over 7 weeks of follow-up in cell-treated animals (pink bars) and a decline in control animals (blue ascending bars). Panel B depicts improvement in regional wall motion in cell-treated animals (pink bars) as early as 4 weeks post-sKMB transplantation, and a correspondent decline in wall motion in control animals (blue ascending bars). Improved contractility after cell therapy persisted through week 7. Abbreviation: SKMBs, skeletal myoblasts. *indicates $p \leq 0.05$, cell-treated versus control animals.

$1.1 \times 10^8 \pm 6.8 \times 10^6$ autologous BMNCs (or saline) into apical, anterior, and lateral regions of the LV, including thinned sections of the scar avoiding ventricular perforation. The procedure was successful in 6 out of 7 cases (one animal suffered ventricular fibrillation with unsuccessful termination attempts). LV function at a 6-week follow-up was significantly better in the cell-treated animals versus the control group (Fig. 3). An important finding of this study was almost a 50% reduction in lung ventilation time (23 versus 44 minutes during a conventionally performed procedure). Further investigations are required to clarify whether the efficacy of robotic cell transplantation equals that of catheter-based injection techniques.

Systemic (i.e. intravenous) administration of cells may ultimately offer therapeutic benefits because of simplicity of administration and homing of cells to the injured myocardium. However, systemic delivery may generate systemic problems. Only if homing is specifically targeted to the injured segments of the myocardium could intravenous delivery become suited for therapeutic application. The exact fate of exogenous cells is not completely understood; however, it is clear from preclinical and some clinical data so far that only a small percentage of the total number of injected cells are retained after the placement into the injured myocardial segment. To begin to dissect the fate of administered cells, we have recently conducted biodistribution of CD34$^+$ cells by utilizing lanthanide-labeling technology (Europium^{TM}, BIOPALs, Worcester, Massachusetts, USA). Early preliminary results indicated that a rapid sequestration of cells into the lungs, spleen, and liver occurred after intravenous administration in a rat model of myocardial infarction (LAD ligation). As seen in Fig. 4, cell homing to the heart and sequestration in the spleen increased over 9 days, as the number of cells counted in the liver and in the lungs gradually fell. Exact signals responsible for that type of behavior are being investigated.

Overall, both preclinical and clinical data show that SKMBs have the potential to become an efficacious therapy for patients with HF, as these cells are capable of reverse remodeling of the LV. The recently reported results of the Myoblast Autologous Grafting in Ischemic Cardiomyopathy phase II trial showed reductions in end-diastolic and end-systolic LV volume, with the absolute increases of LVEF ranging from 3.0% to 14.0%.[24] Although the trial was terminated prematurely due to slower-than-expected

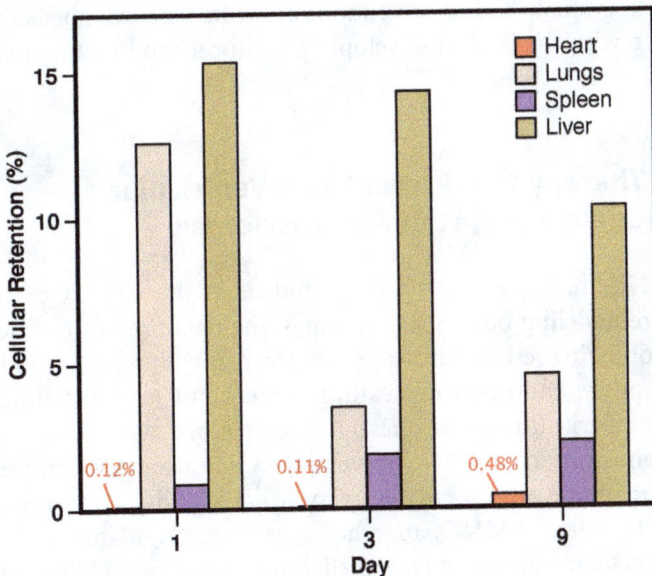

Figure 4. Biodistribution of intravenously administered CD34$^+$ cells in a model of acute myocardial infarction in a rat. Cellular retention is represented for the heart, lungs, liver, and spleen. "Day" indicates day administration of CD34$^+$ cells after LAD ligation and creation of the infarcted zone. Abbreviation: LAD, left anterior descending (artery).

recruitment, Kaplan–Meier analysis of survival free of major adverse cardiac events (MACEs) showed no differences between high dose (800×10^6 SKMBs) versus placebo, and low dose (400×10^6 cells) versus placebo at 30 days and 6 months following the procedure ($p = 0.12$ and 0.87 for the high-dose group and $p = 0.43$ and 0.09 for the low-dose group, respectively). MACE curves separated early in the course of follow-up (at 1 month), although the trial was not powered to detect these differences statistically. Nevertheless, patients in the high-dose SKMB group appeared to have survived equivalently compared to placebo, indicating no increase in MACEs attributable to SKMB transplantation. Time-to-first ventricular arrhythmia was not different among the three groups at both 30 days and 6 months ($p = 0.30$ and 0.12 for the high-dose group, and $p = 0.20$ and 0.23 for the low-dose group). This trial showed that SKMB administration had some effects on the LV remodeling process in patients with LVEF between 15% and 35% and a history of AMI and myocardial akinesia in at least 3 contiguous segments unresponsive to administration of dobutamine. Most importantly, this outcome was achieved without compromising the safety of patients. Therefore, at the very least, SKMBs deserve a more in-depth evaluation, and reiterations between bench and bedside should continue with the goal of developing a clinical product for treatment of HF patients.

3. Cell Therapy Can Prevent Left Ventricular Remodeling and Halt Atherosclerosis

BMNCs have been extensively studied in the context of prevention of LV remodeling post-AMI. Although the data are heterogeneous (Fig. 1), encouraging results have appeared recently with regard to reduction in the composite end-point of death, recurrence of myocardial infarction, and revascularizations in a phase II trial.[10] Along with reduction of angina, administration of BMNCs several days following AMI reduced the size of the infarcted area, increased myocardial viability, and improved regional contractility.[16] BMNCs are a heterogeneous population of hematopoietic precursors, containing endothelial progenitor cells (EPCs) and their subsets (including CD34[+] progenitors), mesenchymal stem cells (MSCs), multiple other progenitors and precursor populations, including monocyte precusors, T and B cell precusors, CD14 cells, etc.[25,26] Administration of unfractionated BMNCs can potentially tip the balance between injury and repair toward repair because multiple reparative pathways are initiated. This beneficial action has been initially seen in patients post-AMI (Fig. 1), but can potentially extend into coronary artery disease to prevent disease-related events, and possibly into HF as well. Studies that compare capabilities of BMNCs in varying stages of disease should help judge functional benefits across the continuum of CHD.

Our recent observation of inhibition of plaque formation in ApoE$^{-/-}$ mice fed a high-fat diet and given injections of BMNCs[27] has fueled further exploration of the hypothesis that administering these cells can induce favorable vascular remodeling and halt progression of atherosclerotic lesions. Atherosclerosis is now viewed as a disease where endothelial inflammation plays a pivotal role in progression of plaque lesions.[28] Once regarded simply as a static barrier between tissue and blood, vascular endothelium is now known to play a key regulatory and integrating role in the initiation and progression of vascular dysfunction.[29] For example, after Ross' seminal work introducing the "response to injury" hypothesis in atherosclerosis, it became apparent that endothelial cells directly regulate vascular function, transport of solutes, and antithrombotic properties of the blood–tissue interface.[30–32] The revised "response to injury" hypothesis focuses on endothelial dysfunction as a trigger for the inflammatory response in atherosclerosis.[32,33] Endothelium-dependent relaxation is impaired early in atherogenesis, often by circulating cytokines mediating vascular inflammation and adhesion of monocytes to the endothelial surface.[32,34] Furthermore, clinical measures of endothelial dysfunction are associated with ischemia in the absence of flow-limiting lesions and predict CHD events, including stroke.[32,35,36] We now understand that vasodilatory, antiplatelet, and antithrombotic processes are primarily regulated at the endothelial level; the loss of normal endothelial function may be the most important driver of the balance in favor of inflammation and thrombosis.[28,29,32]

Therefore, reduction of inflammation combined with reendothelialization seems to be a promising avenue for interfering with the core of the disease pathophysiology. However, in clinical practice, the time of onset and the rate of progression of CHD differ between men and women, as do changes in plaque that lead to AMI (a more frequent rupture in men versus more erosion in women).[37] Our previous data together with these clinical observations empowered us to begin to evaluate the impacts of gender, age, and the extent of disease on BMNC-mediated repair. It is highly likely that the core elements involved in CHD initiation and progression in men and women are similar; however, the mechanisms of endogenous repair are likely to be different.

In our most recent experiments,[38] we administered sex-matched and mismatched BMNCs from wild type mice to apolipoprotein E knockout (ApoE$^{-/-}$) mice (fed on a high-fat diet) every two weeks for four weeks, beginning at the age of 14 weeks. Measurements of atherosclerotic plaque at 1 week following the final injection revealed a significant inhibition of plaque formation in male mice that received female BMNCs [Fig. 5(a)], while no other group benefited. Analysis of progenitor cell profiles in the animals' bone marrow showed sex-based differences in the number of immature EPCs (defined as AC133$^+$/CD34$^+$ cells) in animals that received vehicle injections [Fig. 5(b)]. To verify this finding, we cultured BMNCs from both male and female C57 mice and obtained a similar result. Therefore, a higher plaque burden in males may be related to a lower number of circulating EPCs when atherosclerotic plaque is initiated, which could make endogenous repair process slower and less efficient. However, male animals that received female BMNCs with higher EPC counts showed inhibition of plaque highly likely because of utilization of exogenous cells for repair. The percentage of CXCR4$^+$ cells in male ApoE$^{-/-}$ mice that received female marrow was not significantly different from the male mice which were administered male cells (Table 2), suggesting that either (1) mediation of reduced plaque formation rests solely on exogenous cells, or (2) the involvement of endogenous repair mechanisms is not solely mediated by CXCR4$^+$ cells, but rather involves interactions of those cells with cytokines expressed by plaque, diseased endothelium, and exogenous cells.

Interestingly, levels of total cholesterol were similar in males that received female cells (and showed reduction in plaque) versus vehicle-treated ones (data not shown), affirming that inhibition of plaque formation may depend on efficiency or repair induced by exogenous BMNCs. We have evaluated cytokine profiles using the Luminex® platform, which allowed us to simultaneously evaluate 22 cytokines/chemokines in treated and untreated animals. In general, BMNC treatment created an inflammatory response in ApoE$^{-/-}$ mice; however, the type and the magnitude of the effect varied by sex. Specifically, administration of either male or female BMNCs increased Th1-type proinflammatory cytokines/chemokines (IL-1β, IL-6, IL-12, TNF-α, RANTES, and MCP-1). Increases of these cytokines/chemokines were numerically larger in male mice treated

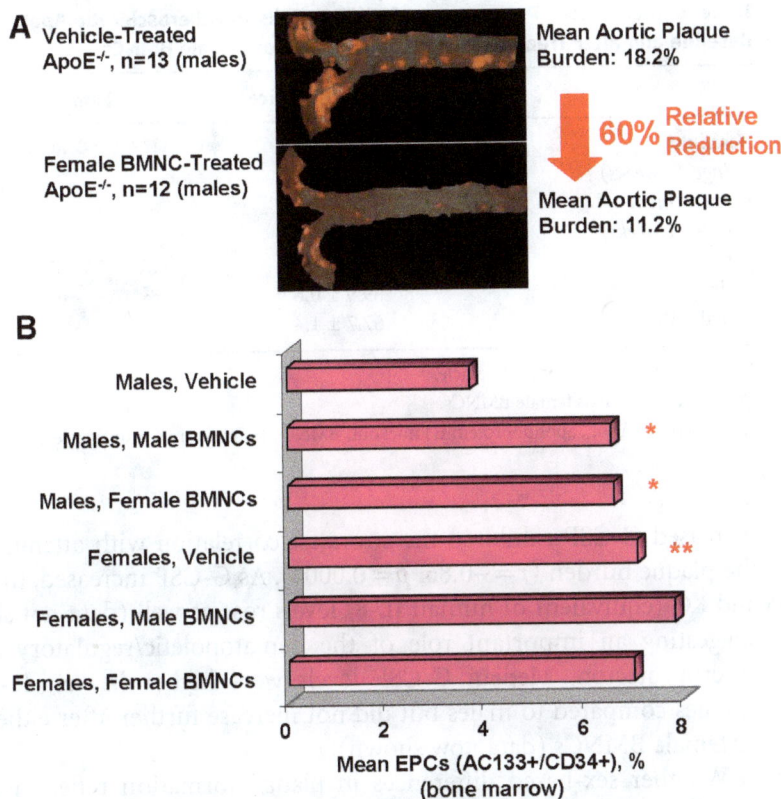

Figure 5. Atheroprotection in atherosclerotic (high-fat diet) ApoE$^{-/-}$ mice after treatment with female but not male BMNCs. Panel A shows a representative *en face* aorta preparation of a male ApoE$^{-/-}$ mouse treated with a vehicle and of a male ApoE$^{-/-}$ mouse that received treatment with female BMNCs. Panel B shows mean percentages of EPCs in treated and untreated male and female ApoE$^{-/-}$ mice. Abbreviations: ApoE$^{-/-}$, apolipoprotein E knockout; BMNCs, bone marrow mononuclear cells; EPCs, endothelial progenitor cells. *indicates $p \leq 0.05$, cell-treated versus vehicle; **indicates $p \leq 0.05$, vehicle-treated males versus females.

with female BMNCs — the group of animals that showed attenuation of plaque formation. Along with the upregulation of the Th1 cytokines, Th2 (anti-inflammatory) cytokines (IL-5, IL-10, and IL-13) levels rose as well; again, larger numerical increases in Th2 cytokines occurred in male mice with a reduced plaque burden. In female mice, chemokines, fewer Th1- and more Th2-type cytokines were upregulated after treatment with either male or female BMNCs. Out of the 22 cytokines/chemokines, only

Table 2. Mean (±SE) Percentages of CXCR4⁺ Cells in Atherosclerotic ApoE⁻/⁻ Mice at Baseline and after Treatment with Vehicle, Male or Female BMNCs

	Male ApoE$^{-/-}$ Mice	Female ApoE$^{-/-}$ Mice
Baseline *(age 14 weeks)*	64.2 ± 1.0	69.3 ± 3.3
Posttreatment *(age 21 weeks)*		
Vehicle	66.7 ± 1.6	67.8 ± 1.3
Male BMNCs	69.9 ± 0.6	61.4 ± 1.3*
Female BMNCs	67.7 ± 1.4	68.3 ± 1.6†

*$p < 0.05$, male BMNCs versus vehicle.
†$p < 0.05$, male versus female BMNCs.
Abbreviations: ApoE, apolipoprotein E knockout; BMNCs, bone marrow mononuclear cells.

increased G-CSF exhibited the strongest correlation with attenuation of the plaque burden ($r = -0.86$, $p = 0.0004$). As G-CSF increased, the IL-15 and KC (equivalent of human IL-8) levels rose as well (data not shown), suggesting an important role of the hematopoietic/regulatory axis in atheroprotecton. Median G-CSF levels were higher in vehicle-treated females compared to males but did not increase further after either male or female BMNCs (data now shown).

Whether sex-based differences in plaque formation reflect a greater reparative capacity of female progenitors or just represent differences in the onset and the rate of progression of atherosclerosis requires further research and confirmation. Investigation into the sex-based differences can not only have an impact on our understanding of the peculiarities in endothelial repair, but most importantly on the ways of screening, diagnosing, and treating CAD in men and women.

Overall, cell therapy with BMNCs appears to have substantial effects on the pathophysiology of atherosclerosis leading to a reduction in plaque formation, at least in our preclinical data. These observations, together with phase II clinical data showing reduction of ischemic events post-BMNC administration immediately post-AMI, show that exogenous BMNCs are capable of shifting the balance between tissue injury and repair toward repair. Although much remains to be learned about the mechanistic side of cell-mediated repair in various states of disease, it is clear that therapy with BMNCs will soon enter definitive phase III trials, and will hopefully become a clinical reality in the near future.

4. Uncommitted Cardiac Precursors — Simply a Better Cell or a New Way Forward?

Contrary to the centuries-old postulates of developmental biology, self-renewal of the organs of the human body is not limited to the blood, the intestines, and the skin. Identification of hematopoietic stem cells and of resident stem cells with a broad capacity to differentiate in injured tissues fueled research on the mechanisms of tissue and organ repair and regeneration — a field which a decade ago was virtually unheard of.[39] Unfortunately, these cells do not protect the heart from AMI or subsequent scar formation and remodeling, or activate the endogenous remodeling process following significant tissue damage.[40] However, specific populations of cells (e.g. lin⁻c-kit⁺) were shown to differentiate into cardiomyocytes, endothelial cells, and smooth muscle cells and take an active part in mediation of repair.[41]

We have recently identified uncommitted cardiac precursor cells (UPCs) in both a neonatal and an adult rat heart by expression of the stage-specific embryonic antigen-1 (SSEA-1) marker.[42] We observed maturation of the original extracted cells into a multipotent, mesodermally committed state (flk-1⁺/CD31⁻) and then into cardiacally committed cells expressing GATA-4, nkx2.5, and isl-1 (Fig. 6). The number of these cells in cultures increased over time (nkx2.5⁺: from 5% at week 2 to 16% at week 5; GATA-4⁺: from 4% at week 2 to 12% at week 5; and isl-1⁺: from 4% at week 2 to 11% at week 5). These cells terminally differentiated further into mature cardiomyocytes, endothelial cells, and smooth muscle cells. We injected SSEA⁺ cells into a failing rat heart to elucidate the capability of contributing to cardiac repair. The animals that received UPCs showed substantial functional and anti-ischemic benefits: the size of infarcted tissue was smaller in the UPC-treated group, LV fractional shortening improved and the LV end-diastolic diameter was smaller in the SSEA+ group, and pressure–volume loops in the UPC-treated animals exhibited improved diastolic function (data not shown). A much lower dose of cells (1×10^6) was required, compared to current doses of SKMBs and BMNCs, to achieve these benefits. Most importantly, these cells did not result in teratoma formation or otherwise inappropriate differentiation.

All these qualities make UPCs the next candidate to be evaluated for cardiac repair. Ultimately, comparisons of SKMBs, BMNCs, and UPCs will tell which cell is the best for repair of the injured myocardium. However, neither SKMBs nor BMNCs offered a similar degree of functional benefits

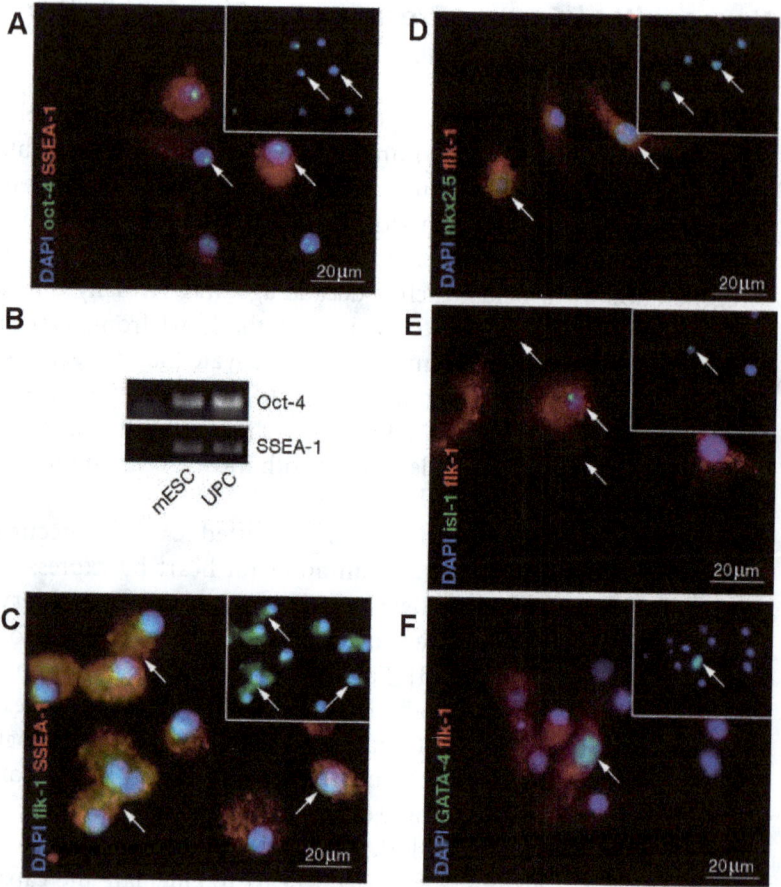

Figure 6. Characterization of adult heart-derived UPCs isolated from primary cultures at day 14. Panel A shows immunostaining of SSEA-1+ (red) and oct-4 (green). Panel B shows a representative comparison of oct-4 and SSEA-1 expression in UPCs and in mESCs by real-time quantitative polymerase chain reaction at 3 weeks. Panels C–F show cardiac transcription factor gene expression in flk+ UPCs. Upon differentiation in cardiac *in vitro* conditions, there is an avid shift toward flk-1 expression concomitant with SSEA-1+ (Panel C), nikx2.5 (Panel D), isl-1 (Panel E), and GATA-4 (Panel F). These images appeared in our original publication in *Nature Clinical Practice — Cardiovascular Medicine* (Ott HC *et al.*, Ref. 41). Abbreviations: mESCs, mouse embryonic stem cells; SSEA-1, stage-specific embryonic antigen 1; UPCs, uncommitted cardiac precursor cells.

coupled with the reduction of infarct size. Administering uncommitted progenitors may result in faster, more efficient, and more sustained repair, and receive more enthusiasm from our clinical colleagues when these cells enter into patient studies.

5. Cell-derived Autologus Organs — Is that What the Future Holds?

Discoveries of undifferentiated cells and recognition of postmitotic capacity within the organs of the body that were once thought to be terminally differentiated fueled a great deal of research efforts geared toward engineering a living tissue using progenitor cells. Recent advances in regulation of the differentiation of uncommitted (or multipotent) progenitors, cell senescence, and repair mechanisms have increased the abilities of bioengineering, moving the field closer to witnessing the arrival of cell-based artificial organs.

However, there are several obstacles in tissue engineering that need to be addressed before leaps of progress can be made. First, before an artificial organ can be constructed, there needs to be a reliable three-dimensional scaffold that would allow seeded cells to attach to the matrix and initiate tissue growth and integration of other tissues into a living system. Both biological and synthetic scaffolds have been proposed.[43,44] For successful tissue engineering, scaffolds should be biocompatible and be able to promote neovascularization. Porous poly(L-lactide-co-glycolide) and collagen-chitosan-hydroxyapatite hydrogel scaffolds have recently been shown to be associated with an increase in leukocyte recruitment, elevated microvascular permeability, and macrophage infiltration.[45] Hydrogel scaffolds did not fare well at all; severe inflammation, as judged by an approximately 15-fold increase in leukocyte adhesion and a pronounced sustained elevation of microvascular permeability, has been observed.[45] It is clear that these types of responses would interfere with a successful outcome.

We have recently taken a distinctly different approach by developing a combination of detergent-based treatments that achieves a very high degree of decellularization of an extracellular matrix (ECM) so that only 5% of quantifiable DNA remains but glycosaminoglycans (GAGs) are left intact.[50] Of note is that the main limitation in decellularization of ECMs has been the loss of GAGs and consequent disruption of the integrity of collagen.[46] Decellularized three-dimensional structure of a native organ could provide a more biocompatible scaffold and a greater degree of successful interactions with exogenously applied cells with regard to interactions with ECMs and development of appropriate phenotypes.

Secondly, with the biocompatibility of scaffold resolved, it is not yet clear which cells would be the best candidate for this delicate job. Should it be embryonic cells? Bone marrow-derived progenitors? Should these be autologous or allogeneic cells? Ethical hurdles notwithstanding, the use of embryonic cells requires development of immune acceptance strategy and minimization of undesirable differentiation. Bone marrow is a rich repository of many kinds of cells, but none of them are created equal. As the quest for the right cells continues, we are beginning to better understand the interactions between the environment and the developing engineered tissue. The use of a bioreactor helps provide controlled conditions and ensure that chemical, mechanical, and other signals are maximized.

As we further embark on the creation of autologous artificial organs, we need to not only underpromise and overdeliver based on rigorous science but also realize that regenerative medicine is becoming global, with research advances all over the world. Appropriate regulation and legislation of this field should be adopted at this time to preserve the attractiveness of this field to scientists, technology partners, and entrepreneurs, as this type of industry has increasingly been developing outside of the United States.

6. Three Important Steps that can Move the Cell Therapy Field Forward

Today, the field of cell therapy is in the iterative stage between bench and bedside. The success of the next generation of clinical trials, as well as of the ultimate "clinical product," relies heavily on these re-iterations, as issues that may hinder or even preclude clinical utilization are being addressed. In the last nine years, the field has started coming out of infancy and maturing into adulthood. Where and, most importantly, *how* do we bring new science forward, advance knowledge, and deliver on the promises? In our opinion, several steps can help ensure future successes.

6.1. Step 1: Creating a registry and biorepository will let us examine existing data and guide the next generation of clinical trials

Decades of CVD research have highlighted the importance of centralized databases in advancing our understanding of a disease process. Cardiovascular medicine would not have advanced as far as it has in the last 25–30 years if the Framingham Study was not initiated and executed in a centralized manner. Large databases provide the power to examine the data retrospectively, while being able to control for numerous covariates — a step not possible to accomplish in a review or even a meta-analysis. Centralized databases also act as testing grounds for new hypotheses, often before clinical trials commence. We believe that a centralized registry of cell therapy trials could not only advance the field but also generate the next set of questions to ask, which, in turn, will greatly advance the science and will move the field closer to creating a cell-based "clinical product." The field of cell therapy has arrived to the point where the next advancement should be creating and employing a large database of all results of clinical trials to serve as a filter for the hypotheses.

The second aspect of the centralized registry is pairing it with a biorepository for blood samples. An initiative like this requires committed funding and resources, centralizing sample collection, storage, flow cytometry, and assays, but could greatly help advance the science. The goal would be to characterize changes within progenitor cell populations in conjunction with clinical data in varying states of disease when different types of cells are given. In that regard, the ability to go back to the samples when new markers, receptors, and pathways emerge will make this effort cost-efficient. Combining the registry for the clinical trials data and the biorepository for the blood and tissue samples seems to be exactly what the field needs to make another decade of major progress and help shape future cell therapy products.

6.2. Step 2: Comparison of different cell types will define the "best cell"

At present, discrepant clinical trial outcomes exist for the same cells in different patients, and for different cells in patients with similar stages of myocardial damage. Only a few side-by-side comparisons of different cell populations have been performed. We clearly lack direct comparisons of different cell types in clearly defined, clinically relevant models of disease.

The major unresolved issues with regard to defining the "best cell" for cardiac repair and regeneration are: (1) whether the mechanism of repair differs when different types of cells are involved in a similar stage of disease; (2) which type of cells delivers the highest functional benefit relative to the context of disease; and (3) whether the delivery technique alters the capability of the cells to home to the site of injury. It is clear that all of these issues are interrelated and are equally important for the advancement of the field. Knowing the critical mechanistic components of how the SKMB- or BMNC- (or other types, such as UPCs) driven repair should enable us to target these and other types of cells to the right pathophysiological contexts, and to achieve efficacy with cells compatible with or better than that of pharmacological therapies, especially when measured by long-term follow-up studies. What is clear from a basic science's point of view is that different environments in the myocardium at the time of injury generate different milieus, and therefore the cells that engraft in one environment may not survive in another. Comparing different cell types in various pathophysiological contexts will help us define how to improve the survival of transplanted cells, which is currently one of the largest hurdles of cell therapy. Approximately 80–90% of all transplanted cells die within the first few days of transplantation into the infarct scar but a subset of the transplanted cells survives and multiplies. Therefore, the question is how to best promote survival — either by genetic expression or by pharmacological means, or by modification of the media. More work is needed in the future to better define the relationship between the microenvironment of the infarct scar and the adjacent myocardial segments and the outcome of the transplanted cells.

It is clear that choosing the best delivery route is a prerequisite for success, right after choosing the right cell for the right environment. A major obstacle to achieving efficacy is the rather poor engraftment seen when cells are administered by intracoronary, intravenous, and intracardiac routes. This limitation is likely to have emerged due to multiple factors, out of which technical difficulties of injecting exactly into the center or the periphery of the scar or precise catheter manipulations in the coronary tree cannot be overemphasized. Therefore, training of the operators becomes pivotal. Alternatively, it may also make sense to restrict the number of centers per region that act as referral centers to utilize cell therapy, at least until the techniques come to solid maturity. We have learned that operator volume and experience were critical determinants of success in CABG and PCI clinical trials and also in routine clinical practice. As the field of cell therapy goes forward, we cannot ignore the importance of appropriately trained specialists. In addition, improved understanding of the biology involved in the delivery route – engraftment interaction could translate into the development of optimal situation-specific delivery systems. These insights can only be obtained when clinical and basic science work together, so that the process can balance between clinically relevant questions and scientifically important observations.

6.3. Step 3: The next generation of clinical trials needs to be based on the consensus regarding protocol design and outcome measurements

At the present time, clinical trials in cell therapy suffer from several major shortcomings primarily involving design and selection of end-points. For example, most studies have been accompanied by an additional revascularization procedure, either by percutaneous coronary intervention or by CABG, making any functional improvement due to exogenous cells nearly impossible to distinguish from the standard of care. In addition, patient characteristics at study entry need to be matched more carefully in prospective trials, which would include baseline comparisons beyond the standard regimen of demographic and basic clinical disease-defining parameters, such as assessments of biomarkers, cytokine profiles, and levels of circulating progenitors, to characterize the milieu and relate the impact of exogenous cells to the outcome appropriately.

What we select to be an "end-point" in cell therapy likely matters a great deal. So far, clinical trials have been geared toward measuring functional improvement of the LV by assessing global LVEF. As we know from the HF trials, improvement of regional contractility may not always translate into better HF numbers because of differences in loading conditions. Since the data with both SKMBs and BMNCs so far suggest that exogenous cells are capable of anti-remodeling effects, measuring those as an end-point in prospective trials will require using a technique with a high sensitivity and specificity in measurements of regional contractile parameters. However, we have begun to appreciate observer dependence of such measurements. Even though cardiac magnetic resonance imaging (CMR) offers the best topographic assessment of the heart, the variability is best minimized by conducting clinical trials at centralized core laboratories where the personnel undergo regular inter- and intraobserver reproducibility assessments. More attention needs to be paid to the peri-infarct zone and the scar volume, and myocardial perfusion quantification. Measuring changes in blood flow was proposed as an end-point for angiogenesis studies,[47] and it is now becoming apparent that cell therapy will need a sensitive measure of blood flow as well. CMR has been used to detect the presence of exogenous cells in the myocardium, as well as to characterize the myocardium prior to transplantation of cells to delineate the areas of myocardial damage.[48]

Finally, medication regimens need to be tracked more carefully throughout the course of the trials, as imbalances in loading conditions and signaling may negatively impact the engraftment of the cells and the overall outcome of the study. We also need to account for the stimulatory effects of drugs, such as statins, PPAR-agonists, erythropoietin, estrogen, angiotensin receptor blockers, and possibly others, in various disease states. Right now, it is completely unclear which, if any, of these combinations of drugs alter the number and the function of progenitor cells available for repair. Clearly, if cell therapy is to be adequately evaluated as a therapy, we will need this type of information for the design of definitive Phase III trials, and also for going forward with clinical applications. In addition, we lack data that evaluate time in disease progression relative to cell transplantation as an additional factor in treatment. Overall, there is a lack of standardization in the current preclinical approach to cell therapy; for example, cell types, doses, preclinical models, and end-points all differ. Attempts to standardize these parameters and to decide on a consensus will move us forward.

Summary

Cell transplantation opened a promising frontier in CVD. The concept of repairing and regenerating damaged myocardium is a real possibility. While many questions still remain, cell therapy has a good chance to eventually become a clinical success. Cell-based therapies have the potential to provide physicians with alternatives for a large patient population that extend beyond revascularization and medical management.

To further advance cell therapy for CVD, we now have to come to a field-wide consensus and standardize future studies. The diversity of cell types, application techniques, and disease stages can be a hurdle and an opportunity — only collaboration will allow us to move forward as a field. Recent clinical trials have shown that cell therapy with SKMBs and BMNCs is able to demonstrate clinical benefits in AMI and HF. Promising results invoked the scientific enthusiasm to warrant large-scale controlled clinical trials to determine the best and safest application of this technology, and to gain a better understanding of its mechanism(s).

As the field progresses, we have a responsibility to promise patients (and the press) only what we can deliver — that is, to tell the truth about cardiac repair. Even though cell-based repair holds out a great promise to modify the pathophysiological process in specific ways, it is crucial for clinicians, patients, and the press to understand that specificity precludes a panacea. As we go forward, some applications of cell therapy will succeed, but some will fail. Disease contexts and not types of cells themselves may come to be the primary determinants of efficacy. We have already experienced a similar process with angiogenic growth factors in CVD, and we now realize that those trials should have more carefully targeted the disease process. As investigators, we need to be realistic in the expectations we place on cell therapy, and ultimately we need to underpromise and overdeliver, based on good, rigorous science. Otherwise, the great potential will be destroyed. Cell therapy is a new and very promising alternative that warrants much more exploration, inspiration, and investment of our time and resources.

Acknowledgments

This work has been supported in part by the NHLBI/NIH award to Dr Doris A. Taylor (R01-HL-063346), the Minnesota Partnership for Biotechnology and Medical Genomics award, the Medtronic Foundation, and by funding from the Center for Cardiovascular Repair, University of Minnesota.

The authors sincerely thank the entire team of the Center for Cardiovascular Repair, as well as Harald C. Ott, MD, and Wendy D. Nelson, PhD, for their continuous, outstanding efforts and their dedication to the advancement of science. The authors also wish to express their appreciation to the Cytokine Reference Laboratory of the University of Minnesota (Director: Angela Panoskaltsis-Mortari, PhD) for ongoing synergistic collaboration.

References

1. Hochman JS, Lamas GA, Buller CE, *et al.* (Occluded Artery Trial Investigators). (2006) Coronary intervention for persistent occlusion after myocardial infarction. *N Engl J Med* 355: 2395–2407.
2. Armstrong PW, Granger CB, Adams PX, *et al.* (APEX AMI Investigators). (2007) Pexelizumab for acute ST-elevation myocardial infarction in patients undergoing primary percutaneous coronary intervention: a randomized controlled trial. *JAMA* 297: 43–51.

3. Thom T, Haase N, Rosamond W, *et al.* (American Heart Association Statistics Committee and Stroke Statistics Subcommittee). (2006) Heart disease and stroke statistics — 2006 update: a report from the American Heart Association Statistics Committee and Stroke Statistics Subcommitte. *Circulation:* 113: e696.

4. Miller LW, Missov ED. (2001) Epidemiology of heart failure. *Cardiol Clin* 1: 547–555.

5. Jong P, Yusuf S, Rousseau MF, *et al.* (2003) Effect of enalapril on 12-year survival and life expectancy in patients with left ventricular systolic dysfunction: a follow-up study. *Lancet* 361: 1843–1848.

6. Jost A, Rauch B, Hochadel M, *et al.* (HELUMA study group). (2005) Beta-blocker treatment of chronic systolic heart failure improves prognosis even in patients meeting one or more exclusion criteria for the MERIT-HF study. *Eur Heart J* 26: 2689–2697.

7. Lenzen MJ, Boersma E, Reimer WJ, *et al.* (2005) Under-utilization of evidence-based drug treatment in patients with heart failure is only partially explained by dissimilarity to patients enrolled in landmark trials: a report from the Euro Heart Survey on Heart Failure. *Eur Heart J* 26: 2706–2713.

8. Pfisterer M, Brunner-La Rocca HP, Buser PT, *et al.* (BASKET-LATE Investigators). (2006) Late clinical events after clopidogrel discontinuation may limit the benefit of drug-eluting stents: an observational study of drug-eluting versus bare-metal stents. *J Am Coll Cardiol* 48: 2584–2591.

9. Simons M. (2005) Angiogenesis: where do we stand now? *Circulation* 111: 1556–1566.

10. Schachinger V, Erbs S, Elsasser A, *et al.* (REPAIR-AMI Investigators). (2006) Intracoronary bone marrow-derived progenitor cells in acute myocardial infarction. *N Engl J Med* 355: 1210–1221.

11. Taylor DA, Atkins BZ, Hungspreugs P, *et al.* (1998) Regenerating functional myocardium: improved performance after skeletal myoblast transplantation. *Nat Med* 4: 929–933.

12. Atkins BZ, Lewis CW, Kraus WE, *et al.* (1999) Intracardiac transplantation of skeletal myoblasts yields two populations of striated cells *in situ. Ann Thor Surg* 67: 124–129.

13. Hutcheson KA, Atkins BZ, Hueman MT, *et al.* (2000) Comparison of benefits on myocardial performance of cellular cardiomyoplasty with skeletal myoblasts and fibroblasts. *Cell Transplant* 9: 359–368.

14. Menasché P, Hagège AA, Vilquin JT, *et al.* (2003) Autologous skeletal myoblast transplantation for severe postinfarction left ventricular dysfunction. *J Am Coll Cardiol* 41: 1078–1083.

15. Taylor DA, Zenovich AG. (2007) Cell therapy for left ventricular remodeling. *Curr Heart Fail Rep* 4: 3–10.

16. Zenovich AG, Davis BH, Taylor DA. (2007) Comparison of intracardiac cell transplantation: autologous skeletal myoblasts versus bone marrow cells. In: Kauser K, Zeiher A (eds.), *Handbook of Experimental Pharmacology*. Springer-Verlag, pp. 117–165.

17. Scorsin M, Hagège A, Vilquin JT, *et al.* (2000) Comparison of the effects of fetal cardiomyocyte and skeletal myoblast transplantation on postinfarction left ventricular function. *J Thorac Cardiovasc Surg* 119: 1169–1175.

18. Suzuki K, Brand NL, Allen S, *et al.* (2001) Overexpression of connexin 43 in skeletal myoblasts: relevance to cell transplantation to the heart. *J Thorac Cardiovasc Surg* 122: 759–766.

19. Moos AJ, Zareba W, Hall WJ, *et al.* (Multicenter Automatic Defibrillator Implantation Trial II Investigators). (2002) Prophylactic implantation of a defibrillator in patients with myocardial infarction and reduced ejection fraction. *N Engl J Med* 346: 877–883.

20. Mark DB, Melson CL, Anstrom KJ, *et al.* (SCD-HeFT Investigators). (2006) Cost-effectiveness of defibrillator therapy or amiodarone in chronic stable heart failure: results from the Sudden Cardiac Death in Heart Failure Trial (SCD-HeFT). *Circulation* 114: 135–142.

21. Siminiak T, Kalawski R, Fiszer D, *et al.* (2004) Autologous skeletal myoblast transplantation for the treatment of postinfarction myocardial injury: phase I clinical study with 12 months of follow-up. *Am Heart J* 148: 531–537.

22. Ott HC, Kroess R, Bonaros N, *et al.* (2005) Intramyocardial microdepot injection increases the efficacy of skeletal myoblast transplantation. *Eur J Cardiothorac Surg* 27: 1017–1021.

23. Ott HC, Brechtken J, Swingen C, *et al.* (2006) Robotic minimally invasive cell transplantation for heart failure. *J Thorac Cardiovasc Surg* 132: 170–173.

24. Menasché P. Alfieri O, Janssens S, *et al.* (2008) Myoblast autologous grafting in ischemic cardiomyopathy (MAGIC) trial: first randomized placebo-controlled study of myoblast transplantation. *Circulation* 117: 1189–1200.

25. Saulnier N, Di Campli C, Zocco M, *et al.* (2005) From stem cell to solid organ: bone marrow, peripheral blood or umbilical cord blood as favorable source? *Eur Rev Med Pharmacol Sci* 9: 315–324.

26. Verfaillie C. (2005) Multipotent adult progenitor cells: an update. *Novartis Foundation Symposium* 265: 55–61.

27. Rauscher FM, Goldschmidt-Clermont PJ, Davis BH, *et al.* (2003) Aging, progenitor cell exhaustion, and atherosclerosis. *Circulation* 108: 457–463.

28. Goldschmidt-Clermont PJ, Creager MA, Losordo DW, *et al.* (2005) Atherosclerosis 2005: recent discoveries and novel hypotheses. *Circulation* 112: 3348–3353.

29. Hansson GK. (2005) Inflammation, atherosclerosis, and coronary artery disease. *N Engl J Med* **352**: 1685–1695.

30. Dimmeler S, Haendeler J, Galle J, Zeiher AM. (1997) Oxidized low-density lipoprotein induces apoptosis of human endothelial cells by activation of CPP32-like proteases. A mechanistic clue to the "response to injury" hypothesis. *Circulation* **95**: 1760–1763.

31. Sun P, Dwyer KM, Merz CN, *et al.* (2000) Blood pressure, LDL cholesterol, and intima-media thickness: a test of the "response to injury" hypothesis of atherosclerosis. *Arterioscler Thromb Vasc Biol* **20**: 2005–2010.

32. Endemann DH, Schiffrin EL. (2004) Endothelial dysfunction. *J Am Soc Nephrol* **15**: 1983–1992.

33. Ross R. (1999) Atherosclerosis — an inflammatory disease. *N Engl J Med* **340**: 115–126.

34. Bhagat K, Vallance P. (1997) Inflammatory cytokines impair endothelium-dependent dilatation in human veins *in vivo*. *Circulation* **96**: 3042–3047.

35. Suwaidi JA, Hamasaki S, Higano ST, *et al.* (2000) Long-term follow-up of patients with mild coronary artery disease and endothelial dysfunction. *Circulation* **101**: 948–954.

36. Targonski PV, Bonetti PO, Pumper GM, *et al.* (2003) Coronary endothelial dysfunction is associated with an increased risk of cerebrovascular events. *Circulation* **107**: 2805–2809.

37. Bairey Merz C, Shaw JL, Reis SE, *et al.* (2006) Insights from the NHLBI-sponsored Women's Ischemia Syndrome Evaluation (WISE) Study. Part II: Gender differences in presentation, diagnosis, and outcome with regard to gender-based pathophysiology of atherosclerosis and macrovascular and microvascular coronary disease. *J Am Coll Cardiol* **47**; S21–29.

38. Nelson WD, Zenovich AG, Ott HC, *et al.* (2007) Sex-dependent attention of plaque growth after treatment with bone marrow mononuclear cells. *Circ Res* **101**: 1319–1327.

39. Anversa P, Kajstura J, Leri A, Bolli R. (2006) Life and death of cardiac stem cells: a paradigm shift in cardiac biology. *Circulation* **113**: 1451–1463.

40. Urbanek K, Rota M, Cascapera S, *et al.* (2005) Cardiac stem cells possess growth factor-receptor systems that after activation regenerate the infarcted myocardium, improving ventricular function and long-term survival. *Cir Res* **97**: 663–673.

41. Beltrami AP, Barlucchi L, Torella D, *et al.* (2003) Adult cardiac stem cells are multipotent and support myocardial regeneration. *Cell* **114**: 763–776.

42. Ott HC, Matthiesen TS, Brechtken J, *et al.* (2007) The adult human heart as a source for stem cells: repair strategies with embryonic-like progenitor cells. *Nat Clin Pract Cardiovasc Med* **4(Suppl 1)**: S27–39.

43. Kim BS, Mooney DJ. (1998) Development of biocompatible synthetic extracellular matrices for tissue engineering. *Trends Biotechnol* **16**: 224–230.

44. Piskin E. (2002) Biodegradable polymeric matrices for bioartificial implants. *Int J Artif Organs* 25: 404–410.
45. Rücker M, Laschke MW, Junker D, *et al.* (2006) Angiogenic and inflammatory response to biodegradable scaffolds in dorsal skinfold chambers of mice. *Biomaterials* 27: 5027–5028.
46. Gilbert TW, Sellaro TL, Badylak SF. (2006) Decellularization of tissues and organs. *Biomaterials* 27: 3675–3683.
47. Wilke N, Zenovich A, Jerosch-Herold M, Henry T. (2001) Cardiac magnetic resonance imaging for the assessment of myocardial angiogenesis. *Curr Interv Cardiol Rep* 3: 205–212.
48. Zhou R, Acton P, Ferrari V. (2006) Imaging stem cells implanted in infarcted myocardium. *J Am Coll Cardiol* 48: 2094–2106.
49. MuCue ID, Swingen C, Feldberg T, *et al.* (2008) The real estate of myoblast cardiac transplantation: negative remodeling is associated with location. *J Heart Lung Transplant* 27: 116–123.
50. Ott HC, Matthiesen TS, Goh SK, *et al.* (2008) Perfusion-decellularized matrix: using nature's platform to engineer a bioartificial heart. *Nat Med* 14: 213–221.

10

Postnatal Stem Cells for Myocardial Repair

Mohammad Nurulqadr Jameel, Peter Eckman, Arthur From,
Catherine Verfaillie and Jianyi Zhang*

1. Introduction

Congestive heart failure (CHF) is an end-stage and often irreversible clinical condition. Although optimal medical management has been shown to limit symptoms and slow clinical progression, such therapy only modestly prolongs life. In failing myocardium, abnormalities of myocardial energy generation commonly parallel structural and contractile abnormalities and may themselves contribute to the development of CHF. It is also known that the quantity of myocardium lost during a myocardial infarction (MI) is inversely related to post-MI left ventricular (LV) function and patient prognosis. A potential therapy for limiting postinfarction LV remodeling and the consequent development of CHF is the replacement of infarcted myocardium with new myocardium generated from transplanted stem cells.

Recent studies have provided evidence that cardiomyocyte regeneration may occur during physiological and pathological states in the heart; these data highlight the possibilities that myocardial regeneration may occur via cardiomyocyte proliferation or proliferation and differentiation of putative cardiac stem cells.[1] In this article, we will first review the studies from experimental and clinical trials that use different types of postnatal stem cells in treating cardiovascular disease, and then discuss methods used for enhancing the efficacy of stem cell treatment. Studies using embryonic stem cells are reviewed elsewhere in this book.

*Correspondence: Department of Medicine/Cardiology, University of Minnesota Medical School, Variety Club Research Center, 401 East River Road, Rm 268, Minneapolis, MN 55455, USA. Tel: 612-624-8970 E-mail: zhang047@umn.edu

2. Cells for Myocardial Repair in Ischemic Heart Disease

2.1. Cardiac stem cells

2.1.1. *Can the cardiomyocytes replicate?*

From 1850 to 1911, myocardial hypertrophy was thought to result from hyperplasia *and* hypertrophy of myocytes.[2] This concept was questioned in subsequent reports from 1921 to 1925, which concluded that the increase in cardiac mass in the pathological heart was solely due to cellular hypertrophy.[3] These data led to the belief that the heart is a postmitotic organ and that myocytes are terminally differentiated cells that lack the ability to replicate. This conclusion was based on the inability to identify mitotic figures in myocytes, as well as the observation that regions of transmural infarction evolved into essentially avascular, thin collagenous scars. This paradigm has been dominant for the past 50 years.

However, recent studies have suggested that cardiomyocyte replication may occur in the post-neonatal heart.[1] Studies have documented activation of the cell cycle machinery, karyokinesis and cytokinesis in a subpopulation of myocytes.[4,5] Figure 1 shows an example of myocyte proliferation in humans. Another interesting observation was the identification of male

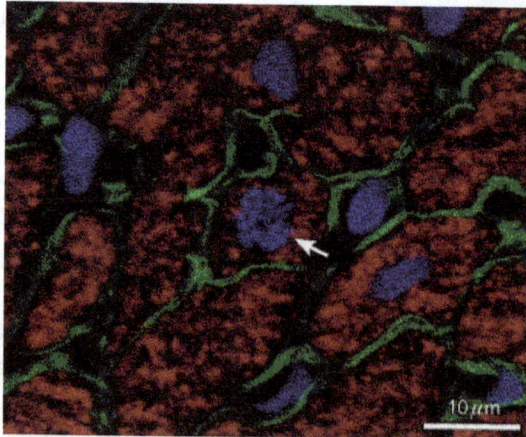

Figure 1. Myocyte proliferation in humans. A small dividing myocyte (alpha-sarcomeric actin; red) with metaphase chromosomes (arrow) is present in the left ventricular myocardium of a patient affected by chronic ischemic cardiomyopathy. Nuclei and metaphase chromosomes are labeled by propidium iodide (blue). Laminin defines the boundary of the cells (green).[159]

myocytes in female hearts transplanted into male recipients.[6] In this study of sex-mismatched cardiac transplants, the female donor heart in the male recipient had a significant number of Y-chromosome positive myocytes and coronary vessels (Fig. 2). Even when cardiac chimerism was taken into account, these results raised the possibility that the male progenitor cells

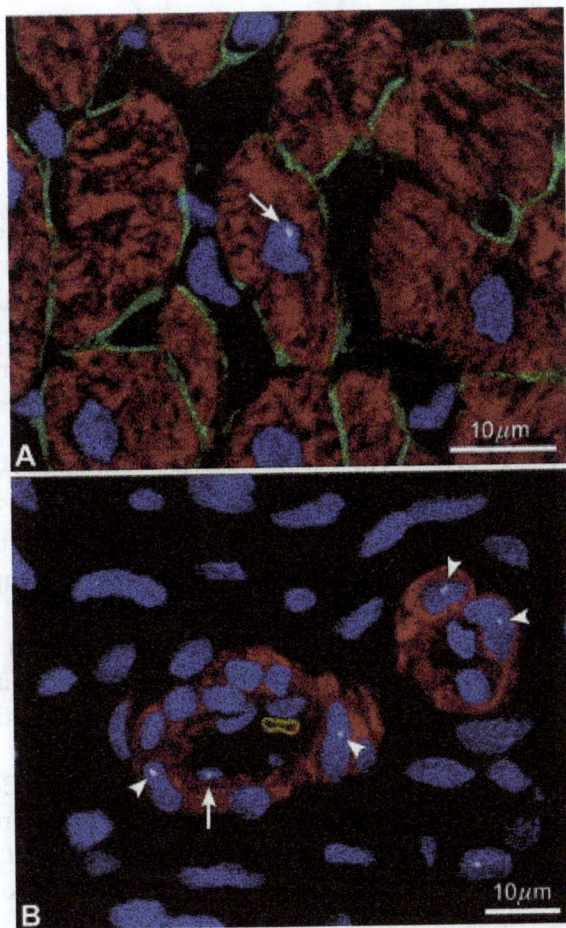

Figure 2. Chimerism of the transplanted female human heart. The localization of the Y chromosome in the nucleus of a myocyte (A; alpha-sarcomeric actin, red; arrow), endothelial cell (B, arrow) and smooth muscle cells (B; α-smooth muscle actin, red; arrowheads) is illustrated in the left ventricle of a female heart transplanted in a male recipient. Laminin defines the boundary of the cells (A, green). A red blood cell is present in the lumen of the coronary arteriole (B; glycophorin; A, yellow).[6]

migrated into the cardiac allografts and gave rise to cardiac progenies. These studies have challenged the established concept of the heart as an organ incapable of regeneration and have opened new experimental and therapeutic vistas.

2.1.2. *Types of cardiac progenitor cells*

If the heart is a self-renewing organ, a stem cell compartment must exist which can replenish all of the different types of cells within the heart. The discovery of niches of adult stem cells within the heart itself supports this possibility. These reports have demonstrated the existence of four types of stem cell-like populations in the adult heart, which have been termed cardiac progenitor cells (CPCs).

Belrami et al.[7] isolated cells expressing the receptor for the stem cell factor c-kit from the interstitial regions of adult myocardium of rats. The highest densities of these lineage negative c-kit[+] stem cells were in the atria and ventricular apex. These cells were self-renewing, clonogenic and multipotent. Furthermore, they had the ability to differentiate into cardiomocytes, endothelial cells and smooth muscle cells. Resident murine CPCs have also been separated on the basis of presence of stem cell antigen 1(Sca-1).[8] A small percentage of these cells activated cardiac genes but did not beat in response to DNA demethylation with 5-azacytidine.[8] However, Oh et al. showed with the help of Cre (Type 1 topoisomerase from P1 bacteriophage) recombinase techniques that the apparent cardiac differentiation was the result of cell fusion in 50% of cases.[8] Matsura et al. also isolated Sca-1[+] cells from adult murine hearts, which differentiated into beating cardiomyocytes in the presence of oxytocin but not 5-azacytidine.[9] Other groups have isolated so-called side population (SP) cells from mouse hearts based on their ability to exclude Hoechst 33342 dye.[10,11] These cells express Abcg2, an ATP-binding cassette (ABC) transporter. They are Sca-1[+] and c-kit low and differentiate into cardiomyocytes after coculture with rat cardiomyocytes. Finally, Laugwitz et al. discovered a group of cells from postnatal mouse hearts using isl-1 (homeobox gene islet-1) transcription factor.[12] These also express the cardiac transcription factors Nkx2.5 and GATA4 but not Sca-1, c-kit or CD31. These cells also have the ability to differentiate into functional cardiomyocytes *in vitro* and *in vivo*. The isl-1[+] cells have not been identified in adult hearts.[12]

Anversa et al. claim that CPCs are undifferentiated cells that express the stem cell-related antigens c-kit, MDR-1 (another ABC transporter) and

Sca-1 in variable combinations.[13] Quantitative data in the mouse, rat, dog and human heart have demonstrated that there is one CPC per approximately 30,000–40,000 myocardial cells. Approximately 65% of all CPCs possess the three stem cell antigens, approximately 20% two stem cell antigens and approximately 15% only one. Approximately 5% of each of these CPCs exclusively express c-kit, MDR1 or Sca-1.[13]

Significantly, none of the above-mentioned reports shows a signature CPC phenotype. This cell population also has significant overlap in the expression of other surface markers. It remains to be elucidated whether these cells are actually the same stem cell type and if differing surface markers reflect differing developmental phases or qualitatively separate subpopulations. These CPCs may participate in myocyte turnover, the rate of which remains to be determined.

2.1.3. *Origin of the cardiac progenitor cell*

The origins of CPCs also remain unclear. The cycling cardiomyocytes might be derived from uncommitted stem-like population cells that reside in the heart and expand and differentiate into cardiomyocytes in response to proper stimulation. Alternatively, these stem-like cells may reside in the bone marrow and then be mobilized into the circulation and induced to home to the heart by signals emanating from the injured heart.

Mouquet *et al.* demonstrated that cardiac side population (SP) cells are maintained by local progenitor cell proliferation under physiological conditions.[14] After MI, this cardiac SP is decreased by as much as 60% in the infarct and to a lesser degree in the noninfarct regions within 1 day. Cardiac SP pools are subsequently reconstituted to baseline levels within 7 days after MI, through both proliferation of resident cardiac SP cells and homing of bone marrow (BM) — derived stem cells to specific areas of myocardial injury. These cells then undergo immunophenotypic conversion and adopt a cardiac SP phenotype (CD45$^+$ to CD45$^-$).[14] BM-derived stem cells accounted for approximately 25% of the SP cells in the heart under pathological conditions as compared to < 1% under physiological conditions.[14] In addition to CD45$^+$ cells, bone marrow also contains CD45$^-$, CXCR4$^+$ and Sca-1$^+$ cells within the nonadherent, nonhematopoietic mononuclear fraction, which express early cardiac markers such as Nkx2.5 and GATA-4.[15] These cells mobilize into blood after MI and home to the infarcted myocardium in mice. Cerisoli *et al.*[16] also showed that (at least in pathological conditions) part of the c-kit$^+$ CPC population may derive from cells that originate in the bone marrow and are able to adopt in the heart the same functions and features of the local CPCs.

2.1.4. *Myocardial regeneration from cardiac progenitor cells*

Some pertinent studies of stem cells in animal models of ischemic heart disease are summarized in Table 1. Anversa *et al.* showed that the direct intramyocardial injection of c-kit⁺ cells into an ischemic rat heart reconstituted well-differentiated myocardium comprising of blood-carrying new vessels and cardiomyocytes with the characteristics of young cells; these cells were present in approximately 70% of the ventricle.[7] Later, it was also shown that intracoronary delivery of these cardiac stem cells in an ischemia/reperfusion rat model resulted in myocardial regeneration, infarct size reduction of 29%, and improvement of left ventricular function.[17] Given intravenously after ischemia/reperfusion, Sca-1⁺ cells also homed to injured myocardium and differentiated into cardiomyocytes.[8] The relative contributions of regenerated cardiomyocytes and preservation of injured native cardiomyocytes in these studies require clarification.

X Wang *et al.* recently reported that heart-derived Sca-1⁺/CD31⁻ cells possess stem cell characteristics and play an important role in cardiac repair.[18] In that study, immunofluorescent staining and fluorescence-activated cell sorter (FACS) analysis indicated that endogenous Sca-1⁺/CD31⁻ cells significantly increased in the infarct and peri-infarct areas at 3 and 7 days after MI. Western blotting confirmed elevated Sca-1 protein expression 7 days after MI. Sca-1⁺/CD31⁻ cells cultured *in vitro* were induced to express both endothelial cell and cardiomyocyte markers. Transplantation of Sca-1⁺/CD31⁻ cells into a murine model of MI led to functional preservation and decreased remodeling after MI.[18] Immunohistochemistry data indicated a significant increase of neovascularization, but a low level of cardiomyocyte regeneration at the infarct border zone. Despite the absence of significant cardiomyocyte regeneration, cell transplantation remarkably improved myocardial bioenergetics.[18] These findings provide evidence that Sca-1⁺/CD31⁻ cells possess both endothelial cell and cardiomyocyte progenitor cell characteristics. However, this study also reported that the regeneration rates of cardiomyocytes or endothelial cells from the engrafted stem cells were very low (only ~80 cells per heart). Hence, trophic effects associated with the transplanted cells were most likely the basis of the beneficial effects of these cells.[18] Expansion of these progenitor cells may have therapeutic applicability to the treatment of MI.

Table 1. Stem Cell Transplantation in Animal Models of Ischemic Heart Disease

Author	Species	Model	Cell Type	Mode of Delivery	Result	Reference
Jackson *et al.*	Mouse	Ischemia/ Reperfusion	BMC c-kit$^+$ Sca-1$^+$	Intracoronary	Cardiomyocyte and endothelial differentiation observed.	28
Orlic *et al.*	Mouse	MI	BMC c-kit$^+$	Intramyocardial	Cardiomyocyte and endothelial differentiation observed.	29
Murry *et a.l*	Mouse	MI	BMC c-kit$^+$	Intramyocardial	No cardiomyocyte differentiation.	30
Balsam *et al.*	Mouse	MI	BMC c-kit$^+$ Sca-1$^+$	Intramyocardial	No cardiomyocyte differentiation.	31
Beltrami *et al.*	Rat	MI	CSC c-kit$^+$	Intramyocardial	New vessel and cardiomyocyte formation (no fusion) and improved EF.	7
Dawn *et al.*	Rat	Ischemia/ Reperfusion	CSC c-kit$^+$	Intracoronary	New vessel and cardiomyocyte formation (no fusion) and improved EF.	17
Oh *et al.*	Mouse	Ischemia/ Reperfusion	CSC Sca-1$^+$	Intravenous via right jugular vein	Cardiomyocyte differentiation, half due to fusion.	8

(Continued)

Table 1. (*Continued*)

Author	Species	Model	Cell Type	Mode of Delivery	Result	Reference
Wang et al.	Mouse	MI	CSC Sca-1$^+$	Intramyocardial	Significant increase in neovascularization and modest myocardial regeneration. Improved EF. Paracrine effect contributory.	18
Johnston et al.	Pig	Ischemia/Reperfusion	CSC c-kit$^+$	Intracoronary	Evidence of engraftment in the border zone and within infarct at 8 weeks.	22
Smith et al.	Mouse	MI	Human CSC c-kit$^+$	Intramyocardial	Cardiac regeneration with improved EF.	26
Tomita et al.	Pig	MI	MSC	Intramyocardial	Formed cardiac tissue, promoted angiogenesis and improved EF.	62
Shake et al.	Pig	MI	MSC	Intramyocardial	Differentiated into cell expressing muscle markers and improved contractile function.	63
Amado et al.	Pig	MI	Allogeneic MSC	Intracoronary	Long-term engraftment with improved EF.	64

BMC, Bone marrow cell; CSC, cardiac stem cell; MSC, mesenchymal stem cell; EF, ejection fraction.

CPCs and early committed cells (ECCs) express c-Met and insulin-like growth factor I receptors (IGF-IR), and synthesize and secrete the corresponding ligands hepatocyte growth factor (HGF) and insulin-like growth factor 1 (IGF-1).[19] HGF mobilizes CSCs-ECCs and IGF-1 promotes their survival and proliferation.[19] Therefore, in another study, HGF and IGF-1 were injected in mice with MI and a growth factor gradient was introduced between the site of storage of primitive cells in the atria and the region bordering the infarct to facilitate homing. The newly formed myocardium contained arterioles, capillaries, and functionally competent myocytes that increased in size over time. This regeneration was associated with improved ventricular performance and induced increased survival. Surprisingly, this intervention rescued animals with infarcts that constituted 86% of ventricular mass. The above findings have been replicated in a dog model, where HGF and IGF-1 were also used to stimulate resident cardiac stem cells after myocardial infarction; growth factor therapy again resulted in improvement of myocardial function.[20]

Before their therapeutic use, CPCs have to be isolated from fragments of myocardium and expanded *in vitro*. This was achieved in a pig model,[21] where c-kit$^+$ cells were isolated and each cell was propagated to form approximately 400,000 cells. Another group performed autologous transplantation of CPCs in an ischemia/reperfusion swine model.[22] Each pig had a biopsy from the right ventricular septum at the time of injury. The biopsies weighed 92 mg and yielded mean cell counts of 14.2×10^6 cells after isolation and expansion (after 2.8 cell passages over 23 days). Intracoronary delivery was performed 4 weeks after injury. Engraftment occurred in the MI border zone, and islands of engrafted cells were present within the scar 8 weeks after coronary delivery.[22]

Human cardiac progenitor cells (hCPCs) have also been isolated from myocardium, expanded *in vitro* and then used for transplantation in animal models of ischemic myocardium. Hosoda *et al.* isolated human cardiac progenitor cells from surgical samples.[23] These c-kit$^+$ hCPCs were injected into the hearts of immunodeficient mice and rats. Foci of myocardial regeneration were identified at 2–3 weeks and consisted of myocytes, resistance arterioles and capillaries.[23] The presence of connexin 43 and N-cadherin in the developing human myocytes strongly suggested that the engrafted human cells were becoming functionally competent. Two-photon microscopy was used to further demonstrate the functional integration of enhanced green fluorescent protein (EGFP) — positive human myocytes with the surrounding myocardium.[23] Torella *et al.*[24] have also isolated hCPCs from myocardial samples from all four chambers of the human heart. These were c-kit⁺, MDR-1⁺ and CD133⁺. One clone could generate over 5×10^9 cells and form functional myocardium after injection into infarcted rat hearts.[24] CPCs from subcultures of postnatal atrial and ventricular human biopsy samples can form multicellular clusters known as cardiospheres.[25] These are capable of long term renewal and formed myocardium when injected into infarcted mouse hearts. Marban et al.[26] performed percutaneous endomyocardial biopsies on 70 adult patients and grew cells from them in primary culture. Cardiospheres-developed in these cultures, which were plated to yield cardiosphere-derived cells (CDCs). Two distinct populations of CDCs were identified by FACS analysis: c-kit⁺ CDCs and CD90⁺ CDCs.[26] The mesenchyme-like CD105⁺ CD90⁺ subpopulation may provide physical or secretory support to the c-kit⁺ population during CDC expansion. After injection into the border zone of immunodeficient mice with acute MI, these human CDCs engrafted and migrated into the infarct region and resulted in improvement of ventricular function.[26]

These studies provide a rationale for the use of hCPCs in patients with ischemic heart disease. These cells seem to be excellent candidates for exogenous stem cell therapy, though they have to be harvested from patients and expanded *ex vivo* to generate numbers sufficient for transplantation. To date, there have been no reported clinical trials of hCPCs.

2.2. Bone marrow-derived stem cells

A number of studies have suggested that cardiomyocytes may be generated from circulating bone marrow stem cells, though more recent studies have questioned this contention. Bittner et al. were the first to suggest that cardiac muscle cells may be derived from BM cells.[27] Goodel et al. showed that following transplantation of murine BM SP cells (c-kit⁺, Sca-1⁺, CD34⁻/ low), donor-derived cells with cardiomyocyte morphology as well as smooth muscle and endothelial cells were found in the heart following LAD ligation.[28] Orlic et al.[29] showed that transplantation of GFP⁺ Lin-c-kit⁺ cells [presumably containing both hematopoietic stem cells (HSCs) and mesenchymal stem cells] into the ventricular wall after LAD ligation resulted in improved function of the ventricle, and detected a large number of GFP⁺ cells with the cardiac phenotype in the myocardium. In contrast to these finding, other laboratories showed that lineage-negative, c-kit-positive cells did not differentiate into cardiomyocytes.[30,31] However, these studies also observed an improvement in the cardiac function of animals receiving stem cell transplantation. This could be attributed to improvement of the passive mechanical properties of the infarct and subsequent amelioration of ventricular remodeling,[32] or to the paracrine effect of injected cells on injured native cardiomyocytes. The term "paracrine effect" refers to the production of local signaling molecules by engrafted cells that may improve perfusion or enhance survival of host cardiomyocytes.[33]

2.3. Mesenchymal stem cells

2.3.1. *Phenotype and differentiation potential*

The notion of mesenchymal stem cells (MSCs) was introduced in 1961 by Friedenstein who documented that marrow stromal cells contain osteogenic progenitors.[34] However, it was Caplan's group in the late 1980s and in the 1990s who identified a subset of cells within the bone marrow that gave rise to osteoblasts and adipocytes and termed them MSCs.[35]

MSCs are present in many different organs of the body, including muscle, skin, adipose tissue and bone marrow. They can be isolated from the bone marrow by a simple process involving Ficoll centrifugation and adhering cell culture in a defined serum-containing medium. They can be expanded for 4–20 population doublings only[36] with preservation of the karyotype, telomerase activity and telomere length.[37,38] Phenotypically, these cells are negative for CD31, CD34 and CD45, unlike hematopoietic progenitors from the bone marrow, and are positive for CD29, CD44, CD71, CD90, CD105, CD106, CD120a, CD124, SH2, SH3 and SH4.[39–41] In the bone marrow, only 0.001–0.01% of the initial unfractionated bone marrow mononuclear cell (BMMNC) population consists of MSCs.[36,39] However, in many recent rodent studies, the adherent fibroblastic cells obtained from the unfractionated mononuclear class of the bone marrow are termed MSCs.[42,43]

MSCs have been reported to have the potential to differentiate into any tissue of mesenchymal origin.[35] MSCs derived from rodent marrow aspiration have been shown to differentiate into cardiomyocytes in the presence of 5-aza-cytidine.[44,45] The morphology of the cells changes from spindle-shaped to ball-shaped and finally to rod-shaped. Thereafter, these cells fuse together to form a syncitium which resembles a myotube.[46] These exhibit markers of fetal cardiomyocytes.[45] Specific transcription factors of the myocyte and cardiac lineage, including GATA4, Nkx2.5 and HAND 1/2, can be detected.[44] There are differences (as compared to native cardiomyocytes) in cardiomyocytes derived from MSCs. Firstly, the β-isoform of the cardiac myosin heavy chain is more abundant than the α-isoform in these cells. Secondly, there is more α-skeletal actin than α-cardiac actinin, and myosin light chain 2v is also present. Thirdly, MEF2A and MEF2D replace MEF2C isoforms from early to late passage. These cells beat spontaneously and synchronously, which is most likely due to the formation of intercalated discs, as has been shown when they are cocultured with neonatal myocytes.[47] They express competent α- and β-adrenergic and muscarinic receptors, as indicated by an increased rate of contraction in response to isoproterenol and a decreased rate of contraction induced by β-adrenergic blockers.[48] This *in vitro* differentiation to myocytes has been replicated using MSCs from pig bone marrow-derived MSCs.[49] However, there is no report thus far of myocyte differentiation from human MSCs.

2.3.2. *Mesenchymal stem cells for myocardial repair*

MSCs have features that make them attractive candidates for cell transplantation. As they are easily accessible and expandable, MSCs could potentially become an "off the shelf" allogeneic product which would be more cost-effective, easier to administer, allow a greater number of cells to be transplanted and, possibly of importance, permit transplantation at the time of urgent interventions to relieve ischemia and injury, such as percutaneous or surgical revascularization procedures. Importantly, these cells appear to avoid rejection by being hypoimmunogenic.[50–52] They lack MHC-II and B-7 costimulatory molecule expression and thus limit T-cell responses.[53,54] They can also directly inhibit inflammatory responses via paracrine mechanisms, including production of transforming growth factor beta1 and hepatocyte growth factor.[55,56] All the above properties make them potentially excellent candidates for cell transplantation.

MSC transplantation was tested in a study in which isogenic adult rats were used as donors and recipients to simulate autologous transplantation clinically. MSC intracoronary injection in these rat hearts after myocardial infarction showed milieu-dependent differentiation of these cells — fibroblastic phenotype within the scar and the cardiomyocyte phenotype outside the infarction area.[57] However, direct intramyocardial injection of autologous MSCs into the scar resulted in differentiation into heart-like myocytes within the scar tissue, induced angiogenesis, and improved myocardial function.[58] Delivery of MSCs via direct left ventricular cavity infusion in a rat MI model resulted in preferential migration and colonization of the cells in the ischemic myocardium at 1 week.[59] MSCs also resulted in increased vascularity and improved cardiac function at 2 months in a canine model of chronic ischemic disease.[60] However, using a rat model of postinfarction LV remodeling, Kloner's group found that the beneficial effects on left ventricular function were short term and absent after 6 months.[61]

MSC transplantation has led to functional improvement in large animal models of MI. Direct intramyocardial injection of 5-azacytadine treated autologous MSCs was performed 4 weeks after MI in a swine model. These cells formed islands of heartlike tissue, induced angiogenesis, prevented thinning and dilatation of the infarct region, and improved regional and global contractile function.[62] In another study, direct intramyocardial injection of autologous MSCs 2 weeks after MI led to engraftment within the host myocardium, expression of muscle specific proteins by a fraction of engrafted cells, and attenuation of the contractile dysfunction and pathological scar thinning.[63] Percutaneous delivery of allogenic MSCs 3 days after MI in a porcine model resulted in long term engraftment at 8 weeks, profound reduction in scar size, and near-normalization of cardiac function.[64]

2.4. Multipotent adult progenitor cells

Verfaillie *et al.*[65-70] defined another set of progenitor cells which are present in adult bone marrow obtained from humans, pigs, rats and mice; these cells were named multipotent adult progenitor cells (MAPCs). MAPCs are isolated from bone marrow aspirates, the brain and other tissues as an adherent culture which shows extensive replicative capacity, sustained telomere length and escape from replicative senescence.[67] This is an important distinction as compared to MSCs, which reach replicative senescence after 25–35 population doublings. MAPCs are maintained on fibronectin in the presence of epidermal growth factor and platelet-derived growth factor supplemented with leukemia inhibitor factor (rodent). They are found only when cells are maintained at low density ($500/cm^2$) and not allowed to grow to semiconfluence or confluence. When maintained under such conditions, MAPCs can be expanded for more than 50–80 population doublings in humans, 100 population doublings in swine, and 80–150 population doublings in rodent.[70–72] This is an important feature because it allows the large scale expansion of MAPCs for clinical use. Thus, MAPCs have the potential to be an off-the-shelf consistent clinical product. They have low levels of Flk-1, Sca-1 and Thy-1 and higher levels of CD13 and stage-specific antigen I. They lack CD34, CD44 and CD45. Unlike MSCs and HSCs, these cells do not express c-kit. As described by Tolar *et al.*,[73] MAPCs share immunological properties with MSCs which could enhance their allogeneic use, although the relatively low to negative expression of class I MHC may make this population sensitive to NK cell rejection.

MAPCs show the tissue regenerative capacity of embryonic stem cells, which is relatively unique in the adult stem cell field, where more stringent tissue regenerative capacity has been described. *In vitro* and *in vivo* animal studies show that in addition to their capacity for mesodermal differentiation, MAPCs can also differentiate into cells with ectodermal and endodermal characteristics.[68,69] Furthermore, MAPCs have been shown to differentiate into hepatocyte-like cells expressing cytokeratin (CK-19), alpha-fetoprotein, albumin, HepPar-1 and CD26.[69] Jiang *et al.*[67] have shown that a single MAPC injected into a mouse blastocyst is capable of generating viable chimeras with engraftment frequencies of up to 40%. This is also true of heart tissue, where similar frequencies of cardiomyocytes derived from the MAPC donor are found, indicating both the lineage potential of this cell population and the safety for use in tissue-regenerative strategies.

Other published reports replicate the finding of pluripotent (mesodermal, ectodermal, endodermal lineage capacity) adherent stem cell populations from bone marrow. D'Ippolito *et al.* described the isolation of pluripotent cells with extensive replicative capacity from marrow of animal and human vertebral bodies.[74–76] Yoon *et al.*[77] also described isolation of pluripotent cells from bone marrow and use of these cells in a rat ligation model of MI. Animals receiving human pluripotent stem cells showed significant recovery with improved left ventricular wall motion and ejection fraction, as well as increased vessel density in the infarct and peri-infarct zones.[77]

This broad tissue-regenerative capacity combined with extensive proliferation capacity thus makes MAPCs an attractive candidate for cellular therapy.[71,73–75,77] There are reports of ectopic tissue or tumor formation by this class of cells. Currently, the effects of direct intramyocardial injection of MAPCs are being studied in a swine heart model of MI.

2.5. Skeletal myoblasts

Skeletal myoblasts are derived from satellite cells, which are precursor cells under the basal membrane of skeletal muscle fibers. These cells are normally in a quiescent state,[78] but after tissue injury, satellite cells are mobilized and proliferate to become skeletal myoblasts and help in regenerating the damaged muscle tissue.

It was first shown in 1993 that skeletal myoblasts form viable, long-term skeletal myotube grafts after transplantation into adult hearts.[79] Later, Taylor *et al.* transplanted autologous skeletal myoblasts in cryoinfarcted rabbit myocardium and showed engraftment at 3 weeks, with subsequent improvement in systolic performance.[78] Since then, there have been numerous studies that have consistently shown that skeletal myoblasts can form multinucleated myoblasts but not differentiate into cardiomyocytes.[80] These cells are not electromechanically coupled to each other or to surrounding cardiomyocytes.[81,82] Despite these findings, left ventricular function improved after transplantation in these animal models of MI. Several mechanisms have been postulated, including paracrine factors secreted by the engrafted cells and changing the elastic properties of the scar tissue, both of which could alter the course of postinfarction remodeling.

2.6. Endothelial progenitor cells

Endothelial progenitor cells (EPCs) were first isolated from blood in 1997.[83] They originate from a common hemangioblast precursor in the bone marrow.[84] However, many other cells, including myeloid/monocyte (CD14$^+$) cells, MAPCs and stem cells from adult organs, can also differentiate into cells with EPC characteristics. Thus, circulating EPCs are a heterogeneous group of cells originating from multiple precursors within the bone marrow and present in different stages of endothelial differentiation in peripheral blood. Therefore, the characterization of these cells is difficult because they share certain surface markers of hematopoietic cells and some adult endothelial cells. Typically, they express CD34 (a hematopoietic cell characteristic), CD-133 (a more specific marker of EPCs) and KDR (kinase insert domain-containing receptor), which is the receptor for vascular endothelial growth factor. In a study of sex-mismatched bone marrow transplant patients by Hebbel and coworkers, 95% of circulating endothelial cells in the peripheral blood of the transplant patients had the recipient genotype, but 5% had the donor genotype.[85] The endothelial cells with the donor phenotype had delayed growth in culture but high proliferative capacity, with 1,023-fold expansion within 1 month, and were termed endothelial outgrowth cells (EOCs). Hebbel and coworkers concluded that the EOCs were of bone marrow origin. In contrast, the cells with the recipient phenotype had only 17-fold expansion within the same period. These circulating endothelial cells most likely originated from the vessel wall.

EPCs have been used in different animal models of cardiovascular disease. Intravenous delivery of CD34[+] cells into athymic nude rats with MI caused angiogenesis in the peri-infarct region, leading to decreased myocyte apoptosis, reduced interstitial fibrosis and improvement of left ventricular function.[33] Similarly, intramyocardial implantation of CD34[+] selected human peripheral blood mononuclear cells into nude rats after MI resulted in neovascularization and improved left ventricular function.[86]

2.7. Umbilical cord blood stem cells

Stem cells are found in higher numbers in the umbilical cord as compared to adult human blood and bone marrow.[87] Human umbilical cord blood (UCB) contains fibroblast-like cells termed unrestricted somatic stem cells, which adhere to culture dishes, are negative for c-kit, CD34 and CD45, and differentiate *in vitro* and *in vivo* into a variety of tissue types, including cardiomyocytes.[88] Direct intramyocardial injection of these human unrestricted somatic cells into the infarcted hearts of immunosuppressed pigs resulted in improved perfusion and wall motion, reduced infarct size and enhanced cardiac function.[89] Further, intravenous injection of human mononuclear UCB cells, a small fraction of which were CD34[+], into NOD/scid mice led to enhanced neovascularization with capillary endothelial cells of both human and mouse origin and reduced infarct size.[90] However, no myocytes of human origin were found, arguing against cardiomyogenic differentiation and regeneration of cardiomyocytes from donor cells. Additionally, direct intramyocardial injection of UCB CD34[+] cells into the peri-infarct rim in a rat model resulted in improved cardiac function.[91] There have been no reported clinical studies of UCB transplantation.

3. Cell Transplantation in Nonischemic Cardiomyopathy

To date, the majority of clinical studies concerned with cellular therapy for cardiovascular disease have been designed to reverse or prevent systolic heart failure in patients with ischemic heart disease. Preclinical studies have also been largely focused on this pathological problem, in part due to the availability of well-established and predictable animal models of progressive postinfarction left ventricular remodeling which culminates in CHF. Considering the large number of patients with heart failure of nonischemic etiology, and despite the infrequent reports of clinical investigations involving this population, a review of nonischemic preclinical models is warranted in anticipation of clinical trials involving patients with nonischemic cardiomyopathy (NICM).

The earliest report of stem cell therapy in a nonclinical, nonischemic setting involved the transplantation of skeletal myoblasts into the hearts of dogs with a canine version of Duchenne muscular dystrophy. Although no functional assessment was performed, the demonstration of dystrophin-positive myogenic cells in the hearts of recipient dogs provided proof of the principle that transplanted cells could engraft into a myopathic heart.[79] Later, Scorsin *et al.*[92] reported functional improvement after fetal cardiomyocyte transplantation in a mouse model of doxorubicin-induced cardiomyopathy. Several groups have subsequently studied the role of cell transplantation in doxorubicin-induced heart failure models. Suzuki *et al.*[93] performed intracoronary infusion of skeletal myoblasts in rats and showed functional improvement as measured *ex vivo* and observed grafted cells to be widely distributed throughout the left ventricle. Agbulut *et al.*[94] studied the effects of unpurified bone marrow cells and purified Sca-1$^+$ cells in a mouse model of doxorubicin-induced cardiomyopathy. They found evidence suggesting myogenic differentiation in the engrafted unpurified bone marrow cells but not in the engrafted purified Sca-1$^+$ cells 2 weeks posttransplant. Transplanted bone marrow cells have also been beneficial in a rabbit model of doxorubicin-induced cardiomyopathy.[95] More recently, freshly aspirated bone marrow cells were injected into the LV free wall of rabbits following a course of doxorubicin administration, and function was assessed 4 weeks posttransplant.[96] Functional improvement was documented despite the fact that immunohistochemistry studies failed to detect myocytes regenerated from engrafted cells. Regardless of the mechanism(s) involved, the consistent functional benefit associated with cell transplantation in these studies suggests that further experimental work utilizing cell transplantation to ameliorate doxorubicin toxicity is warranted.

The TO-2 hamster strain is considered a representative model of human hereditary dilated cardiomyathy. In this strain, the promoter region of the delta-sarcoglycan gene is deleted and cardiomyocytes do not express alpha, beta, gamma or delta sarcoglycan. The number of cardiomyocytes declines after birth because of apoptosis, and progressive cardiac remodeling leads to CHF. Yoo *et al.*[97] procured ventricular cells from young asymptomatic hamsters and transplanted them after BrdU labeling into older symptomatic animals. Functional improvement was seen along with evidence of myocyte-like cells which stained positive for BrdU, troponin I and the myosin light chain. Transplanted smooth muscle cells and skeletal myoblasts yielded similar results.[98,99]

Cell transplantation has also been performed in rat models of myosin immunization-induced myocarditis.[100,101] Intramyocardial injection of MSCs resulted in functional improvement, and intravenous delivery of spleen-derived EPCs caused a decreased scar area and improved fractional shortening. Inflammatory infiltrate and fibrosis were also noted to be decreased in a mouse model of Chagas cardiomyopathy in which bone marrow cells were administered intravenously.[102] Similarly cotransplanted MSCs and skeletal myoblasts were also found to ameliorate ventricular dysfunction in rats with Chagas disease.[103] Wang *et al.*[104] recently reported functional benefit in a large animal model of nonischemic cardioyopathy. VEGF-transfected MSCs were transplanted in a porcine model of pressure overload heart failure induced by ascending aortic banding. Transplantation resulted in improved perfusion, bioenergetics and contractile function of the hypertrophying left ventricles.

Taken together, the available preclinical studies of cell therapy for non-ischemic cardiomyopathy have shown it to be beneficial in multiple pathological models in multiple species, despite the use of differing cell sources and routes of delivery. These findings suggest that this approach should be further explored in animal models with the goal of developing clinically useful treatments.

4. Clinical Trials of Stem Cell Treatment in Heart Disease

Tables 2 and 3 summarize the important clinical trials of stem cells in ischemic cardiac disease. Skeletal myoblasts were chosen as the initial cell

Table 2. Clinical Trials of Stem Cell Treatment in Acute Myocardial Infarction

Author and Reference	Cell Type	Mean Cell Dose ($\times 10^6$)	Number Treated	Days after MI	Randomized	Primary End-Point	P Value	Result
Strauer et al.[112]	BMMNC CD34+ AC133+	28	10	8	No	Safety	—	EF unchanged, regional contractility increased
Assmus et al., TOPCARE-AMI[113]	1. BMMNC CD34, CD45 5.5×10^6 CD34, CD133 0.28×10^6	16	29	4.9	No	Safety and feasibility	—	EF increased
	2. Blood derived progenitor cells (endothelial characteristics)	213	30					
Wollert et al., BOOST[115]	BMMNC CD34+ 9.5×10^6	2,460	30	4.8	Yes	LVEF	0.0026	Short-term benefit at 6 months, but no difference between groups at 18 months

(Continued)

Table 2. (*Continued*)

Author and Reference	Cell Type	Mean Cell Dose (×10⁶)	Number Treated	Days after MI	Randomized	Primary End-Point	P Value	Result
Janssens et al.[118]	BMMNC 2·8 × 10⁶ CD34⁺ 2·0 × 10⁶ CD133⁺ 0·2 × 10⁶ CD90⁺/Thy-1⁺ 2·5 × 10⁶ CD105⁺/endoglin 7·0 × 10⁶ CD117⁺/c-kit 32 × 10⁶ CD73⁺	304 nucleated cells and 172 mononuclear cells	33	<1	Yes	LVEF	0.36	Decreased infarct size
Chen et al.[125]	MSCs	48,000–60,000	34	18	Yes	LVEF	0.01	EF improved 14% above controls
REPAIR-AMI[116]	BMMNC CD34⁺CD45⁺ 2.5 × 10⁶ CD133, CD45 1.9 × 10⁶ CD34⁺CD133⁺ CD45⁺ 1.8 × 10⁶	198	101	3–7	Yes	LVEF	0.01	Improved EF at 4 months

(*Continued*)

Table 2. (Continued)

Author and Reference	Cell Type	Mean Cell Dose ($\times 10^6$)	Number Treated	Days after MI	Randomized	Primary End-Point	PValue	Result
ASTAMI[117]	BMMNC CD34$^+$ 0.7 × 10^6 CD133$^+$ 1.7 × 10^6	68	50	4–8	Yes	LVEF	0.7	No benefit at 6 months
MAGIC Cell 3 DES[129]	GCSF-mobilized PBSC 9.3% CD34$^+$ 15.1% KDR$^+$ 2.2% AC133$^+$ 5.7% CD34/KDR	14,00 leukocytes	25 16 (old MI)	4	Yes	LVEF	<0.05	EF improved at 6 months in AMI, no improvement in old MI patients
STEMI[133]	GCSF mobilized stem cells. (CD34$^+$, CD34$^-$CXCR4$^+$)	None	39	<2	Yes	Systolic wall thickening	1.0	No benefit at 6 months from GCSF treatment alone

BMMNC, bone marrow mononuclear cells; MSCs, mesenchymal stem cells; PBSC, peripheral blood stem cell; GCSF, granulocyte colony stimulating factor; EF, ejection fraction.

Note: All are autologous cells delivered via intracoronary infusion. BMMNCs are a mixed population of cells containing hematopoetic, mesenchymal and other progenitor cells.

Table 3. Clinical Trials of Stem Cells for Chronic Ischemic Disease

Author	Cell Type	Mode	Number Treated	Randomized	Primary End-Point	P Value	Result
Menasche et al.[105]	Skeletal myoblasts	Open heart surgery	10	No	Safety and feasibility	—	Improved NYHA class and improved EF
Dib et al.[160]	Skeletal myoblasts	Open heart surgery	30	No	Safety and feasibility	—	Improved EF
POZNAN[110]	Skeletal myoblasts	Transcoronary venous	10	No	Safety and feasibility	—	Improved NYHA class and improved EF
Perin et al.[120]	BMMNC	Endomyocardial	14	No	Safety	—	Improved EF and reduced symptoms
Fuchs et al.[121]	BMMNC	Endomyocardial	10	No	Safety and feasibility	—	Improved angina score, no change in EF
Tse et al.[122]	BMMNC	Endomyocardial	8	No	Safety and feasibility	—	No change in EF, Improved wall thickening
TOPCARE-CHD[123]	CPC/BMMNC	Intracoronary	75	Yes	LVEF	0.003	EF improved after 3 months, more in BMC than in CPC
Erbs et al.[127]	GCSF mobilized CPC	Intracoronary	13	Yes	CFR and LVEF	<0.05	LVEF and CFR increased at 3 months

BMMNC, bone marrow mononuclear cells; CPC, circulating progenitor cells; GCSF, granulocyte colony stimulating factor; EF, ejection fraction; CFR, coronary flow reserve.

type to be used in these clinical trials, because they: (1) are at an advanced stage of myogenic differentiation which markedly decreases the chances of tumorogenesis; (2) have high proliferative potential and the initial muscle biopsy can be easily expanded *in vitro*; (3) allow autologous transplantation, thus eliminating the risk of rejection and the need for immunosuppression; and (4) are very resistant to ischemia and can survive if injected around the scar tissue with minimal perfusion. Beginning in 2000, there have been 6 phase I safety and feasibility studies of skeletal myoblast transplantation in patients with severe left ventricular dysfunction caused by MI. Four of these studies[105–108] were surgical and entailed myoblast implantation at the time of coronary artery bypass grafting or left ventricular assist device implantation, and two were catheter-based trials[109,110] using an endoventricular or a coronary sinus transvenous approach. Long-term engraftment of myoblasts has been documented in pathological specimens which showed skeletal myocytes aligned parallel to host cardiomyocytes and embedded in scar tissue for up to 18 months after transplantation.[107,111] There has been concern over the development of ventricular tachycardia in some of these studies, which has prompted the use of intracardiac defibrillators (ICDs) in protocols for skeletal myoblast transplantation. Left ventricular function was not the primary end-point in these studies; however, it tended to improve following transplantation of skeletal myoblasts. Ongoing large, randomized, placebo-controlled, double-blind studies will determine the efficacy and safety of this treatment, as well as its role in the clinic.

Following skeletal myoblast transplantation studies, the next wave of clinical trials examined the efficacy and safety of bone marrow cell (BMC) transplantation. In the first published human trial of BMC transplantation, Strauer *et al.*[112] aspirated BMMNCs and reinfused them into the infarct-related artery 7 days after MI in 10 patients and a control group of 10 patients who refused the treatment. The treated group showed significant improvements of myocardial perfusion and wall motion indexes. The Transplantation of Progenitor Cells and Regeneration Enhancement in Acute Myocardial Infarction (TOPCARE-AMI) trial revealed significant improvement in the LV ejection fraction as well as significantly enhanced myocardial viability and regional wall motion in the infarct area following transplantation of BMMNCs or blood-derived progenitor cells.[113,114] In the "bone marrow transfer to enhance ST-elevation infarct regeneration" (BOOST) study,[115] 60 patients were randomized after successful percutaneous coronary intervention for acute MI to receive either BMMNCs or standard therapy. The treated group showed significant short-term increase of the LV ejection fraction at 6 months, but no difference between the two groups was present at 18 months. The recently published Reinfusion of Enriched Progenitor Cells and Infarct Remodeling in Acute Myocardial Infarction (REPAIR-AMI) trial[116] randomized 204 patients with acute MI to receive an intracoronary infusion of progenitor cells derived from bone marrow or placebo medium into the infarct artery 3–7 days after successful reperfusion therapy. The absolute increase in LVEF was significantly greater (2.5%) in the BMC group than in the placebo group at 4 months. However, the Autologous Stem Cell Transplantation in Acute Myocardial Infarction (ASTAMI) study[117] found no effects of intracoronary injection of autologous mononuclear BMCs on global left ventricular function at 6 months. Similarly, the ST-Elevation Acute Myocardial Infarction (STEMI) trial[118] failed to show any improvement of left ventricular function. Differences in cell preparation[119] and numbers may have contributed to these conflicting results.

BMCs have also been employed as therapeutic agents in patients with chronic heart failure. Perin *et al.*[120] enrolled 14 subjects and 7 control subjects with ischemic cardiomyopathy. The treatment group received endomyocardial injection of approximately 30 million BMMNCs. Transplanted hearts showed a significant reduction in reversible myocardial perfusion defects and a significant improvement in overall LV function. Other studies have reported improved angina scores[121] and improved wall thickening and radial shortening in transplanted hearts.[122] In the recently published Transplantation of Progenitor Cells and Recovery of LV Function in Patients with Chronic Ischemic Heart Disease (TOPCARE-CHD) trial,[123] 75 patients with stable ischemic heart disease who had MI at least 3 months previously were assigned to receive either no cell infusion or infusion of circulating progenitor cells or BMCs into the patent coronary artery supplying the most dyskinetic left ventricular area. The transplantation of BMCs was associated with a moderate (2.9 percentage points) but significant improvement in left ventricular function 3 months posttransplantation.

MSC transplantation has also moved to the clinical arena based on the promising results of the preclinical studies. Chen *et al.* randomized 69 patients who received a primary percutaneous intervention within 12 hours of acute MI to receive intracoronary injection of autologous MSCs or saline 18 days after percutaneous intervention.[124,125] They reported a significant improvement in left ventricular function at 3 months, which was maintained at 6 months. A significant reduction in the size of the perfusion defect (measured by positron emission tomography at 3 months) was also seen in these patients.[124,125] Currently, a large randomized multicenter Phase I trial using allogenic MSCs is being conducted in the United States by Osiris Therapeutics in patients with acute MI.

The clinical application of EPCs is limited by several factors. Firstly, there are few circulating EPCs and it is difficult to expand them into sufficient numbers without inducing a change in the phenotype or the development of cell senescence. Secondly, EPCs in patients with cardiovascular diseases and diabetes and in elderly patients are functionally impaired, thus making autologous transplantation with these cells problematic. In an uncontrolled clinical study, autologous AC133[+] cells were injected into the infarct border at the time of bypass surgery in 6 patients; improved perfusion in the infarct region and better left ventricular function were reported to be present 3–9 months after surgery.[126] Erb *et al.* randomized patients with chronically occluded coronary arteries to receive intracoronary progenitor cells or a placebo. They mobilized BMCs using GCSF, harvested them from peripheral blood and expanded them *ex vivo*. The intracoronary delivery of these cells led to improvement in coronary flow reserve and cardiac function at 3 months posttransplantation.[127] Currently, clinical trials using CD34+ cells from bone marrow that are enriched in EPC content are underway.

5. Techniques for Enhancing the Efficacy of Stem Cell Therapy

Although it is a consistent observation in the literature that cellular transplantation improves LV contractile function, the cell engraftment rate a few weeks after the transplantation is usually very low. Therefore, it is clear that the majority of cells transplanted to the heart do not achieve durable engraftment. Furthermore, the majority of transplanted cells that do engraft remain as spindle-shaped stem cells and do not fully differentiate into host cardiac cell phenotypes. Therefore, methods to enhance the rates of stem cell engraftment, the proliferation of engrafted cells, and the differentiation of the engrafted cells into functioning cardiomyocytes are urgently needed.

5.1. Differentiation and proliferation

Factors that promote differentiation into cardiomyocytes are under investigation. 5-azacytidine is known to cause differentiation of CSCs and MSCs into cardiomyocytes,[8,45] and oxytocin also enhances differentiation of CSCs.[9] Transforming growth factor (TGF) beta stimulates the myogenic differentiation of CD117+ stem cells by upregulating GATA-4 and NKx-2.5 expression. Therefore, the intramyocardial implantation of TGF-beta-preprogrammed CD117+ cells in mice hearts after acute MI effectively assisted myocardial regeneration and induced therapeutic angiogenesis, contributing to functional cardiac regeneration.[128]

5.2. Mobilization

Many therapies have been used to mobilize stem cells from bone marrow to the systemic circulation in order to enhance endogenous repair and to facilitate collection of these cells for *ex vivo* expansion and use in cell therapy. These include granulocyte colony stimulating factor (GCSF), vascular endothelial growth factor (VEGF), stromal cell-derived factor 1 (SDF-1), angiopoeitin 1, placental growth factor and erythropoietin. Most studies confirm an improvement in endothelial regeneration or angiogenesis. G-CSF has probably been the most studied factor. In the MAGIC Cell-3-DES randomized controlled trial,[129] intracoronary infusion of peripheral blood stem cells mobilized by GCSF resulted in improvement of left ventricular function and no increased risk of restenosis as reported in the previous MAGIC Cell trial.[130] More recently, cytokine therapy alone has been used to facilitate regeneration of cardiac tissue. In an ischemia/reperfusion model in mice, GCSF+ Flt-3 ligand (FL), GCSF+ stem cell factor (SCF) or GCSF was administered, and it was noted that postinfarct cytokine therapy with GCSF+ FL or GCSF+ SCF caused mobilization and homing of BMCs to the heart and limited adverse left ventricular remodeling.[131] Kanellakis *et al.*[132] used GCSF and SCF therapy, which improved left ventricular function of MI, but the newly formed cells were of cardiac rather than bone marrow origin. In the recently published double-blind, randomized, placebo-controlled stem cells in MI (STEMMI) trial,[133] subcutaneous GCSF was found to be safe but did not lead to improvement in ventricular function.

5.3. Homing

One of the foremost challenges for the medical field currently is to devise ways to enhance the homing of stem cells to the injured region of the heart. It has been shown that recruitment of CXCR4-positive progenitor cells to regenerating tissues is mediated by hypoxia gradients via hypoxia inducible factor 1 (HIF-1) — induced expression of stromal cell-derived growth factor 1 (SDF-1).[134] Furthermore, transplantation of syngeneic cardiac fibroblasts transfected to express SDF-1 induced homing of CD-117[+] stem cells to the injured myocardium and led to improved ventricular function.[135] High mobility group box protein 1 (HMGB1), released after cell necrosis, is another protein which causes progenitor cell homing. Limana *et al.* injected 200 ng of purified HMGB1 into the peri-infarct region of murine hearts 4 hours after permanent coronary artery occlusion.[136] This caused proliferation and differentiation of endogenous cardiac c-kit[+] stem cells into cardiomyocytes and led to improvement in ventricular function. Integrins are another class of molecules which play a role in the adhesion and transmigration of stem cells.

It is also known that the microenvironment after acute MI is favorable to cell homing as compared to chronically infarcted myocardium. While BM-derived cells are known to home to the infarct zone within days of acute MI, mobilization or infusion of these cells 2 months after MI does not result in stem cell engraftment into the infarct zone.[135] Using a rat model of permanent coronary artery ligation, Lu *et al.*[137] examined the local conditions requisite for cell homing and migration, including the temporal expression of intercellular adhesion molecule (ICAM-1), monocyte chemoattractant protein 1 (MCP-1) and matrix metalloproteinase (MMP) in the infarcted myocardium; they concluded that the optimal period for cell homing and migration is the 2-week period following MI.

5.4. Function and survival

Assuming that the number of transplanted cells that survive is critical to therapeutic benefit, multiple groups are working on methods to increase the survival engraftment of transplanted cells; a great deal of experimental evidence suggests that this strategy may be effective. Apoptosis can be decreased by constitutive expression of Akt (a serine threonine kinase with potent prosurvival activity) or by heat shock prior to transplantation.[138] Rat MSCs transduced to overexpress Akt1 (encoding the Akt protein) transplanted into ischemic myocardium were found to inhibit cardiac remodeling by reducing inflammation, collagen deposition and myocyte hypertrophy in a dose-dependent fashion.[139] MSCs transduced to express Akt were also studied in an ischemic porcine model which showed an improvement in EF as compared to nontransduced MSCs. In order to determine the mechanism of the beneficial effect, the effects of the apoptotic stimulus, H_2O_2, on MSCs transduced with Akt were studied *in vitro*. Akt-MSCs were found to be more resistant to apoptosis; this was related to higher levels of extracellular signal-regulated protein kinase (ERK) activation and VEGF.[140] The precise mechanism of benefit is unclear, because medium conditioned with Akt-transduced MSCs was found to markedly inhibit hypoxia-induced apoptosis and trigger vigorous spontaneous contraction of adult rat cardiomyocytes *in vitro*. Injection of conditioned medium into infarcted hearts also significantly limits infarct size and improves ventricular function.[141] Several genes that are potential mediators of these effects (VEGF, FGF-2, HGF, IGF-I, TB4) are significantly upregulated in the Akt-MSCs, particularly in response to hypoxia.[141,142] A significant concern also exists regarding the potential tumorogenecity of Akt-transduced cells, particularly when Akt is constitutively expressed, because Akt has been shown to be sufficient to induce oncogenic transformation of cells and tumor formation. Therapeutic efforts are underway to target the Akt pathway for treatment of malignancies.[143] Another strategy that has been widely tested involves attempts to increase vasculogenesis with VEGF. Transfection with VEGF and IGF-1 improved survival of transplanted bone marrow cells in a rat model of MI.[144] the delivery of which had undergone adenoviral transduction. Overexpression of VEGF also resulted in improved LV function and neovascularization,[145] but the addition of VEGF protein alone to cells did not show any benefit in a rat model of fetal cardiomyocyte transplantation.[146] MSCs overexpressing VEGF also improved myocardial bioenergetics and left ventricular function in a porcine model of pressure-overloaded hearts, that this approach is effective in a nonischemic setting.[104]

Augmentation of the expressions of other gene products, including cardiotrophin 1 (CT-1), heme oxygenase 1 (HO-1), an IL-1 inhibitor, and CuZn-superoxide dismutase, has also been tested and found effective. Myoblasts expressing CT-1 (known to have potent hypertrophic and survival effects on myocytes) retard the transition to heart failure in Dahl salt-sensitive hypertensive rats. MB-CT1-treated animals showed a significant alleviation of LV dilation and contractile dysfunction compared with the sham group.[147] Hypoxia-inducible factor 1 alpha (HIF-1a) has been studied in the setting of skeletal myoblast transplantation in a rat coronary artery ligation model and found to increase EF dramatically (27%) above the baseline and result in a significantly greater degree of angiogenesis cell engraftment, and cell survival.[148] MSCs transfected with a hypoxia-regulated HO-1 vector are more tolerant to hypoxia-reoxygen injury *in vitro* and result in improved viability in ischemic hearts.[149] Skeletal myoblasts secreting an IL-1 inhibitor have been found to modulate adverse remodeling in infarcted murine myocardium. IL-1 influences post-MI hypertrophy and collagen turnover. Modulation of adverse remodeling has been postulated as a key mechanism by which transplanted cells might improve cardiac function following MI.[150] Treatment with CuZn-superoxide dismutase has been shown to attenuate the initial rapid cell death following transplantation, leaving a twofold increase in the total number of engrafted cells at 72 hours compared with controls.[151]

Use of viruses for gene expression cannot be translated into clinical studies because of the risk of mutagenesis, carcinogenesis and induction of an immune response. Jo *et al.*[152] developed a nonviral carrier of cationized polysaccharide for genetic engineering of MSCs. Spermine-introduced dextran of cationized polysaccharide (spermine-dextran) was internalized into MSCs by way of a sugar-recognizable receptor to enhance the expression level of plasmid deoxyribonucleic acid (DNA). When genetically engineered by the spermine-dextran complex with plasmid DNA of adrenomedullin (AM), MSCs secreted a large amount of AM, an anti-apoptotic and angiogenic peptide. Transplantation of AM gene-engineered MSCs improved cardiac function after MI significantly more than did nontransduced MSCs. Thus, this genetic engineering technology using the nonviral spermine-dextran (and other promising new methods) is a promising strategy for improving MSC therapy for ischemic heart disease.

5.5. Use of biomaterials to design the microenvironment

It seems that the microenvironment in which the cells are injected is of extreme importance to their survival and subsequent beneficial effects. Biomaterials offer a means of providing quantitative adhesion, growth, or migration signals.[153] They can be designed to regulate quantitative timed release of factors which direct cellular differentiation pathways, such as angiogenesis and vascular maturation. Moreover, smart biomaterials even respond to the local environment, such as protease activity or mechanical forces, with controlled release or activation.[153]

Cell sheet technology has been used to transplant a monolayer of MSCs in an ischemic myocardium with functional improvement.[154] The success of a sheet type graft has been extended to a hamster model of dilated cardiomyopathy in which a reduction in LV dilatation, increased wall thickness, decreased myocardial fibrosis and prolonged life expectancy were all observed.[155]

Cardiac patches are three-dimensional matrices composed of natural or synthetic scaffold materials that can host the cells to allow longer cell viability and enhance differentiation and integration. They may also contain factors that enhance neovascularization to allow the patch to survive in ischemic tissue. G Zhang *et al.* showed that controlled release of stromal-cell-derived factor 1α (SDF-1α) *in situ* increases stem cell homing to the infarcted heart.[156] Recombinant mouse SDF-1α was covalently bound to the PEGylated fibrinogen, as evidenced by immunoprecipitation and western blotting. The PEGylated fibrinogen bound with recombinant mouse SDF-1α was mixed with thrombin to form the PEGylated fibrin patch. *In vitro* results showed that SDF-1α was successfully bound to the PEGylated fibrin patch and could be released from the patch for up to 10 days. This PEGylated fibrin patch with bound SDF-1α (100 ng) was placed on the surface of the infarct region of the left ventricle (LV) after acute MI. Two weeks after infarction, the myocardial recruitment of c-kit$^+$ cells was significantly higher in the SDF-1α PEGylated fibrin patch–treated group. At day 28 post MI, unlike the control group, the SDF-1α releasing patch group maintained stable release of SDF-1α concurrent with additional stem cell homing. Moreover, the LV function was significantly improved compared with the control groups. To enhance the engraftment rate and to direct cell differentiation, 3D porous PEGylated biomaterial was used to deliver MSCs in mouse hearts with AMI.[157] This significantly enhanced the engraftment rate (by 10-fold) with associated improvement in LV EF (significantly more than with direct MSCs injection). It is interesting to note that LV functional improvement was only observed at week 4 postinfarction and was not proportionate to the dramatically increased engraftment rate, suggesting that additional interventions may be needed (such as directing myocyte differentiation by HGF) in order to achieve significant myocardial regeneration and more improvement of LV function.

Lee *et al.*[158] designed self-assembling peptide nanofibers for prolonged delivery of insulin-like growth factor 1 (IGF-1), a cardiomyoyte growth and differentiation factor, to the myocardium using a "biotin sandwich" approach. Biotinylated IGF-1 was complexed with tetravalent streptavidin and then bound to biotinylated self-assembling peptides. This biotin sandwich strategy allowed binding of IGF-1 but did not prevent self-assembly of the peptides into nanofibers within the myocardium. IGF-1 that was bound to peptide nanofibers activated Akt, decreased activation of caspase 3, and increased expression of cardiac troponin I in cardiomyocytes. After injection into rat myocardium, biotinylated nanofibers provided sustained IGF-1 delivery for 28 days, and targeted delivery of IGF-1 *in vivo* increased activation of Akt in the myocardium. When combined with transplanted cardiomyocytes, IGF-1 delivery by biotinylated nanofibers decreased caspase 3 cleavage by 28% and increased the myocyte cross-sectional area by 25% compared with cells embedded within nanofibers alone or with untethered IGF-1. Finally, cell therapy with IGF-1 delivery by biotinylated nanofibers improved systolic function after experimental MI, demonstrating how engineering the local cellular microenvironment can improve cell therapy.

Most of these new biomaterials provide much greater flexibility for regenerating tissues *ex vivo*, but emerging technologies like self-assembling nanofibers can now establish intramyocardial cellular microenvironments by injection. This may allow percutaneous cardiac regeneration and repair approaches, i.e. injectable-tissue engineering. Finally, materials can be made to multifunction by providing sequential signals with custom design of differential release kinetics for individual factors. Thus, new rationally designed biomaterials no longer simply coexist with tissues, but can provide precision bioactive control of the microenvironment that may be required for cardiac regeneration and repair.

References

1. Anversa P, Nadal-Ginard B. (2002) Myocyte renewal and ventricular remodeling. *Nature* **415:** 240–243.
2. Kollicker A. (1852) *Handbuch der Gewebelehre des Menschen*, 6th edn. (Engelmann, Leipzig).
3. Aschoff L. (1921) *Pathologische Anatomie* (Jena, Germany).
4. Beltrami AP, Urbanek K, Kajstura J, *et al.* (2001) Evidence that human cardiac myocytes divide after myocardial infarction. *N Engl J Med* **344:** 1750–1757.
5. Urbanek K, Quaini F, Tasca G, *et al.* (2003) Intense myocyte formation from cardiac stem cells in human cardiac hypertrophy. *Proc Natl Acad Sci USA* **100:** 10440–10445.

6. Quaini F, Urbanek K, Beltrami AP, *et al.* (2002) Chimerism of the transplanted heart. *N Eng J Med* **346**: 5–15.
7. Beltrami AP, Barlucchi L, Torella D, *et al.* (2003) Adult cardiac stem cells are multipotent and support myocardial regeneration. *Cell* **114**: 763–776.
8. Oh H, Bradfute SB, Gallardo TD, *et al.* (2003) Cardiac progenitor cells from adult myocardium: homing, differentiation, and fusion after infarction. *Proc Natl Acad Sci USA* **100**: 12313–12318.
9. Matsuura K, Nagai T, Nishigaki N, *et al.* (2004) Adult cardiac Sca-1 positive cells differentiate into beating cardiomyocytes. *J Biol Chem* **279**: 11384–11391.
10. Martin CM, Meeson AP, Robertson SM, *et al.* (2004) Persistent expression of the ATP-binding cassette transporter, Abcg2, identifies cardiac SP cells in the developing and adult heart. *Dev Biol* **265**: 262–275.
11. Pfister O, Mouquet F, Jain M, *et al.* (2005) CD31– but not CD31+ cardiac side population cells exhibit functional cardiomyogenic differentiation. *Circ Res* **97**: 52–61.
12. Laugwitz KL, Moretti A, Lam J, *et al.* (2005) Postnatal isl1+ cardioblasts enter fully differentiated cardiomyocyte lineages. *Nature* **433**: 647–653.
13. Leri A, Kajstura J, Anversa P. (2005) Cardiac stem cells and mechanisms of myocardial regeneration. *Physiol Rev* **85**: 1373–1416.
14. Mouquet F, Pfister O, Jain M, *et al.* (2005) Restoration of cardiac progenitor cells after myocardial infarction by self-proliferation and selective homing of bone marrow-derived stem cells. *Circ Res* **97**: 1090–1092.
15. Kucia M, Dawn B, Hunt G, *et al.* (2004) Cells expressing early cardiac markers reside in the bone marrow and are mobilized into the peripheral blood after myocardial infarction. *Circ Res* **95**: 1191–1199,
16. Cerisoli F, Chimenti I, Gaetani R. (2006) Kit-positive cardiac stem cells (CSCs) can be generated in damaged heart from bone marrow-derived cells. *Circulation* **114**: II-164.
17. Dawn B, Stein AB, Urbanek K, *et al.* (2005) Cardiac stem cells delivered intravascularly traverse the vessel barrier, regenerate infarcted myocardium, and improve cardiac function. *Proc Natl Acad Sci USA* **102**: 3766–3771.
18. Wang X, Hu Q, Nakamura Y, *et al.* (2006) The role of the Sca-1$^+$/CD31$^-$ cardiac progenitor cell population in postinfarction left ventricular remodeling. *Stem Cells.* **24**: 1779–1788.
19. Urbanek K, Rota M, Cascapera S, *et al.* (2005) Cardiac stem cells possess growth factor–receptor systems that after activation regenerate the infarcted myocardium, improving ventricular function and long-term survival. *Circ Res* **97**: 663–673.
20. Linke A, Müller P, Nurzynska D, *et al.* (2005) Stem cells in the dog heart are self-renewing, clonogenis, and multipotent and regenerate infarcted

myocardium, improving cardiac function. *Proc Natl Acad Sci USA* **102:** 8966–8971.

21. Bearzi C, Muller P, Amano K. (2006) Identification and characterization of cardiac stem cells in the pig heart. *Circulation* **114:** II-125.

22. Johnston P, Sasano T, Mills K. (2006) Isolation, expansion and delivery of cardiac derived stem cells in a porcine model of myocardial infarction. *Circulation* **114:** II-125.

23. Hosoda T, Bearzi C, Amano S. (2006) Human cardiac progenitor cells regenerate cardiomyocytes and coronary vessels repairing the infarcted myocardium. *Circulation* **114:** II-51.

24. Torella D, Elliso GM, Karakikes I. (2006) Biological properties and regenerative potential, *in vitro* and *in vivo*, of human cardiac stem cells isolated from each of the four chambers of the adult human heart. *Circulation* **114:** II-87.

25. Messina E, De Angelis L, Frati G, *et al.* (2004) Isolation and expansion of adult cardiac stem cells from human and murine heart. *Circ Res* **95:** 911–921.

26. Smith RR, Barile L, Cho HC. (2006) Cardiogenicity and regenerative potential of cardiac-derived stem cells isolated from adult human and porcine endomyocardial biopsy specimens. *Circulation* **114:** II-51.

27. Bittner RE, Schöfer C, Weipoltshammer K, *et al.* (1999) Recruitment of bone marrow-derived cells by skeletal and cardiac muscle in adult dystrophic mdx mice. *Anat Embryol* **199:** 391–396.

28. Jackson KA, Majka SM, Wang H, *et al.* (2001) Regeneration of ischemic cardiac muscle and vascular endothelium by adult stem cells. *J Clin Invest* **107:** 1395–1402.

29. Orlic D, Kajstura J, Chimenti S, *et al.* (2001) Bone marrow cells regenerate infarcted myocardium. *Nature* **410:** 701–705.

30. Murry CE, Soonpaa MH, Reinecke H, *et al.* (2004) Haematopoietic stem cells do not transdifferentiate into cardiac myocytes in myocardial infarcts. *Nature* **428:** 664–668.

31. Balsam LB, Wagers AJ, Christensen JL, *et al.* (2001) Haematopoietic stem cells adopt mature haematopoietic fates in ischaemic myocardium. *Nature* **428:** 668–673.

32. Jain M, DerSimonian H, Brenner DA, *et al.* (2001) Cell therapy attenuates deleterious ventricular remodeling and improves cardiac performance after myocardial infarction. *Circulation* **103:** 1920–1927.

33. Kocher AA, Schuster MD, Szabolcs MJ, *et al.* (2001) Neovascularization of ischemic myocardium by human bone marrow-derived angioblasts prevents cardiomyocyte apoptosis, reduces remodeling and improves cardiac function. *Nat Med* **7:** 430–436.

34. Friedenstein AJ. (1961) Osteogenetic activity of transplanted transitional epithelium. *Acta Anat* **45:** 31–59.

35. Caplan AI. (1991) Mesenchymal stem cells. *J Orthop Res* **9**: 641–650.
36. Prockop DJ. (1997) Marrow stromal cells as stem cells for nonhematopoietic tissues. *Science* **276**: 71–74.
37. Phinney DG, Kopen G, Righter W, *et al.* (1999) Donor variation in the growth properties and osteogenic potential of human marrow stromal cells. *J Cell Biochem* **75**: 424–436.
38. Pittenger MF, Mackay AM, Beck SC, *et al.* (1999) Multilineage potential of adult human mesenchymal stem cells. *Science* **284**: 143–147.
39. Pittenger MF, Martin BJ. (2004) Mesenchymal stem cells and their potential as cardiac therapeutics. *Circ Res* **95**: 9–20.
40. Majumdar MK, Keane-Moore M, Buyaner D, *et al.* (2003) Characterization and functionality of cell surface molecules on human mesenchymal stem cells. *J Biomed Sci* **10**: 228–241.
41. Haynesworth SE, Baber MA, Caplan AI. (1992) Cell surface antigens on human marrow-derived mesenchymal cells are detected by monoclonal antibodies. *Bone* **13**: 69–80.
42. Alhadlaq A, Mao JJ. (2004) Mesenchymal stem cells: isolation and therapeutics. *Stem Cells Dev* **13**: 436–448.
43. Minguell JJ, Erices A, Conget P. (2001) Mesenchymal stem cells. *Exp Biol Med* **226**: 507–520.
44. Fukuda K. (2002) Molecular characterization of regenerated cardiomyocytes derived from adult mesenchymal stem cells. *Congenit Anom (Kyoto)* **42**: 1–9.
45. Makino S, Fukuda K, Miyoshi S, *et al.* (1999) Cardiomyocytes can be generated from marrow stromal cells *in vitro*. *J Clin Invest* **103**: 697–705.
46. Fukuda K. (2003) Use of adult marrow mesenchymal stem cells for regeneration of cardiomyocutes. *Bone Marrow Transplant* **32 Suppl 1**: S25–27.
47. Tomita S, Nakatani T, Fukuhara S, *et al.* (2002) Bone marrow stromal cells contract synchronously with cardiomyocytes in a coculture system. *Jpn J Thorac Cardiovasc Surg* **50**: 321–324.
48. Hakuno D, Fukuda K, Makino S, *et al.* (2002) Bone marrow-derived regenerated cardiomyocytes (CMG cells) express functional adrenergic and muscarinic receptors. *Circulation* **105**: 380–386.
49. Liu J, Hu Q, Wang Z, *et al.* (2004) Autologous stem cell transplantation for myocardial repair. *Am J Physiol Heart Circ Physiol* **287**: H501–511.
50. Bartholomew A, Sturgeon C, Siatskas M, *et al.* (2002) Mesenchymal stem cells suppress lymphocyte proliferation *in vitro* and prolong skin graft survival *in vivo*. *Exp Hematol* **30**: 42–48.
51. Le Blanc K, Tammik L, Sundberg B, *et al.* (2003) Mesenchymal stem cells inhibit and stimulate mixed lymphocyte cultures and mitogenic responses independently of the major histocompatibility complex. *Scand J Immunol* **57**: 11–20.

52. Tse WT, Pendleton JD, Beyer WM, *et al.* (2003) Suppression of allogeneic T-cell proliferation by human marrow stromal cells: implications in transplantation. *Transplantation* 75: 389–397.

53. Zimmet JM, Hare JM. (2005) Emerging role for bone marrow derived mesenchymal stem cells in myocardial regenerative therapy. *Basic Res Cardiol* 100: 471–481.

54. Ryan JM, Barry FP, Murphy JM, Mahon BP. (2005) Mesenchymal stem cells avoid allogeneic rejection. *J Inflam (Lond)* 2: 8.

55. Le Blanc K, Tammik C, Rosendahl K, *et al.* (2003) HLA expression and immunologic properties of differentiated and undifferentiated mesenchymal stem cells. *Exp Hematol* 31: 890–896.

56. Di Nicola M, Carlo-Stella C, Magni M, *et al.* (2002) Human bone marrow stromal cells suppress T-lymphocyte proliferation induced by cellular or nonspecific mitogenic stimuli. *Blood* 99: 3838–3843.

57. Wang JS, Shum-Tim D, Chedrawy E, Chiu RC. (2001) The coronary delivery of marrow stromal cells for myocardial regeneration: pathophysiologic and therapeutic implications. *J Thorac Cardiovasc Surg* 122: 699–705.

58. Tomita S, Li RK, Weisel RD, *et al.* (1999) Autologous transplantation of bone marrow cells improves damaged heart function. *Circulation* 100: II247–265.

59. Barbash IM, Chouraqui P, Baron J, *et al.* (2003) Systemic delivery of bone marrow-derived mesenchymal stem cells to the infarcted myocardium: feasibility, cell migration, and body distribution. *Circulation* 108: 863–868.

60. Silva GV, Litovsky S, Assad JA, *et al.* (2005) Mesenchymal stem cells differentiate into an endothelial phenotype, enhance vascular density, and improve heart function in a canine chronic ischemia model. *Circulation* 111: 150–156.

61. Dai W, Hale SL, Martin BJ, *et al.* (2005) Allogeneic mesenchymal stem cell transplantation in postinfarcted rat myocardium: short- and long-term effects. *Circulation* 112: 214–223.

62. Tomia S, Mickle DA, Weisel JD, *et al.* (2002) Improved heart function with myogenesis and angiogenesis after autologous porcine bone marrow stromal cell transplantation. *J Thorac Cardiovasc Surg* 123: 1132–1140.

63. Shake JG, Gruber PJ, Baumgartner WA, *et al.* (2002) Mesenchymal stem cell implantation in a swine myocardial infarct model: engraftment and functional effects. *Ann Thorac Surg* 73: 1919–1925.

64. Amado LC, Saliaris AP, Schuleri KH, *et al.* (2005) Cardiac repair with intramyocardial injection of allogeneic mesenchymal stem cells after myocardial infarction. *Proc Natl Acad Sci USA* 102: 11474–11479.

65. Reyes M, Dudek A, Jahagirdar B, *et al.* (2002) Origin of endothelial progenitors in human postnatal bone marrow. *J Clin Invest* 109: 337–346.

66. Jiang Y, Jahagirdar BN, Reinhardt RL, *et al.* (2002) Pluripotency of mesenchymal stem cells derived from adult marrow. *Nature* 418: 41–49.

67. Jiang Y, Vaessen B, Lenvik T, *et al.* (2002) Multipotent progenitor cells can be isolated from postnatal murine bone marrow, muscle, and brain. *Exp Hematol* 30: 896–904.

68. Reyes M, Lund T, Lenvik T, *et al.* (2001) Purification and *ex vivo* expansion of postnatal human marrow mesodermal progenitor cells. *Blood* 98: 2615–2625.

69. Schwartz RE, Reyes M, Koodie L, *et al.* (2002) Multipotent adult progenitor cells from bone marrow differentiate into functional hepatocyte-like cells. *J Clin Invest* 109: 1291–1302.

70. Zeng L, Rahrmann E, Hu Q, *et al.* (2006) Multipotent adult progenitor cells from swine bone marrow. *Stem Cells* 24: 2355–2366.

71. Verfaillie CM, Schwartz R, Reyes M, Jiang Y. (2003) Unexpected potential of adult stem cells. *Ann N Y Acad Sci* 996: 231–234.

72. Lakshmipathy U, Verfaillie C. (2005) Stem cell plasticity. *Blood* 19: 29–38.

73. Tolar J, O'Shaughnessy MJ, Panoskaltsis-Mortari A, *et al.* (2006) Host factors that impact the biodistribution and persistence of multipotent adult progenitor cells. *Blood* 107: 4182–4188.

74. D'Ippolito G, Diabira S, Howard GA, *et al.* (2004) Marrow-isolated adult multilineage inducible (MIAMI) cells, a unique population of postnatal young and old human cells with extensive expansion and differentiation potential. *J Cell Sci* 117: 2971–2981.

75. D'Ippolito G, Howard GA, Roos BA, Schiller PC. (2006) Sustained stromal stem cell self-renewal and osteoblastic differentiation during aging. *Rejuvenation Res* 9: 10–19.

76. D'Ippolito G, Howard GA, Roos BA, Schiller PC. (2006) Isolation and characterization of marrow-isolated adult multilineage inducible (MIAMI) cells. *Exp Hematol* 34: 1608–1610.

77. Yoon YS, Wecker A, Heyd L, *et al.* (2005) Clonally expanded novel multipotent stem cells from human bone marrow regenerate myocardium after myocardial infarction. *J Clin Invest* 115: 326–338.

78. Taylor DA, Atkins BZ, Hungspreugs P, *et al.* (1998) Regenerating functional myocardium: improved performance after skeletal myoblast transplantation. *Nat Med* 4: 929–933.

79. Koh GY, Klug MG, Soonpaa MH, Field LJ. (1993) Differentiation and long-term survival of C2C12 myoblast grafts in heart. *J Clin Invest* 92: 1548–1554.

80. Dowell JD, Rubart M, Pasumarthi KB, *et al.* (2003) Myocyte and myogenic stem cell transplantation in the heart. *Cardiovasc Res* 58: 336–350.

81. Murry CE, Wiseman RW, Schwartz SM, Hauschka SD. (1996) Skeletal myoblast transplantation for repair of myocardial necrosis. *J Clin Invest* 98: 2512–2523.

82. Leobon B, Garcin I, Menasché P, *et al.* (2003) Myoblasts transplanted into rat infarcted myocardium are functionally isolated from their host. *Proc Natl Acad Sci USA* 100: 7808–7811.

83. Asahara T, Murohara T, Sullivan A, *et al.* (1997) Isolation of putative progenitor endothelial cells for angiogenesis. *Science* 275: 964–967.
84. Masuda H, Asahara T. (2003) Post-natal endothelial progenitor cells for neovascularization in tissue regeneration. *Cardiovasc Res* 58: 390–398.
85. Lin Y, Weisdorf DJ, Solovey A, Hebbel RP. (2000) Origins of circulating endothelial cells and endothelial outgrowth from blood. *J Clin Invest* 105: 71–77.
86. Kawamoto A, Tkebuchava T, Yamaguchi J, *et al.* (2003) Intramyocardial transplantation of autologous endothelial progenitor cells for therapeutic neovascularization of myocardial ischemia. *Circulation* 107: 461–468.
87. Mayani H, Lansdorp PM. (1998) Biology of human umbilical cord blood-derived hematopoietic stem/progenitor cells. *Stem Cells* 16: 153–165.
88. Kogler G, Sensken S, Airey JA, *et al.* (2004) A new human somatic stem cell from placental cord blood with intrinsic pluripotent differentiation potential. *J Exp Med* 200: 123–135.
89. Kim BO, Tian H, Prasongsukarn K, *et al.* (2005) Cell transplantation improves ventricular function after a myocardial infarction: a preclinical study of human unrestricted somatic stem cells in a porcine model. *Circulation* 112: I96-104.
90. Ma N, Stamm C, Kaminski A, *et al.* (2005) Human cord blood cells induce angiogenesis following myocardial infarction in NOD/scid-mice. *Cardiovasc Res* 66: 45–54.
91. Hirata Y, Sata M, Motomura N, *et al.* (2005) Human umbilical cord blood cells improve cardiac function after myocardial infarction. *Biochem Biophys Res Commun* 327: 609–614.
92. Scorsin M, Hagège AA, Dolizy I, *et al.* (1998) Can cellular transplantation improve function in doxorubicin-induced heart failure? *Circulation* 98: II151–155.
93. Suzuki K, Murtuza B, Suzuki N, *et al.* (2001) Intracoronary infusion of skeletal myoblasts improves cardiac function in doxorubicin-induced heart failure. *Circulation* 104: I213–217.
94. Agbulut O, Menot ML, Li Z, *et al.* (2003) Temporal patterns of bone marrow cell differentiation following transplantation in doxorubicin-induced cardiomyopathy. *Cardiovasc Res* 58: 451–459.
95. Zhang J, Li GS, Li GC, *et al.* (2005) Autologous mesenchymal stem cells transplantation in adriamycin-induced cardiomyopathy. *Chin Med J (Engl)* 118: 73–76.
96. Lu C, Arai M, Misao Y, *et al.* (2006) Autologous bone marrow cell transplantation improves left ventricular function in rabbit hearts with cardiomyopathy via myocardial regeneration-unrelated mechanisms. *Heart Vessels* 21: 180–187.
97. Yoo KJ, Li RK, Weisel RD, *et al.* (2000) Heart cell transplantation improves heart function in dilates cardiomyopathic hamsters. *Circulation* 102: III204–209.

98. Yoo KL, Li RA, Weisel RD, *et al.* (2000) Autologous smooth muscle cell transplantation improved heart function in dilated cardiomyopathy. *Ann Thorac Surg* **70**: 859–865.

99. Pouly J, Hagège AA, Vilquin JT, *et al.* (2004) Does the functional efficacy of skeletal myoblast transplantation extend to nonischemic cardiomyopathy? *Circulation* **110**: 1626–1631.

100. Nagaya N, Kangawa K, Itoh T, *et al.* (2005) Transplantation of mesenchymal stem cells improves cardiac function in a rat model of dilated cardiomyopathy. *Circulation* **112**: 1128–1135.

101. Werner L, Deutsch V, Barshack I, *et al.* (2005) Transfer of endothelial progenitor cells improves myocardial performance in rats with dilated cardiomyopathy induced following experimental myocarditis. *J Mol Cell Cardiol* **39**: 691–697.

102. Soares MB, Lima RS, Rocha LL, *et al.* (2004) Transplanted bone marrow cells repair heart tissue and reduce myocarditis in chronic chagasic mice. *Amer J Pathol* **164**: 441–447.

103. Guarita-Souza LC, Carvalho KA, Woitowicz V, *et al.* (2006) Simultaneous autologous transplantation of cocultured mesenchymal stem cells and skeletal myoblasts improves ventricular function in a murin model of Chagas disease. *Circulation* **114**: I120–124.

104. Wang X, Hu Q, Mansoor A, *et al.* (2003) Autologous skeletal myoblast transplantation for severe postinfarction left ventricular dysfunction. *Amer J Physiol* **290**: H1393–1405.

105. Menasché P, Hagège AA, Vilquin JT, *et al.* (2003) Autologous skeletal myoblast transplantation for severe postinfarction left ventricular dysfunction. *J Amer Coll Cardiol* **41**: 1078–1083.

106. Herreros J, Prósper F, Perez A, *et al.* (2003) Autologous intramyocardial injection of cultured skeletal muscle-derived stem cells in patients with nonacute myocardial infarction. *Eur Heart J* **24**: 2012–2020.

107. Pagani FD, DerSimonian H, Zawadzka A, *et al.* (2003) Autologous skeletal myoblasts transplanted to ischemia-damaged myocardium in humans. Histological analysis of cell survival and differentiation. *J Amer Coll Cardiol* **41**: 879–888.

108. Siminiak T, Kalawski R, Fiszer D, *et al.* (2004) Autologous skeletal myoblast transplantation for the treatment of postinfarction myocardial injury: phase I clinical study with 12 months of follow-up. *Amer Heart J* **148**: 531–537.

109. Smits PC, van Geuns RJ, Poldermans D, *et al.* (2003) Catheter-based intramyocardial injection of autologous skeletal myoblasts as a primary treatment of ischemic heart failure: clinical experience with six-month follow-up. *J Amer Coll Cardiol* **42**: 2063–2069.

110. Siminiak T, Fiszer D, Jerzykowska O, *et al.* (2005) Percutaneous transcoronary-venous transplantation of autologous skeletal myoblasts in the

treatment of post-infarction myocardial contractility impairment: the POZNAN trial. *Eur Heart J* **26**: 1188–1195.

111. Hagège AA, Carrion C, Menasché P, *et al.* (2003) Viability and differentiation of autologous skeletal myoblast grafts in ischaemic cardiomyopathy. *Lancet* **361**: 491–492.

112. Strauer BE, Brehm M, Zeus T, *et al.* (2002) Repair of infarcted myocardium by autologous intracoronary mononuclear bone marrow cell transplantation in humans. *Circulation* **106**: 1913–1918.

113. Assmus B, Schachinger V, Teupe C, *et al.* (2002) Transplantation of progenitor cells and regeneration enhancement in acute myocardial infarction (TOPCARE-AMI). *Circulation* **106**: 3009–3017.

114. Schächinger V, Assmus B, Britten MB, *et al.* (2004) Transplantation of progenitor cells and regeneration enhancement in acute myocardial infarction: final one-year results of the TOPCARE-AMI trial. *J Amer Coll Cardiol* **44**: 1690–1699.

115. Wolkert KC, Meyer GP, Lotz J, *et al.* (2004) Intracoronary autologous bone-marrow cell transfer after myocardial infarction: the BOOST randomized controlled clinical trial. *Lancet* **364**: 141–148.

116. Schächinger V, Erbs S, Elsasser A, *et al.* (REPAIR-AIM Investigators). (2006) Intracoronary bone marrow-derived progenitor cells in acute myocardial infarction. *N Engl J Med* **355**: 1210–1221.

117. Lunde K, Solheim S, Aakhus S, *et al.* (2006) Intracoronary injection of mononuclear bone marrow cells in acute myocardial infarction. *N Engl J Med* **355**: 1199–1209.

118. Janssens S, Dubois C, Bogaert J, *et al.* (2006) Autologous bone marrow-derived stem-cell transfer in patients with ST-segment elevation myocardial infarction: double-blind, randomised control trial. *Lancet* **367**: 113–121.

119. Seeger FH, Tonn T, Krzossok N, *et al.* (2006) Cell isolation procedures matter: a comparison of different isolation protocols of bone marrow mononuclear cells used for cell therapy in patients with acute myocardial infarction. *Circulation* **114**: II-51.

120. Perin EC, Dohmann HF, Borojevic R, *et al.* (2003) Transendocardial, autologous bone marrow cell transplantation for severe, chronic ischemic heart failure. *Circulation* **107**: 2294–2302.

121. Fuchs S, Satler LF, Kornowski R, *et al.* (2003) Catheter-based autologous bone marrow myocardial injection in no-option patients with advanced coronary artery disease: a feasibility study. *J Amer Coll Cardiol* **41**: 1721–1724.

122. Tse HF, Kwong YL, Chan JK, *et al.* (2003) Angiogenesis in ischaemic myocardium by intramyocardial autologous bone marrow mononuclear cell implantation. *Lancet* **361**: 47–49.

123. Assmus B, Honold J, Schachinger V, *et al.* (2006) Transcoronary transplantation of progenitor cells after myocardial infarction. *N Eng J Med* **355**: 1222–1232.

124. Chen SL, Fang WW, Qian J, *et al.* (2004) Improvement of cardiac function after transplantation of autologous bone marrow mesenchymal stem cells in patients with acute myocardial infarction. *Amer J Cardiol* 117: 1443–1448.

125. Cjhen SL, Fang WW, Ye F, *et al.* (2004) Effect on left ventricular function of intracoronary transplantation of autologous bone marrow mesenchymal stem cell in patients with acute myocardial infarction. *Amer J Cardiol* 94: 92–95.

126. Stamm C, Westphal B, Kleine HD, *et al.* (2003) Autologous bone marrow stem-cell transplantation for myocardial regeneration. *Lancet* 361: 45–46.

127. Erbs S, Linke A, Adams V, *et al.* (2005) Transplantation of blood-derived progenitor cells after recanalization of chronic coronary artery occlusion: first randomized and placebo-controlled study. *Circ Res* 97: 756–762.

128. Li TS, Hayashi M, Ito H, *et al.* (2005) Regeneration of infarcted myocardium by intramyocardial implantation of *ex vivo* transforming growth factor beta-preprogrammed bone marrow stem cells. *Circulation* 111: 2438–2445.

129. Kang HJ, Lee HY, Na SH, *et al.* (2006) Differential effect of intracoronary infusion of mobilized peripheral blood stem cells by granulocyte colony-stimulating factor on left ventricular function and remodeling in patients with acute myocardial infarction versus old myocardial infarction: the MAGIC Cell-3-DES randomized, controlled trial. *Circulation* 114: I145–151.

130. Kang HJ, Kim HS, Zhang SY, *et al.* (2004) Effects of intracoronary infusion of peripheral blood stem cells mobilised with granulocyte-colony stimulating factor on left ventricular systolic function and restenosis after coronary stenting in myocardial infarction: the MAGIC cell randomised clinical trial. *Lancet* 363: 751–756.

131. Dawn B, Guo Y, Rezazadeh A, *et al.* (2006) Postinfarct cytokine therapy regenerates cardiac tissue and improves left ventricular function. *Circ Res* 98: 1098–1105.

132. Kanellakis P, Slater NJ, Du XJ, *et al.* (2006) Granulocyte colony-stimulating factor and stem cell factor improve endogenous repair after myocardial infarction. *Cardiovasc Res* 70: 117–125.

133. Ripa RS, Jorgensen E, Wang Y, *et al.* (2006) Stem cell mobilization induced by subcutaneous granulocyte-colony stimulating factor to improve cardiac regeneration after acute ST-elevation myocardial infarction: result of the double-blind, randomized, placebo-controlled stem cells in myocardial infarction (STEMMI) trial. *Circulation* 113: 1983–1992.

134. Ceradini DJ, Kulkarni AR, Callaghan MJ, *et al.* (2004) Progenitor cell trafficking is regulated by hypoxic gradients through HIF-1 induction of SDF-1. *Nat Med* 10: 858–864.

135. Askari AT, Unzek S, Popvic ZB, *et al.* (2003) Effect of stromal cell-derived factor 1 on stem-cell homing and tissue regeneration in ischaemic cardiomyopathy. *Lancet* 362: 697–703.

136. Limana F, Germani A, Zacheo A, *et al.* (2005) Exogenous high-mobility group box 1 protein induces myocardial regeneration after infarction via enhanced cardiac C-kit+ cell proliferation and differentiation. *Circ Res* 97: e78–83.
137. Lu L, Zhang JQ, Ramires FJ, Sun Y. (2004) Molecular and cellular events at the site of myocardial infarction: from the perspective of rebuilding myocardial tissue. *Biochem Biophys Res Comm* 320: 907–913.
138. Zhang M, Methot D, Poppa V, *et al.* (2001) Cardiomyocyte grafting for cardiac repair: graft cell death and anti-death strategies. *J Mol Cell Cardiol* 33: 907–921.
139. Mangi AA, Noiseux N, Kong D, *et al.* (2003) Mesenchymal stem cells modified with Akt prevent remodeling and restore performance of infarcted heart. *Nat Med* 9: 1195–1201.
140. Lim SY, Kim YS, Ahn Y, *et al.* (2006) The effects of mesenchymal stem cells transduced with Akt in a porcine myocardial infarction model. *Cardiovasc Res* 70: 530–542.
141. Gnecchi M, He H, Liang OD, *et al.* (2005) Paracrine action accounts for marked protection of ischemic heart by Akt-modified mesenchymal stem cells. *Nat Med* 11: 367–368.
142. Gnecchi M, He H, Noiseux N, *et al.* (2006) Evidence supporting paracrine hypothesis for Akt-modified mesenchymal stem cell-mediated cardiac protection and functional improvement. *FASEB J* 20: 661–669.
143. Cheng JQ, Lindsley CW, Cheng GZ, *et al.* (2005) The Akt/PKB pathway: molecular target for cancer drug discovery. *Oncogene* 24: 7482–7492.
144. Yau TM, Kim C, Li G, *et al.* (2005) Maximizing ventricular function with multimodal cell-based gene therapy. *Circulation* 112: 1123–128.
145. Askari A, Unzek S, Goldman CK, *et al.* (2004) Cellular, but not direct, adenoviral delivery of vascular endothelial growth factor results in improved left ventricular function and neovascularization in dilated ischemic cardiomyopathy. *J Amer Coll Cardiol* 43: 1908–1914.
146. Schuh A, Breuer S, Al Dashti R, *et al.* (2005) Administration of vascular endothelial growth factor adjunctive to fetal cardiomyocyte transplantation and improvement of cardiac function in the rat model. *J Cardiovasc Pharmacol Ther* 10: 55–66.
147. Toh R, Kawashima S, Kawai M, *et al.* (2004) Transplantation of cardiotrophin-1-expressing myoblasts to the left ventricular wall alleviates the transition from compensatory hypertrophy to congestive heard failure in Dahl salt-sensitive hypertensive rats. *J Amer Coll Cardiol* 43: 2337–2347.
148. Azarnoush K, Maurel A, Sebbah L, *et al.* (2005) Enhancement of the functional benefits of skeletal myoblast transplantation by means of coadministration of hypoxia-inducible factor 1 alpha. *J Thorac Cardiovasc Surg* 130: 173–179.

149. Tang YL, Tang Y, Zhang YC, *et al.* (2005) Improved graft mesenchymal stem cell survival in ischemic heart with a hypoxia-regulated heme oxygenase-1 vector. *J Amer Coll Cardiol* **46:** 1339–1350.

150. Murtuza B, Suzuki K, Bou-Gharios G, *et al.* (2004) Transplantation of skeletal myoblasts secreting an IL-1 inhibitor modulates adverse remodeling in infarcted murine myocardium. *Proc Natl Acad Sci USA* **101:** 4216–4221.

151. Suzuki K, Murtuza B, Beauchamp JR, *et al.* (2004) Dynamics and mediators of acute graft attrition after myoblast transplantation to the heart. *FASEB J* **18:** 1153–1155.

152. Jo J, Nagaya N, Miyahara Y, *et al.* (2006) Transplantation of genetically engineered mesenchymal stem cells improves cardiac function in rats with myocardial infarction: benefit of a novel nonviral vector, cationized dextran. *Tissue Eng* **13:** 313–322.

153. Davis ME, Hsieh PC, Grodzinsky AJ, Lee RT. (2005) Custom design of the cardiac microenvironment with biomaterials. *Circ Res* **97:** 8–15.

154. Miyahara Y, Nagaya N, Kataoka M, *et al.* (2006) Monolayered mesenchymal stem cells repair scarred myocardium after myocardial infarction. *Nat Med* **12:** 459–465.

155. Kondoh H, Sawa Y, Miyagawa S, *et al.* (2006) Longer preservation of cardiac performance by sheet-shaped myoblast implantation in dilated cardiomyopathic hamsters. *Cardiovasc Res* **69:** 466–475.

156. Zhang G, Wang X, Wang Z, *et al.* (2006) A PEGylated fibrin patch for mesenchymal stem cell delivery. *Tissue Eng* **12:** 9–19.

157. Zhang G, Hu Q, Braunlin EA. (2006) Enhancing efficacy of cell transplantation in hearts with post-infarction LV remodeling by an injectable biomatrix. *Circulation* **114:** II-239.

158. Davis ME, Hsieh PC, Takahasi T, *et al.* (2006) Local myocardial insulin-like growth factor 1 infarction. *Proc Natl Acad Sci USA* **103:** 8155–8160.

159. Anversa P, Leri A, Kajstura J. (2006) Cardiac regeneration. *J Amer Coll Cardiol* **47:** 1769–1776.

160. Dib N, Michler RE, Pagani FD, *et al.* (2005) Safety and feasibility of autologous myoblast transplantation in patients with ischemic cardiomyopathy: four-year follow-up. *Circulation* **112:** 1748–1755.

STEM CELLS FOR THE NERVOUS SYSTEM

11

Use of a β-Galactosidase Reporter Coupled to Cell-Specific Promoters to Examine Differentiation of Neural Progenitor Cells *In Vivo* and *In Vitro*

Dale S. Gregerson*

1. Introduction

Several retinal diseases, including retinitis pigmentosa (RP) and other inherited retinal diseases, age-related macular degeneration (ARMD), inflammatory diseases and infections, diabetic retinopathy and retinopathy of prematurity (ROP), result in severe visual loss from the degeneration of retinal photoreceptors, neurons and ganglion cells. Although some amphibians and fish display the capacity for retinal growth and regeneration after birth from cells at the ciliary margin,[1] the adult mammalian retina ceases to grow shortly after birth and does not regenerate. The loss of these cells from disease, injury or inflammation in the mammalian retina is particularly detrimental due to the absence of regeneration. No effective long term treatments for retinal/photoreceptor degeneration or damage exist. Over two million residents of the US alone are currently affected by these diseases. Many of these diseases are slowly progressive, with a continual, increasing loss of vision.

The ability to regenerate and repair damage to critical, terminally differentiated tissues is at the center of hopes for the therapeutic use of stem cells. Embryonic stem (ES)[2] cells have unlimited self-renewal and multipotent differentiation potential. Conversely, tissue-specific progenitor cells have less self-renewal ability than ES cells and, although they

*Correspondence: Department of Ophthalmology, University of Minnesota, Lions Research Building, Rm 314, 2001 6th St SE, Minneapolis, MN 55431-2404, USA. Tel: 612-626-0772; Fax: 612-626-0781; E-mail: grege001@umn.edu

differentiate into multiple lineages, are not multipotent.[3–8] Stem or progenitor cells have been identified in many tissues, including some collected from adults, and include hematopoietic,[9,10] neural,[11,12] gastrointestinal,[13] epidermal,[14] hepatic[15] and mesenchymal stem cells.[16–18] Their availability provides some advantages.

1.1. Stem/progenitor cells

Several studies have reported that transplantation of stem/progenitor cells from various sources into or around the retina led to expression of retina-specific proteins. In particular, evidence for cells that produce proteins specific for photoreceptor cells has been sought. These populations were largely derived from the sources described below.

1.1.1. *Retinal progenitor cells (RPCs)*

RPC have been isolated from the embryonic[19] and adult[20] neural retina. Progenitor cells from the adult mouse ciliary margin were able to differentiate into cells expressing retina-specific proteins *in vivo*.[21,22] Green fluorescent protein (GFP)–marked RPCs inoculated into the subretinal space of mice with spontaneous retinal degeneration were found widely dispersed in the retina. Some donor cells found in the retinal outer layers expressed photoreceptor cell proteins,[23] suggestive of site-specific cues for differentiation. Evidence of rescuing light-mediated behavior was proposed to be a neuroprotective effect rather than replacement of photoreceptor cells, as there was little evidence for repopulation of photoreceptors by donor cells.[23]

1.1.2. *Neural progenitor cells (NPCs)*

The isolation and culture of brain-derived neural stem cells in a serum-free medium with EGF and b-FGF was initially described over 10 years ago.[24,25] NPCs have been tested extensively in studies seeking evidence of repair and restoration of function in the brain as well as the retina. Human NPCs treated with TGF-β3 expressed opsin in culture.[26] Inoculation into the rat vitreous cavity after a needle injury to the retina yielded rare opsin-positive cells in the retina 30 days later. Donor cells were concentrated around the lesions.[26] As some cells were opsin-expressing after 30 days, this suggests that cues from the surrounding tissue induced and sustained the opsin production. NPCs isolated from the brain have been inoculated into adult retinal degeneration RCS rats and retinal degeneration (rd) mice. Use of GFP transgenic mice as donors permitted tracking, and showed widespread incorporation into the retina following subretinal inoculation.[27] Adult rat brain progenitor cells transplanted into the vitreous cavity of newborn rats were able to incorporate into the retina, but did not express markers specific for retinal cells.[28] Although extensive incorporation has been found in some models, generation of functional photoreceptor cells has been quite limited.[29–32]

1.1.3. *Bone marrow*

Adult CD90[+] marrow stromal cells (MSCs) from RCS retinal degeneration rats have been injected into the subretinal space of adult RCS rats. Use of GFP- or BrdU-labeled MSCs allowed visualization of cells integrated into the retina that appeared to make synapse-like interactions and opsin, but no outer segments, in the initial bleb area.[33] A recent finding by Otani *et al.*[34] showed that inoculation of adult lineage-negative (Lin[−]) hematopoietic stem cells rescued the retinal blood vessels in rd mice from complete degeneration, preserved limited evidence of an electroretinogram, and promoted survival of host cones.

1.1.4. *Embryonic stem cells*

Meyer *et al.*[35] injected *in vitro* "neuralized" ES cells into the vitreous of rd mice. After six weeks, the vast majority of GFP-labeled ES cells were spread out on top of the retina; a few processes were extended into the retina, and some expressed neuronal and synaptic markers. Rare cells were found in the inner plexiform layer. Hara *et al.*[36] injected ES cells into the vitreous. Examination at days 5 and 30 revealed that most donor cells were found on the surface of the retina. Some cells sent processes into the inner plexiform layer (IPL), and expressed retinal and neuronal markers.

Tissue-specific cues that direct the migration and integration of progenitor cells into the retina are critical and elusive elements in the search for the successful application of stem cells to tissue repair. NPCs were found to commit to a neuronal fate when cocultured with brain-derived astrocytes.[37] The role of astrocytes may not be the same for differentiation in the retina, as the limited number of astrocytes that populate the eye are generated outside of the eye, and which migrate in via the optic nerve. The major population of retinal glia, Müller cells, could conceivably perform a similar function, but Müller cells were found to inhibit rod development in cultures of the neonatal mouse retina.[38] *In vitro* studies of differentiation show that several factors, such as retinoic acid[39] and taurine,[40] induce a photoreceptor cell fate. As noted above, human NPCs treated with TGF-β3 and inoculated into the rat vitreous cavity of the needle-injured retina produced some opsin-positive cells that were found in the retina after 30 days.[26] Relatively few studies involve grafting of stem/progenitor cells into the retina after initiating differentiation *in vitro*. Although specific cues for differentiation are relatively uncertain, a striking dependency on the developmental state of the precursor cells has been shown. MacLaren *et al.*[41] demonstrated that retinal cell suspensions prepared from postnatal mice at the time when substantial numbers of postmitotic photoreceptor cell precursors are present (P3–P7) were able to generate functional photoreceptor cells in the adult mouse retina if inoculated into the subretinal space. These cells could also be incorporated into mice with retinal degeneration.

Restoration and maintenance of the precise interactions and organization of neural and nonneural retinal cells is crucial for recovery of visual function. It is not currently known what role other retinal cell types may have in determining the fate of grafted progenitor cells, and studies are required to reveal important activities and interactions. Results to date demonstrate that cues for differentiation are provided by the surrounding cells and tissue. Visualizing the activity of these cues is critical for progress in promoting restorative differentiation and regeneration by grafted stem cells. A difficult scientific obstacle to realization of the potential of stem or progenitor cells is the lack of understanding of how their incorporation into the damaged tissue can be promoted, and how differentiation into the necessary cells can be induced to restore function to the damaged tissue or organ. Inducing neural precursor properties prior to subretinal inoculation yielded small numbers of integrated cells with the ability to express a few retinal proteins.[42] Progress in this area could lead to breakthroughs in the use of stem/progenitor cells.

Studies of amphibians, fish and birds, which demonstrate continued retinal growth and repair to varying degrees,[1,43] together with the finding of retinal progenitor cells in both the neural retina and the ciliary epithelium,[20,21] suggest that retinal stem/progenitor cells may be locally present in mammals, but that their activity is constrained by unknown factors. This defines two broad strategies for exploiting stem or progenitor cells:

(1) discover how to activate local progenitor cells, or (2) administer exogenous stem cells induced to enter a differentiation program that will lead to regeneration of the tissue according to the signals provided by the tissue, and/or provided exogenously by the practitioner. If the results of MacLaren *et al.*[41] apply to the human retina, this requires procuring sufficient numbers of postmitotic cells from donors, or learning how to expand and differentiate precursors *in vitro* to give sufficient numbers of the necessary postmitotic cells. We describe here some of our efforts with the second strategy, using NPCs isolated from β-gal-expressing transgenic mice, including hi-arr-β-gal (retinal photoreceptor cell expression), GFAP-β-gal (astrocyte expression), and ROSA26 donors. Each NPC donor type may yield a distinct pattern of Xgal staining based on the inductive cues, and provide a means to assess *in vitro* conditioning strategies.

1.1.5. *A new strategy — NPCs from transgenic progenitor cell donors expressing cell-specific reporters*

We hypothesized that NPCs cultured or grafted under conditions that would promote differentiation may exhibit markers of mature retinal photoreceptor cells, or other cells of the retina or brain. Evidence of differentiation was sought by using NPCs from transgenic mice that express β-gal based on the activities of one of three specific promoters. This includes the arrestin promoter (hi-arr-β-gal mice), the GFAP promoter (GFAP-β-gal mice) and the ROSA26 mice. The β-gal reporter gene used in the hi-arr-β-gal mice and the GFAP-β-gal mice is not to be confused with the more frequent applications of β-gal-transduced/transfected cells, in which it simply marks the presence of transferred cells due to constitutive expression, often via CMV or retroviral LTR promoters. As described below, the NPCs used in these studies were isolated from transgenic mice that express β-gal under the control of specific promoters. *In vitro* studies were done to explore the conditions that would promote or inhibit expression of the β-gal by the various Tg promoter–reporter constructs.

1.1.6. *Hi-arr-β-gal Tg mice*

Under the control of the arrestin promoter, hi-arr-β-gal mice exhibit high levels of β-gal expression in the retinal photoreceptor cells (neurons), and some pineal gland expression, as well as slight histochemically detectable expression in a few unidentified brain cells.[44–47] No other systemic expression

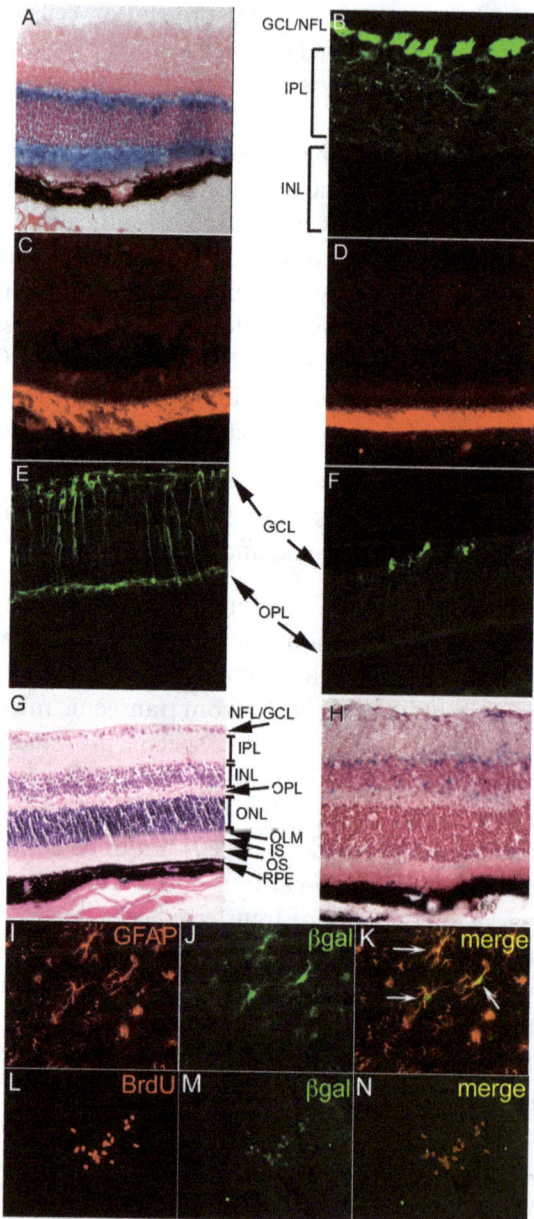

Figure 1. (A–H) Expression patterns of transgenic and endogenous markers of differentiated neurons, photoreceptors, and glia in retinal sections. (**A**) Xgal stain of hi-arr-β-gal mouse retina; (**B**) anti-β3-tubulin stain of normal retina; (**C**) antiarrestin stain of normal retina; (**D**) antirhodopsin stain of normal retina; (**E**) anti-GFAP stain of normal retina; (**F**) anti-β-gal

of β-gal has been found in these mice. The expression in the rod photoreceptor cell layer includes the entire cell cytoplasm from the synapses in the outer plexiform layer to the tips of the outer segments adjacent to the retinal pigment epithelium (RPE) (Fig. 1A).

1.1.7. GFAP-β-gal mice

These mice express β-gal on the GFAP promoter, resulting in expression in a subset of astrocytes in the retina (Fig. 1F) and the brain (Fig. 1J).[48,49] The GFAP-β-gal promoter does not drive expression in the outer layers of the retina in normal GFAP-β-gal mice. Even though other neural glia, including retinal Müller cells, express GFAP (Fig. 1E), only a limited set of astrocytes from these transgenic mice express β-gal, and this does not include Müller cells (Figs. 1I vs 1J). The β-gal in these mice contains a nuclear localizing signal, resulting in a substantial concentration of β-gal in the nucleus.

1.1.8. ROSA26 mice

ROSA26 mice express β-gal based on the activity of an unidentified embryonic promoter.[50,51] While expression in embryonic tissue is widespread, the scope of expression is more limited on the adult B10.A background (Figs. 1H), and shows that the vast majority of highly differentiated cells in the B10.A retina do not express Xgal detectable β-gal from the ROSA26 promoter. Expression of β-gal is found in the GCL, and at the interface between the IPL and the INL (Fig. 1H). While we do not know the developmental state at which ROSA26 expression of *lacZ* is turned off in the majority of retinal cells, it is clear from the examination of embryos and newborns that it is expressed in many more immature cells, and may help to identify donor cells in an immature state. We found bright BrdU and β-gal costaining cells (Figs. 1L and 1M) following local grafting of BrdU-labeled ROSA26 NPCs.

---

stain of retina from GFAP-β-gal transgenic mouse retina; **(G)** H&E stain of normal retina; **(H)** Xgal stain for β-gal expression in ROSA26 mouse retina. **(I–K)** A subpopulation of GFAP⁺ astrocytes from the GFAP-β-gal Tg mouse also expresses β-gal on the GFAP promoter. Immunofluorescence staining for GFAP and β-gal in GFAP-β-gal Tg mouse brain (15 μm frozen sections). **(I)** Anti-GFAP (red); **(J)** anti-β-gal (green); and **(K)** merge. **(L–N)** BrdU-labeled ROSA26 NPCs on the surface of the retina. **(L)** anti-BrdU (red); **(M)** anti-β-gal; and **(N)** merge. Abbreviations: NFL, nerve fiber layer; GCL, ganglion cell layer; IPL, inner plexiform layer; INL, inner nuclear layer; OPL, outer plexiform layer; ONL, outer nuclear layer; OLM, outer limiting membrane; IS, photoreceptor inner segment; OS, photoreceptor outer segment; RPE, retinal pigment epithelium.

This makes it a useful positive control for the other β-gal transgenic mice, BrdU staining procedures and NPC engraftment.

1.1.9. *Preparation of NPCs*

The NPCs used in our studies were isolated and cultured largely as described by Richards *et al.*[25] and Reynolds and Weiss[24] in a serum-free medium supplemented with EGF and b-FGF. The maintenance and induction media are described in Table 1. E14 mouse embryos provided the brain tissue which was minced and dissociated in 0.05% trypsin-EDTA. The dissociated brain cells were washed and seeded to a density of 1×10^6 cells in 5 ml of media per well in a 6-well plate. The culture medium was a serum-free maintenance medium described in Table 1, and by others.[52–54]

2. *In Vitro* Studies

2.1. Characterization and differentiation of the β-gal-expressing NPCs *in vitro*

To determine the ability of the NPCs to differentiate into neural cells and express markers expected of this lineage, we subjected them to three different differentiation protocols. They were maintained and grown as

Table 1. Media Composition for NPC Cultures

Component	Maintenance Medium	Induction Media FCS	RTT	RTT & FCS
DMEM/F12; 1:1	×	×	×	×
N-2 supplement	×	×	×	×
Glucose; 0.2%	×	×	×	×
β-mercaptoethanol; 5 μM	×	×	×	×
Pyruvate; 100 μg/ml	×	×	×	×
Glutamine; 2 mM	×	×	×	×
EGF; 10 ng/ml	×	×	×	×
b-FGF; 10 ng/ml	×	×	×	×
Heparin; 0.002%	×	×	×	×
FCS; 10%		×		×
Retinoic acid; 0.5 μM			×	×
Taurine; 50 μM			×	×
TGF-β3; 10 ng/ml			×	×

neurospheres in a maintenance medium in the undifferentiated, mitotic state (Table 1). To induce *in vitro* differentiation, the NPCs were placed in 4- or 8-well chamber slides at a high or low density with one of the three differentiation media (Table 1). Induced neurospheres were attached to the slide by the next day. By day 2, most of the cells expressed nestin and a few cells expressed glial fibrillary acidic protein (GFAP).[55] The cultures were maintained for up to six weeks, with media changes every 4–5 days. The slides were fixed with 4% paraformaldehyde at days 2, 7, 14, 21 and 42 in culture and processed for immunofluorescence.

2.1.1. *NPCs from hi-arr-β-gal mice after 1 and 2 weeks*

NPCs isolated from hi-arr-β-gal mice and cultured in a maintenance medium did not express detectable levels of β-gal by immunofluorescence (not shown).[55] After one week at high density in an FCS-containing medium, β-gal staining was detectable and expressed in a population morphologically distinct from that expressing GFAP (Figs. 2A and 2B). The β-gal staining was found to closely coincide with the β3-tubulin+ cells (Figs. 2D and 2E). In the normal adult retina, β3-tubulin staining was only detectable in the GCL/NFL and the IPL (Fig. 1B). A culture in an RTT medium gave strong induction of β3-tubulin, but not all cells were double-positive (Figs. 2G and 2H). Culture of the cells in an RTT–FCS combined medium led to more intense β-gal expression in β3-tubulin+ cells after two weeks (Figs. 2M and 2H). Staining for β-gal and GFAP revealed that their expression was beginning to overlap (Figs. 2J and 2K). Seeding the NPCs at high versus low density changed the morphology of the β-gal+ cells, but distinct neural and glial populations remained. Through two weeks in an RTT-free differentiation medium, the distinction between glia and neuron was maintained, i.e. β-gal and β3-tubulin were coexpressed, but β-gal and GFAP were not.

In other experiments, coculture of hi-arr-β-gal NPCs on monolayers of RPE cells did not induce expression of β-gal, nor did coculture of the NPCs with enzymatically dissociated retinal cells, or with retinal explants in cultures up to one week after initiation of the cultures (data not shown).

2.1.2. *NPCs from hi-arr-β-gal mice after 3 and 6 weeks*

In the normal adult retina, arrestin is concentrated in the photoreceptor IS and OS (Fig. 1C). After three weeks in an RTT medium,

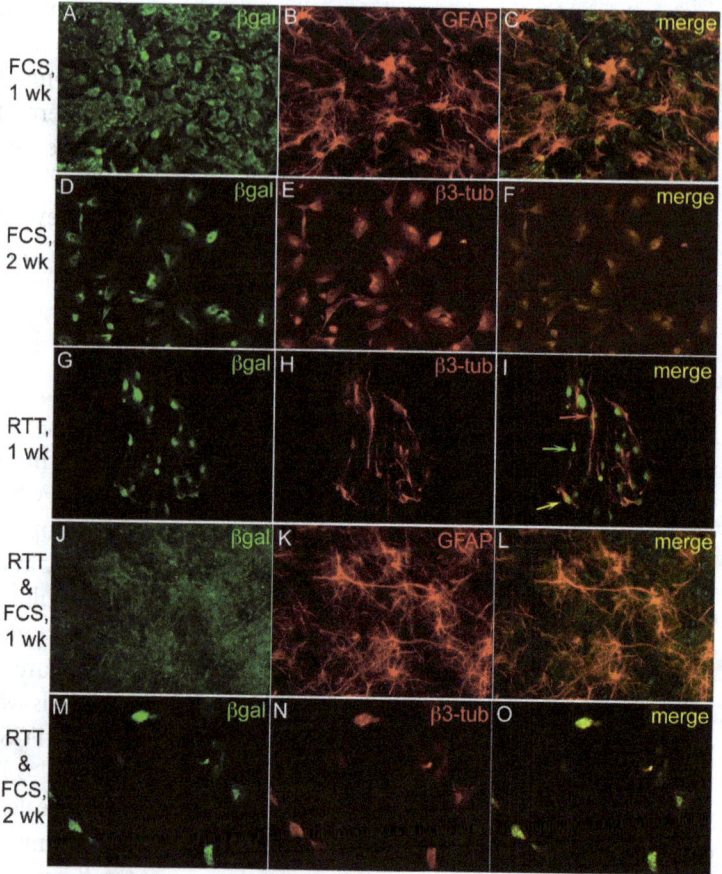

Figure 2. NPCs from hi-arr-β-gal mice after 1 and 2 weeks in culture with differentiation-inducing media. Panels on the left (**A, D, G, J, M**) were stained with FITC-labeled anti-β-gal (green). Panels in the center column are from cultures stained with anti-GFAP (**B, K**) (red) or anti-β3-tubulin (**E, H, N**) (red), as indicated by the labels on the figures. Panels on the right side are the merged images. Culture conditions (medium and duration) are shown in the figure.

β3-tubulin and arrestin were coexpressed (Figs. 3A and 3B), but GFAP and arrestin were also coexpressed (Figs. 3E and 3F), indicating that the two previously distinct lineages, neuron (β3-tubulin$^+$) vs glia (GFAP$^+$), were lost. Six weeks in culture with an RTT and FCS medium led to a near-complete overlap between β3-tubulin and GFAP (Figs. 3M and 3H), indicating a loss of specific expression. By six weeks in culture, arrestin and β-gal, β3-tubulin and GFAP were coexpressed

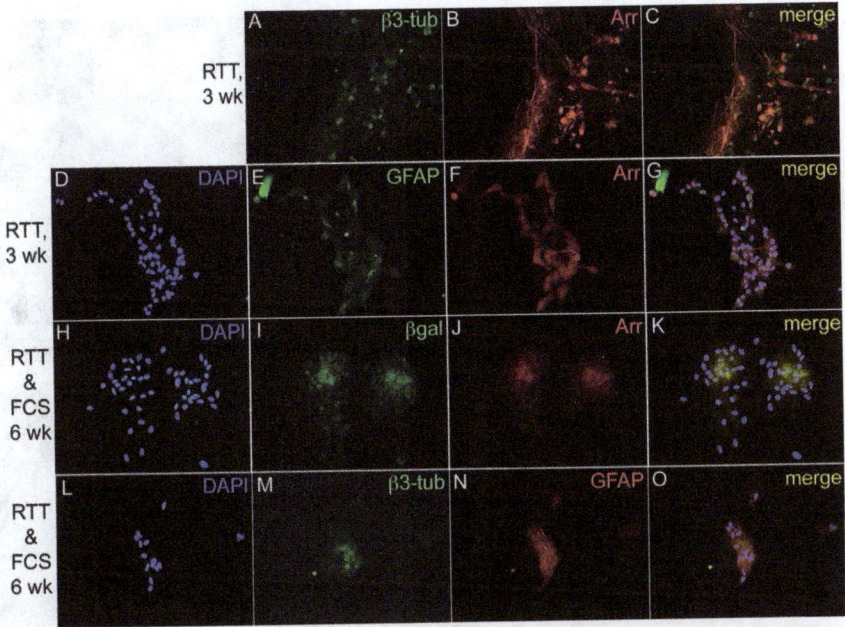

Figure 3. NPCs from hi-arr-β-gal mice after 3 and 6 weeks in culture with differentiation-inducing media. Panels on the left (**D, H, L**) were stained with the nuclear stain DAPI (blue). Panels on the left, center column are from cultures stained with anti-β3-tubulin (**A, M**), anti-GFAP (**E**) (red) and anti-β-gal (**I**) (green). Panels on the right, center column were stained with antiarrestin (**B, F, J**) and anti-GFAP (**N**) (red), as indicated by the labels on the figures. Panels on the right side are the merged images. Culture conditions (medium and duration) are shown in the figure.

by virtually all cells in these cultures, and the RTT–FCS cultures (Figs. 3I and 3J).

2.1.3. NPCs from GFAP-β-gal mice after 2 and 6 weeks

The β-gal expressed in the GFAP-β-gal transgenic mice contains a nuclear localizing signal, so that the nuclei are most heavily stained, giving the β-gal⁺ cells a different appearance than seen in the NPCs from the hi-arr-β-gal mice described above. After two weeks in an FCS-containing medium, GFAP expression was found to be in cells distinct from β3-tubulin⁺ cells, as expected (Figs. 4B and 4C). As observed *in vivo* in these Tg mice, some, but not all, of the GFAP⁺ cells were β-gal⁺ (Figs. 4F and 4G). Inclusion of RTT with the FCS led to nearly all nuclei staining for β-gal (Figs. 4I–4K).

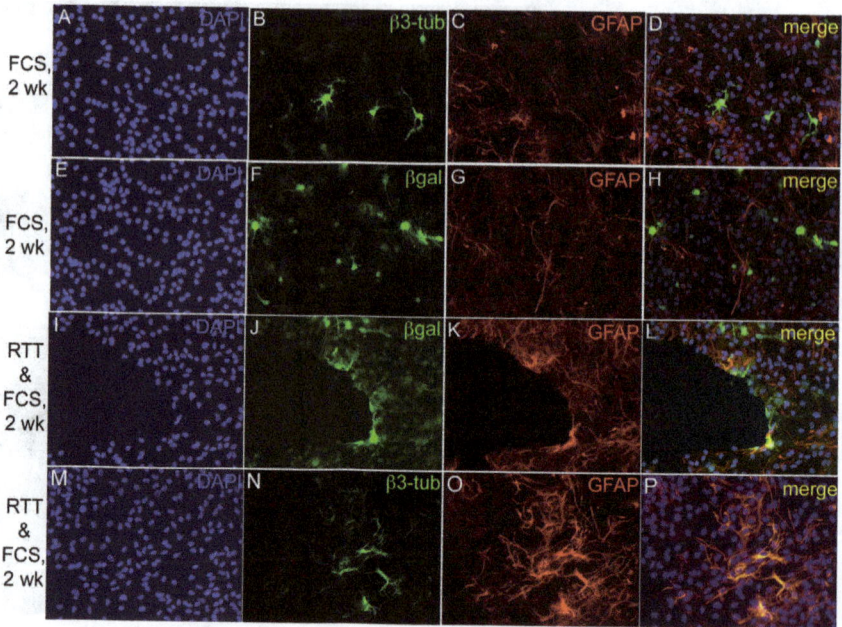

Figure 4. NPCs from GFAP-β-gal mice after 2 weeks in culture with differentiation-inducing media. Panels on the left (**A, E, I, M**) were stained with the nuclear stain DAPI (blue). Panels on the left, center column are from cultures stained with anti-β3-tubulin (**B, N**) and anti-β-gal (**F, J**) (green). Panels on the right, center column were stained with anti-GFAP (**C, G, K, O**) (red), as indicated by the labels on the figures. Panels on the right side are the merged images. Culture conditions (medium and duration) are shown in the figure.

As with the hi-arr-β-gal-derived NPC, inclusion of the RTT supplements led to the near-complete overlap of β3-tubulin and GFAP expression (Figs. 4N and 4O).

2.1.4. NPCs from ROSA26 mice at 1, 2 and 6 weeks

After one week in an FCS medium, virtually all the cells are β-gal$^+$; some are also GFAP$^+$ or β-tubulin$^+$ (Figs. 5A–5E). A different morphology is found after one week in RTT; β-gal now labels only a subset of cells distinct in part from GFAP$^+$ cells (Figs. 5G and 5H). By two weeks in RTT, β3-tubulin and arrestin staining identifies discreet populations (Figs. 5J and 5K). By six weeks, most of the nuclei are β-gal$^+$, but few glial cells (GFAP$^+$) are still distinct in the ROSA26 populations (Figs. 5N and 5O).

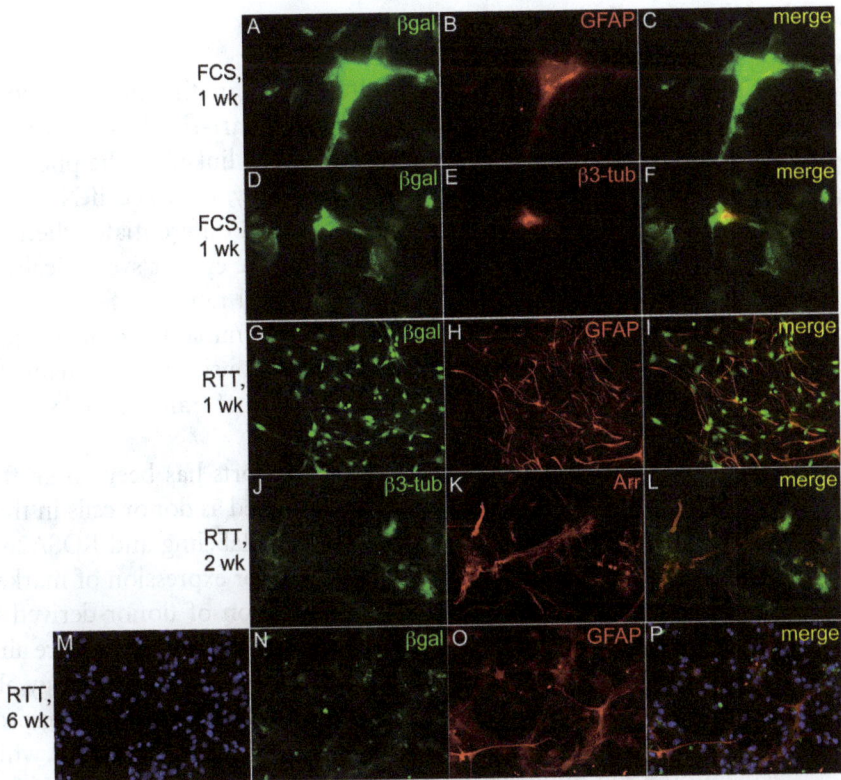

Figure 5. NPCs from ROSA26 mice after 1, 2 and 6 weeks in culture with differentiation-inducing media. The panel on the left (**M**) was stained with the nuclear stain DAPI (blue). Panels on the left, center column are from cultures stained with anti-β-gal (**A, D, G, N**), or anti-β3-tubulin (**J**) (green). Panels on the right, center column were stained with anti-GFAP (**B, H, O**), anti-β3-tubulin (**E**) and antiarrestin (**K**) (red), as indicated by the labels on the figures. Panels on the right side are the merged images. Culture conditions (medium and duration) are shown in the figure.

2.2. Summary

Although the RTT medium initially stimulated differentiation into populations that expressed β3-tubulin or GFAP, this distinct staining pattern was generally lost by three weeks. NPCs cultured in the FCS-containing medium maintained the two populations for the duration of the study (six weeks). These results demonstrated the utility of employing NPCs expressing cell(or lineage)-specific promoter/reporter combinations to search for conditions that might allow *in vitro* expansion and differentiation of precursor cells for tissue repair.

3. *In Vivo* Studies

The goal of the *in vivo* studies was to investigate the incorporation and differentiation of transplanted NPCs from hi-arr-β-gal mice by assaying for their expression of the β-gal reporter gene linked to the photoreceptor cell-specific arrestin promoter. Accordingly, we asked if NPCs could survive, proliferate, localize to the retina, and differentiate when transplanted into the anterior chamber (AC) of the eye, or systemically (IP). Detection was based on staining for the β-gal reporter. Some stem cell recipients were pretreated to produce an inflammatory retinal injury in order to provide an injured environment considered by many to be required for the survival and/or differentiation of transplanted stem/progenitor cells.

The strategy taken in many published reports has been to graft cells bearing markers that allow them to be identified as donor cells in the target tissue (constitutive GFP expression, BrdU labeling and ROSA26 lacZ expression, for example), and then test them for expression of markers of neural differentiation. This allows identification of donor-derived cells, but raises some difficulties when the markers of differentiation are already present in other cells in the recipient retina. Further, this strategy allows transfer of syngeneic cells, diminishing issues of rejection. In our approach, detection is based on β-gal produced from the arrestin promoter, which is active in the rod photoreceptor cells of the retina[44,46,56] (Fig. 1). If induction of this promoter occurred in the NPCs or their progeny in the environment of the retina, it would indicate differentiation toward a mature photoreceptor phenotype. As a result, it is the "differentiated" cells that stand out against a negative background, as shown below. Cells that did not enter this lineage would not be seen.

3.1. NPC transplantation into the brain

As a control for the activity of the arrestin/β-gal promoter/reporter, and to confirm that the NPC preparation would engraft and survive as expected, BrdU-labeled hi-arr-β-gal NPCs (5×10^4 in 2 μl) were inoculated into the striatal area of the brains of adult B10.A mice. No specific lesions were made to promote engraftment. At two days, BrdU-positive cells were found predominantly in the needle track of the transplant site. No β-gal positive cells were observed. At 14 days after transplantation, many BrdU-positive cells had migrated away from the transplant site, but there was no detectable expression of β-gal.[55] The absence of β-gal positive cells suggests that the environment of the striatum did not induce differentiation of the neural stem cells into retinal cells so that the rod photoreceptor cell arrestin promoter was not active in CNS tissue, confirming the specificity of the promoter.

3.2. Injury-conditioning of retinas of recipient mice

It has been reported that migration of hippocampal progenitor cells transplanted to the vitreous cavity is greater in developing, diseased and injured eyes than in adult, normal eyes.[28,57,58] Given the consensus that tissue injury promotes stem cell engraftment, some mice were immunized with a self-peptide of interphotoreceptor retinoid binding protein (IRBP) to induce retinal autoimmune disease (experimental autoimmune uveoretinitis, EAU).[59,60] The disease was limited in severity by the use of this mildly pathogenic peptide, and through whole body irradiation (800 R) at 14 days post-EAU induction to terminate the active response and eliminate the responding T lymphocytes. In these mice, pathology due to EAU was mild (data not shown).

3.3. Anterior chamber inoculation of NPCs and analysis for Xgal⁺ cells

Three days after the irradiation used to terminate the retinal autoimmune response, the mice were given an AC inoculation of 5×10^4 NPCs in 2 μl. After 14 weeks, the eyes were collected for study. Cryosections and retinal wholemounts were examined by microscopy for Xgal⁺ cells. Xgal-positive cells were found in the retinal wholemounts isolated from one of two mice that were not immunized, and in wholemounts from four of six mice that were immunized to induce the retinal autoimmune disease, irradiated, and then given NPCs. While most reports of stem cell incorporation into the retina describe inoculation into the vitreous cavity or subretinal space by transcleral or transcorneal routes, the AC injection of cells is less traumatic to the retina, and avoids a depot effect in it. A representative region from an NPC-inoculated, Xgal-stained retinal wholemount is shown in Fig. 6. Panels A through E are the same field at five different focal planes in the outer nuclear layer. It is clear that most of the cells have a perinuclear stain, and processes extend out in all directions. All the Xgal⁺ cells were found in the peripheral retina.

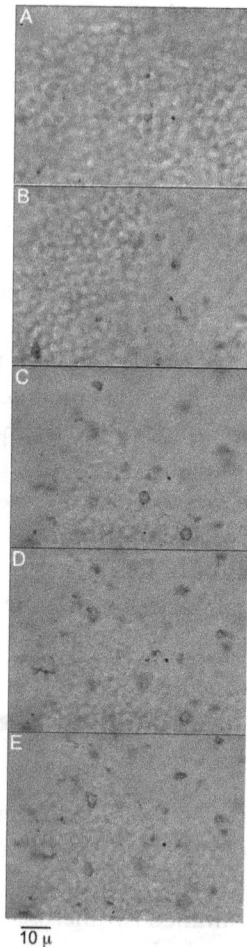

10 μ

Figure 6. Xgal staining of a retinal wholemount taken from an AC-inoculated eye. Consecutive high power views of Xgal-stained cells in a retinal wholemount are the same field at five different focal planes (panels **A–E**), and show that the cells are vertically distributed in the outer nuclear layer of the retina. To prepare retinal wholemounts, the retina was removed from the choroid, and flat mounts were then incubated overnight in Xgal at 37°C. Retinas were then washed in 1 X DPBS and mounted onto a slide. Tissue was air-dried and dehydrated in ethanol and xylene before being coverslipped in Permount.

Eyes were also taken for sectioning of the retinas from NPC-grafted mice to examine the distribution of the Xgal-positive cells in the various layers of the retina. Cells were found in all the layers of the retina, but 66% were found in the photoreceptor cell layer.[55] Altogether, the results showed that the Xgal-positive cells were dispersed, that there was no accumulation of cells on the surface of the retina, and that there was no wholesale proliferation in the eye. Xgal-positive cells were present in all the layers of the retina, but only rare cells were found on the inner surface. These results contrast with a number of reports in which the vast majority of cells were found spread on the inner surface following inoculation into the vitreous cavity. Even with the "injury" of retinal degeneration, the vast majority of ES cells,[36] or retinoic acid-neuralized ES cells[35] inoculated into the vitreous cavity remained on the surface of the retina after six weeks, although there was evidence for expression of neuronal and synaptic markers that entered the IPL. Conversely, it was reported that inoculation of three different ocular stem cell preparations into the retina led to donor cells being found distributed throughout the layers of the retina, and provided evidence for layer-specific cues that altered expression of specific proteins by the donor cells.[61]

3.4. IP inoculation of NPCs

As an alternative to AC inoculation, we also tested the outcome of IP inoculation of the NPCs into mice without EAU, thus avoiding all inoculation-associated effects on the eye. For the IP protocol, 5×10^6 stem cells in 100 μl saline were inoculated. The IP inoculation of NPCs was intended to test the multipotency of the NPCs by assaying for evidence that they might differentiate along other paths, perhaps forming elements of the immune system. It is widely thought that NPCs, or neurally induced ES cells, do not integrate into the normal adult retina,[35,57] and we expected that IP-inoculated NPCs would have little chance of appearing in the normal adult mouse retina, thus serving as a negative control for subsequent manipulations. Previous studies reporting that neural stem/progenitor cells failed to migrate into the normal adult retina used local inoculations into either the subretinal space[33] or the vitreous cavity.[36] In contrast, neither local injection nor the need for retinal injury was found to be essential in our experiments in which the NPCs were administered IP. The two mice given NPCs via the IP route both gave evidence of retinal incorporation without induction of EAU and irradiation (Table 2). The results show that some of these cells migrated from the circulation into the retina and established a stable presence that resulted in expression of the β-gal reporter gene. Finding Xgal-positive cells in significant numbers in the retina following IP inoculation was somewhat unexpected. It is, however, consistent with

Table 2. Integration and Xgal Expression of hi-arr-β-gal NPCs

Exp.	Pretreatment	Injection Site	# of NPCs	Xgal⁺ Retinas
1	Imm & Irrad	AC	5×10^4	4 of 6
2	None	AC	5×10^4	0 of 2
3	None	IP	5×10^6	2 of 2

Imm — immunized; Irrad — irradiated; AC — anterior chamber; IP — intraperitoneal.

Table 3. NPCs or Their Progeny Migrate to Remote Tissue Sites and Survive Long Term

Inoculation site	Number of cells	Treatment	β-gal PCR⁺ Spleen	Brain
AC	$5{-}10 \times 10^4$	None	9 of 16*	
IP	$1{-}5 \times 10^6$	None	6 of 8	1 of 3
Organs†	$5{-}20 \times 10^4$	None	8 of 9	

*Incidence: number of positive samples vs total samples.
†Brain, kidney, thymus.

the report that murine MAPCs inoculated IV showed more widespread engraftment if the recipients were lightly irradiated (250 R) prior to transfer, but that irradiation was not strictly required.[52]

3.5. Engraftment of donor cells in other sites

The plasticity of stem cells is well known[62] and extends to various tissue-specific stem/progenitor cells.[4] For example, BM stem cells gave rise to cells expressing neuron-specific markers in the CNS.[8] As a result, we looked for evidence of NPC-specific DNA in other tissues by PCR. The majority of NPC-transferred mice, both IP- and AC-inoculated, were PCR-positive for β-gal in tests of DNA isolated from the spleen at the termination of the experiments (Table 3). After 16 weeks, two of three mice receiving inoculations into the brain were PCR-positive in the spleen, and one of three IP-inoculated recipients was PCR-positive in the brain (Table 3). Various control mice that did not receive the NPCs were reliably PCR-negative. Engraftment of NPCs in nonretinal tissues was stable, given the 14–16 weeks that passed between NSC inoculation and PCR analysis.

To date, we have found no evidence that the cells in the brain, spleen, kidney or thymus expressed β-gal, demonstrating the tissue-specific expression of the photoreceptor-cell-specific protein, and indicating that inductive cues for differentiation and expression of the reporter gene were present and received in the retina, but not elsewhere.

Concerning the multipotency of the IP-inoculated NPCs, there was no evidence for lymphoid cells derived from the NPCs, even though β-gal DNA was found in the spleen. There seems little functional reason for neural stem cells to take up residence in the spleen as neural cells, raising the possibility that the NPCs are pluripotent and entered another lineage in the splenic environment. However, no donor-derived immune cells were detected. We have made IP inoculations of NPCs into sublethally irradiated recombinase-deficient B10.A mice. These mice should be highly favorable to repopulation by syngeneic cells with hematopoietic potential, but no reconstitution of the lymphoid compartment was found to date (data not shown), although the spleens were positive for the β-gal DNA. The NPCs we have prepared appear to have little ability to produce immune cells under these conditions. The NPC-derived cells from the spleen have not been further characterized.

3.6. Summary

We found that NPC-derived cells expressed a photoreceptor cell-specific protein in the retina. Since interactions with other cells/factors acting in concert with intrinsic cues determine the fate of cells during development, the milieu is expected to strongly impact the differentiation of the NPCs in whatever tissue they may enter. The behavior of transplanted NPCs appears to be strongly influenced by their microenvironment. NPCs carrying the arrestin/β-gal reporter construct inoculated into the brain did not become β-gal-positive by our assays, even though BrdU-positive cells were readily found. This suggests that while both the retinal and brain environments are "neural," only the retina contained the signals that induced the NPCs to activate the arrestin promoter and produce β-gal. Our experience with the retinoic acid,[39] taurine[40] and TGF-β[26] supplemented media raises a note of caution regarding their use *in vitro* to induce photoreceptor cell-like properties. In our hands, these supplements induced the expression of arrestin and rhodopsin, but led to the loss of distinct neuron/glia properties through concurrent expression of GFAP.

In these studies, the expression of a rod photoreceptor cell protein/reporter, arrestin-promoted β-gal, was the principal criterion by which we evaluated most retinal samples for evidence of stem cell engraftment, survival and differentiation. It was also part of our strategy to provide ample time for the cells to respond to the environment — 14 weeks. Their presence in the retina at 14 weeks postgrafting indicates that the cells had taken up residence, and were not simply migrating through the tissue. The presence of Xgal+ cells in all the layers suggests that at least some inductive cues for the arrestin promoter are present throughout the retina. No Xgal+ cells were found in other tissues of the eye. Our observations that both AC-inoculated and IP-inoculated NPCs survive, migrate into the retina, and differentiate to produce a photoreceptor cell protein suggest that these NPCs would be useful for future studies of the applications of stem cells to the damaged retina.

A clear dependency on retinal tissue disruption to promote stem cell engraftment was not found. It has been reported that migration of hippocampal progenitor cells transplanted to the vitreous cavity is much greater in developing,[28] diseased[58] and injured eyes[57] than in adult, normal eyes. Retinal injury was also required for incorporation and differentiation of ocular stem cell preparations.[57] The degree of required injury in other studies appeared to be substantial, since deletion of retinal ganglion cells only was not a sufficient stimulus for engraftment of C17.2 NPCs.[63] In the majority of sections taken from eyes that were nominally "injured" by EAU, there was no evidence of EAU. Perhaps the nature of the injury is an important factor. It is possible that the large number of IP-inoculated IPNPCs, 5×10^6, was a factor that compensated for the lack of injury. Inoculation of 5×10^4 NPCs into the AC of inflammation-injured eyes gave more frequent engraftment than inoculation into unmanipulated eyes.

3.7. Prospective studies/interpretation

The studies to date using this strategy provide a proof of principle. The contrasts in β-gal expression that we may find in the retina between mice receiving NPCs from the arrestin vs GFAP promoters, which are expressed in differentiated cells, and the ROSA26 promoter, which is expressed in immature cells, will, in future studies, provide the opportunity to indirectly visualize the activity of specific cues that drive differentiation. The total number of NSC-derived cells and their distribution should not be affected by the presence or absence of any of the β-gal promoters used here. The ROSA26 mice have been widely used for tracking studies, and we have never seen any evidence of pathology or adverse effects in the hi-arr-β-gal or GFAP-β-gal mice.

Acknowledgments

The author thanks Thien Sam, Heidi Roehrich and Jing Xiao for their assistance in making the studies. This work was supported by a University of Minnesota AHC Faculty Research Development grant, MN Lions and Lionesses Clubs, and Research to Prevent Blindness.

References

1. Reh TA, Levine EM. (1998) Multipotential stem cells and progenitors in the vertebrate retina. *J Neurobiol* **36**: 206–220.
2. Thomson JA, Kalishman J, Golos TG, *et al.* (1995) Isolation of a primate embryonic stem cell line. *Proc Natl Acad Sci USA* **92**: 7844–7848.
3. Ferrari G, Cusella-De Angelis G, Coletta M, *et al.* (1998) Muscle regeneration by bone marrow-derived myogenic progenitors. *Science* **279**: 1528–1530.
4. Jackson KA, Mi T, Goodell MA. (1999) Hematopoietic potential of stem cells isolated from murine skeletal muscle. *Proc Natl Acad Sci USA* **96**: 14482–14486.
5. Takahashi T, Kalka C, Masuda H, *et al.* (1999) Ischemia- and cytokine-induced mobilization of bone marrow-derived endothelial progenitor cells for neovascularization. *Nat Med* **5**: 434–438.
6. Petersen BE, Bowen WC, Patrene KD, *et al.* (1999) Bone marrow as a potential source of hepatic oval cells. *Science* **284**: 1168–1170.
7. Theise ND, Badve S, Saxena R, *et al.* (2000) Derivation of hepatocytes from bone marrow cells in mice after radiation-induced myeloablation. *Hepatology* **31**: 235–240.
8. Mezey E, Chandross KJ, Harta G, *et al.* (2000) Turning blood into brain: cells bearing neuronal antigens generated *in vivo* from bone marrow. *Science* **290**: 1779–1782.
9. Akashi K, Traver D, Miyamoto T, Weissman IL. (2000) A clonogenic common myeloid progenitor that gives rise to all myeloid lineages. *Nature* **404**: 193–197.
10. Akashi K, Reya T, Dalma-Weiszhausz D, Weissman IL. (2000) Lymphoid precursors. *Curr Opin Immunol* **12**: 144–150.
11. Gage FH. (2000) Mammalian neural stem cells. *Science* **287**: 1433–1438.

12. Sakakibara S, Imai T, Hamaguchi K, *et al.* (1996) Mouse-musashi-1, a neural rna-binding protein highly enriched in the mammalian cns stem cell. *Dev Biol* **176**: 230–242.

13. Potten CS, Booth C, Pritchard DM. (1997) The intestinal epithelial stem cell: the mucosal governor. *Int J Exp Pathol* **78**: 219–243.

14. Moles JP, Watt FM. (1997) The epidermal stem cell compartment: variation in expression levels of e-cadherin and catenins within the basal layer of human epidermis. *J Histochem Cytochem* **45**: 867–874.

15. Alison M, Sarraf C. (1998) Hepatic stem cells. *J Hepatol* **29**: 676–682.

16. Pittenger MF, Mackay AM, Beck SC, *et al.* (1999) Multilineage potential of adult human mesenchymal stem cells. *Science* **284**: 143–147.

17. Haynesworth SE, Goshima J, Goldberg VM, Caplan AI. (1992) Characterization of cells with osteogenic potential from human marrow. *Bone* **13**: 81–88.

18. Gronthos S, Zannettino AC, Graves SE, *et al.* (1999) Differential cell surface expression of the stro-1 and alkaline phosphatase antigens on discrete developmental stages in primary cultures of human bone cells. *J Bone Miner Res* **14**: 47–56.

19. Ahmad I, Dooley CM, Thoreson WB, *et al.* (1999) *In vitro* analysis of a mammalian retinal progenitor that gives rise to neurons and glia. *Brain Res* **831**: 1–10.

20. Zhao X, Das AV, Soto-Leon F, Ahmad I. (2005) Growth factor-responsive progenitors in the postnatal mammalian retina. *Dev Dyn* **6**: 6.

21. Tropepe V, Coles BL, Chiasson BJ, *et al.* (2000) Retinal stem cells in the adult mammalian eye. *Science* **287**: 2032–2036.

22. Ahmad I, Tang L, Pham H. (2000) Identification of neural progenitors in the adult mammalian eye. *Biochem Biophys Res Commun* **270**: 517–521.

23. Klassen HJ, Ng TF, Kurimoto Y, *et al.* (2004) Multipotent retinal progenitors express developmental markers, differentiate into retinal neurons, and preserve light-mediated behavior. *Invest Ophthalmol Vis Sci* **45**: 4167–4173.

24. Reynolds BA, Weiss S. (1992) Generation of neurons and astrocytes from isolated cells of the adult mammalian central nervous system. *Science* **255**: 1707–1710.

25. Richards LJ, Kilpatrick TJ, Bartlett PF. (1992) *De novo* generation of neuronal cells from the adult mouse brain. *Proc Natl Acad Sci USA* **89**: 8591–8595.

26. Dong X, Pulido JS, Qu T, Sugaya K. (2003) Differentiation of human neural stem cells into retinal cells. *Neuroreports* **14**: 143–146.

27. Mizumoto H, Mizumoto K, Shatos MA, *et al.* (2003) Retinal transplantation of neural progenitor cells derived from the brain of GFP transgenic mice. *Vision Res* **43**: 1699–1708.

28. Takahashi M, Palmer TD, Takahashi J, Gage FH. (1998) Widespread integration and survival of adult-derived neural progenitor cells in the developing optic retina. *Mol Cell Neurosci* 12: 340–348.

29. Sakaguchi DS, Van Hoffelen SJ, Theusch E, *et al.* (2004) Transplantation of neural progenitor cells into the developing retina of the Brazilian opossum: an *in vivo* system for studying stem/progenitor cell plasticity. *Dev Neurosci* 26: 336–345.

30. Van Hoffelen SJ, Young MJ, Shatos MA, Sakaguchi DS. (2003) Incorporation of murine brain progenitor cells into the developing mammalian retina. *Invest Ophthalmol Vis Sci* 44: 426–434.

31. Chacko DM, Das AV, Zhao X, *et al.* (2003) Transplantation of ocular stem cells: the role of injury in incorporation and differentiation of grafted cells in the retina. *Vision Res* 43: 937–946.

32. Chacko DM, Rogers JA, Turner JE, Ahmad I. (2000) Survival and differentiation of cultured retinal progenitors transplanted in the subretinal space of the rat. *Biochem Biophys Res Commun* 268: 842–846.

33. Kicic A, Shen WY, Wilson AS, *et al.* (2003) Differentiation of marrow stromal cells into photoreceptors in the rat eye. *J Neurosci* 23: 7742–7749.

34. Otani A, Dorrell MI, Kinder K, *et al.* (2004) Rescue of retinal degeneration by intravitreally injected adult bone marrow-derived lineage-negative hematopoietic stem cells. *J Clin Invest* 114: 765–774.

35. Meyer JS, Katz ML, Maruniak JA, Kirk MD. (2004) Neural differentiation of mouse embryonic stem cells *in vitro* and after transplantation into eyes of mutant mice with rapid retinal degeneration. *Brain Res* 1014: 131–144.

36. Hara A, Niwa M, Kunisada T, *et al.* (2004) Embryonic stem cells are capable of generating a neuronal network in the adult mouse retina. *Brain Res* 999: 216–221.

37. Song H, Stevens CF, Gage FH. (2002) Astroglia induce neurogenesis from adult neural stem cells. *Nature* 417: 39–44.

38. Neophytou C, Vernallis AB, Smith A, Raff MC. (1997) Muller-cell-derived leukaemia inhibitory factor arrests rod photoreceptor differentiation at a postmitotic pre-rod stage of development. *Development* 124: 2345–2354.

39. Kelley MW, Turner JK, Reh TA. (1994) Retinoic acid promotes differentiation of photoreceptors *in vitro*. *Development* 120: 2091–2102.

40. Altshuler D, Lo Turco JJ, Rush J, Cepko C. (1993) Taurine promotes the differentiation of a vertebrate retinal cell type *in vitro*. *Development* 119: 1317–1328.

41. MacLaren RE, Pearson RA, MacNeil A, *et al.* (2006) Retinal repair by transplantation of photoreceptor precursors. *Nature* 444: 203–207.

42. Banin E, Obolensky A, Idelson M, *et al.* (2006) Retinal incorporation and differentiation of neural precursors derived from human embryonic stem cells. *Stem Cells* 24: 246–257.

43. Coulombre JL, Coulombre AJ. (1965) Regeneration of neural retina from the pigmented epithelium in the chick embryo. *Dev Biol* **12**: 79–92.

44. Gregerson DS, Dou C. (2002) Spontaneous induction of immunoregulation by an endogenous retinal antigen. *Invest Ophthalmol Vis Sci* **43**: 2984–2991.

45. McPherson SW, Yang J, Chan CC, *et al.* (2003) Resting CD8 T cells recognize beta-galactosidase expressed in the immune-privileged retina and mediate autoimmune disease when activated. *Immunology* **110**: 386–396.

46. Gregerson DS, Torseth JW, McPherson SW, *et al.* (1999) Retinal expression of a neo-self antigen, beta-galactosidase, is not tolerogenic and creates a target for autoimmune uveoretinitis. *J Immunol* **163**: 1073–1080.

47. Gregerson DS, Xiao J. (2001) Failure of memory (CD44 high) CD4 T cells to recognize their target antigen in retina. *J Neuroimmunol* **120**: 34–41.

48. Verderber L, Johnson W, Mucke L, Sarthy V. (1995) Differential regulation of a glial fibrillary acidic protein-*lacz* transgene in retinal astrocytes and muller cells. *Invest Ophthalmol Vis Sci* **36**:1137–1143.

49. Johnson WB, Ruppe MD, Rockenstein EM, *et al.* (1995) Indicator expression directed by regulatory sequences of the glial fibrillary acidic protein (GFAP) gene: *in vivo* comparison of distinct GFAP-*lacz* transgenes. *Glia* **13**: 174–184.

50. Friedrich G, Soriano P. (1991) Promoter traps in embryonic stem cells: a genetic screen to identify and mutate developmental genes in mice. *Genes Develop* **5**: 1513–1523.

51. Zambrowicz BP, Imamoto A, Fiering S, *et al.* (1997) Disruption of overlapping transcripts in the rosa beta geo 26 gene trap strain leads to widespread expression of beta-galactosidase in mouse embryos and hematopoietic cells. *Proc Nat Acad Sci USA* **94**: 3789–3794.

52. Jiang Y, Jahagirdar BN, Reinhardt RL, *et al.* (2002) Pluripotency of mesenchymal stem cells derived from adult marrow. *Nature* **418**: 41–49.

53. Zhao LR, Duan WM, Reyes M, *et al.* (2002) Human bone marrow stem cells exhibit neural phenotypes and ameliorate neurological deficits after grafting into the ischemic brain of rats. *Exp Neurol* **174**: 11–20.

54. Keene CD, Ortiz-Gonzalez XR, Jiang Y, *et al.* (2003) Neural differentiation and incorporation of bone marrow-derived multipotent adult progenitor cells after single cell transplantation into blastocyst stage mouse embryos. *Cell Transplant* **12**: 201–213.

55. Sam TN, Xiao J, Roehrich H, *et al.* (2006) Engrafted neural progenitor cells express a tissue-restricted reporter gene associated with differentiated retinal photoreceptor cells. *Cell Transplant* **15**: 147–160.

56. Sunayashiki-Kusuzaki K, Kikuchi T, Wawrousek EF, Shinohara T. (1997) Arrestin and phosducin are expressed in a small number of brain cells. *Molec Brain Res* **52**: 112–120.

57. Nishida A, Takahashi M, Tanihara H, *et al.* (2000) Incorporation and differentiation of hippocampus-derived neural stem cells transplanted in injured adult rat retina. *Invest Ophthalmol Vis Sci* **41**: 4268–4274.

58. Young MJ, Ray J, Whiteley SJ, *et al.* (2000) Neuronal differentiation and morphological integration of hippocampal progenitor cells transplanted to the retina of immature and mature dystrophic rats. *Mol Cell Neurosci* **16**: 197–205.

59. Avichezer D, Liou GI, Chan CC, *et al.* (2003) Interphotoreceptor retinoid-binding protein (IRBP)-deficient C57BL/6 mice have enhanced immunological and immunopathogenic responses to IRBP and an altered recognition of IRBP epitopes. *J Autoimmun* **21**: 185–194.

60. Namba K, Ogasawara K, Kitaichi N, *et al.* (1998) Identification of a peptide inducing experimental autoimmune uveoretinitis (eau) in H-2A^k-carrying mice. *Clin Exp Immunol* **111**: 442–449.

61. Chacko DM, Das AV, Zhao X, *et al.* (2003) Transplantation of ocular stem cells: the role of injury in incorporation and differentiation of grafted cells in the retina. *Vis Res* **43**: 937–946.

62. Fuchs E, Segre JA. (2000) Stem cells: a new lease on life. *Cell* **100**: 143–155.

63. Mellough CB, Cui Q, Spalding KL, *et al.* (2004) Fate of multipotent neural precursor cells transplanted into mouse retina selectively depleted of retinal ganglion cells. *Exp Neurol* **186**: 6–19.

12

From Neural Stem Cells to Neuroregeneration

Terry C. Burns*, Walter C. Low and Catherine M. Verfaillie

1. Introduction

In many parts of the body, cells lost due to wear and tear or injury are swiftly replaced by new functional cells derived from tissue-specific stem cells. The adult brain, however, fails to regenerate after damage caused by neurodegenerative disease or injury, leading to irreversible loss of function. In the past 15 years it has been found that the adult mammalian brain contains neural stem cells (NSCs) that generate new neurons in specific regions throughout life. These may be involved in certain normal brain functions such as learning and memory. In this article, we follow mammalian NSCs from development, during which they give rise to all neural cells of the central nervous system, to adulthood, when they continue to generate functional neurons in discrete adult neurogenic regions. We then critically examine the idea that NSCs from multiple sources may, under certain conditions, be induced to participate in functionally meaningful neuroregenerative processes.

2. NSC Biology

2.1. The development of NSCs

The first markers of neural identity appear after gastrulation in medial epiblast cells in response to FGF signaling. These neuroepithelial cells represent the earliest neural stem cells, and can be cultured *in vitro* in the

---

*Correspondence: Department of Neurosurgery, Graduate Program in Neuroscience, University of Minnesota Medical School, Minneapolis, MN 55455, USA. E-mail: burn0280@umn.edu

presence of bFGF. The flat neuroepithelial sheet then invaginates to form a tube, the hollow center of which gives rise to the CNS ventricular system. The neural tube is then patterned into the putative spinal cord caudally and the various brain regions rostrally.

Neuroepithelial cells initially proliferate via symmetrical self-renewing cell divisions. However, as proliferation in the neural tube proceeds, notch signaling promotes transition into radial glia cells (RGCs) that maintain both ventricular and pial contacts.[1] Unlike neuroepithelial cells, RGCs demonstrate features of astrocytes and appear coincident with the onset of neurogenesis.[2] RGCs have long been recognized as guides necessary for newly born neurons to migrate to the appropriate cell layer in the telencephalon, after which RGCs were believed to differentiate into astrocytes. Only recently have RGCs themselves been recognized as the direct precursors to neurons and later glia. While maintaining their pial and ventricular connections, RGCs undergo asymmetric divisions giving rise to neuroblasts that then use the radial glial fiber as a guide for migration away from the ventricle.[3]

In the rodent neocortex, RGCs undergo repeated asymmetric divisions giving rise to neurons or neuronal precursors,[4] followed by glia. The transition from neural to glial production results in an upregulation of EGF receptors on RGCs such that isolated NSC cultures can subsequently be maintained *in vitro* in the presence of EGF without need for FGF.[5] RGCs in certain brain regions, such as the cortex, give rise to multiple types of neurons in a specific order before giving rise to glia.[6] This organized temporal generation of progeny is preserved even *in vitro*, highlighting the need to understand developmental processes in order to efficiently generate desired neuronal subtypes for clinical applications.

The final cell division of radial glia is generally symmetric, giving rise to either two neurons or two glia. As such, radial glia disappear during the early postnatal period in most brain regions, with the exception of the retina and cerebellum, where radial glia persist as nonproliferative Müller glia and Bergman-glia, respectively. In the striatal subventricular zone, some radial glia give rise to ependymal cells, while certain others maintain their ventricular end-foot and persist as "subventricular zone NSCs" throughout adult life.[7] A second population of adult GFAP+ precursor cells is found in the subgranular zone of the hippocampal dentate gyrus. These adult neural progenitor populations will be discussed later in this chapter.

2.2. Defining features of neural stem and progenitor cells

Stem Cell Glossary

Multipotent: Capable of clonally generating multiple functional cell types. Adult stem cells are considered multipotent.

Pluripotent: A multipotent cell capable of clonally generating all functional cell types of the entire organism. Germ cells and embryonic stem cells meet these criteria.

Self-renewing: Capable of cell division that gives rise to progeny identical to the parent. In this way, numerous cell divisions should be possible after which cells indistinguishable from the original stem cell still exist. Although the proliferative potential of stem cells should therefore be "unlimited," the accumulation of genetic alternations can confound the study of stem cells after extensive passaging. Telomerase activity, though present, is lower in adult stem cells than in pluripotent or germ cells, leading ultimately to senescence.

Symmetric division: A cell division that gives rise to two identical daughter cells. The daughters may be different than the parent, exemplified by a final division of an NPC that may generate two neurons or two astrocytes. However, if the two daughters are identical to the parent, the division is called a self-renewing symmetric division.

Asymmetric division: A cell division that gives rise to two different cell types. If one is identical to the parent, this is a self-renewing asymmetric division.

Senescence: Most progenitor cells ultimately cease dividing in response to mitogenic stimuli. This may reflect shortened telomere length, preventing the cell from continued division without risk of chromosomal damage.

Stem cell: A cell that can undergo extensive self-renewal, is multipotent and can reconstitute a tissue *in vivo*.

Progenitor: A proliferative cell that gives rise to a restricted set of progeny and that is generally capable of a finite number of cell divisions. This term is sometimes used interchangeably with stem cells.

Precursor: A general term used to describe any cell that gives rise to another cell, including stem cells and progenitors.

Stem cells are by definition multipotent, self-renewing, and capable of extensive proliferation, though they are normally relatively quiescent *in vivo*. Progenitor cells by comparison, have a more restricted fate and are capable of only limited proliferation. A spectrum of stem cells exist, with varying potencies. Pluripotent stem cells have the potential to generate any cell type in any organ. Embryonic stem cells meet the criteria for pluripotency — a feature that can be tested by implantation of cells into the developing blastocyst. A pluripotent cell should be capable of generating a chimera, from which graft-derived germ cells can go on to generate an entire functioning organism. During normal development, stem cells, with greater potential (e.g. zygote) generally give rise to cells with more restricted potential. In this way, the organized generation of the required number and diversity of functional cell types is facilitated. NSCs are an example of multipotent stem cells in that they are capable of generating multiple functional cell types, namely neurons, oligodendrocytes, and astrocytes. More committed progenitor cells also exist in the CNS. NSCs *in vivo* give rise to transit amplifying cells that can proliferate extensively for a limited period of time and subsequently give rise to large numbers of neurons. Likewise, oligodendrocyte progenitor cells (OPCs) exist throughout the parenchyma and give rise to mature functioning oligodendrocytes throughout life, but do not spontaneously generate neurons. In recent years, the divisions between stem cells and progenitors have become blurred, as have the distinctions between different types of adult stem cells, with numerous reports of adult cells and progenitors demonstrating unexpected plasticity. The solid lines in Fig. 1 represent the generally accepted lineages of NSCs and progenitors, while the dashed lines represent examples of progenitors giving rise to progeny outside of their normal repertoire — transitions that often involve specific cytokines such as FGF, SHH, and BMPs.[8,9]

2.3. NSCs in culture

The *in vitro* culture of NSCs was first described in 1992 by Reynolds and Weiss.[10] In this study, the authors described the ability of cells from adult striatal tissue to proliferate as nonadherent spheres in the presence of EGF, and to differentiate into neurons and glia upon growth factor withdrawal. These spherical clusters of cells, now popularly termed neurospheres,

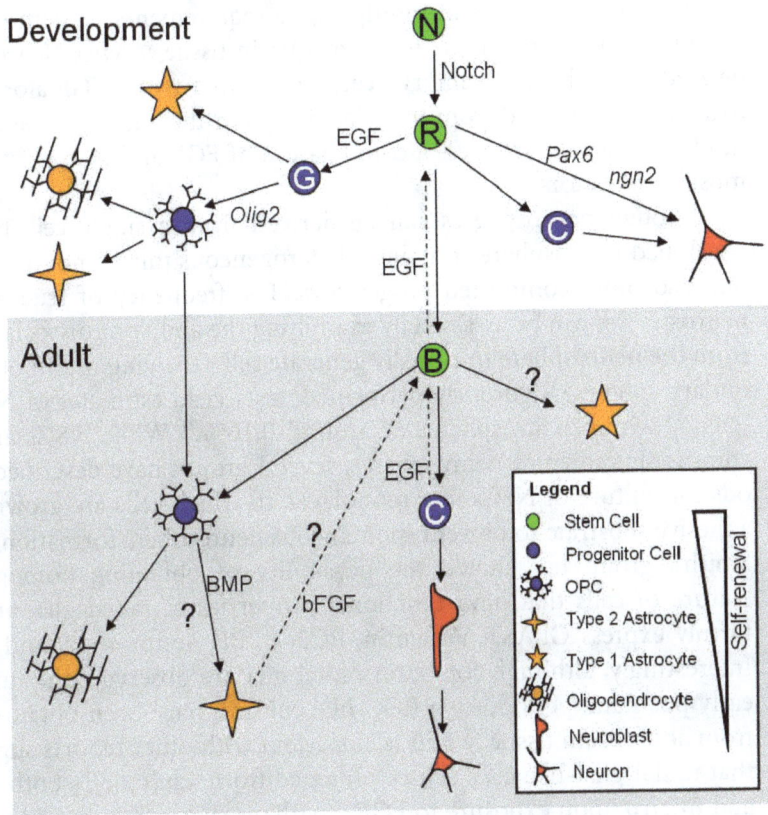

Figure 1. Lineage tree for neural stem cell differentiation. During development, neuroepithelial cells **(N)** give rise to radial glia **(R)**, from which all neural cells of the brain develop. Radial glia may give rise to neurons either directly, or via neural progenitor cells **(C)**. Glial progenitors may give rise to type 1 astrocytes, or to bipotent oligodendrocyte progenitor cells **(OPCs)**. These may give rise to type 2 astrocytes or oligodendrocytes. Most radial glia do not persist after birth; however, some give rise to subventricular zone astrocytes (type B cells), **(B)**, which serve as neural stem cells throughout adult life. Oligodendrocyte precursor cells also persist throughout life. Although SVZ astrocytes can give rise to neurons, oligodendrocytes and astrocytes *in vitro*, they mostly give rise to neurons *in vivo*, via type C cells (transit amplifying cells), and neuroblasts. Demyelinating lesions can promote their differentiation into oligodendrocytes *in vivo*; however, it remains unclear if they give rise to astrocytes *in vivo*. OPCs can give rise to oligodendrocytes throughout life. Isolated OPCs can be induced to generate type 2 astrocytes, which, upon exposure to bFGF, can exhibit properties of multipotent neural stem cells. Whether such transformation is possible *in vivo* remains unclear. Dashed lines represent movement up the differentiation hierarchy toward a more potent cell type. ?: Not yet confirmed *in vivo*.

fulfill the self-renewal and multipotency requirements of stem cells and can functionally contribute to normal brain tissue *in vivo*. Neurospheres derived from the early embryo can be cultured with bFGF alone, while neurospheres derived from the late embryo or the adult may be cultured in EGF alone. Nevertheless a combination of EGF and FGF is favored by most investigators.

Although neurospheres can be derived from a single cell, the fully developed neurosphere contains a heterogeneous mix of not only NSCs, but also more committed progenitors. The frequency of true NSCs in neurospheres can be assessed by examining the ability of dissociated cells from the neurosphere to clonally generate self-renewing multipotent secondary spheres. Rigorously performed, tests yield estimates of NSC frequency within neurospheres of around 0.16%.[11] While NSCs are most commonly grown as neurospheres, several groups have described methods for culture of NSCs in a monolayer, in which cells are grown on an adhesive substrate to prevent spontaneous neurosphere formation. Austin Smith's group has shown the possibility of obtaining homogeneous culture of cells that have functional properties of radial glia, and uniformly express GLAST, Vimentin, RC2, BLBP, nestin, sox2, and pax6.[12] Interestingly, although forebrain radial glia are observed only until the early postnatal period, such RGC-like cultures have been obtained even from adult brain tissue.[13] This is consistent with other reports suggesting that radial-glial-like cells can be obtained from adult NSCs both *in vitro* and *in vivo* upon exposure to EGF,[14] perhaps suggesting that adult NSCs and radial glia are different states of the same cell.

What fates do NSCs adopt after *in vitro* differentiation? Although neurons are the most common progeny of subventricular zone (SVZ) NSCs *in vivo*, NSCs derived from the adult brain differentiate mostly into astrocytes, a modest percentage of neurons, and very few oligodendrocytes, though various *in vitro* manipulations can alter the relative yields of neurons and glia. The most common neurotransmitter phenotype generated from NSCs *in vitro* is GABAergic, in keeping with the normal *in vivo* progeny of adult NSCs. However, cholinergic, glutamatergic and various neuroendocrine cells are also spontaneously generated.[15] Generally the yield of brainstem or spinal cord subtypes is very low compared to those of the forebrain. NSCs have not proven to be a reliable source of generating midbrain dopaminergic neurons or motorneurons. However, occasional studies have indicated that good yields of MN[16] and DA[17] neurons can be obtained, via either priming[16] or overexpression of appropriate combinations of transcription factors.[17] Thus, although the array of neural subtypes achievable *in vitro* is substantially greater than that observed from adult neural stem cells *in vivo*, using current culture techniques, the range of neural phenotypes that can be generated efficiently from NSCs is still limited.

2.4. **Where are NSCs found in the brain?**

To date, no specific markers have been identified for definitive prospective identification of NSCs. As such, stem cells are defined based on their self-renewing and multipotent behavior and can therefore only be identified retrospectively. Areas of ongoing neurogenesis in the adult brain have been identified mostly based on BrdU incorporation by dividing cells, with subsequent identification of BrdU in neuronal cells. However, such methodology does not definitively identify where the new BrdU-labeled neuron originated, nor does it identify quiescent stem cells with latent neurogenic potential. Nevertheless, work based heavily on BrdU labeling has established the subventricular zone and hippocampal dentate gyrus as two regions that undergo neurogenesis throughout life. An alternative method for identifying NSCs is to cultivate cells from a given brain region *in vitro* and demonstrate the appearance of multipotent self-renewing cells as discussed above. *In vitro* neurogenesis has been observed after culture of virtually all adult brain regions, which could mean that all brain regions may harbor NSCs. Alternatively, *in vitro* manipulations may impart neurogenic capabilities to progenitors that would not normally generate neurons *in vivo*.

Though exposure to specific factors may enable certain progenitors from diverse brain regions, the remainder of this section will discuss only adult brain regions found to harbor NSCs, defined as multipotent, self-renewing cells that can be cultured in bFGF, EGF, or both. Based on this criterion, NSCs have been found to exist in the SVZ as well as other regions of the ventricular neuroaxis including the spinal cord.[18] Regions developmentally derived from periventricular regions, including the rostral extension (RE) of the SVZ along the RMS and into the olfactory bulb,[19] as well as a caudal extension of the SVZ called the subcallosal zone (SCZ),[20] also contain multipotent NSCs capable of self-renewal in EGF and/or bFGF. Although NSCs from all ventricular regions are multipotent *in vitro*, SVZ and RMS NSCs appear to give rise mostly to neurons *in vivo*, while the SCZ appears to give rise only to oligodendrocytes.[20]

Several reports have suggested that the hippocampus also contains multipotent NSCs that could be cultured in the presence of bFGF. However, this view has been challenged by other investigators.[21] Technical differences such as the inclusion of ventricular regions in hippocampal microdissection or initial plating in serum-containing media in some cases may help explain why some groups but not others have been able to find cells with NSC properties in the hippocampus. The apparent difficulty in consistently obtaining NSC cultures from such a highly neurogenic region is surprising. Nevertheless, for the purposes of this chapter, neurogenic cells in the hippocampus will be referred to as neural progenitor cells (NPCs). Finally, cells have been isolated from the retina that can be clonally expanded and give rise to photoreceptors, bipolar neurons, and Müller glia. Unlike forebrain NSCs, these cells could self-renew even in the absence of exogenous bFGF or EGF, though paracrine release of bFGF appears to be in part responsible for the retinal stem cell self-renewal.[22] Neural precursors in the SVZ and DG possess properties of astrocytes and express GFAP, though retinal NSCs are GFAP-negative.

3. Adult Neurogenesis

3.1. *In vivo* SVZ neurogenesis

Altman and Dass provided evidence in 1965 that newly born neurons could be found in the olfactory bulb of postnatal rats.[23] It is now recognized that thousands of new neurons are added to the rodent olfactory bulb each day. Two main classes of interneurons exist in the olfactory bulb: granule neurons and periglomerular neurons. Both are GABAergic, though periglomerular neurons additionally express dopamine. Recent studies have revealed different spatial locations of progenitors for these two neural subtypes. Granule cells constitute the vast majority of newly born cells in the olfactory bulb, with less than 3% becoming periglomerular cells.[24]

Although early experiments suggested that ependymal cells were the neural stem cells responsible for olfactory bulb neurogenesis, this conclusion was not supported in further studies. Instead, it is now believed that self-renewing subventricular zone astrocytes (type B cells) reside in the wall of the lateral ventricles, where they divide infrequently, giving rise

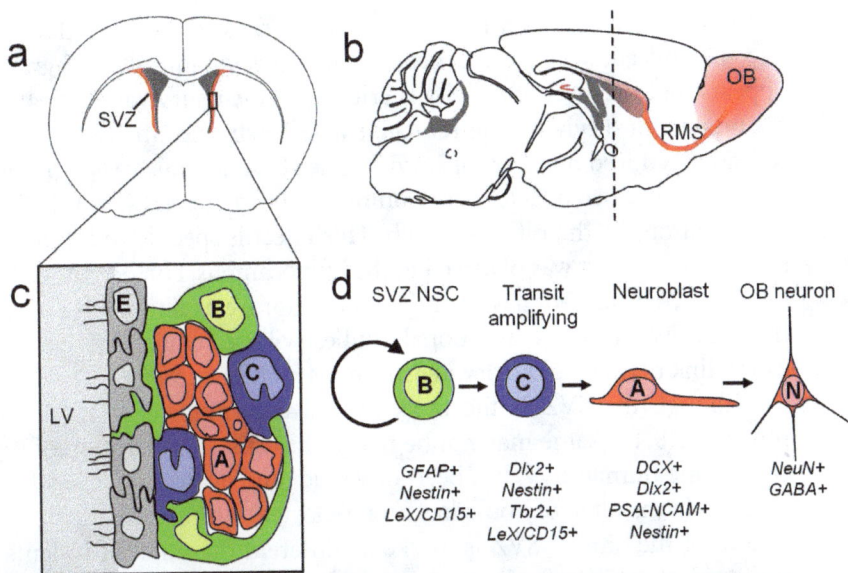

Figure 2. Subventricular zone neurogenesis. **(a)** Coronal section through the subventricular zone at the approximate location of the dashed line in **(b)**. Regions shown in red in **(a)** and **(b)** represent adult brain regions involved in neurogenesis throughout life. **(c)** Diagram of the subventricular zone showing the self-renewing SVZ NSCs (type B cells) in green. These give rise to highly proliferative type C cells (blue), which in turn give rise to migratory neuroblasts (type A cells) which migrate along the rostral migratory zone to the olfactory bulb. **(d)** Schematic representation of the neurogenic lineage of adult neural stem cells. Markers used to identify the various cell populations are shown in italics. Panels **(a)** and **(b)** adapted from *the mouse brain in stereotaxic coordinates*[25]; panels **(c)** and **(d)** are modeled in part after Doetch *et al.*[26]

to rapidly dividing transit amplifying cells termed type C cells (Fig. 2). These in turn give rise to migratory neuroblasts which migrate in chains through an extensive network of specialized glial tunnels in the SVZ toward the RMS and olfactory bulb.

3.2. Functional implications of SVZ neurogenesis

What is the purpose of adding more of these granule and periglomerular cells throughout life? A complete answer to this question is still emerging. However, preliminary studies suggest that olfactory bulb neurogenesis is actively regulated by olfactory stimuli and is associated with functional effects relating to olfactory discrimination and memory.

In normal mice, about half of granule neurons die between day 15 and day 45. Indeed, when these neurons become synaptically integrated, it appears that the level of activity is critical for their survival. After 45 days, the number of surviving cells remains stable. Nostril closure has been shown to lead to reduced numbers of surviving newborn granule cells, while exposure to an enriched olfactory environment led to increased survival of newly born neurons in the olfactory bulb. This effect is specific to the olfactory system, as no effect was observed in the hippocampus. Further, the enriched olfactory environment has been shown to improve olfactory but not spatial memory. Nevertheless, additional studies will be required to establish a causal link between olfactory neurogenesis and function.[24]

Although the SVZ is the most extensive germinal area in the adult rodent CNS, the same may not be true in humans. Proliferation of neural cells in the human SVZ has been observed, and evidence for newly born neurons in the olfactory bulb has been reported. However, the basic organization of the human SVZ is markedly different than that of rodents and is devoid of chains of migrating neuroblasts.[27] In the adult human brain, only occasional cells with neuroblast morphology are observed, and the route via which they migrate to the olfactory bulb, if indeed this occurs, remains to be demonstrated. Such findings may reflect the relative importance of olfaction in rodents and humans, but may also have clinical importance, given that the SVZ appears to be the main source of new neurons mobilized in response to injury (see "Neuroregeneration," below).

3.3. Hippocampal NSCs

A second region in which neurogenesis occurs throughout life is the hippocampal dentate gyrus. Certain RGCs migrate from the ventricle to the dentate gyrus around the perinatal period, after which they serve as precursor cells throughout life. Although hippocampal progenitors are derived from radial glia, this neurogenic region differs from other adult CNS regions containing neural stem cells as it does not maintain direct contact with the ventricle. GFAP[+] hippocampal progenitor cells reside in the subgranular layer, directly below the granule cell layer of the dentate gyrus, and have certain features in common with radial glia, such as the ability to undergo mitosis without retracting complex projections and guidance of migrating progeny. These give rise to "type D" cells, which express similar markers as type A cells in the SVZ/RMS, including PSA-NCAM and DCX. Type D cells divide once before migrating into the granule cell layer to form a new granule cell.[28] Adult humans have an extensive population of hippocampal neural progenitors which give rise to large numbers of neurons *in vivo* and *in vitro*.[29]

Newly born neurons in the adult rodent dentate gyrus have been shown via retroviral labeling to mature and to synaptically and functionally integrate into the hippocampal circuitry. Ultimately, the functional properties of adult born dentate granule neurons become indistinguishable from those born during development. Unlike the olfactory bulb, where most cell death occurs after synaptic integration, many new hippocampal neurons are lost throughout the first three weeks.[30] Increased

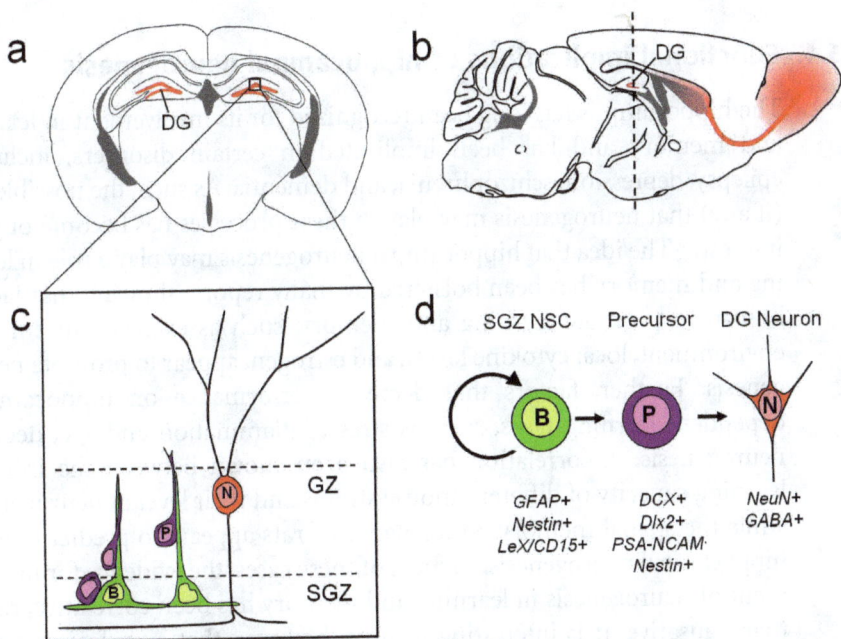

Figure 3. Subgranular zone neurogenesis. **(a)** Coronal section through the hippocampus at the approximate location shown by the dashed line in **(b)**. Regions involved in adult neurogenesis are shown in red in panels **(a)** and **(b)**. **(c)** Diagrammatic representation of neurogenesis in the dentate gyrus. NSCs (type B cells) in the subgranular zone (SGZ) give rise to precursor cells which migrate up the radial extension of the type B astrocyte. Progenitor cells then detach from the radial fiber and differentiate into a granule neuron. Markers used to identify the various cell populations are shown in italics. This lineage is illustrated in **(d)**, which also shows markers used to identify the various cell populations in italics. Panels **(a)** and **(b)** adapted from *the mouse brain atlas in stereotaxic coordinates*[25]; panels **(c)** and **(d)** modeled in part after Doetch *et al.*[26]

hippocampal activity, induced for example by hippocampus dependent learning tasks, housing in enriched environments, or even experimental induction of long term potentiation[31] promotes survival of new neurons via both NMDA-receptor-mediated synaptic[30] and other trophic[32] effects.

The number of newly born neurons in the hippocampus drops precipitously with age, a change due almost completely to a drop in cell production with no significant change in the percentage of newly born cells that survive. This finding is consistent with increased basal levels of inflammation and cortisol with age, both of which are known to inhibit neurogenesis, along with decreased levels of mitogens including FGF-2, IGF-1, and VEGF. Adrenalectomy, growth factor infusion and exercise can restore neurogenesis in aged animals to levels similar to those of younger animals.[33]

3.4. Functional implications of hippocampal neurogenesis

The hippocampus has long been recognized for its involvement in learning and memory and has been implicated in certain disorders, including epilepsy, depression, schizophrenia, and dementia. As such, the possible role (if any) that neurogenesis may play in these processes has become of great interest.[33] The idea that hippocampal neurogenesis may play a role in learning and memory has been bolstered by many reports showing that factors known to promote learning and memory, such as exercise, an enriched environment, local cytokine levels, and estrogen, appear to promote neurogenesis. Further, factors that decrease performance on hippocampus-dependent learning tasks, such as stress, inflammation and age, decrease neurogenesis. A correlation has also been shown between the inherent learning capacity of different mouse strains and their level of neurogenesis, while the spatial memory performance of rats appears to predict levels of hippocampal neurogenesis. In most of these cases, the evidence for involvement of neurogenesis in learning and memory has been correlative, rather than causative. It is interesting to note, however, that even learning itself appears to directly induce neurogenesis, while *in vivo* induction of long term potentiation, an experimental model of synaptic plasticity, was further shown to enhance neurogenesis through both increased progenitor proliferation and enhanced neuronal survival.[31] However, these findings are not without dissent. For example, stress can in some cases enhance memory formation, while estrogens in certain situations have been argued to have negative effects on learning, in spite of effects on neurogenesis that would predict the opposite. Further, certain learning tasks either do not affect or decrease neurogenesis. Technical and methodological considerations may account for some of these discrepancies, but these issues would be best resolved by experiments that directly investigate possible causal connections between learning and neurogenesis.[34]

Unfortunately, however, such experiments have been inherently difficult to design, as any intervention aimed to enhance or reduce hippocampal neurogenesis may have additional effects on the rest of the hippocampus. For example, the major strategies employed to date to inhibit neurogenesis have included irradiation and chemotherapeutic agents to inhibit cell proliferation. Although reduced neurogenesis has been shown to inhibit hippocampus-dependent learning, the mitotic inhibitors and irradiation used to observe these effects may also affect the overall health of the animal and alter the local hippocampal microenvironment, respectively, and conclusions of such experiments have been conflicting.[34,35] Thus, further studies employing more selective techniques are needed in order to definitively assess functional implications of hippocampal neurogenesis.

4. Neuroregeneration

4.1. Introduction to neuroregeneration

Almost 90 years ago, Ramon y Cajal wrote: "In adult centres the nerve paths are something fixed, ended, immutable. Everything may die, nothing may be regenerated. It is for the science of the future to change, if possible, this harsh decree."[36] Though we now appreciate that limited neurogenesis occurs in specific brain regions throughout life, Cajal's statement still accurately reflects the sobering challenge of reconstructing damaged CNS circuitry. Much hope has been placed on the putative capacity of stem cell technology to one day cure Parkinson's disease and to make paraplegics walk. However, we are only now beginning to understand even the basic principles underlying NSC function. Nevertheless, there are hints that under specific circumstances, the CNS may be receptive to the integration of new neurons, even outside of neurogenic areas. Further, some lines of evidence suggest that the brain itself may re-express certain developmental signals in response to injury, lending credence to the idea that small regenerative steps may eventually be achieved. In this section, we discuss the potential of NSCs to generate specific cell types, then explore the two primary approaches to NSC-mediated brain repair: mobilization of endogenous NSCs and cell transplantation. First, however, a few words are needed about methodologies for evaluating neuroregeneration.

4.2. Methodological considerations for the study of neuroregeneration

The most widely employed method for labeling newly born cells is with thymidine analogs such as BrdU[37] which are incorporated into DNA during the S phase of mitosis. Newly born neurons have now been reported in diverse regions of the brain, including hippocampal CA1, cortex, striatum, amygdala, subcortical white matter, around the third ventricle, the dorsal vagus complex, and the substantia nigra. In many cases, however, these findings have been hotly disputed, resulting in uncertainty regarding basal levels of neurogenesis outside of the DG and SVZ/OB.[33,38]

Some of the concerns surrounding the use of BrdU as a marker for newly born neurons are as follows:[33,37]

(1) BrdU is taken up by postmitotic cells undergoing DNA repair, as occurs at low levels in all normal cells. However, a number of other methods have demonstrated that under basal conditions, the risk of misidentifying newly born cells based on routine DNA repair is extremely low.

(2) Satellite cells are mitotic cells often found in close association with neurons. Hence, careful confocal microscopy and 3D reconstruction is important for confirming the colocalization of BrdU and neuronal markers.

(3) Perhaps the most significant risk for misidentification of newly born cells is in areas of injury. Reports have demonstrated that cells damaged after injury may re-enter the S phase in an attempt to repair damage, though in such cases the cell is generally believed to be severely damaged, and dies within the next 28 days. Such problems appear to be most common in areas of combined ischemia and hypoxia.

(4) BrdU has been shown to have toxic effects when used at high doses, which may therefore alter the behavior of the progenitor cells being studied. Conversely, very low doses may yield false negative results.

Additional challenges have arisen in regard to definitive identification of transplanted cells. A high percentage of cells die shortly after transplantation, making identification with cytoplasmic beads and membrane dyes unreliable due to possible leakage and uptake by host cells. Therefore, labels that theoretically can not be transferred to host cells such as BrdU, which is incorporated into the cell's DNA, or transgenic labels have been recommended.[39] The relative ease of labeling with BrdU and the avoidance of silencing concerns has led to widespread use of BrdU to label cells prior to transplantation. Recent studies, however, have suggested that nucleotides released from cells that die after transplantation can be taken up by dividing host cells, leading to misidentification of host cells as transplanted cells.[37,40] Thus, studies in which BrdU was used as the sole label should be interpreted with caution. An additional layer of complexity is that transplanted cells may fuse with endogenous host cells, leading to expression of exogenous transgenes in host cells.[41] Proof of neurogenesis from either endogenous or transplanted cells thus requires a battery of appropriate controls to rule out a considerable number of potential pitfalls in the quest for accurate interpretation of experimental results.

4.3. NSC plasticity

A dogma of developmental biology has stated that precursor cells of greater developmental potential give rise to precursors of progressively more restricted potential. By the time of birth, only tissue-specific stem/progenitor cells exist which have traditionally been considered restricted to a specific repertoire of progeny relevant for maintenance and repair of a specific organ. The past several years have seen an abundance of reports that stem cells may possess significantly greater potential than previously thought possible, with numerous examples of somatic stem cells crossing tissue and even germ layer boundaries to generate cell types from unrelated organs. This behavior is commonly termed "transdifferentiation." NSCs, for example, were in quick succession shown to generate blood, muscle, and even to contribute to all three germ layers after implantation into the developing blastocyst,[42] suggesting pluripotency.

Meanwhile, numerous other papers suggested that nonectodermal cells could give rise to neural cells.[43] A second wave of papers, however, quickly followed suggesting explanations other than transdifferentiation to explain the findings, such as culture artifact, genetic transformation and cell fusion, while still other reports were simply unable to replicate the original claims of transdifferentiation.[44] As such, although fusion-independent differentiation of NSCs has recently been demonstrated *in vitro*,[45] widespread differentiation of NSCs to nonneural lineages is probably unlikely, and remains to be conclusively demonstrated *in vivo*.

4.4. Mobilization of endogenous NSCs

4.4.1. *Regulating NSC behavior in vivo*

A burgeoning number of intrinsic and extrinsic factors are being identified that regulate the behavior of NSCs both *in vitro* and *in vivo*, many of which are being investigated for potential neuroregenerative applications. Regulation of regenerative processes may occur at several levels, including stem cell self-renewal, fate specification, migration, functional integration, and survival. Thus, effective mobilization strategies may require specific interventions to guide each of these steps via infusion of cytokines, viral transduction with regulatable constructs to provide appropriate environmental support, and genetic manipulation of the progenitor cells themselves, all in concert with appropriate rehabilitative programs to capitalize on innate brain plasticity.[46]

The environment surrounding the endogenous adult stem cell is critical for its maintenance as a self-renewing entity and is termed the "stem cell niche."[47] This niche is composed of specific cell types that deposit specific extracellular matrix molecules and cytokines that together ensure that self-renewing stem cells are retained. Endothelial cells, astrocytes, and bone-marrow-derived cells are especially critical in the regulation of maintenance and differentiation of adult NSCs.[48] Table 1 provides examples of factors involved in regulation of NSC proliferation and differentiation. Of note is that certain cytokines, such as bFGF, appear to expand numbers of NSCs, while others, such as EGF and BDNF, promote migration of progenitors into adjacent parenchyma. BDNF, particularly in combination with Noggin, promotes not only migration, but also neural differentiation into region-appropriate neural subtypes in the striatum.

Several injury models spontaneously result in increased proliferation in both the SVZ and the DG, possibly via upregulation of endogenous mitogens. Increased proliferation in the DG generally leads to robust increases in neurogenesis in the DG only, though SVZ-derived cells have been shown to migrate widely throughout the brain, differentiating into neurons and glia in areas of injury. In most cases, the level of spontaneous neurogenesis is very low, and the vast majority of newly born neurons fail to survive, yielding estimates of approximately 0.2% replacement of lost neurons.[96] Recent reports, however, suggest that such neurogenesis may continue for months after injury,[97] perhaps leading cumulatively to meaningful numbers of new neurons. An additional challenge is that although some new neurons may be found reliably after damage to the striatum,

Table 1. Extrinsic Regulators of Adult Neurogenesis

Factor	Cell Proliferation	Additional Effects	Comments	References
Hormones & Peptides				
Corticosterone	↓	↓ neurogenesis	DG only; ↑ w/stress	49, 50
Estrogen	↑		DG only	51
Prolactin	↑	↑ neurogenesis	SVZ/OB only	52
Neurotransmitters				
Glutamate	↓			53
Serotonin	↑	↑ neurogenesis	↓ in depression	54–56
GABA	↓ (SVZ)	↑ integration (DG)		32, 53, 57
Norepinephrine	↑			52
Dopamine	↑			58–61
Cytokines				
bFGF	↑	↑ neurogenesis	SVZ/OB only	62, 63
EGF	↑	↓ neurogenesis		14, 62, 64
TGF-α	↑			64, 65
IGF-1	↑		Role in ischemia	51, 66–68
VEGF	↑	↑ neurogenesis & survival	Linked to DG activity	69
BDNF	↑ (SVZ only)	↑ neurogenesis		70
CNTF	↑			71
SHH		↓ neurogenesis		72, 73
BMP		↑ neurogenesis		74
Noggin		↑ neurogenesis	Made by ependymal cells	74
Wnt	↑			75

(Continued)

Table 1. (*Continued*)

Factor	Cell Proliferation	Additional Effects	Comments	References
Intercellular Interactions				
Notch		↑ NSC survival; ↓ neurogenesis	See Ref. 76	76–79
Eph/Ephrin signaling	→	↓ NSC survival; guide migration		80, 81
Environmental stimuli				
Environmental enrichment	←	↑ neuronal survival	See Ref. 35	82, 83
Learning	←	↑ neuronal survival		31, 84
Exercise	←			85, 86
Ischemia	←			66, 87
Epilepsy	←			88–90
Depression	→			54, 56
Stress	→		↓ serotonin	49, 50
Aging	→		Via corticosterone	91, 92
Inflammation	→		Multiple mechanisms	87, 94, 95
			See Ref. 93	

which is directly adjacent to the SVZ, such as in Huntington's disease or striatal ischemia, new neurons have not consistently been observed after ischemia in the cortex, which is further from the SVZ. Nevertheless, several reports have now indicated that this basal level of spontaneous neuroregeneration can be substantially augmented by the use of cytokines. Table 2 offers a sampling of methods and findings from studies that have investigated the mobilization of SVZ-derived cells into adjacent parenchyma via administration of various factors and/or injury stimuli.

4.4.2. *Functional benefits of injury-induced neurogenesis?*

To date, very little proof of principle has been provided to suggest that even if an adequate number of new neurons could be generated, they would be able to functionally integrate into host circuitry and yield behaviorally meaningful benefits. Although certain treatments have been shown to promote functional recovery after injury and the time course of such recovery in some cases parallels the gradual production of new neurons over time,[76] clear evidence that the new neurons are actually responsible for the functional gains is still lacking. With few exceptions, the vast majority of studies have not investigated the electrophysiological properties of newly born neurons in areas of injury, or assessed the formation of synaptic connections with surrounding cells. Indeed, the fact that the vast majority of newly born neurons die may suggest failure to integrate into existing circuitry. Two groups, however, have suggested that under the right circumstances, functional integration of newly born neurons may be possible in nonneurogenic regions after selective loss.

The CA1 region of the hippocampus is selectively susceptible to hypoxia, and pyramindal neurons in this region are specifically ablated after transient global forebrain ischemia. Nakatomi *et al.*[109] elegantly demonstrated that a low level of spontaneous regeneration occurs in this region, wherein new neurons appear to be derived from the posterior periventricular region — a region likely similar to the SCZ described by Seri *et al.*[20] Infusion of bFGF and EGF into the lateral ventricles for five days significantly increased this baseline level of recovery, leading to almost complete regeneration of the CA1 area. New neurons were shown to make appropriate synaptic connections and develop electrophysiological properties typical of CA1 neurons. Further, behavioral analysis following regeneration revealed virtually complete resolution of ischemia-induced memory deficits, in marked contrast to the untreated animals. In a second

Table 2. Mobilization of Newly Born Cells to Nonneurogenic Brain Regions

Factors administered	Injury Stimulus	Findings	Refs
Cytokines only			
bFGF (ICV)	—	↑new neurons in OB; no new cells in striatum	62
EGF (ICV)	—	↓new neurons but ↑ astrocytes in OB; SVZ expansion; migration into striatum; minimal neural differentiation	14, 62, 64, 98
BDNF (ICV)	—	BrdU+ cells in striatum, septum, thalamus, hypothalamus; 27–42% β3-tub+ at d16	99
BDNF AV (ICV)	—	↑OB neurogenesis: new GAD+; DARPP-32+ calbindin-D28+ cells in striatum at 5–8 weeks	100
BDNF + Noggin AV (ICV)	—	Higher % new striatal neurons than BDNF alone at 8 weeks; projections formed to globus pallidus	101
BDNF + Noggin AV (into striatum)	—	Very few new neurons near SVZ at 3 weeks	101
Notch ligand (ICV)	—	Hu+ cells in cortex at day 45	76
Injury only			
—	6-OHDA	No migration	102
—	Stroke (cortex)	New BrdU+ neurons in cortex	103
—	Stroke (cortex + striatum)	Neurons in striatum + cortex at 2 weeks (dcx+)[104] and NeuN+ at 30 & 60 days[249]	104, 105
—	Stroke (cortex + striatum)	New neurons in striatum only	96, 106, 107
—	Global Ischemia	Increased DG proliferation; New CA1 neurons	108, 109
—	Huntington's	Increase B cells; Migration into striatum	110, 111

(Continued)

Table 2. *(Continued)*

Factors administered	Injury Stimulus	Findings	Refs
Cytokines only			
—	Demyelination	Migration of SVZ and RMS-derived cells into white matter + oligo/astrocyte differentiation	112, 113
—	Synchronous apoptotic degeneration in cortex	Layer- and region-specific regeneration of corticothalamic or corticospinal neurons that reform appropriate connections	114–117
Injury + cytokines			
TGFα (into striatum)	6-OHDA	Proliferation and migration toward TGF-α + neural differentiation; behavioral improvement	102
TGF-α (ICV)	6-OHDA	SVZ expansion; migration into striatum: minimal neural differentiation	65
bFGF (AAV into penumbra)	Stroke (cortex)	Progressive neurogenesis over 90 days with gradual behavioral recovery	118
bFGF (AV ICV)	Global Ischemia	Widely distributed neurons, including in cortex at 2 weeks	119
EGF + albumin (ICV)	Stroke (striatum)	200-fold increase in striatal neural replacement from 0.1 to 20%	120
EGF + EPO (ICV)	Stroke (cortex)	Robust cortical "regeneration." Functional improvement reversed upon removal of new cortical tissue.	121
Notch ligand + bFGF	Stroke (cortex + striatum)	Progressive improvement in function over 45 days.	76
EGF + bFGF (ICV)	Global Ischemia	Robust regeneration of functional CA1 neurons	109

ICV: Intracerebroventricular; AV: adenovirus; AAV: Adeno-associated virus example, select populations of cortical pyramidal neurons can be specifically ablated by retrograde transport of nanospheres containing a chromophore that releases a cytotoxic singlet oxygen upon exposure to 670 nm laser illumination, leading to apoptotic degeneration without inflammation. This model has been used to study spontaneous regeneration of cortical projection neurons.[114] Substantial numbers of new neurons were born after injury, which appeared to migrate from the SVZ through the CC to the area of injury, where they integrated into appropriate cortical layers. In the case of the corticospinal neuronal regeneration, a retrograde label injected into the cervical spinal cord could be demonstrated in some of the newly born neurons starting at eight weeks after injury, suggesting that in spite of the inhibitory environment of the adult CNS, at least some new projections could extend long distances from the cortex to the spinal cord. Furthermore, although at least half of the newly born neurons died within 12 weeks, some surviving cells, including some with spinal projections, survived over one year.[114] Such robust regeneration as shown by these two examples has not been observed in injury models featuring widespread necrosis, suggesting that preservation of the local cytoarchitecture may greatly facilitate endogenous mechanisms for spontaneous neural replacement in the adult brain. The development of new animal models in which endogenous neural stem cells can be prospectively labeled prior to injury will greatly facilitate the functional assessment of newly born neurons in areas of injury.[122] Further, tools for specifically ablating newly born neurons will be required to assess the direct involvement of such neurons in observed functional recovery.

4.4.3. Evidence for human NSC mobilization

For such neuroregenerative strategies to be meaningful for human medicine, it is important to know whether neurogenesis in response to injury also occurs in aged animals, in which the basal level of neurogenesis may be only a fraction of that in younger animals. Indeed, if basal levels of neurogenesis predict the level of repair possible after injury, then the relevance of strategies that appear to be effective in young adult experimental animals to human medicine, where most patients are in their later years, may be grim. However, in spite of decreased SVZ and DG proliferation under basal conditions in aged animals, induction of stroke in 15 month-old rats yielded numbers of new neurons in the ischemic striatum that were not significantly different than in three-month-old animals, with surviving cells observed seven weeks after injury.[123] Further, increased cell proliferation has been observed in the SVZ of humans with Huntington's disease, where proliferation rates correlated with disease severity and numbers of CAG repeats.[111] Recently, newly born neurons have also been found adjacent to areas of ischemia in human post-mortem brains.[124] As such, although the distance new neurons would need to migrate may be much greater in the human brain, augmentation of spontaneous regeneration by the use of cytokines may prove to be a clinically relevant strategy.

4.4.4. NSC-derived tumors: Too much of a good thing?

Infusion of mitogens into the brain is not without potential risks. It is well known that excessive proliferation is associated with increased risk of genetic alteration. Thus, stem cells in the adult brain (along with most other adult tissues) are relatively quiescent, with much of the proliferative burden borne by transit amplifying cells. The most common brain tumors in adults are glioblastomas, and some evidence suggests that these may be derived from neural stem or progenitor cells.[125] The tumor-forming cells of glioblastoma cultures *in vitro* have properties of NSCs, and strategies to induce differentiation of these cells may reduce their tumorgenicity after transplantation. PDGF is a mitogen that induces proliferation of transient amplifying cells. Multiple publications have revealed that chronic infusion of PDGF can lead to highly invasive glioblastoma-like growths.[126] As such, the mobilization of endogenous NSCs, especially in situations involving expansion of the progenitor pool, may be associated with risks that must be taken into consideration in the development of regenerative strategies. Careful evaluation of tumorgenic risks associated with putative thera-peutic strategies will be essential before translation of such approaches to clinical trials.

4.5. Oligodendrocyte precursors: Stem cells hiding in the parenchyma?

Oligodendrocyte precursor cells (OPCs) represent the largest pool of actively proliferating cells in the adult brain, accounting for about 70% of labeled cells after an injection of BrdU.[127] OPCs express NG2 proteogly-can, PDGFRα, A2B5, CNPase, and olig2. Although traditionally regarded solely as progenitor cells, recent studies suggest that these cells, which are distributed throughout the brain, may have broader potential than previously realized. Many NG2+ OPCs exist in both white and gray matter throughout life. In gray matter their numbers equal those of oligodendrocytes, while in white matter they are outnumbered by oligodendrocytes four to one.[127]

True to their name, OPCs can be activated *in vivo* in response to demyelinating lesions and give rise to new oligodendrocytes, though this process eventually fails in multiple sclerosis for reasons that are still under investigation. A certain number of new oligodendrocytes are also made throughout adult life even in the absence of injury, and retroviral tracing studies suggest that adult OPCs may additionally give rise to astrocytes *in vivo*. The functions of OPCs in the brain are not limited to their role as progenitors. OPCs have a dynamic and complex morphology, making contact with the nodes of ranvier and synaptic terminals. Indeed, they are likely quite active participants in synaptic function. OPCs also react rapidly to injury and constitute a major component of the glial scar that forms in response to CNS damage.[127]

Upon isolation, OPCs can give rise to oligodendrocytes under low serum conditions, and to type 2 astrocytes under high serum conditions or in the presence of BMPs. Alternatively, OPCs can be propagated without differentiation in the presence of PDGF. Remarkably, OPC-derived type 2 astrocytes have been found to assume properties of multipotent NSCs when expanded in the presence of bFGF, with potential to generate neurons and type 1 astrocytes as well as oligodendrocytes.[9]

OPCs from human subcortical white matter have further been shown, even without the use of serum or BMPs, to give rise to functional neurons, astrocytes and oligodendrocytes.[128] Whether or not these cells could correspond to the multipotent progenitors of the SCZ[20] (see Section 3.41), remains unclear, as negative selection against GFAP+ cells was not performed. Indeed, the expression profile of type 2 astrocytes overlaps with that of OPCs; type 2 astrocytes express OPC markers such as A2B5 in addition to GFAP. Further, NG2, a common marker of OPCs, has been observed on NSCs in adult germinal areas.[129] To date, the only reports describing isolation of multipotent NSCs from adult cortical gray matter have employed serum as part of the isolation or culture procedure,[130] leaving open the possibility that these cultures may have derived from reprogrammed oligodendrocyte precursors. Future studies will be required to determine definitively whether or not known adult germinal zone multipotent adults are in fact type 2 astrocytes, and if so, whether all adult type 2 astrocytes in white matter are multipotent.

Can OPCs be mobilized *in vivo* to generate neurons? Buffo *et al.* recently demonstrated that retroviral knockdown of olig2 or overexpression of pax6 following cortical injury resulted in the acquisition of neuronal characteristics such as DCX expression morphology in 12% of transduced cells. This suggested *in situ* differentiation of proliferating cells, most likely including OPCs, into neurons.[131] The number of surviving cells at 30 days after injection, however, was quite low, in keeping with the idea that additional signals may be required to promote survival of newly born neurons. A similar retroviral strategy by Ohori *et al.*[132] compared the overexpression of neurogenin2 and Mash1 in combination with EGF and bFGF treatment in an injured spinal cord. Although cytokines alone could induce the differentiation of proliferating cells into immature neurons, concomitant overexpression of ngn2 and Mash1 resulted in both increased numbers and maturation of new neurons and oligodendrocytes, respectively, from cells with phenotypic characteristics of OPCs. As such, whether reprogrammed by extrinsic factors or modulated via genetic methods, OPCs may represent an additional source of endogenous cells for neuroregeneration.

4.6. Neuroregeneration using exogenous cells

In most studies to date, the number of endogenous neurons generated in response to injury or mobilization has been modest. Further, the repertoire of neuronal phenotypes generated has primarily been restricted to GABAergic, though cortical projection neurons may be generated under specific conditions. Thus, the introduction of exogenous cells may provide an alternate or complementary neuroregenerative strategy for certain CNS disorders.[133] *In vitro* culture of stem cells theoretically allows for the generation of unlimited numbers of cells, and generally permits greater control over cell fate.[15,16] Further, cell types that are challenging to obtain from NSCs, such as dopaminergic neurons, may be generated from more primitive cells, such as embryonic stem cells. Proof of principle for the potential clinical value of transplanted neural cells has been obtained from pioneering studies in which fetal human brain tissue was transplanted into patients with neurodegenerative disease.[134] Although significant work remains to optimize transplantation techniques and minimize side effects, some patients with Parkinson's disease have experienced substantial lasting benefit from fetal cell transplantation. Practical and ethical limitations on the availability of fetal brain tissue will likely require the availability of stem-cell-derived neurons before such therapies could be considered practical for widespread use. Additionally, stem cells are becoming increasingly recognized as an important source of secreted molecules that may further promote functional recovery.[135]

Cell replacement strategies may be particularly appropriate for CNS conditions in which a specific subset of cells are lost. Such is the case in at least the early stages of several neurodegenerative diseases, such as Parkinson's disease, Huntington's disease (HD), Amyotrophic Lateral Sclerosis (ALS), and Multiple Sclerosis (MS). What is the best source of cells for CNS transplantation? NSCs have been employed successfully to generate functional cell types lost in HD and MS, whereas embryonic stem cells have proven a more efficient source of DA neurons for animal studies of cell replacement. Synaptic integration of neurons from fetal, NSC and ESC sources has been demonstrated, though care to avoid teratoma formation is needed when using cells derived from ESCs. Although cell replacement may reasonably be expected to yield functional benefits in situations where release of a single neurotransmitter can improve function, or where replacement of experimentally ablated myelin is observable, reconstruction of entire ablated brain regions comprising multiple cell types will likely remain a substantial challenge. Indeed, it remains to be determined if cell replacement strategies, whether via endogenous or exogenous sources of progenitors, can ever fully restore the original brain circuitry after such extensive damage as is observed following stroke or in late stages of Alzheimer's disease. As such, there is strong impetus to preserve as many of the original connections as possible during neurodegenerative disease and brain injury to circumvent the need for replacement.

Remarkably, it is being increasingly found that stem cells from multiple sources possess inherent trophic properties that appear to protect neurons from death in regions of injury. Although the mechanisms underlying these observations remain to be fully elucidated, release of neurotrophic molecules, modulation of inflammation, and promotion of endogenous angiogenesis and neurogenesis have all been proposed. Further, NSCs have been shown to migrate to areas of injury, potentially helping to target neurotrophic activity to appropriate locations. Such migration may also allow specific delivery of engineered neuroprotective molecules to areas of neurodegeneration, or chemotherapeutic agents to tumors, while spontaneous expression[136] or engineered transgene expression[137] of specific enzymes may be beneficial for treatment of lysosomal storage disorders. Interestingly, some of the trophic cell transplantation studies that have yielded the most promising results have included transplantation of nonneural cells such as bone marrow-derived stem cells, even though in many cases, very few of these cells are identifiable in the brain following either local or systemic administration. The idea that the transplanted cells may, in some cases, be directly responsible for functional improvement has been demonstrated by transplantation of human cells that can then be selectively killed by administration of diphtheria toxin. However, the exact nature of the beneficial effect, whether due to synaptic integration or some other long term supportive role, will require further study. Ultimately, optimized neuroregenerative strategies may employ a combination of cell replacement from either endogenous or exogenous sources and delivery of trophic factors to promote both survival of damaged neurons and functional integration of new neurons.

Conclusion

It is now undisputed that new functional neurons are continually added to the mammalian brain throughout life. The idea of harnessing this innate neurogenic substrate for neuroregenerative purposes is attractive in light of the current paucity of effective treatment for both acute and chronic neurodegenerative conditions. Over the past several years, proof of principle has been obtained that endogenous NSCs can give rise to new neurons in nonneurogenic areas in response to injury and/or mobilization. However, in most cases, the number of new neurons is small, their survival is poor, and both their functional properties and the behavioral implications of their presence remain ill-defined.

The complexity of adult CNS circuitry will likely continue to pose regenerative challenges well after methods have been established to generate suitable numbers of neurons of all desired subtypes. As such, the importance of research efforts to prevent or halt the progression of neurodegenerative conditions cannot be overstated. Until such goals are realized, however, continued efforts to understand the regulation of constitutive adult neurogenesis and the signaling pathways underlying stem cell plasticity and cell fate specification seek to obtain clinically viable strategies for regeneration of damaged CNS circuitry. Studies indicating that adult-born neurons can, under appropriate circumstances, generate region-appropriate neurons capable of forming long distance connections are encouraging, suggesting that developmental guidance signals may remain or even be reactivated in the injured adult brain. Some of the many challenges facing the field of neuroregeneration will include preservation or recreation of guidance cues in regions of widespread injury such as stroke or Alzheimer's disease, recruitment of appropriate progenitors to ostensibly nonneurogenic brain regions such as the ventral midbrain for Parkinson's disease, and translation of these strategies to the clinical arena.

References

1. Gaiano N, Nye JS, Fishell G. (2000) Radial glial identity is promoted by Notch1 signaling in the murine forebrain. *Neuron* 26: 395–404.
2. Gotz M, Barde YA. (2005) Radial glial cells defined and major intermediates between embryonic stem cells and CNS neurons. *Neuron* 46: 369–372.
3. Noctor SC, Flint AC, Weissman TA, *et al.* (2001) Neurons derived from radial glial cells establish radial units in neocortex. *Nature* 409: 714–720.
4. Anthony TE, Klein C, Fishell G, Heintz N. (2004) Radial glia serve as neuronal progenitors in all regions of the central nervous system. *Neuron* 41: 881–890.
5. Burrows RC, Wancio D, Levitt P, Lillien L. (1977) Response diversity and the timing of progenitor cell maturation are regulated by developmental changes in EGFR expression in the cortex. *Neuron* 19: 251–267.
6. Shen Q, Wang Y, Dimos JT, *et al.* (2006) The timing of cortical neurogenesis is encoded within lineages of individual progenitor cells. *Nat Neurosci* 9: 743–751.
7. Merkle FT, Tramontin AD, Garcia-Verdugo JM, Alvarez-Buylla A. (2004) Radial glia give rise to adult neural stem cells in the subventricular zone. *Proc Natl Acad Sci USA* 101: 17528–17532.
8. Gabay L, Lowell S, Rubin LL, Anderson DJ. (2003) Deregulation of dorsoventral patterning by FGF confers trilineage differentiation capacity on CNS stem cells *in vitro*. *Neuron* 40: 485–499.
9. Kondo T, Raff M. (2000) Oligodendrocyte precursor cells reprogrammed to become multipotential CNS stem cells. *Science* 289: 1754–1757.
10. Reynolds BA, Weiss S. (1992) Generation of neurons and astrocytes from isolated cells of the adult mammalian central nervous system. *Science* 255: 1707–1710.
11. Reynolds BA, Rietze RL. (2005) Neural stem cells and neurospheres — re-evaluating the relationship. *Nat Methods* 2: 333–336.

12. Conti L, Pollard SM, Gorba T, *et al.* (2005) Niche-independent symmetrical self-renewal of a mammalian tissue stem cell. *PLoS Biol* **3**: e283.
13. Pollard SM, Conti L, Sun Y, *et al.* (2006) Adherent neural stem (NS) cells from fetal and adult forebrain. *Cereb Cortex* **16 Suppl 1**: i112–120.
14. Gregg C, Weiss S. (2003) Generation of functional radial glial cells by embryonic and adult forebrain neural stem cells. *J Neurosci* **23**: 11587–11601.
15. Markakis EA, Palmer TD, Randolph-Moore L, *et al.* (2004) Novel neuronal phenotypes from neural progenitor cells. *J Neurosci* **24**: 2886–2897.
16. Wu P, Tarasenko YI, Gu Y, *et al.* (2002) Region-specific generation of cholinergic neurons from fetal human neural stem cells grafted in adult rat. *Nat Neurosci* **5**: 1271–1278.
17. Shim JW, Park CH, Bae YC, *et al.* (2007) Generation of functional dopamine neurons from neural precursor cells isolated from the subventricular zone and white matter of the adult rat brain using Nurr1 overexpression. *Stem Cells* **25**: 1252–1262.
18. Weiss S, Dunne C, Hewson J, *et al.* (1996) Multipotent CNS stem cells are present in the adult mammalian spinal cord and ventricular neuroaxis. *J Neurosci* **16**: 7599–7609.
19. Gritti A, Bonfanti L, Doetsch F, *et al.* (2002) Multipotent neural stem cells reside into the rostral extension and olfactory bulb of adult rodents. *J Neurosci* **22**: 437–445.
20. Seri B, Herrera DG, Gritti A, *et al.* (2006) Composition and organization of the SCZ: a large germinal layer containing neural stem cells in the adult mammalian brain. *Cereb Cortex* **16 Suppl 1**: i103–111.
21. Seaberg RM, van der Kooy D. (2003) Stem and progenitor cells: the premature desertion of rigorous definitions. *Trends Neurosci* **26**: 125–131.
22. Tropepe V, Coles BL, Chiasson BJ, *et al.* (2000) Retinal stem cells in the adult mammalian eye. *Science* **287**: 2032–2036.
23. Altman J, Das GD. (1965) Post-natal origin of microneurones in the rat brain. *Nature* **207**: 953–956.
24. Doetsch F, Hen R. (2005) Young and excitable: the function of new neurons in the adult mammalian brain. *Curr Opin Neurobiol* **15**: 121–128.
25. Paxinos, GF, Keith BJ. *The Mouse Brain in Stereotaxic Coordinates.* Academic Press, Sydney, 2001.
26. Doetsch, F. (2003) The glial identity of neural stem cells. *Nat Neurosci* **6**: 1127–1134.
27. Sanai N, Tramontin AD, Quiñones-Hinojosa A, *et al.* (2004) Unique astrocyte ribbon in adult human brain contains neural stem cells but lacks chain migration. *Nature* **427**: 740–744.
28. Seri B, Garcia-Verdugo JM, McEwen BS, Alvarez-Buylla A. (2001) Astrocytes give rise to new neurons in the adult mammalian hippocampus. *J Neurosci* **21**: 7153–7160.

29. Roy NS, Wang S, Jiang L, *et al.* (2000) *In vitro* neurogenesis by progenitor cells isolated from the adult human hippocampus. *Nat Med* 6: 271–277.

30. Tashiro A, Sandler VM, Toni N, *et al.* (2006) NMDA-receptor-mediated, cell-specific integration of new neurons in adult dentate gyrus. *Nature* 442: 929–933.

31. Bruel-Jungerman E, Davis S, Rampon C, Laroche S. (2006) Long-term potentiation enhances neurogenesis in the adult dentate gyrus. *J Neurosci* 26: 5888–5893.

32. Ge S, Goh EL, Sailor KA, *et al.* (2006) GABA regulates synaptic integration of newly generated neurons in the adult brain. *Nature* 439: 589–593.

33. Abrous DN, Koehl M, Le Moal M. (2005) Adult neurogenesis: from precursors to network and physiology. *Physiol Rev* 85: 523–569.

34. Leuner B, Gould E, Shors TJ. (2006) Is there a link between adult neurogenesis and learning? *Hippocampus* 16: 216–224.

35. Meshi D, Drew MR, Saxe M, *et al.* (2006) Hippocampal neurogenesis is not required for behavioral effects of environmental enrichment. *Nat Neurosci* 9: 729–731.

36. Ramon y Cajal S. (1928) *Degeneration and Regeneration of the Nervous System.* Haffner, New York.

37. Taupin P. (2007) BrdU immunohistochemistry for studying adult neurogenesis: Paradigms, pitfalls, limitations, and validation. *Brain Res Rev* 53: 198–214.

38. Taupin P. (2006) Neurogenesis in the adult central nervous system. *C R Biol* 329: 465–475.

39. Cao QL, Onifer SM, Whittemore SR. (2002) Labeling stem cells *in vitro* for identification of their differentiated phenotypes after grafting into the CNS. *Methods Mol Biol* 198: 307–318.

40. Burns TC, Ortiz-González XR, Gutiérrez-Pérez M, *et al.* (2006) Thymidine analogs are transferred from prelabeled donor to host cells in the central nervous system after transplantation: a word of caution. *Stem Cells* 24: 1121–1127.

41. Wurmser AE, Gage FH. (2002) Stem cells: cell fusion causes confusion. *Nature* 416: 485–487.

42. Clarke DL, Johansson CB, Wilbertz J, *et al.* (2000) Generalized potential of adult neural stem cells. *Science* 288: 1660–1663.

43. Jiang Y, Henderson D, Blackstad M, *et al.* (2003) Neuroectodermal differentiation from mouse multipotent adult progenitor cells. *Proc Natl Acad Sci USA* 100 Suppl 1: 11854–11860.

44. Verfaillie CM. (2002) Adult stem cells: assessing the case for pluripotency. *Trends Cell Biol* 12: 502–508.

45. Wurmser AE, Nakashima K, Summers RG, *et al.* (2004) Cell fusion-independent differentiation of neural stem cells to the endothelial lineage. *Nature* 430: 350–356.

46. Hallbergson AF, Gnatenco C, Peterson DA. (2003) Neurogenesis and brain injury: managing a renewable resource for repair. *J Clin Invest* **112**: 1128–1133.
47. Scadden DT. (2006) The stem-cell niche as an entity of action. *Nature* **441**: 1075–1079.
48. Hagg T. (2005) Molecular regulation of adult CNS neurogenesis: an integrated view. *Trends Neurosci* **28**: 589–595.
49. Montaron MF, Drapeau E, Dupret D, *et al.* (2006) Lifelong corticosterone level determines age-related decline in neurogenesis and memory. *Neurobiol Aging* **27**: 645–654.
50. Gould E, Tanapat P, McEwen BS, *et al.* (1998) Proliferation of granule cell precursors in the dentate gyrus of adult monkeys is diminished by stress. *Proc Natl Acad Sci USA* **95**: 3168–3171.
51. Perez-Martin M, Azcoitia I, Trejo JL, *et al.* (2003) An antagonist of estrogen receptors blocks the induction of adult neurogenesis by insulin-like growth factor-I in the dentate gyrus of adult female rat. *Eur J Neurosci* **18**: 923–930.
52. Shingo T, Gregg C, Enwere E, *et al.* (2003) Pregnancy-stimulated neurogenesis in the adult female forebrain mediated by prolactin. *Science* **299**: 117–120.
53. Deisseroth K, Singla S, Toda H, *et al.* (2004) Excitation–neurogenesis coupling in adult neural stem/progenitor cells. *Neuron* **42**: 535–552.
54. Santarelli L, Saxe M, Gross C, *et al.* (2003) Requirement of hippocampal neurogenesis for the behavioral effects of antidepressants. *Science* **301**: 805–809.
55. Jacobs BL, Praag H, Gage FH. (2000) Adult brain neurogenesis and psychiatry: a novel theory of depression. *Mol Psychiatry* **5**: 262–269.
56. Grote HE, Bull ND, Howard ML, *et al.* (2005) Cognitive disorders and neurogenesis deficits in Huntington's disease mice are rescued by fluoxetine. *Eur J Neurosci* **22**: 2081–2088.
57. Liu X, Wang Q, Haydar TF, Bordey A. (2005) Nonsynaptic GABA signaling in postnatal subventricular zone controls proliferation of GFAP-expressing progenitors. *Nat Neurosci* **8**: 1179–1187.
58. Kulkarni VA, Jha S, Vaidya VA. (2002) Depletion of norepinephrine decreases the proliferation, but does not influence the survival and differentiation, of granule cell progenitors in the adult rat hippocampus. *Eur J Neurosci* **16**: 2008–2012.
59. Baker SA, Baker KA, Hagg T. (2004) Dopaminergic nigrostriatal projections regulate neural precursor proliferation in the adult mouse subventricular zone. *Eur J Neurosci* **20**: 575–579.
60. Kippin TE, Kapur S, van der Kooy D. (2005) Dopamine specifically inhibits forebrain neural stem cell proliferation, suggesting a novel effect of antipsychotic drugs. *J Neurosci* **25**: 5815–5823.
61. Höglinger GU, Rizk P, Muriel MP, *et al.* (2004) Dopamine depletion impairs precursor cell proliferation in Parkinson's disease. *Nat Neurosci* **7**: 726–735.

62. Kuhn HG, Winkler J, Kempermann G, *et al.* (1997) Epidermal growth factor and fibroblast growth factor-2 have different effects on neural progenitors in the adult rat brain. *J Neurosci* **17**: 5820–5829.

63. Jin K, Sun Y, Xie L, *et al.* (2003) Neurogenesis and aging: FGF-2 and HB-EGF restore neurogenesis in hippocampus and subventricular zone of aged mice. *Aging Cell* **2**: 175–183.

64. Craig CG, Tropepe V, Morshead CM, *et al.* (1996) *In vivo* growth factor expansion of endogenous subependymal neural precursor cell populations in the adult mouse brain. *J Neurosci* **16**: 2649–2658.

65. Cooper O, Isacson O. (2004) Intrastriatal transforming growth factor alpha delivery to a model of Parkinson's disease induces proliferation and migration of endogenous adult neural progenitor cells without differentiation into dopaminergic neurons. *J Neurosci* **24**: 8924–8931.

66. Yan YP, Sailor KA, Vemuganti R, Dempsey RJ. (2006) Insulin-like growth factor-1 is an endogenous mediator of focal ischemia-induced neural progenitor proliferation. *Eur J Neurosci* **24**: 45–54.

67. Aberg MA, Aberg ND, Palmer TD, *et al.* (2003) IGF-I has a direct proliferative effect in adult hippocampal progenitor cells. *Mol Cell Neurosci* **24**: 23–40.

68. Trejo JL, Carro E, Torres-Aleman I. (2001) Circulating insulin-like growth factor I mediates exercise-induced increases in the number of new neurons in the adult hippocampus. *J Neurosci* **21**: 1628–1634.

69. Cao L, Jiao X, Zuzga DS, *et al.* (2004) VEGF links hippocampal activity with neurogenesis, learning and memory. *Nat Genet* **36**: 827–835.

70. Bull ND, Bartlett PF. (2005) The adult mouse hippocampal progenitor is neurogenic but not a stem cell. *J Neurosci* **25**: 10815–10821.

71. Emsley JG, Hagg T. (2003) Endogenous and exogenous cillary neurotrophic factor enhances forebrain neurogenesis in adult mice. *Exp Neurol* **183**: 298–310.

72. Ahn S, Joyner AL. (2005) *In vivo* analysis of quiescent adult neural stem cells responding to Sonic hedgehog. *Nature* **437**: 894–897.

73. Machold R, Hayashi S, Rutlin M, *et al.* (2003) Sonic hedgehog is required for progenitor cell maintenance in telencephalic stem cell niches. *Neuron* **39**: 937–950.

74. Lim DA, Tramontin AD, Trevejo JM, *et al.* (2000) Noggin antagonizes BMP signaling to create a niche for adult neurogenesis. *Neuron* **28**: 713–726.

75. Lie DC, Colamarino SA, Song HG, *et al.* (2005) Wnt signalling regulates adult hippocampal neurogenesis. *Nature* **437**: 1370–1375.

76. Androutsellis-Theotokis A, Leker RR, Soldner F, *et al.* (2006) Notch signalling regulates stem cell numbers *in vitro* and *in vivo*. *Nature* **442**: 823–826.

77. Hitoshi S, Alexson T, Tropepe V, *et al.* (2002) Notch pathway molecules are essential for the maintenance, but not the generation, of mammalian neural stem cells. *Genes Dev* **16**: 846–858.

78. Patten BA, Sardi SP, Koirala S, *et al.* (2006) Notch1 signaling regulates radial glia differentiation through multiple transcriptional mechanisms. *J Neurosci* 26: 3102–3108.

79. Louvi A, Artavanis Tsakonas S. (2006) Notch signalling in vertebrate neural development. *Nat Rev Neurosci* 7: 93–102.

80. Depaepe V, Suarez-Gonzalez N, Dufour A, *et al.* (2005) Ephrin signalling controls brain size by regulating apoptosis of neural progenitors. *Nature* 435: 1244–1250.

81. Conover JC, Doetsch F, Garcia-Verdugo JM, *et al.* (2000) Disruption of Eph/ephrin signaling affects migration and proliferation in the adult sub-ventricular zone. *Nat Neurosci* 3: 1091–1097.

82. Kempermann G, Kuhn HG, Gage FH. (1998) Experience-induced neurogenesis in the senescent dentate gyrus. *J Neurosci* 18: 3206–3212.

83. Kempermann G, Kuhn HG, Gage FH. (1997) More hippocampal neurons in adult mice living in an enriched environment. *Nature* 386: 493–495.

84. Schmidt-Hieber C, Jonas P, Bischofberger J. (2004) Enhanced synaptic plasticity in newly generated granule cells of the adult hippocampus. *Nature* 429: 184–187.

85. van Praag H, Shubert T, Zhao C, Gage FH. (2005) Exercise enhances learning and hippocampal neurogenesis in aged mice. *J Neurosci* 25: 8680–8685.

86. van Praag H, Kempermann G, Gage FH. (1999) Running increases cell proliferation and neurogenesis in the adult mouse dentate gyrus. *Nat Neurosci* 2: 266–270.

87. Hoehn BD, Palmer TD, Steinberg GK. (2005) Neurogenesis in rats after focal cerebral ischemia is enhanced by indomethacin. *Stroke* 36: 2718–2724.

88. Scharfman HE, Sollas AL, Goodman JH. (2002) Spontaneous recurrent seizures after pilocarpine-induced status epilepticus activate calbindin-immunoreactive hilar cells of the rat dentate gyrus. *Neuroscience* 111: 71–81.

89. Jakubs K, Nanobashvili A, Bonde S, *et al.* (2006) Environment matters: synaptic properties of neurons born in the epileptic adult brain develop to reduce excitability. *Neuron* 52: 1047–1059.

90. Parent JM, Yu TW, Leibowitz RT, *et al.* (1997) Dentate granule cell neurogenesis is increased by seizures and contributes to aberrant network reorganization in the adult rat hippocampus. *J Neurosci* 17: 3727–3738.

91. Cameron HA, McKay RD. (1992) Restoring production of hippocampal neurons in old age. *Nat Neurosci* 2: 894–897.

92. Seki T, Arai Y. (1995) Age-related production of new granule cells in the adult dentate gyrus. *Neuroreport* 6: 2479–2482.

93. Ziv Y, Ron N, Butovsky O, *et al.* (2006) Immune cells contribute to the maintenance of neurogenesis and spatial learning abilities in adulthood. *Nat Neurosci* 9: 268–275.

94. Monje ML, Toda H, Palmer TD. (2003) Inflammatory blockade restores adult hippocampal neurogenesis. *Science* 302: 1760–1765.

95. Ekdahl CT, Claasen JH, Bonde S, *et al.* (2003) Inflammation is detrimental for neurogenesis in adult brain. *Proc Natl Acad Sci USA* 100: 13632–13637.

96. Arvidsson A, Collin T, Kirik D, *et al.* (2002) Neuronal replacement from endogenous precursors in the adult brain after stroke. *Nat Med* 8: 963–970.

97. Thored P, Arvidsson A, Cacci E, *et al.* (2006) Persistent production of neurons from adult brain stem cells during recovery after stroke. *Stem Cells* 24: 739–747.

98. Doetsch F, Petreanu L, Caille I, Garcia-Verdugo JM, Alvarez-Buylla A. (2002) EGF converts transit-amplifying neurogenic precursors in the adult brain into multipotent stem cells. *Neuron* 36: 1021–1034.

99. Pencea V, Bingaman KD, Wiegand SJ, Luskin MB. (2001) Infusion of brain-derived neurotrophic factor into the lateral ventricle of the adult rat leads to new neurons in the parenchyma of the striatum, septum, thalamus, and hypothalamus. *J Neurosci* 21: 6706–6717.

100. Benraiss A, Chmielnicki E, Lerner K, *et al.* (2001) Adenoviral brain-derived neurotrophic factor induces both neostriatal and olfactory neuronal recruitment from endogenous progenitor cells in the adult forebrain. *J Neurosci* 21: 6718–6731.

101. Chmielnicki E, Benraiss A, Economides AN, Goldman SA. (2004) Adenovirally expressed noggin and brain-derived neurotrophic factor cooperate to induce new medium spiny neurons from resident progenitor cells in the adult striatal ventricular zone. *J Neurosci* 24: 2133–2142.

102. Fallon J, *et al.* (2000) In vivo induction of massive proliferation, directed migration, and differentiation of neural cells in the adult mammalian brain. *Proc Natl Acad Sci USA* 97: 14686–14691.

103. Gu W, Brannstrom T, Wester P. (2000) Cortical neurogenesis in adult rats after reversible photothrombotic stroke. *J Cereb Blood Flow Metab* 20: 1166–1173.

104. Jin K, *et al.* (2003) Directed migration of neuronal precursors into the ischemic cerebral cortex and striatum. *Mol Cell Neurosci* 24: 171–189.

105. Jiang W, Gu W, Brannstrom T, *et al.* (2001) Cortical neurogenesis in adult rats after transient middle cerebral artery occlusion. *Stroke* 32: 1201–1207.

106. Zhang RL, Zhang ZG, Zhang L, Chopp M. (2001) Proliferation and differentiation of progenitor cells in the cortex and the subventricular zone in the adult rat after focal cerebral ischemia. *Neuroscience* 105: 33–41.

107. Parent JM, Vexler ZS, Gong C, *et al.* (2002) Rat forebrain neurogenesis and striatal neuron replacement after focal stroke. *Ann Neurol* 52: 802–813.

108. Bendel O, *et al.* (2005) Reappearance of hippocampal CA1 neurons after ischemia is associated with recovery of learning and memory. *J Cereb Blood Flow Metab* 25: 1586–1595.
109. Nakatomi H, Kuiru T, Okabe S, *et al.* (2002) Regeneration of hippocampal pyramidal neurons after ischemic brain injury by recruitment of endogenous neural progenitors. *Cell* 110: 429–441.
110. Batista CM, *et al.* (2006) A progressive and cell non-autonomous increase in striatal neural stem cells in the Huntington's disease R6/2 mouse. *J Neurosci* 26: 10452–10460.
111. Curtis MA, Penney EB, Pearson AG, *et al.* (2003) Increased cell proliferation and neurogenesis in the adult human Huntington's disease brain. *Proc Natl Acad Sci USA* 100: 9023–9027.
112. Picard-Riera N, *et al.* (2002) Experimental autoimmune encephalomyelitis mobilizes neural progenitors from the subventricular zone to undergo oligodendrogenesis in adult mice. *Proc Natl Acad Sci USA* 99: 13211–13216.
113. Nait-Oumesmar B, *et al.* (1999) Progenitor cells of the adult mouse subventricular zone proliferate, migrate and differentiate into oligodendrocytes after demyelination. *Eur J Neurosci* 11: 4357–4366.
114. Chen J, Magavi SS, Macklis JD. (2004) Neurogenesis of corticospinal motor neurons extending spinal projections in adult mice. *Proc Natl Acad Sci USA* 101: 16357–16362.
115. Catapano LA, Arlotta P, Cage TA, Macklis JD. (2004) Stage-specific and opposing roles of BDNF, NT-3 and bFGF in differentiation of purified callosal projection neurons toward cellular repair of complex circuitry. *Eur J Neurosci* 19: 2421–2434.
116. Magavi SS, Macklis JD. (2002) Induction of neuronal type-specific neurogenesis in the cerebral cortex of adult mice: manipulation of neural precursors in situ. *Brain Res Dev Brain Res* 134: 57–76.
117. Magavi SS, Leavitt BR, Macklis JD. (2000) Induction of neurogenesis in the neocortex of adult mice. *Nature* 405: 951–955.
118. Leker RR, *et al.* (2007) Long-lasting regeneration after ischemia in the cerebral cortex. *Stroke* 38: 153–161.
119. Matsuoka N, *et al.* (2003) Adenovirus-mediated gene transfer of fibroblast growth factor-2 increases BrdU-positive cells after forebrain ischemia in gerbils. *Stroke* 34: 1519–1525.
120. Teramoto T, Qiu J, Plumier JC, Moskowitz MA. (2003) EGF amplifies the replacement of parvalbumin-expressing striatal interneurons after ischemia. *J Clin Invest* 111: 1125–1132.
121. Kolb B, *et al.* (2006) Growth factor-stimulated generation of new cortical tissue and functional recovery after stroke damage to the motor cortex of rats. *J Cereb Blood Flow Metab.*

122. Carlen M, Meletis K, Barnabe-Heider F, Frisen J. (2006) Genetic visualization of neurogenesis. *Exp Cell Res* **312**: 2851–2859.

123. Darsalia V, Heldmann U, Lindvall O, Kokaia Z. (2005) Stroke-induced neurogenesis in aged brain. *Stroke* **36**: 1790–1795.

124. Jin K, Wang X, Xie L, *et al.* (2006) Evidence for stroke-induced neurogenesis in the human brain. *Proc Natl Acad Sci USA* **103**: 13198–13202.

125. Nicolis SK. (2006) Cancer stem cells and "stemness" genes in neuro-oncology. *Neurobiol Dis.*

126. Jackson EL, Garcia-Verdugo JM, Gil-Perotin S, *et al.* (2006) PDGFR alpha-positive B cells are neural stem cells in the adult SVZ that form glioma-like growths in response to increased PDGF signaling. *Neuron* **51**: 187–199.

127. Levine JM, Reynolds R, Fawcett JW. (2001) The oligodendrocyte precursor cell in health and disease. *Trends Neurosci* **24**: 39–47.

128. Nunes MC, Roy NS, Keyoung HM, *et al.* (2003) Identification and isolation of multipotential neural progenitor cells from the subcortical white matter of the adult human brain. *Nat Med* **9**: 439–447.

129. Aguirre AA, Chittajallu R, Belachew S, Gallo V. (2004) NG2-expressing cells in the subventricular zone are type C-like cells and contribute to interneuron generation in the postnatal hippocampus. *J Cell Biol* **165**: 575–589.

130. Palmer TD, Ray J, Gage FH. (1995) FGF-2-responsive neuronal progenitors reside in proliferative and quiescent regions of the adult rodent brain. *Mol Cell Neurosci* **6**: 474–486.

131. Buffo A, Vosko MR, Ertürk D, *et al.* (2005) Expression pattern of the transcription factor Olig2 in response to brain injuries: implications for neuronal repair. *Proc Natl Acad Sci USA* **102**: 18183–18188.

132. Ohori Y, Yamamoto S, Nagao M, *et al.* (2006) Growth factor treatment and genetic manipulation stimulate neurogenesis and oligodendrogenesis by endogenous neural progenitors in the injured adult spinal cord. *J Neurosci* **26**: 11948–11960.

133. Goldman S. (2005) Stem and progenitor cell-based therapy of the human central nervous system. *Nat Biotechnol* **23**: 862–871.

134. Bjorklund A, Dunnett SB, Brundin P, *et al.* (2003) Neural transplantation for the treatment of Parkinson's disease. *Lancet Neurol* **2**: 437–445.

135. Ourednik J, Ourednik V, Lynch WP, *et al.* (2002) Neural stem cells display an inherent mechanism for rescuing dysfunctional neurons. *Nat Biotechnol* **20**: 1103–1110.

136. Snyder EY, Taylor RM, Wolfe JH. (1995) Neural progenitor cell engraftment corrects lysosomal storage throughout the MPS VII mouse brain. *Nature* **374**: 367–370.

137. Muller FJ, Snyder EY, Loring JF. (2006) Gene therapy: can neural stem cells deliver? *Nat Rev Neurosci* **7**: 75–84.

13

Cochlear Stem Cells/Progenitors

Jizhen Lin*, Water Low and Catherine Verfaillie

1. Introduction

Cochlear stem cells/progenitors hold promise for cell replacement therapy of sensorineural hearing loss (SNHL), a degenerative hearing disorder that affects approximately 30 million Americans and 250 million people globally. The major cause of SNHL is the loss of hair cells and spiral ganglion neurons due to aging, antibiotic use, noise exposure, and genetic defects. There is no effective remedy for deafness at the present time. Electric devices such as cochlear implants and hearing aids help but do not restore hearing biologically. This has highlighted the need for innovative therapeutics for SNHL, and cell replacement therapy is a promising possibility.

Cochlear stem cells/progenitors are present in the embryonic stage of the mammalian otocyst from which multiple cells such as hair cells, neurons, and supporting cells develop. Recent studies have indicated the presence of cochlear stem cells/progenitors in the postnatal days, up to 21 days, in the organ of Corti in mice. It is unclear whether or not cochlear stem cells/progenitors remain in the adult mammalian organ of Corti; if any exist, their numbers must be very limited and it is difficult to cultivate them using current cell culture methods.

Efforts have recently been made to find alternative stem cells for cell replacement therapy. Cochlear stem cells/progenitors are derived from stem cells in the placode from which the otocyst originates. The stem cells in the placode descend from pluripotent embryonic stem cells (ESCs) that are pluripotent and capable of forming all the body's cell lineages. Therefore, in theory, ESCs should be capable of becoming cochlear hair cells/progenitors under certain circumstances. Recent works by Li *et al.*

*Correspondence: 216 Lions Research Building, University of Minnesota, 2001 6th St SE, Minneapolis, MN 55455, USA. Tel: (612) 626-9872; Fax: (612) 626-9871; E-mail: linxx004@tc.umn.edu

have demonstrated that ESCs develop into cochlear hair cells when induced properly *in vitro* and then injected into a chick embryonic otic sac, where the developmental cues guided the differentiation of injected ESCs.[1] Therefore, it is apparent that ESCs can covert into cochlear stem cells/progenitors given the proper environment.

Bone marrow stem cells, referred to as multipotent adult progenitor cells (MAPCs), are similar to ESCs in nature and may potentially become cochlear stem cells/progenitors as well under certain conditions. Recent studies have shown that murine MAPCs are capable of becoming neurons and glial cells (neuroectoderm) *in vitro* when induced properly.[2,3] Mesenchymal stem cells (MSCs) from mouse bone marrow, similar to MAPCs, have been recently reported to become auditory hair cells under the influence of *Math1* (*Atoh1*) *in vitro* or under the guidance of the chick inner ear developmental cues *in vivo*.[4] This is significant, because it opens a new door to biological restoration of hearing in an autograft manner. Neural stem cells (NSCs), derived from brain tissue, are similar to cochlear stem cells/progenitors in nature. Recent studies have demonstrated that NSCs share many cellular markers with cochlear stem cells/progenitors.[5] Therefore, it is not surprising that NSCs can readily be converted into cochlear stem cells/progenitors and used for cell replacement therapy in SNHL. Lastly, nasal stem cells, which are derived from the olfactory mucosal epithelium, are another type of NSCs which have been effective in helping paraplegic patients.[6] The availability of MAPCs, MSCs, and nasal stem cells greatly increases the cellular sources of cochlear stem cells/progenitors for cell replacement therapy of deafness in an autograft manner. Recently, inducible pluripotent stem cells (iPS) appear to be another resource for cochlear stem cells/progenitors.

Research on cochlear stem cells/progenitors is still in its infancy. First, it is not clear whether cochlear stem cells/progenitors remain in the mature organ of Corti. Second, if they are in the adult organ of Corti, it is poorly understood why they do not proliferate and replace lost hair cells and neurons after cochlear damage. Finally, the growth, proliferation, and differentiation processes of these stem cells/progenitors, as well as the necessary growth factors for them to become hair cells and/or neurons, are yet to be determined.

In this chapter, we focus on the fundamental issues, such as isolation, growth, proliferation, and differentiation of cochlear stem cells/progenitors. Isolation of cochlear stem cells/progenitors is the first step in studying the biology of cochlear stem cells/progenitors and identifying cochlear tissue-specific growth factors that guide the differentiation of cochlear stem cells/progenitors. A fundamental understanding of the biology of cochlear stem cells/progenitors and the determination of cochlear tissue-specific growth factors will provide us with powerful tools to convert other stem cells such as ESCs, MAPCs, and NSCs, into cochlear stem cells/progenitors or to stimulate the proliferation and differentiation of endogenous cochlear stem cells/progenitors. The study of cochlear stem cells/progenitors not only increases our understanding of the biology of these cells, it also opens the possibility for using a variety of stem cells/progenitors as cell replacement therapy for SNHL.

2. The Discovery of Cochlear Stem Cells/Progenitors in the Mammalian Organ of Corti

Most nonmammalian vertebrates retain the capability to generate new sensory epithelial cells throughout their lives. In birds and amphibians, the loss of hair cells triggers the production of new hair cells,[7-9] and thus hearing is restored. Recent studies have shown that in these animals, supporting cells proliferate and differentiate into hair cells in response to cochlear damage, and transdifferentiation is thought to be a potential mechanism for hair cell regeneration.[8,10]

It had been controversial in the past whether or not cochlear stem cells remain in the adult mammalian organ of Corti. In rodents and mammals, cell types in the organ of Corti have been studied and classified. There are hair cells, pillar cells, Hensen cells, Deiters cells, inner/outer sulcus cells, and Claudius cells, but there have been no stem or stem-like cells identified from the adult organ of Corti. Loss of hair cells in the mammalian organ of Corti leads to very limited or no production of new hair cells, which further strengthens the doubt that cochlear stem cells/progenitors remain after hair cells develop. However, the regeneration of hair cells does occur in the mammalian organ of Corti under certain circumstances.[10,11] In 1999, Kalinec *et al.* established organ-of-Corti cell lines from the postnatal (P14) organ of Corti in mice.[12] In 2002, Malgrange *et al.* isolated auditory hair cells from the postnatal (P0) organ of Corti in rats.[13] In 2003, Ozeki *et al.* established several progenitor hair cell lines.[14] Accumulating evidence suggests that cochlear stem cells/progenitors remain in the postnatal organ of Corti in mice and rats.

Recently, cochlear stem cells/progenitors have been isolated from post-natal (P1) C57B/6J mice.[5] Procedures for the isolation of cochlear stem cells/progenitors include five steps: (1) precise dissection of the postnatal organ of Corti under an operating microscope[14]; (2) physical dissociation of cells from the organ of Corti; (3) culture of physically dissociated cells in the following growth media: DMEM/F12 supplemented with 1% N2, 10 ng/mL EGF, and 10 ng/mL basic fibroblast growth factor (bFGF) in an atmosphere of 5% CO_2 at 37°C; (4) harvest of floating cochlear stem cells/progenitors from primary cell cultures by transferring them to new dishes; and (5) maintenance of cochlear stem cells/progenitors at a low cell density in the above-mentioned growth media.

Morphologically, cochlear stem cells/progenitors appear to be small, round, and bright under a contrast microscope [Fig. 1(a)]. Unlike adher-ent primary cultures, they grow in suspension or loosely attached to the above-mentioned adherent primary culture cells, expressing nestin (a marker for immature neuroepithelial cells), inhibitor of differentiation (Id1, a marker for proliferating neuroepithelial cells), Sox2 (a pan-neuroepithelial marker), and glial fibrillary acid protein (GFAP, a neural stem cell marker) [Figs. 1(b)–1(f)].

Cochlear stem cells/progenitors grow in the above-mentioned growth media, and they rapidly become cellular spheres [Fig. 1(g)] when plated at a low cell density of $1–2 \times 10^2$ single cells per mm^2 or 7–10 small cellular spheres per mm^2. Spheres derived from cochlear stem cells/progenitors express nestin, Sox2, *Math1*, and espin [Figs. 1(h) and 1(I)]. It has been noted that cochlear stem cells/progenitors at a high cell density grow in an adherent manner due to cell–cell interactions and/or rapid consumption of EGF and bFGF in the growth media. Adherent cochlear stem cells/pro-genitors are morphologically similar to primary cells but genetically dif-ferent from them.

In the P1 organ of Corti, nestin is positive for some cells [Fig. 1(j)], sug-gesting the source of cochlear stem cells/progenitors.

3. The Markers of Cochlear Stem Cells/Progenitors

Cellular markers for cochlear stem cells/progenitors have not been studied until recently. Reverse transcription-polymerase chain reaction (RT-PCR) and immunohistochemistry data demonstrate that cochlear stem cells/progenitors highly express neuroepithelial cell markers Id1, nestin,

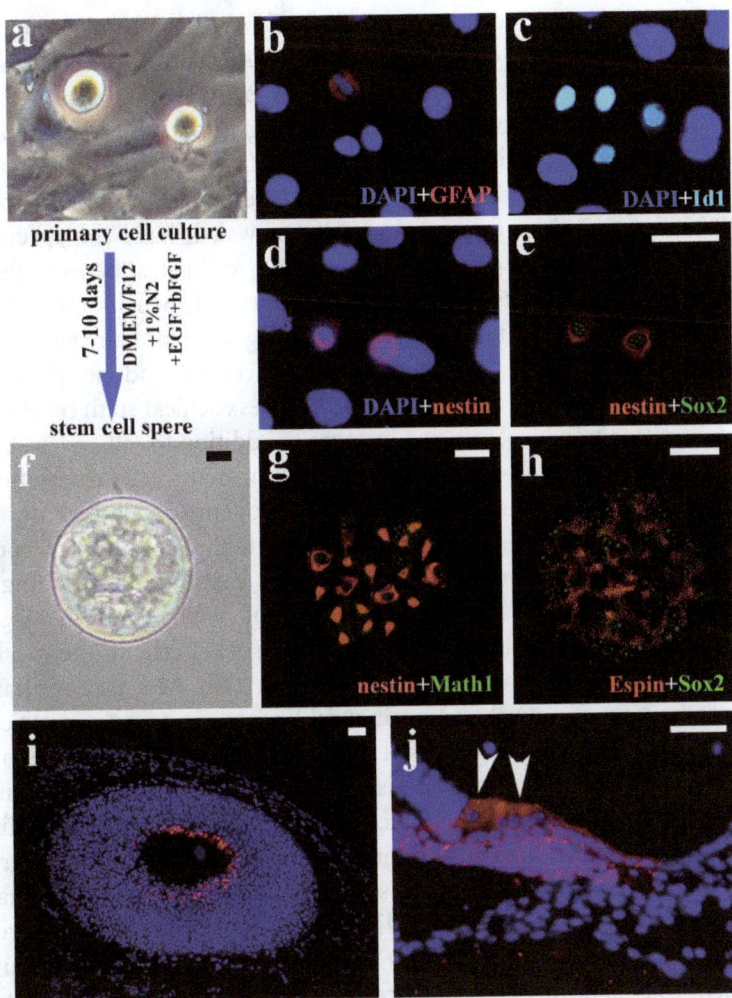

Figure 1. Cochlear stem cells/progenitors are small, round, bright, and yellowish, growing in suspension (a), positive for GFAP (b), Id1 (c), nestin (d), and Sox2 (e). Within 7–10 days, a single cochlear stem cell grows into a sphere (f) in which some cells express nestin/*Math1* (g) and some espin/Sox2 (h). Note that nonstem cells are "adherent" (a) and negative from GFAP (b), nestin (d), and Sox2 (e) in a primary cell culture from the postnatal day 1 organ of Corti in mice. Bar = 10 μm.

Musashi (Msi), and paired box gene 2 (*Pax2*). Id1, the first member of the Id family, is a positive regulator of neural epithelial cell proliferation.[15,16] Recently, it has been suggested that Id1 regulates the proliferation of progenitor hair cells (sensory epithelial cells) by sequestering basic helix-loop-helix (bHLH) transcription factors such as *Math1*.[15,17] The downregulation of Id1 *in vitro* reduces cellular DNA synthesis, cell cycle progression, and cell counts of progenitor hair cells, which proves that Id1 is essential for proliferation.[15] In cochlear stem cells/progenitors, abundant expression of Id1 makes them active in proliferation but inactive in differentiation. This is an important mechanism for cochlear stem cells/progenitors to maintain their fast-growing property and self-renewal. It has been speculated that cochlear stem cells/progenitors grow into cellular spheres with few cells differentiated because the majority express Id1, which acts to antagonize *Math1*, an important differentiation driving force for hair cells.[18,19] Since Id1 plays a role in the neurogenesis of the central and peripheral nervous systems,[15,16] it is also expected that Id1 makes cochlear stem cells/progenitors committed to neuroepithelial lineages in addition to self-renewal.

Nestin is a well-known marker for immature neuroepithelial cells. Similar to NSCs, cochlear stem cells/progenitors express nestin abundantly. Msi, another important neuroepithelial cell marker, is originally found in neural precursor cells that are capable of becoming neurons and glial cells. Loss-of-function mutations in Msi result in a loss of neurons and glial cells but gain of shaft and socket cells in the mechanosensory organs of *Drosophila*,[20] suggesting that Msi is essential for the production of neural and nonneural lineages in an asymmetric manner. It suggests that Msi is involved in the asymmetric cell division of sensory organ precursors that give rise to neuron/glial progenitors and shaft/socket progenitors during neurogenesis. Asymmetric cell division is observed in the nervous system for the production of cellular mosaics such as neurons vs glial cells. It is, therefore, speculated that Msi is involved in asymmetric cell division of cochlear stem cells/progenitors for the production of cellular mosaics such as hair cells vs supporting cells in collaboration with patterning genes such as *Pax2* and lateral inhibition (the Delta–Notch inhibitory system). Similar to nestin, Msi is required for maintenance of stem cell identity.[21]

Pax2 as a patterning gene of cochlear development[22] is highly expressed in cochlear stem cells/progenitors but is weakly expressed in NSCs. It is becoming clear from a recent study that NSCs are rich in Pax6, whereas cochlear stem cells/progenitors are rich in Pax2.[5] The former is responsible for the patterning of the brain tissue whereas the latter is responsible for the patterning of the cochlear tissue. Sox1, Sox2, Pax6, and GFAP, highly expressed in NSCs, are weakly expressed in cochlear stem cells/progenitors [Fig. 2(a)]. Sox1 is an early marker for specified neural cells[23] whereas Sox2

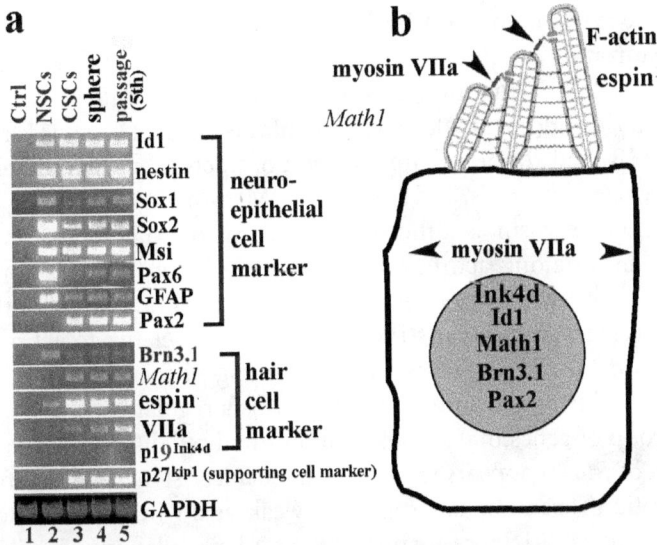

Figure 2. Cochlear stem cells/progenitors (espin) are rich in neural stem cell markers (Id1, nestin, Musashi) and hair cells markers (Pax2, espin) and relatively weak in neural stem cell markers (Pax6, GFAP, Sox1 & 2). Myosin VIIa, *Math1*, and Pax2 are postive in cochlear stem cell but negative in neural stem cells (NSCs). The gene expression profile in the passaged cultures of CSCs remains unchanged compared with CSCs and spheres (a). Hair cell markers are expressed in hair cells with Id1, *Math1*, and Brn3.1 being located in the nucleus, myosin VIIa, and p16^{Ink4d} being located in the cytosol, and myosin VIIa (arrowheads), espin, and F-actin being expressed in the stereocilia (b).

is a protein to maintain neural progenitor identity.[24] GFAP is a classic NSC marker but is weakly expressed in cochlear stem cells/progenitors.

In addition, cochlear stem cells/progenitors weakly express myosin VIIa, *Math1*, Brn3.1 (POU4f3, Brn3c), and espin, which are important hair cell markers [Fig. 2(a)]. These hair cell markers are abundant in mature hair cells but are barely detectable in cochlear stem cells/progenitors. Myosin VIIa is a protein expressed in the cytosol and stereocilia of a hair cell. In the stereocilia, it anchors transduction channels to the stereocilia membrane. In addition, it is an adaptor protein between actin (a predominant protein of the stereocilia) and cadherin 23 (a major structural protein of the tip link).[25] Some investigators believe that myosin VIIa acts as a molecular motor to maintain the resting tension of tip links [Fig. 2(b)] and therefore contributes to the sensitivity of channel proteins to deflection of stereocilia in response to sound vibration. Mutations of the myosin VIIa protein make the tension of tip links loose and thus result in hearing loss.[26]

In the cytosol, myosin VIIa is an important protein that is involved in the organization of microtubules and the hair cell cytoskeleton. It is well known that a mutation of myosin VIIa in the human chromosone 11q results in widespread degeneration of the organ of Corti (Usher syndrome, type IB) and abnormal organization of microtubules in the photoreceptor cells, nasal cilia cells, and sperm cells. Myosin VIIa, belonging to an unconventional myosin, is a motor molecule with structurally conserved heads that move along actin filaments when energy (ATP) is consumed. The highly divergent tails of myosin VIIa are presumed to be tethered to different macromolecular structures that move relative to actin filaments.

Math1 is recognized as a molecular trigger for progenitor hair cell generation during the embryonic stage. It is a regulatory protein that controls a group of genes that are relevant to initial cell differentiation. The expression of *Math1* appears to be weak in cochlear stem cells but strong in postmitotic progenitor hair cells and weak again in differentiated hair cells. Without *Math1*, there are no progenitor hair cells appearing in the organ of Corti,[27] no granule neurons in the cerebellum,[28] and null production of goblet cells in the intestine of mice.[29] Brn3.1 was originally found in the neurons of the central nervous system. Its role in hearing was recognized in a genetic study in which the mutation of Brn3.1 results in progressive SNHL.[30] Knockout of this gene in mice resulted in no auditory hair cell differentiation and maturation.[31] espin is a cross-linker of F-actin in hair cell stereocilia. Mutations in espin are implicated in deafness in mice[32] and humans.

To our knowledge, none of these proteins alone are specific to cochlear hair cells. Brn3.1 and *Math1* are expressed in some neurons whereas myosin VIIa and espin are expressed in some mucosal epithelial cells, but myosin VIIa, *Math1*, Brn3.1, and espin in combination are restricted to cochlear hair cells.

4. The Quiescent State of Cochlear Stem Cells/Progenitors in the Adult Mammalian Organ of Corti

The existence of cochlear stem cells/progenitors has been elegantly proven in recent years. One may wonder why they are quiescent in the adult mammalian organ of Corti. This is a question that puzzles many otological researchers. In the mammalian organ of Corti, there is no to little cellular proliferation activity found in the late-term embryonic cochlea and postnatal days. This suggests that cellular proliferation activity of cochlear stem cells/progenitors is inhibited.

If so, what makes cochlear stem cells/progenitors stop proliferating and renewing themselves? The mechanism is not understood, but a possible explanation is that the developed hair cells in the organ of Corti highly express inhibitors for the cell cycle progression (p19^{Ink4d} and retinoblastoma, Rb1), which prevents them from entering into the cell cycle. Evidence supporting this theory is that the knockout of p19^{Ink4d} results in the cell cycle progression of developed hair cells[33] and the knockout of Rb1 results in the generation of extra hair cells in mice.[34] Interestingly, p19^{Ink4d} acts to block cell cycle progression and protects cells from apoptosis (removal of p19^{Ink4d} resulting in activation of caspase 3),[35] whereas Rb1 inhibits cell cycle progression but apparently has nothing to do with apoptosis.

Another explanation is that a local inhibitory signaling system (Delta–Notch) in the organ of Corti, so called lateral inhibition,[36] suppresses proliferation and differentiation of supporting cells. By definition, lateral inhibition is a cell–cell interaction involved in the production of cellular mosaics [Fig. 3(a)]. In a multiple cellular organ, a specified cell prohibits the development of similar cells in its immediate neighborhood. The Delta–Notch system is believed to be involved in the production of unique hair cell vs supporting cell mosaics. In the developing organ of Corti, all the cells derived from cochlear stem cells/progenitors initially express Delta ligands and Notch receptors. Under this situation, Delta binds to adjacent Notch receptors and activates them.

By cell-to-cell interaction of Delta and Notch, a competition occurs with the winning cell expressing Delta strongly and inhibiting its immediate neighbors from doing likewise through Notch receptors. The winning cell becomes a hair cell (high Delta but low Notch) and the others become supporting cells [e.g. high Notch but low Delta cells, Fig. 3(b)]. Then, Notch is inactivated in the winning hair cell because Delta suppresses the Notch signaling pathway [Fig. 3(c)], but is activated in the surrounding cells to prevent them from differentiation and to keep them in a default state. For this reason, supporting cells are thought of as candidate progenitors for hair cells in a sense. All cells in mosaic structure lose their capability to proliferate unless the inhibitory system is removed. Hair cells are prohibited from entering into the cell cycle due to the expression of p19^{Ink4d} and Rb1, whereas supporting cells are inhibited by the Notch–Hes signaling pathway [Fig. 3(d)] and perhaps p27^{Kip1}.

When hair cells are damaged or lost, supporting cells are liberated from this cell–cell lateral inhibition and undergo cellular proliferation in order to fill the cellular vacancy. Unfortunately, the initial developmental cues

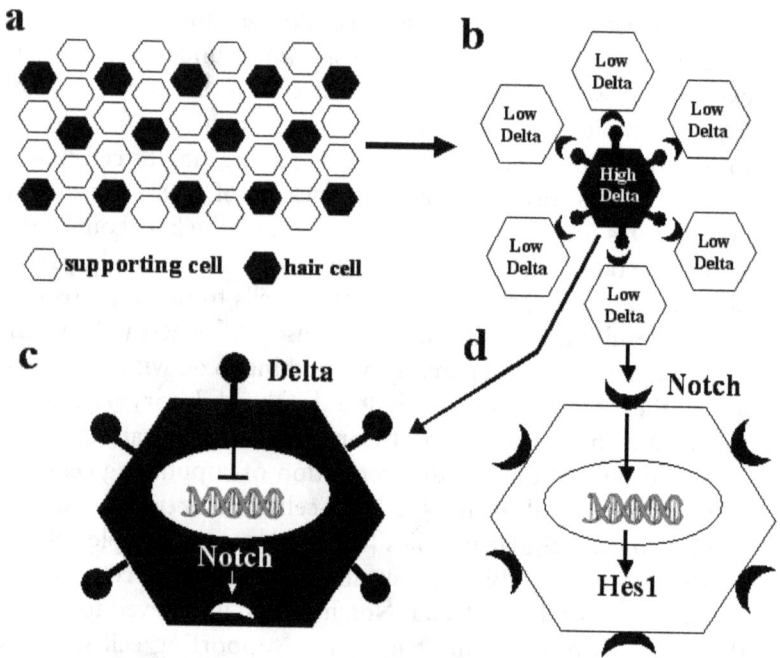

Figure 3. Mosaic features of the cochlear hair vs supporting cells within the developed organ of Corti (a) with a lateral inhibition mechanism by which a winner cell expresses a high level of Delta that activates the Notch receptor of the winner cell's neighbors and results in a low level of Delta in them (b). The high level of Delta ligand in the winner cells negatively regulates the Notch signaling activity within the cell and makes it a hair cell (c). Activation of the Notch receptor in the winner cell's immediate neighbors triggers the expression of Hes1, which suppresses the transcription of mRNAs, including Delta, making it a supporting cell (d).

are no longer present in the adult organ of Corti and, therefore, no supporting cells spontaneously compete to become hair cells unless the developmental cues are provided. What the developmental cues consist of remains to be elucidated.

In addition, growth factors such as EGF and bFGF may not be sufficient for cellular growth and proliferation in the developed mammalian organ of Corti. There should be some EGF and bFGF in the bloodstream, but they may not be able to pass through the blood-labyrinth barrier because the growth factors are either positively or negatively charged, which makes penetration difficult. This assumption is supported by the recent *in vitro* studies in which cochlear stem cells/progenitors isolated from postnatal animals restore their proliferation and renewal capabilities when EGF and bFGF are supplied. It suggests that cochlear stem cells/progenitors do not lose their capabilities to proliferate and self-renewal in the postnatal days because they do not express $p19^{Ink4d}$ as shown above [referred to in Fig. 2(a)]. It is their environments (cell–cell contact inhibition, lateral inhibition system, or loss of developmental cues, etc.) which do not allow them to proliferate and self-renew.

5. The Maintenance of Cochlear Stem Cells/Progenitors

A fundamental feature of stem cells is their lasting ability to multiply. Like most stem cells, cochlear stem cells/progenitors have such capability to grow on a long-term basis when isolated from the cochlear tissue of mice and supplied with the appropriate growth factors. In growth media, DMEM/F12 provides basic electrolytes, salts, and essential amino acids, N2 is a supplement for neural and/or neuronal growth *in vitro*, and EGF and bFGF jointly provide the growth and proliferation power of cochlear stem cells/progenitors. EGF and bFGF alone or in combination increase DNA synthesis of ESCs,[1] cell cycle progression, and cell counts of cochlear stem cells/progenitors.[5] Without EGF and bFGF, the cochlear stem cells/progenitors lose their growth and proliferation power and do not grow into cellular spheres or maintain their self-renewal capability *in vitro*.

Due to the availability of mutant mice, it has become increasingly clear in recent years that cellular proliferation and differentiation are controlled by a group of transcription factors bHLH and dominant negative bHLH.[27,37–39] bHLH family members such as *Math1* and E-box protein (E47) form a bHLH heterodimer that binds to the E-box promoter region and initiates gene transcription, triggering initial cell differentiation of hair cells. Dominant negative bHLH family members such as Id1 sequester bHLH family members (*Math1* or E47) and form a bHLH-dominant negative HLH heterodimer that cannot bind to the E-box promoter and thus switch cells from a differentiation state to a proliferation state (Fig. 4). Therefore, Id1 and *Math1* in cochlear stem cells/progenitors act as cellular switches for proliferation and differentiation.

Id1 is highly expressed in the progenitor hair cells in the early embryonic stage but is downregulated in the late embryonic days.[15,17] In progenitor hair cells, Id1 increases the transcription of nuclear factor kappa B (NF-κB) and translocates NF-κB into the nucleus, which activates NF-κB. Activated NF-κB then binds to the promoter of cyclin D1 and increases the

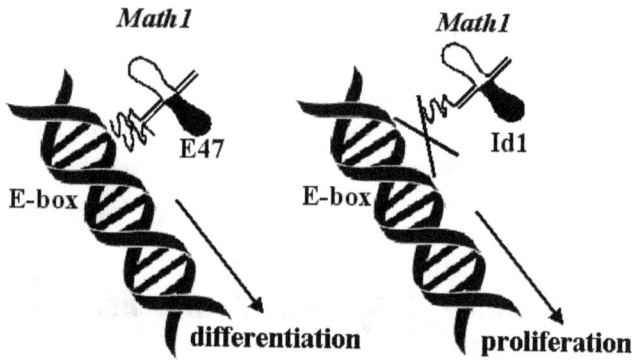

Figure 4. Id1 and *Math1* serve as molecular switches for cochlear stem cell proliferation and differentiation. Cells differentiate when *Math1* and E-box protein (E47) form a heterodimer which binds to the E-box promoter area and activates the transcription of a battery of genes whereas cells proliferate when Id1 sequesters *Math1* and prevents it from binding to the E-box promoter area due to the defective *Math1*–Id1 heterodimer in binding capability.

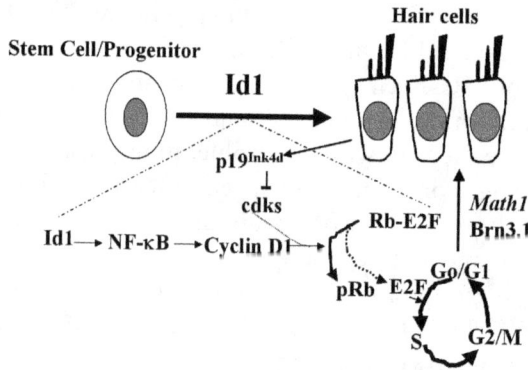

Figure 5. Id1 induces the proliferation of cochlear stem cells/progenitors via an NF-κB/cyclin D1/cdks/E2F-dependent mechanism. Id1, a highly expressed transcription factor in cochlear stem cells/progenitors, increases the activity of NF-κB. Activated NF-κB binds to the promoter of cyclin D1 and increases the expression of cyclin D1. Cyclin D1, with the help of cdks, phosphorylates the Rb-E2F complex, release E2F to drive a cell into cell cycles for proliferation. With the expression of *Math1* and Brn3.1, cells differentiate into hair cells that express p16^{Ink4d}, which blocks cell cycles in a negative manner. Upregulated *Math1* then increases cellular differentiation by sequestering Id1.

transcription of cyclin D1.[15] Cyclin D1, together with cyclin-dependent kinases 4/6 (cdk4/6), phosphorylates retinoblastoma (Rb) which releases E2F transcription factor from the Rb-E2F complex. E2F then drives cells from the G0/G1 phase to the S phase, where they then undergo proliferation (Fig. 5).

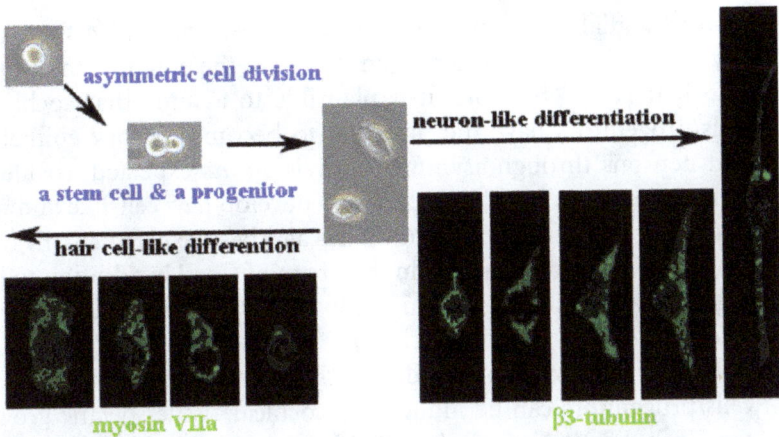

Figure 6. A cochlear stem cell (CSC) has the potential to renew itself by asymmetric cell division through which a stem cell and a progenitor are produced. Similarly, a progenitor cell can generate two daughter cells through an asymmetric cell division mechanism, with one becoming neuron-like and the other becoming hair cell-like.

During the maintenance of cochlear stem cells/progenitors, cells undergo asymmetric division. A parent cell produces two daughter cells, one of which becomes identical to the parent cell (stem cell) and the other becomes a slightly different phenotype or tentative progenitor (Fig. 6). Therefore, in stem cell culture, there are always some stem cells and some progenitors, which is the reason why we use the term cochlear stem cells/progenitors in this article. The organ of Corti consists of a variety of cell types such as hair cells, supporting cells, and neurons. To generate this diversity, cells must form two different cell types from one. This can be achieved in two ways: either two initially identical daughter cells become different because they encounter different environments, or cell fate determinants are segregated into only one of the two daughter cells to make one cell different from its sister cell. Asymmetric division is observed in NSCs and is probably used by cochlear stem cells/progenitors to generate distinct phenotypes (sensory hair cells and neurons) under the influence of induction media (see below).

6. The Oligopotency of Cochlear Stem Cells/Progenitors

Nestin, Pax6, Sox2, GFAP, and Msi are typical NSC markers. Cells expressing these NSC markers have the potential to differentiate into neurons, glial cells, and oligodendrocytes, whereas cells expressing *Math1*, Brn3.1, espin, and myosin VIIa have the potential to differentiate into hair cells. Therefore, it is plausible to assume that cochlear stem cells/progenitors have the potential to become sensory epithelial cells and neurons through asymmetric division. As expected, cochlear stem cells/progenitors have been shown to develop hair cell-like and neuron-like phenotypes *in vitro* (Fig. 6). It has also been experimentally observed that nascent neurons appear in the organ of Corti in addition to hair cell regeneration following the injection of ESCs and vestibular stem cells into chick embryonic otics.[1,40]

It has recently been found that the differentiation of cochlear stem cells/progenitors can be induced by cochlear tissue-specific growth factors, including sonic hedgehog (Shh), EGF, retinoic acid (RA), and brain-derived neurotrophic factor (BDNF), namely SERB.[5] Induction of cochlear stem cell/progenitor differentiation is performed in a two-step protocol. In step 1, the cochlear stem cells/progenitors are incubated with SERB for 14 days to "prime" or guide cellular proliferation. In step 2, SERB is withdrawn for 7 days (step 2) to allow cells to exit cell cycles and commit to cellular differentiation.

As described in the section "The Discovery of Cochlear Stem Cells/Progenitors in the Mammalian Organ of Corti," cochlear stem cells/progenitors are derived from floating cells that may contain NSCs and cochlear stem cells. How do we know that neuron-like and hair cell-like phenotypes are from the same subpopulation? To answer this question, clonal analysis of cochlear stem cells/progenitors has recently been performed. The results indicate that both hair cell-like and neuron-like cells are derived from a single clone,[5] which suggests that cochlear stem cells/progenitors indeed have the potential to become hair cells and neurons via an asymmetric division mechanism.

Theoretically, cochlear stem cells/progenitors should be able to differentiate into glial cells and oligodendrocytes because they express a full set of NSC markers such as nestin, GFAP, Msi, Sox2, and Pax6. They should be able to become supporting cells because they also express supporting cell markers such as $p27^{Kip1}$, as shown above [referred to as Fig. 2(a)]. However, the multipotency of cochlear stem cells/progenitors remains to be studied and further elucidated.

7. The Specific Factors for "Priming" Cochlear Stem Cells/Progenitors

It has been recognized that the growth and proliferation of cochlear stem cells in the presence of cochlear tissue-specific growth factors are not only involved in changes of its size, shape, and quantity, but it is also related to the upregulation of critical hair cell markers. Without appropriate "priming" of growth factors, newly proliferated cells remain to be cochlear stem cells/progenitors and rarely differentiate into terminal phenotypes spontaneously. For example, the progenitor hair cell line (OC1) proliferates actively with DMEM/F12 + 10% fetal bovine serum (FBS) + 10 ng/mL EGF, but it does not differentiate into hair cells and neurons[14] without the "priming" of cochlear tissue-specific growth factors. Cochlear stem cells/ progenitors grow well in growth media (DMEM/F12 + 1% N2 + EGF + bFGF), but they also do not differentiate into hair cells and neurons without "priming" of cochlear tissue-specific growth factors (SERB). Individual factors have some effect on the expression of hair cell markers but not as much as SERB in combination. Cochlear tissue-specific growth factors are essential for the induction of certain proteins that are related to the structure and/or function of hair cells. In cochlear stem cells/progenitors, the process is associated with the induction of *Math1*, myosin VIIa, Brn3.1, and espin. These four molecules are considered to be as important as hair cell markers because their combination is necessary for hair cells to form and function.

There are many growth factors that are involved in the growth and proliferation of cochlear stem cells/progenitors, but not all of them are capable of upregulating the expression of *Math1*, Brn3.1, myosin VIIa, and espin. It has recently been found that the SERB formula is capable of guiding cochlear stem cell/progenitor differentiation into hair cells and neurons. This is due to their synergistic effects on cellular proliferation and differentiation. To better understand the roles of SERB combination in proliferation and differentiation of cochlear stem cells/progenitors, individual factors are addressed below:

- Shh is involved in the development of the inner ear.[41] Blockage of Shh bioactivity with a specific antibody results in the loss of the ventral inner ear structure.[42] Together with retinoic acid (RA), Shh can turn bone marrow stromal cells into sensory neurons.[43]

- EGF has been shown to stimulate the replacement of hair cells following animoglycoside ototoxic damage in rat cochlear organotypic cultures[44] and to induce cochlear hair cell differentiation *in vitro.*[45] It is also one of the important growth factors that induce ESCs to express *Math1*, myosin 7a, Brn3.1, espin, p27[Kip1], and AchRα9.[1,40]
- RA is known as a key factor for neurons in the central nervous system and hair cells in the peripheral nervous system.[11] Recent studies indicate that RA stimulates neurogenesis in adult hippocampal NSC cultures,[46] making it one of the candidate factors for induction of sensorineural epithelial cells.
- BDNF is an important neurotrophin in the central and peripheral nervous systems[47,48] and contributes to cell differentiation, neurogenesis,[49] and the survival of auditory neurons.[48]

The possibility exists that individual factors play critical roles in the "priming" of cochlear stem cells/progenitors *in vivo* or in organ-like cultures, but the synergy between individual factors is more important than individual actions. The possible synergistic effects of Shh, EGF, RA, and BDNF (SERB) on the cellular growth, proliferation, differentiation, and survival of cochlear stem cells/progenitors are summarized in Fig. 7.

8. The Differentiation of Cochlear Stem Cells/Progenitors

Differentiation is a process in which cells become specialized via expression of genes that are necessary for cellular structure and function. For differentiation to start, cells have to exit the cell cycle. Only those cells that are out of the cell cycle or at the G0 phase are able to differentiate. Cell cycle analysis by flow cytometry confirms that in step 1 the majority of the cells are in the S and G2/M phases, whereas in step 2 the majority of the cells are in the G0 or G1 phase of a cell cycle (Fig. 8). From the morphology point of view, cells in step 1 increase in number which some begin to upregulate hair cell markers and neuronal cell markers and preliminarily possess some morphologic and genetic features of hair cells or neurons, though they are far from mature. Therefore, the withdrawal of growth factors in step 2 further promotes cellular differentiation. At the end of step 2, differentiated cells increase in number and cellular morphology begins to change dramatically (Fig. 9).

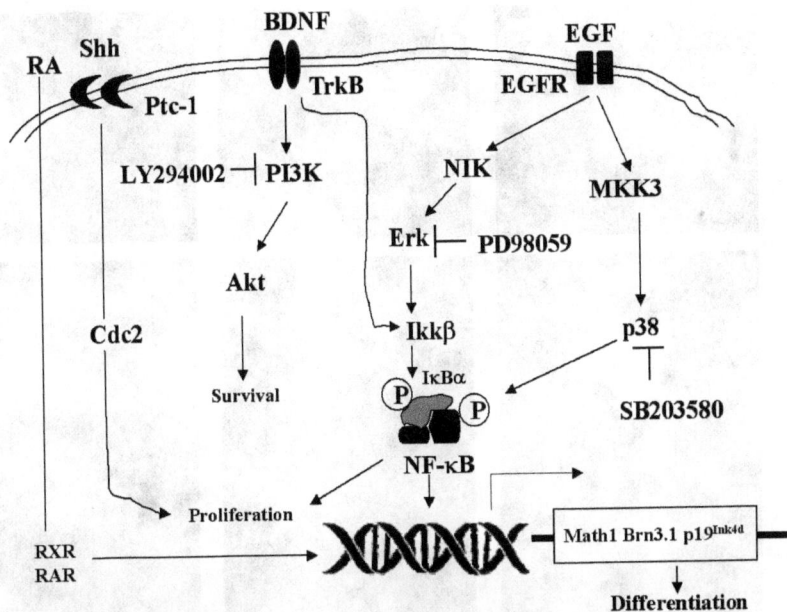

Figure 7. Induction of CSC proliferation by EGF via the Erk pathway and Shh via the *cdc2* gene (Barnes *et al.*, 1997), cellular survival by BDNF via the PI3K-Akt pathway, and expression of *Math1* and Brn3.1 by RA via RXR (Inoue *et al.*, 2006) may in part explain the synergy of SERB in the generation of hair cell- and neuron-like cells *in vitro* in this study.

Increased expression of *Math1*, Brn3.1, and p19^{Ink4d} is a sign that cochlear stem cells/progenitors have committed to a hair cell lineage. It has been noted that *Math1* is induced in step 1 and is strengthened in step 2. Recently, it has been shown that *Math1* induces the expression of multiple hair cell markers myosin VIIa, espin, Brn3.1, and jagged2, as well as supporting cell marker p27^{Kip1} in MSCs.[4] Introduction of *Math1* cDNA using the adenoassociated virus (AAV) as a carrier has been shown to generate new hair cells in the deafened cochlea of guinea pigs,[18,19,50] which raises hope that upregulation of the *Math1* gene by other methods, such as cochlear tissue-specific growth factors, may lead to the generation of new hair cells in damaged cochleas. Increased expression of Brn3.1 strengthens the differentiation of cochlear stem cells/progenitors toward hair cells, whereas increased expression of p19^{Ink4d} prevents differentiated cells from entering the cell cycle again.

In the presence of SERB, cochlear stem cells/progenitors face a fate-determining process: to be hair cells or neurons. How this decision is made

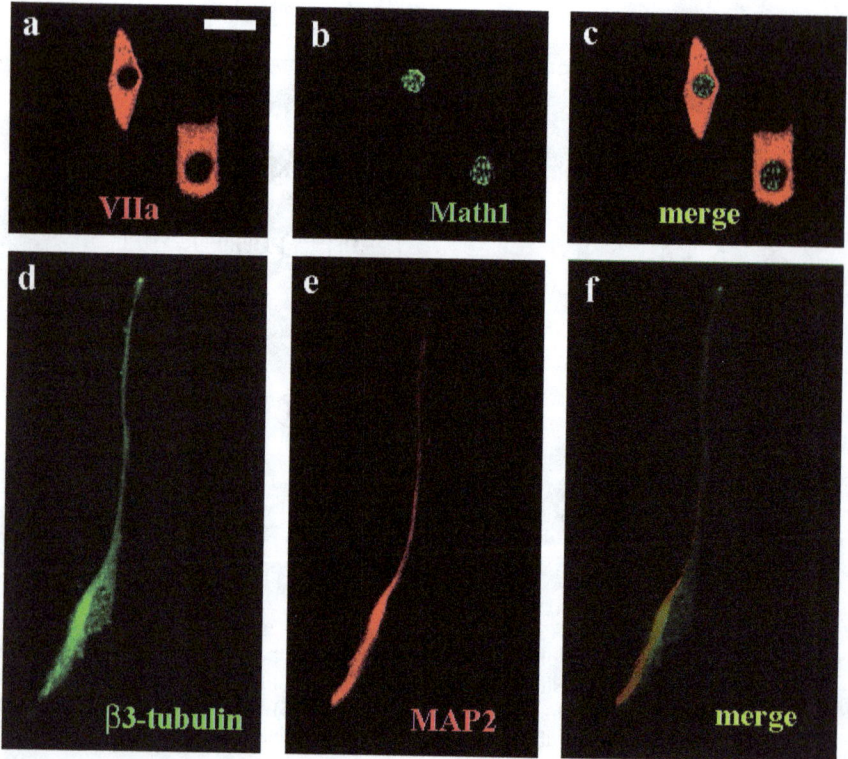

Figure 8. Incubation of cochlear stem cells/progenitors with SERB induces the differentiation of two distinct phenotypes. One is the hair cell-like phenotype that expresses both myosin VIIa and *Math1*, which define nascent hair cells (a–c), and the other is the neuron-like phenotype that expresses both β3-tubulin and MAP2 (d–f), which define nascent neurons. At least 35.8% of cells express both myosin VIIa and *Math1* following SERB treatment, whereas at least 12.1% of cells express both β3-tubulin and MAP2 following SERB treatment (FACS data).

was largely unknown until recently. *In vitro*, activation of the extracellular signal-regulated kinase (Erk, or mitogen-activated protein kinase — MAPK) pathway in cochlear stem cells/progenitors appears to be critical. Blockage of the Erk pathway with a specific inhibitor, PD98059, significantly reduces the expression of hair cell markers *Math1*, myosin VIIa, and espin but not the expression of neuronal cell markers β3-tubulin (a protein for nascent neurons) and microtubule-associated protein 2 (MAP2, a protein for nascent or preliminarily differentiated neurons) by fluorescence-activated cell sorting (FACS) (Fig. 10). This suggests that the Erk

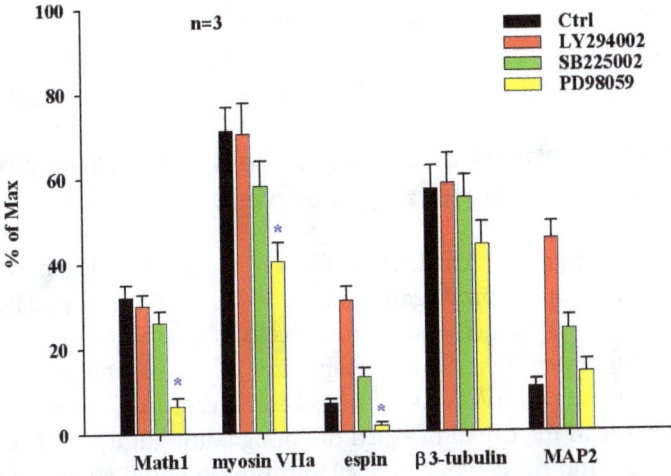

Figure 9. Induction of the *Math1*, myosin VIIa, and espin expression by SERB is significantly inhibited by an Erk signaling pathway inhibitor, PD98059, whereas induction of the β3-tubulin and MAP2 expression by SERB is not, suggesting that the Erk signaling pathway controls the differentiation of sensory epithelial cells but not neuronal cells in the cochlear organ of Corti.

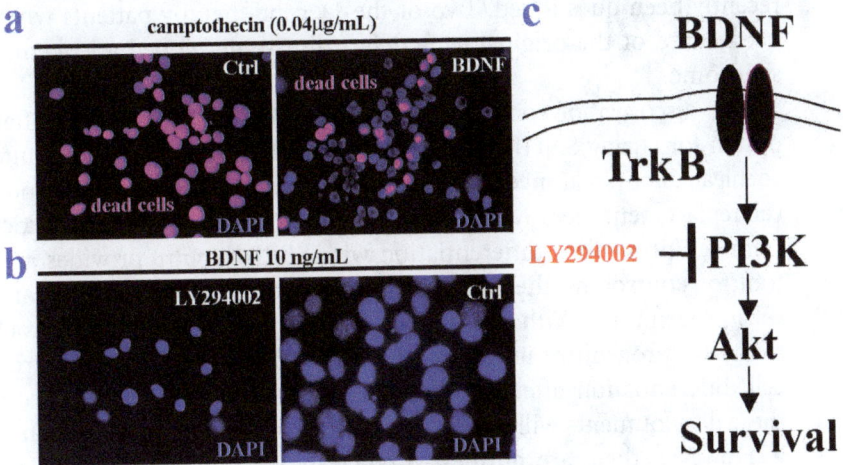

Figure 10. BDNF protects cochlear stem cells/progenitors from cell death induced by camptothecin, a universal inducer of cellular apoptosis (a). A specific inhibitor for the PI3K kinase remarkably reduces the cell numbers of cochlear stem cells/progenitors (b). BDNF binds to the receptor TrkB, which in turn activates the kinase PI3K and the protein kinase B (Akt), contributing to the survival of cochlear stem cells/progenitors (c).

pathway acts as a molecular switch to determine whether a cell becomes a hair cell or a neuron.

9. The Applications of Cochlear Stem Cells/Progenitors in Degenerative Hearing Disorders

One of the greatest challenges in SNHL is restoring degenerative cochlear hair cells and spiral ganglion neurons biologically. There are two approaches by which cochlear hair cells and spiral ganglion neurons can be regenerated, here gene therapy and cell replacement. In recent years, gene therapy with AAV-*Math1* has been successful in regenerating nascent hair cells in the organ of Corti of young adult guinea pigs, as reported by Raphael and his coworkers.[18,50] However, potential concerns remain. One of the concerns in gene therapy is the use of virus-derived vectors that are potentially risky. The commercially available AAV vector itself is immunogen-reduced, but the antibodies invoked by AAV infection in the upper respiratory tract are capable of destroying the cells that contain AAV vectors. For this reason, lentivirus vectors may be a better choice. In addition to the immunologic concern, the safety of using virus-derived vectors has recently been questioned. Two of the 11 gene therapy patients who have been cured of the original gene defect disease developed a leukemia-like syndrome.[51]

The discovery of cochlear stem cells/progenitors in neonates offers an option for curing SNHL. Success in the injection of *Math1* cDNA into the cochlear duct (scala media) in gene therapy provides technical support for cell replacement therapy in SNHL. More importantly, induction of cochlear stem cell/progenitor differentiation with SERB *in vitro* provides a useful tool for controlling the differentiation process of injected cochlear stem cells/progenitors. With these new developments, injection of cochlear stem cells/progenitors into the cochlear duct (scala media) and controllable cell differentiation after injection are theoretically feasible. If successful, these developments will have a tremendous impact on public health.

However, there are hurdles to be cleared before cell replacement can become a clinical choice for treatment of SNHL. First, what should be done to prepare stem cells before injection? Efforts have been made to inject fetal NSCs into the modiolus of cochlea in mice, but no cochlear hair cells have been regenerated in the organ of Corti.[52] Ito and his coworkers have reported that an injection of mouse ESCs into the inner ear did not result in the regeneration of hair cells.[53] However, injected NSCs and ESCs survive in the cochlear tissues.[52,53] The question here is how to direct or guide the stem cells initial differentiation *in vitro* and ensure their differentiation after injection. Second, the embryonic developmental cues are no longer present in the developed organ of Corti. What factors are necessary for fostering and promoting the differentiation of injected stem cells *in vivo*? Is the SERB formula able to direct stem cell differentiation *in vivo*? Finally, what would be appropriate vehicles for carrying developmental cues into the cochlea?

10. The Future Directions of Cochlear Stem Cell/ Progenitor Research

10.1. Determination of the cochlear stem cell niche

It has been demonstrated that the growth media for cochlear stem cells/ progenitors are able to maintain the identity of cochlear stem cells/ progenitors, but the detailed cellular microenvironment providing support and stimuli necessary for sustaining self-renewal on a long-term basis remains to be elucidated. We are not certain whether cochlear stem cells/progenitors maintained with the aforementioned growth media are "immortal strand" or "*in vitro* stem cell." By definition, "immortal strand" selectively retains parental DNA strands during asymmetric self-renewal, a potential mechanism for protection of stem cells from the mutations associated with replication. "*In vitro* stem cell" stands for self-renewal *ex vivo* in cells that do not overtly behave as stem cells *in vivo*. It occurs due to liberation from inductive commitment signals or by creation of a synthetic stem cell state. This is a fundamental issue for the maintenance of cochlear stem cells/progenitors and will influence whether we are able to duplicate cochlear stem cells/progenitors with fidelity for clinical use.

10.2. Promotion of endogenous cochlear stem cell/progenitor proliferation and differentiation

With the finding of cochlear stem cells/progenitors in the neonates of mice, the possibility arises of repairing damaged and lost hair cells and spiral ganglion neurons by stimulating the residual cochlear stem cells/progenitors. The question here is how to deliver cochlear tissue-specific growth factors such as SERB into the cochlear duct (scala media) and direct the differentiation of residual cochlear stem cells/progenitors into hair cells and spiral ganglion neurons. First, direct injection of SERB into the scala media is not acceptable because repeated trauma to the cochlear duct would damage the vulnerable organ-of-Corti structure. Second, intravenous administration of SERB is not practical because the blood-labyrinth barrier would prevent SERB from entering the inner ear based upon the fact that growth factors are either positively or negatively charged. It is known that the basement membrane does not allow negatively or positively charged substances to freely gain access to the brain and inner ear.

10.3. Use of nanoparticles to deliver factors

Theoretically, nanoparticles are ideal carriers for factors because they are capable of freely penetrating cells and biological membranes due to their ability to "neutralize" electronically charged factors. Furthermore, nanoparticles such as iron–cobalt (FeCo) are highly magnetic and can be guided to an organ or tissue using a magnetic device outside of the body. This provides an opportunity to aggregate nanoparticles in the cochlear tissue. In addition, specific antibodies can be attached to nanoparticles' surfaces via biological "glue" such as poly-L-lysine to endow nanoparticles with the ability to target cell subpopulation. Construction of such an antibody–antigen recognition system for magnetic nanoparticles is important and technically innovative. Antibody-attached nanoparticles could deliver drugs and factors directly to targeted cells and have extensive application prospects in research. This "smart" drug/factor delivery system will enhance treatment efficiency and avoid side effects of drugs and factors. Simultaneously, nanoparticles such as FeCo nanoparticles can be visualized by magnetic resonance imaging (MRI) due to their high magnetic momentum.

An aurum (gold) surface on nanoparticles is favored because gold atoms are inert and negatively charged.[54] This provides a surface for positively charged peptides ($-NH_3^+$ group of Shh, EGF, and BDNF) to attach in an electrostatic interaction manner. They will neutralize positively charged peptides and help them go through some biological membranes or barriers, namely the blood-labyrinth barrier. It has been shown that the electrostatic interaction between the positive charge of the $-NH_3^+$ group and the negative charge of the gold surface yields a stable attachment that is more than 600 piconewtons.[55] The gold surface property can be used for carrying peptides and antibodies via poly-L-lysine. Also, the gold surface protects FeCo cores from corrosion and oxidation in a biological environment.

Are nanoparticles cytotoxic? Some are and some are not. Based upon the recent reports, gold nanoparticles are not cytotoxic[56] and nanodiamonds are not toxic to a variety of cell types either.[57] However, cerium oxide nanoparticles appear to have low cytotoxicity toward human lung cancer cells due to their oxidative stress in the cells.[58]

Nanoparticles, in combination with stem cell replacement therapy, provide an opportunity to establish a completely new method for generation of sensory hair cells and neurons from exogenously injected cochlear stem cells/progenitors and/or endogenously residual cochlear stem cells/progenitors. If successful, that will open a new door to treatment of SNHL.

Acknowledgments

We are grateful to Drs. Samuel Levine, Tina Huang, Yehoash Raphael, and Vladimir Tsuprun for their valuable comments and suggestions during the preparation of this book chapter. This study is in part supported by NIDCD R01008165, P30CD04660, and NCI R03 CA107989, as well as private foundations such as the Lions 5M International Hearing Foundation.

References

1. Li H, Roblin G, Liu H, Heller S. (2003) Generation of hair cells by stepwise differentiation of embryonic stem cells. *Proc Natl Acad Sci USA* **260:** 13495–13500.
2. Jiang Y, Henderson D, Blackstad M, *et al.* (2003) Neuroectodermal differentiation from mouse multipotent adult progenitor cells. *Proc Natl Acad Sci USA* **100 Suppl 1:** 11854–11860.
3. Jiang Y, Jahagirdar BN, Reinhardt RL, *et al.* (2002) Pluripotency of mesenchymal stem cells derived from adult marrow. *Nature* **418:** 41–49.
4. Jeon SJ, Oshima K, Heller S, Edge AS. (2006) Bone marrow mesenchymal stem cells are progenitors *in vitro* for inner ear hair cells. *Mol Cell Neurosci* **34:** 59–68.
5. Lin J, Feng L, Hamajima Y, *et al.* (2007) Cochlear stem cells/progenitors and degenerative hearing disorders. *Curr Med Chem* **14:** 2937–2943.
6. Marhsall CT, Lu C, Winstead W, *et al.* (2006) The therapeutic potential of human olfactory-derived stem cells. *Histol Histopathol* **21:** 633–643.

7. Ryals BM, Rubel EW. (1998) Hair cell regeneration after acoustic trauma in adult Cortunix quail. *Science* **240**: 1774–1776.

8. Stone JS, Rubel EW. (2000) Cellular studies of auditory hair cell regeneration in birds. *Proc Natl Acad Sci USA* **97**: 11714–11721.

9. Woolley SM, Wissman AM, Rubel EW. (2001) Hair cell regeneration and recovery of auditory thresholds following aminoglycoside ototoxicity in Bengalese finches. *Hear Res* **153**: 181–195.

10. Corwin JT, Contanche DA. (1988) Regeneration of sensory hair cells after acoustic trauma. *Science* **240**: 1772–1774.

11. Lefebvre PP, Malgrange B, Staecker H, *et al.* (2003) Retinoic acid stimulates regeneration of mammalian auditory hair cells. *Science* **260**: 692–695.

12. Kalinec F, Kalinec G, Boukhvalova M, Kachar B. (1999) Establishment and characterization of conditionally immortalized organ of Corti. *Cell Biol Intl* **23**: 175–184.

13. Malgrange B, Belachew S, Thiry M, *et al.* (2002) Proliferative generation of mammalian auditory hair cells in culture. *Mech Dev* **112**: 79–88.

14. Ozeki M, Duan L, Obritch W, Lin J. (2003) Establishment and characterization of progenitory hair cell lines in rats. *Hear Res* **179**: 43–52.

15. Ozeki M, Hamajima Y, Feng L, *et al.* (2006) Id1 induces the proliferation of cochlear sensorineural epithelial cells via the NF-kB/cyclin D1 pathway *in vitro. J Neurosci Res* **85**: 515–524.

16. Tzeng SF. (2003) Inhibitors of DNA binding in neural cell proliferation and differentiation. *Neurochem Res* **28**: 45–52.

17. Jones JM, Montcouquiol M, Dabdoub A, *et al.* (2006) Inhibitors of differentiation and DNA binding (Ids) regulate Math1 and hair cell formation during the development of the organ of Corti. *J Neurosci* **26**: 550–558.

18. Kawamoto K, Ishimoto S, Minoda R, *et al.* (2003) Math1 gene transfer generates new cochlear hair cells in mature guinea pigs *in vivo. J Neurosci* **23**: 4395–4400.

19. Zheng JL, Gao WQ. (2000) Overexpression of Math1 induces robust production of extra hair cells in postnatal rat inner ears. *Nat Neurosci* **3**: 580–586.

20. Sakakibara S, Imai T, Hamaguchi K, *et al.* (1996) Mouse-Musashi-1, a neural RNA-binding protein highly enriched in the mammalian CNS stem cell. *Dev Biol* **176**: 23–242.

21. Siddall NA, McLaughlin EA, Marriner NL, Hime GR. (2006) The RNA-binding protein Musashi is required intrinsically to maintain stem cell identity. *Proc Natl Acad Sci USA* **103**: 8402–8407.

22. Burton Q, Cole LK, Mulheisen M, *et al.* (2004) The role of Pax2 in mouse inner ear development. *Dev Biol* **272**: 161–175.

23. Aubert J, Stavridis MP, Tweedie S, *et al.* (2003) Screening for mammalian neural genes via fluorescence-activated cell sorter purification of neural

precursors fro mSox1-gfp knock-in mice. *Proc Natl Acad Sci USA* **100 Suppl 1:** 11836–11841.

24. Graham V, Khudyakov J, Ellis P, Pevny L. (2003) SOX2 functions to maintain neural progenitor identity. *Neuron* **39:** 749–765,

25. Sollner C, Rauch GJ, Siemens J, *et al.* (2004) Mutations in cadherin 23 affect tip links in zebrafish sensory hair cells. *Nature* **428:** 955–959.

26. Kros CJ, Marcotti W, van Netten SM, *et al.* (2002) Reduced climbing and increased slipping adaptation in cochlear hair cells of mice with Myo7a mutations. *Nat Neurosci* **5:** 41–47.

27. Bermingham NA, Hassan BA, Price SD, *et al.* (1999) Math1: an essential gene for the generation of inner ear hair cells. *Science* **283:** 1837–1841.

28. Ben-Arie N, Bellen HJ, Armstrong DL, *et al.* (1997) Math1 is essential for genesis of cerebellar granule neurons. *Nature* **390:** 169–172.

29. Yang Q, Bermingham NA, Finegold MJ, Zoghbi HW. (2001) Requirement of Math1 for secretory cell lineage commitment in the mouse intenstine. *Science* **294:** 2155–2158.

30. Vahava O, Morell R, Lynch ED, *et al.* (1998) Mutation in transcription factor POU4F3 associated with inherited progressive hearing loss in humans. *Science* **279:** 1950–1954.

31. Xiang M, Gao WQ, Hasson T, Shin JJ. (1998) Requirement for Brn-3c in maturation and survival, but not in fate determination of inner ear hair cells. *Development* **125:** 3935–3946.

32. Zheng L, Sekerkova G, Vranich K, *et al.* (2000) The deaf jerker mouse has a mutation in the gene encoding the espin actin-bundling proteins of hair cell stereocilia and lacks espins. *Cell* **102:** 377–385.

33. Chen P, Zindy F, Abdala C, *et al.* (2003) Progressive hearing loss in mice lacking the cyclin-dependent kinase inhibitor Ink4d. *Natil Cell Biol* **5:** 422–426.

34. Sage C, Huang M, Karimi K, *et al.* (2005) Proliferation of functional hair cells *in vivo* in the absence of the retinoblastoma protein. *Science* **307:** 1114–1118.

35. Tavera-Mendoza LE, Wang TT, White JH. (2006) p19INK4D and cell death. *Cell Cycle* **5:** 596–598.

36. Lanford PJ, Lan Y, Jiang R, *et al.* (1999) Notch signaling pathway mediates hair cell development in mammalian cochlea. *Nat Genet* **21:** 289–292.

37. Stump G, Durrer A, Klein A, *et al.* (2002) Notch1 and its ligands Delta-like and Jagged are expressed and active in distinct cell populations in the postnatal mouse brain. *Mech Dev* **114:** 153.

38. Zheng JL, Shou J, Guillemot F, *et al.* (2000) Hes1 is a negative regulator of inner ear hair cell differentiation. *Development* **127:** 4551–4560.

39. Zine A, Aubert A, Qiu J, *et al.* (2001) Hes1 and Hes5 activities are required for the normal development of the hair cells in the mammalian inner ear. *J Neurosci* **21:** 4712–4720.

40. Li H, Liu H, Heller S. (2003) Pluripotent stem cells from the adult mouse inner ear. *Nat Med* **9:** 1293–1239.
41. Liu W, Li G, Chien JS, *et al.* (2002) Sonic hedgehog regulates otic capsule chondrogenesis and inner ear development in the mouse embryo. *Dev Biol* **248:** 240–250.
42. Bok J, Bronner-Fraser M, Wu DK. (2005) Role of the hindbrain in dorsoventral but not anteroposterior axial specification of the inner ear. *Development* **132:** 2115–2124.
43. Kondo T, Johnson SA, Yoder MC, *et al.* (2005) Sonic hedgehog and retinoic acid synergistically promote sensory fate specification from bone marrow-derived pluripotent stem cells. *Proc Natl Acad Sci USA* **102:** 4789–4794.
44. Zine A, de Ribaupierre F. (1998) Replacement of mammalian auditory hair cells. *Neuroreport* **9:** 263–268.
45. Doetzlhofer A, White PM, Johnson JE, *et al.* (2004) *In vitro* growth and differentiation of mammalian sensory hair cell progenitors: a requirement for EGF and periotic mesenchyme. *Dev Biol* **272:** 432–447.
46. Takahashi J, Palmer TD, Gage FH. (1999) Retinoic acid and neurotrophins collaborate to regulate neurogenesis in adult-derived neural stem cell cultures. *J Neurobiol* **38:** 65–81.
47. Lu J, Wu Y, Sousa N, Almeida OF. (2005) SMAD pathway mediation of BDNF and TGF beta 2 regulation of proliferation and differentiation of hippocampal granule neurons. *Development* **132:** 3231–3242.
48. Staecker H, Kopke R, Malgrange B, *et al.* (1996) NT-3 and/or BDNF therapy prevents loss of auditory neurons following loss of hair cells. *Neuroreport* **7:** 889–894.
49. Pencea V, Bingaman KD, Wiegand SJ, Luskin MB. (2001) Infusion of brain-derived neurotrophic factor into the lateral ventricle of the adult rat leads to new neurons in the parenchyma of the striatum, septum, thalamus, and hypothalamus. *J Neurosci* **21:** 6707–6717.
50. Izumikawa M, Minoda R, Kawamoto K, *et al.* (2005) Auditory hair cell replacement and hearing improvement by Atoh1 gene therapy in deaf mammals. *Nat Med* **11:** 271–276.
51. Thomas CE, Ehrhardt A, Kay MA. (2003) Progress and problems with the use of viral vectors for gene therapy. *Nat Rev Genet* **4:** 346–358.
52. Tamura T, Nakagawa T, Iguchi F, *et al.* (2004) Transplantation of neural stem cells into the mediolus of mouse cochleae injured by cisplatin. *Acta Otolaryngol* **551:** 65–68.
53. Sakamoto T, Nakagawa T, Endo T, *et al.* (2004) Fates of mouse embryonic stem cells transplanted into the inner ears of adult mice and embryonic chickens. *Acta Otolaryngol* **551:** 48–52.

54. Zhao H, Yuan B, Dou X. (2004) The effects of electrostatic interaction between biological molecules and nano-metal colloid on near-infrared surface-enhanced Raman scattering. *J Opt A* **6:** 900–905.
55. Montemagno C, Bachand G, Stelick S, Bachand M. (1999) Construction biological motor powered nanomechanical devices. *Nanotechnology* **10:** 225–231.
56. Connor EE, Mwamuka J, Gole A, *et al.* (2005) Gold nanoparticles are taken up by human cells but do not cause acute cytotoxicity. *Small* **1:** 325–327.
57. Schrand AM, Huang H, Carlson C, *et al.* (2007) Are diamond nanoparticles cytotoxic? *J Phys Chem B* **111:** 2–7.
58. Lin W, Huang YW, Zhou XD, Ma Y. (2006) Toxicity of cerium oxide nanoparticles in human lung cancer cells. *Int J Toxicol* **25:** 451–457.

14

Intravascular Delivery Systems for Stem Cell Transplantation in Neurologic Disorders

Vallabh Janardhan*, Adnan I. Qureshi and Walter C. Low

1. Introduction

Stem cells are primordial cells that retain the ability to renew themselves, can differentiate into a wide range of cell types, and when transplanted can settle in and function fully in their new environment. In 1998, Thomson and Jones along with their colleagues at the University of Wisconsin–Madison isolated the first human embryonic stem cell line derived from human blastocysts (Fig. 1).[1] In 2002, Jiang et al. identified a rare type of nonhematopoietic stem cell in the adult bone marrow that was pluripotent and could transdifferentiate into a range of specialized tissues including neurons. They called these adult stem cells "multipotent adult progenitor cells" (MAPCs) (Fig. 2).[2] In 2005, Xiao et al. showed that transplantation of a similar type of nonhematopoietic stem cell that exists in human umbilical cord blood ameliorates neurologic deficits postischemic brain injury in rats (Fig. 3).[3] In 2007, De Coppi et al. identified a novel type of pluripotent stem cells from amniotic fluid with therapeutic potential (Fig. 4).[4] Clearly, there are a wide variety of sources of pluripotent stem cells that have been identified in the laboratory that can be useful in a range of diseases. The next big question in attempting to translate the exciting advances in the laboratory to the bedside is: What is the best route of delivery for these stem cells in patients with various diseases?

*Correspondence: 420 Delaware Street SE, 12-158 PWB, Department of Neurology, University of Minnesota, Minneapolis, MN 55455, USA. Tel: 612-626-9517; Fax: 612-626-9464; e-mail: vallabh@umn.edu.

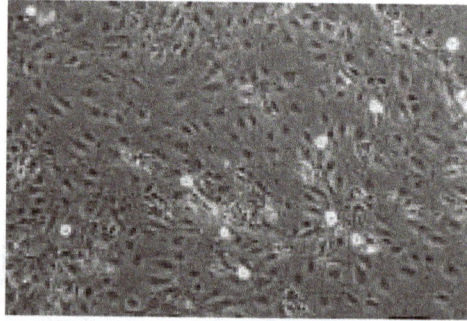

Figure 1. Human embryonic stem cells cultured from human blastocysts. (Reprinted with permission from *Science*. 1998; **282**: 1145–1147.)

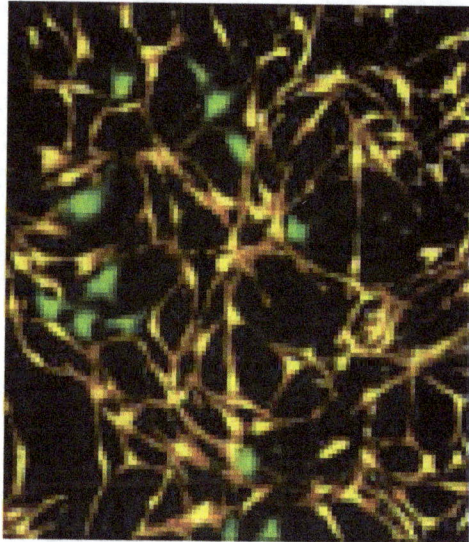

Figure 2. Multipotent adult progenitor cells from mouse bone marrow. (Reprinted with permission from *Nature*. 2002; **418**: 41–49.)

1.1. Nonvascular routes of delivery of stem cells

Surgical transplantation of stem cells is a viable route of delivery for certain diseases, such as neurologic disorders with very small focal lesions. In a Phase II open-label, observer-blind clinical trial, Kondziolka *et al.* demonstrated the safety and feasibility of stereotactic neurosurgical

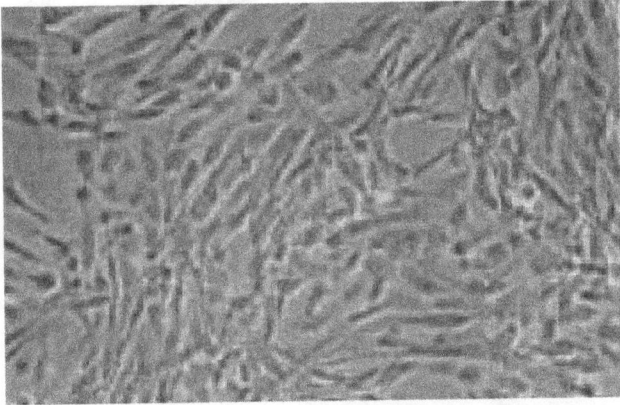

Figure 3. Nonhematopoietic human umbilical cord blood stem cells. (Reprinted with permission from *Stem Cells Dev.* 2005; **14**: 722–733.)

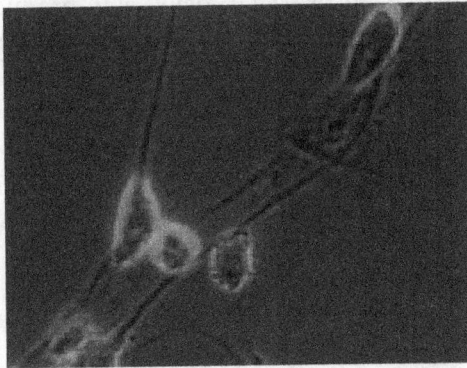

Figure 4. Amniotic fluid stem cells differentiating into pyramid-l shaped cells under conditions for producing dopaminergic neurons. (Reprinted with permission from *Nat Biotechnol.* 2007; **25**: 100–106.)

transplantation of human neuronal stem cells in patients with small subcortical motor strokes.[5] However, stereotactic neurosurgical transplantation has limitations. In neurologic diseases that have larger lesions or affect large parts of the brain, a focal intraparenchymal injection of stem cells does not allow enough stem cell migration for the entire lesion. To be effective, there would have to be multiple focal injections, which would increase the risk of the neurosurgical procedure. Intrathecal or intraventricular injection of stem cells is another viable route of delivery for certain diseases,[6-8] but has similar limitations when it comes to the extent of adequate stem cell migration to lesions that are not near the ventricles or spinal canal.

1.2. Intravascular routes of delivery of stem cells

All cells in the body are in close proximity to blood vessels and receive blood supply for oxygen and nutrients. Therefore, intravascular delivery of stem cells has the theoretical advantage of reaching all parts of the body, including the nervous system.

1.2.1. Intravenous route of delivery

Sanberg *et al.* have shown that intravascular (intravenous) delivery of human umbilical cord blood stem cells may be more effective than intraparenchymal (intrastriatal) delivery in producing long-term functional benefits in a rodent model of stroke.[9] However, stem cells delivered via an intravenous route have to first pass through the pulmonary circulation, where the size of the pulmonary capillaries are around 5.5 microns, prior to reaching extrapulmonary organs such as the brain.[10] In comparison with the size of the pulmonary capillaries, the average size of stem cells such as MAPCs is around 10–12 microns.[11] Therefore, if stem cells delivered via an intravenous route are to improve functional recovery after stroke by cell replacement, they would have to be able to deform while passing through the pulmonary capillaries, similar to blood leukocytes and neutrophils, so that they are not sequestered in the thorax. Tolar *et al.* have shown that MAPCs administered intravenously are predominantly sequestered in the upper thorax, suggesting that these stem cells are limited by the size of the pulmonary capillaries (Fig. 5).[11] Therefore, the beneficial effect of intravenous delivery of human umbilical cord stem cells noted by Sanford *et al.*[9] is probably due to trophic mediators that aid neural plasticity rather than cell replacement.

1.2.2. Intra-arterial route of delivery

In contrast to the intravenous route of delivery, intra-arterial delivery of stem cells eliminates the risk of sequestration in the pulmonary capillaries and could play a role in cell replacement therapy. Tolar *et al.* studied

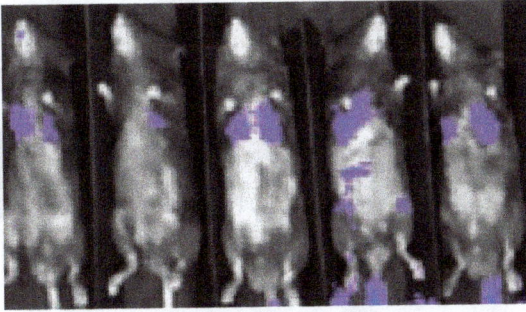

Figure 5. Whole body imaging was performed using luciferase bioluminescence imaging in mice that had intravenous tail vein injection of multipotent adult progenitor cells and showed that the majority of the MAPCs were sequestered in the upper thorax due to the trapping in the pulmonary capillaries. (Reprinted with permission from *Blood*. 2006; **107**: 4182–4188.)

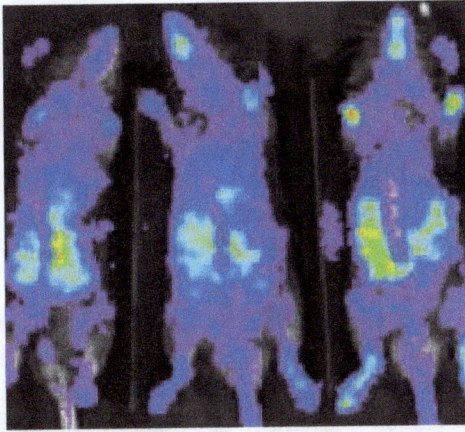

Figure 6. Whole body imaging was performed using luciferase bioluminescence imaging in mice that had intra-arterial injection of multipotent adult progenitor cells and showed that MAPCs were distributed throughout the whole body and had significantly higher distribution in the liver, kidney and brain compared to intravenous injection. (Reprinted with permission from *Blood*. 2006; **107**: 4182–4188.)

the biodistribution of MAPCs via an intravenous route of delivery compared to an intra-arterial route of delivery, and showed that if extrapulmonary organs are the targeted location of delivery, then intra-arterial delivery of stem cells is probably a better alternative to intravenous delivery (Fig. 6).[11] Similar results were obtained in a randomized

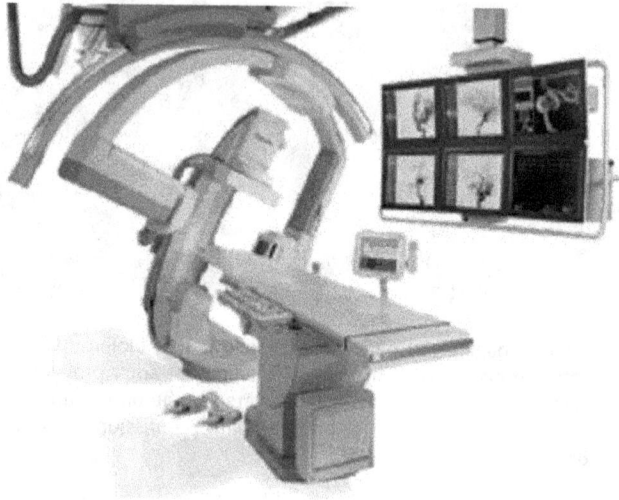

Figure 7. Biplane three-dimensional rotational digital subtraction angiography system with flat-panel detectors (Reprinted with permission from *Neurotherapeutics*. 2007; **4**: 414–419.)

study of three routes of delivery of mesenchymal stem cells in a porcine model of myocardial infarction.[12] The intra-arterial intracoronary route of delivery resulted in increased engraftment within myocardial-infarcted tissue compared to endocardial as well as intravenous routes of delivery.[12]

Advances in interventional neurology and neuroimaging have significantly improved our ability to visualize as well as safely catheterize the distal small arteries supplying areas of the head, neck, brain and spinal cord that were previously difficult to approach.[13] Digital subtraction angiography with flat-panel detector systems has been developed that is superior in image quality for visualizing small intracranial vessels with less radiation exposure compared to digital subtraction angiography with conventional image intensifiers (Fig. 7).[13] Two-dimensional and three-dimensional roadmap capabilities have significantly improved our ability to safely and selectively catheterize distal small intracranial vessels (Fig. 8).[13] The common femoral artery is the most common access site for the arterial sheath for most interventional procedures, followed by the radial artery approach. Advances in steerable and flow-directed microcatheters along with advances in microwires (0.008th–0.014th of an inch in diameter) have enabled safe and selective catheterization of the smallest of blood vessels

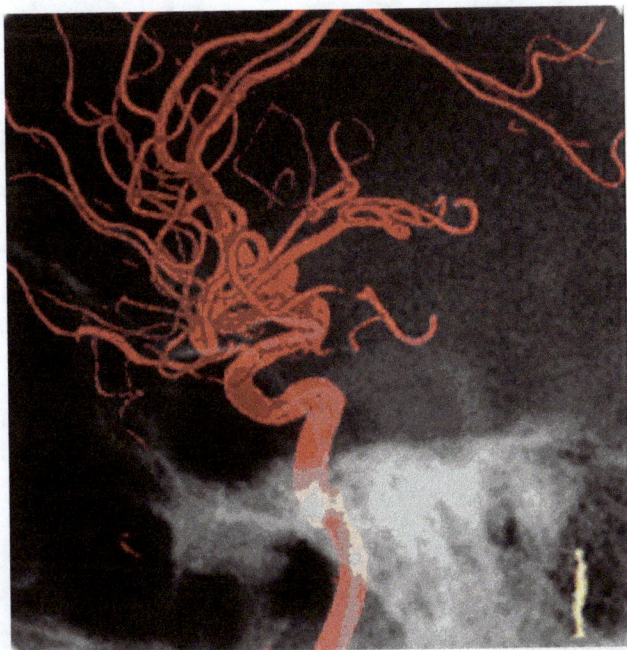

Figure 8. Three-dimensional roadmap with a three-dimensional reconstructed image superimposed on a two-dimensional fluoroscopic image. (Reprinted with permission from *Neurotherapeutics.* 2007; **4**: 414–419.)

in the head, neck, brain and spinal cord (Figs. 9 and 10). The ability to safely reach the deepest parts of the brain and spinal cord with intravascular delivery systems and the availability of various types of pluripotent stem cells make it an exciting time in the field of regenerative medicine to translate the advances in the laboratory to clinical improvement at the bedside. This article will focus on the various neurological disorders that have shown promise with stem cell therapies and how intravascular delivery of stem cells could change the future treatment paradigms in these conditions.

1.2.3. *Optimization of intravascular delivery systems for stem cells*

The effects on stem cell viability by injecting stem cells through intravenous catheters and intra-arterial microcatheters are not well understood. At the University of Minnesota, we have studied the effects of using

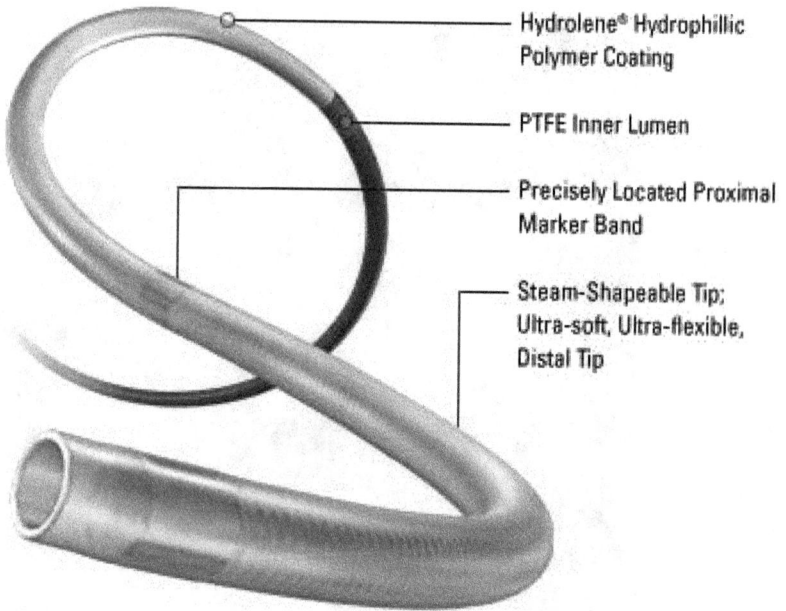

Hydrolene® Hydrophillic Polymer Coating

PTFE Inner Lumen

Precisely Located Proximal Marker Band

Steam-Shapeable Tip; Ultra-soft, Ultra-flexible, Distal Tip

Figure 9. Steerable microcatheter used for selective catheterization of small cerebral blood vessels. (Courtesy of Boston Scientific Corporation, Fremont, CA, USA.)

intravenous angiocatheters (length: 1½ inch; diameter: 18-gauge, 20-gauge and 22-gauge) and endovascular intra-arterial microcatheters in the lengths needed to catheterize the cerebral circulation (Boston Scientific Inc, Excelsior SL-10 microcatheter of 150 cm length and with internal lumen diameter of 0.0165th inch) on the viability of cord blood stem cells. Infusion of cells at densities of 10 million cells/ml via the 1½-inch-long intravenous catheters (18-, 20- or 22-gauge) did not affect cell viability. We found that the density of the stem cells passing through the 300 μL dead space of the 150-cm-long microcatheter greatly affected cell viability. Infusion of cells at a density of 10 million cells/ml resulted in a decrease in cell viability by 27% (Fig. 11) while infusion at a density of 1 million cells/ml did not significantly diminish viability (Fig. 12). Alteration of the infusion rate with a density of 1 million cells/ml did not affect cell viability either. Infusion of 300 μl of cells within 5 min or 1 min and 10 s had no effect. These preliminary studies begin to outline the important parameters that are needed to optimize the endovascular delivery of stem cells.

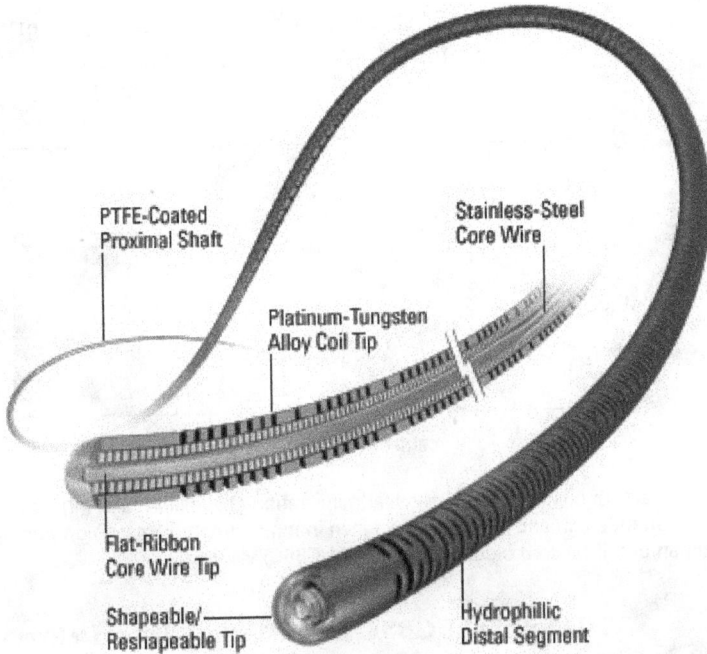

Figure 10. Specialized microwire used for selective catheterization of small cerebral blood vessels. (Courtesy of Boston Scientific Corporation, Fremont, CA, USA.)

2. Traumatic Brain Injury

Traumatic brain injury (TBI) is a leading cause of morbidity and mortality in the United States; an estimated 1.5 million Americans sustain a TBI every year.[14] Nearly 50,000 people die annually as a result of TBI, approximately 230,000 people are hospitalized and survive, and an estimated 80,000–90,000 of the hospitalized people experience the onset of significant long-term disability.[14] In addition, more than 1 million people visit the emergency department every year without requiring hospitalization. Although these people are often considered to have mild TBI, these injuries can lead to significant cognitive and emotional impairment and long-term morbidity.[15]

Currently, no clinical treatment is available to reverse the pathologic cellular cascade underlying progression of cell death and to improve neurobehavioral outcome in people with TBI. Stem cell transplantation is one option to repair the injured CNS by aiming to replace the function of cells lost following the injury.

Viability of nhUCBSC after Microcatheter delivery

cell density = 1 x 10e7 /mL, infusion speed = 120 uL/min

p = 0.008 <0.01

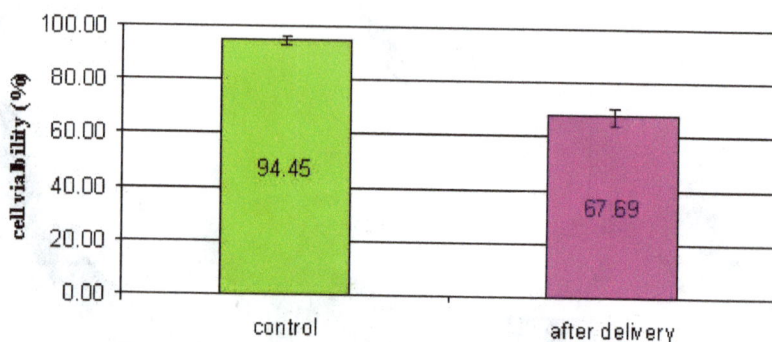

Figure 11. Effects of high density stem cell administration (10 million cells/ml) via a microcatheter (150 cm in length and 0.0165th of an inch in inner lumen diameter) on non-hematopoietic human umbilical cord blood stem cell (nhUCBSC) viability.

Viability of nhUCBSC after Microcatheter Delivery

cell density = 1x10e6/ mL

p > 0.3

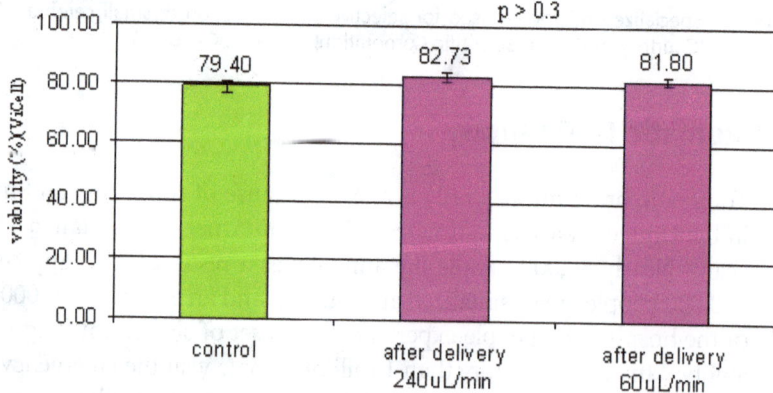

Figure 12. Effects of low density stem cell administration (1 million cells/ml) via a microcatheter (150 cm in length and 0.0165th of an inch in inner lumen diameter) on non-hematopoietic human umbilical cord blood stem cell (nhUCBSC) viability.

2.1.1. *Experimental studies of intravascular stem cell therapies in TBI*

Several experimental studies have tried to evaluate the role of various intravascular stem cell therapies in small animal models of TBI, and they are summarized below. The most common rat models for TBI are the lateral fluid percussion model,[16] and the controlled cortical impact model.[17]

2.1.2. *Choice of stem cell type in TBI*

Rat bone marrow stem cells (stromal cells) delivered intravascularly have been studied in rat models of TBI and are associated with improved functional outcomes.[18–22] They have been shown to display neuronal as well as astrocytic markers at 2 weeks post-intravenous-delivery.[22] At 1–2 weeks post-intravenous-delivery, these bone marrow stromal cells have also been shown to increase the expression of nerve growth factor (NGF) and brain-derived neurotrophic factor (BDNF),[20] as well as induce endogenous cellular proliferation in the brain,[19] which in turn might account for the improved functional outcomes. These bone marrow stromal cells delivered intravenously were noted to be distributed not just in the brain but also into various other organs; however, there were no adverse effects from the systemic distribution of these stem cells.[22] Bone marrow stem cells along with adjunctive medications such as statins, or genetically modified bone marrow stem cells such as those cells expressing nerve growth factors, might help further improve survival and outcome.[18,21] Intravenous delivery of rat bone marrow stem cells combined with atorvastatin given for 2 weeks after TBI is associated with increased survival and improved functional recovery compared to monotherapy with stem cells.[21] Intra-arterial delivery of rat bone marrow stem cells (cultured with NGF and BDNF) is associated with slightly improved engraftment rates of stem cells in the hemisphere ipsilateral to the injection as well as increased expression of neuronal protein MAP-2 compared to intra-arterial delivery of rat bone marrow stromal cells (cultured without NGF and BDNF). Despite these differences, however, there was significant functional improvement in both groups.[18]

Human adult bone marrow stem cells (stromal cells) delivered intravenously have also been studied in rat models of TBI and have been associated with improved functional outcomes.[16,17,23–25] They have been shown to survive and express neuronal, astrocytic and oligodendrocytic markers from 2 weeks[16] to as long as 3 months[17] post-intravenous-delivery of stem cells. At 3 months post intravenous-delivery, human adult bone marrow stem cells are associated with an increased expression of BDNF, but the short-lived peak of NGF noted at 1–2 weeks post intravenous-delivery of stem cells does not persist at 3 months, suggesting that different neurotrophins may have different peaks and periods of expression in the neurorestorative phase after stem cell transplantation.[17]

Human umbilical cord blood stem cells delivered intravenously have also been associated with improved functional outcome in rat models of TBI.[26] At 4 weeks postintravenous delivery, these cells have been noted to express neuronal as well as astrocytic markers.[26] Even though they were injected intravenously, there were significantly greater numbers of these cells in the cerebral hemisphere ipsilateral to the maximal brain injury, suggesting that these cells preferentially entered the brain and migrated into the cerebral hemisphere of maximal TBI.[26] Human umbilical cord blood also contains endothelial progenitor cells, and results from animal studies have shown that these cells also tend to integrate into the vascular structures surrounding the area of brain injury, suggesting that these cells may also be associated with angiogenesis of cerebral blood vessels in TBI.[26]

2.1.3. *Optimal dosage of stem cells in TBI*

Most experimental studies in rat models of TBI have evaluated the effects of cell doses in the range of $1-2 \times 10^6$ bone marrow stem cells (stromal cells) or umbilical cord blood stem cells.[16,19,23–26] Mahmood *et al.* compared the effects of intravenous delivery of human adult bone marrow stem cells (stromal cells) in doses of 1×10^6 cells versus 2×10^6 cells in a rat model of TBI.[24] There were significantly improved functional outcomes at 1 month post-intravenous-delivery of 2×10^6 cells compared to 1×10^6 cells.[24] Mahmood *et al.* also evaluated dose escalation studies of human adult bone marrow stem cells in a rat model of TBI (2×10^6 cells versus 4×10^6 cells versus 8×10^6 cells).[17,25] There were higher numbers of engrafted stem cells in the injury site at 3 months in the higher dosage groups (4×10^6 cells and 8×10^6 cells).[17,25] However, there were improved functional outcomes in all three dosage groups (2×10^6 cells versus 4×10^6 cells versus 8×10^6 cells) without any significant difference between groups, suggesting that there may be a ceiling effect at the 2×10^6 cells dosage level and that this may be the optimal dose for bone marrow stem cells in a rat model of TBI.[17,25]

2.1.4. *Optimal timing of stem cell therapy in TBI*

The timing of intravascular stem cell delivery post-TBI, namely in the acute or chronic phase, is an important question. Most experimental studies in rat models of TBI have evaluated the role of intravascular (intravenous and intra-arterial) delivery of bone marrow (stromal cells) or umbilical cord blood stem cells in cell survival when injected at 24 hours after TBI.[18-26] Lu *et al.* evaluated the role of intravenous delivery of human bone marrow stem cells (stromal cells) in survival of engrafted cells when injected at 72 hours after TBI.[16] Mahmood *et al.* extended this time window and evaluated the role of intravenous delivery of human bone marrow stem cells (stromal cells) in cell survival when injected at 1 week after TBI.[17] Several of these studies have shown that these engrafted stem cells delivered intravascularly at either 24 hours,[25] 72 hours[16] or 1 week after TBI[17] can survive up to 3 months or longer posttransplantation, suggesting that any time between 24 hours and 1 week after the TBI could represent an optimal time window in the acute phase after TBI for intravascular delivery of stem cells. Further studies are needed to evaluate if the same results are applicable for intravascular delivery of stem cells in the chronic phase after TBI.

2.1.5. *Optimal intravascular route of delivery in TBI*

Most of the experimental studies of intravascular delivery of stem cells in small animal models of TBI have been on the intravenous route of delivery and have shown good stem cell survival as well as improved functional outcomes after TBI.[16,17,19-26] Mahmood *et al.* evaluated the role of intracerebral injection (1×10^6 cells) of rat bone marrow stem cells (stromal cells) compared to the intravenous route of delivery (2×10^6 cells) in a rat model of TBI.[19] Progenitor cell proliferation was significantly increased in the subventricular zone and the boundary zone of the injury site in the intracerebral injection group. In comparison, progenitor cell proliferation was significantly increased in those regions as well as the hippocampal regions in the intravenous delivery group, suggesting more widespread migration and survival with the intravascular route of delivery.[19] The functional outcomes were similar in the two routes of delivery.[19]

TBI is a heterogeneous disease with areas of focal as well as diffuse brain injury along with areas of contusions. In the setting of these contusions, the need for multiple intracerebral injections for stem cell delivery is not optimal. Intra-arterial delivery of stem cells in TBI can be performed directly into the internal carotid artery (for lesions in the carotid distribution) or vertebral arteries (for lesions in the vertebrobasilar distribution). For small lesions, selective catheterization of anterior, middle or posterior cerebral arteries and their branches can be performed and cells can be injected directly at the site of maximal TBI. Lu *et al.* have shown the feasibility of ipsilateral intracarotid injection of rat bone marrow stem cells (stromal cells) in a rat model of TBI.[18] They have shown that there is increased survival and migration of stem cells in the hemisphere ipsilateral to the injection.[18] At the present time, both the intravenous and the intra-arterial route appear to be feasible routes of delivery of stem cells after TBI. Further studies are needed to compare the intravenous with the intra-arterial route of delivery of stem cells in order to understand the optimal route of intravascular delivery of stem cells after TBI.

3. Spinal Cord Injury

Spinal cord injury (SCI) affects more than 250,000 individuals in the United States. The incidence of SCI is around 11,000 individuals every year, most of whom are young.[27] There is significant morbidity and mortality associated with SCI, as well as a huge economic burden. The average cost of treating SCI for a quadriplegic patient (C1–C4 level injury) is US$741,425 for the first year, and the approximate lifetime cost for an adult 25 years of age at the time of injury is close to US$3 million.[27] The average cost of treating SCI for a paraplegic patient is US$270,913 for the first year, and approximately US$1 million over the lifetime of an adult 25 years of age at the time of injury.[27]

Steroids are the mainstay for acute SCI to decrease spinal cord edema, and spinal surgery can be helpful when there is evidence of spinal cord compression. However, there are no current treatments to salvage neuronal injury in the spinal cord. In addition, there are no current treatments for spinal cord infarction. Stems cells have the potential to salvage the injured or ischemic spinal cord and help recover function.

3.1. Experimental studies of intravascular stem cell therapies in SCI

3.1.1. *Choice of stem cell type in SCI*

Bone marrow stem cells can help improve functional recovery after SCI.[28–30] In a balloon compression model of SCI in rats, Urdzikova *et al.* compared the role of intravenous delivery of rat mesenchymal stem cells versus intravenous delivery of rat mononuclear bone marrow stem cells versus granulocyte-colony-stimulating factor administration (to mobilize marrow stem cells) in improving hind limb sensitivity and functional recovery after SCI.[28] They found that the most promising results in terms of improved functional recovery as well as better engraftment of stem cells were obtained with rat mesenchymal stem cells.[28] Liu *et al.* have recently shown that intravenous delivery of rat bone marrow stem cells (stromal cells) 24 hours after a stroke is associated with increased axonal sprouting in the denervated corticospinal tracts from the cortex to the spinal motor neurons.[31] In a rat model of SCI using an aneurysm clip compression technique, Saporta *et al.* have shown that human umbilical cord blood stem cells are significantly associated with improved functional outcomes.[32] In the first human pilot study of intravascular delivery (intravenous or intra-arterial delivery) of stem cells in acute (10–30 days) and chronic (2–17 months) SCI, Sykova *et al.* used autologous adult bone marrow stem cells and showed significant functional improvement in the SCI patients treated in the acute phase.[30]

Experimental studies in rat models of SCI have also used intravenous delivery of other types of stem cells, such as neural progenitor cells (obtained from the E15 fetal hippocampus of transgenic rats),[33] as well as oligodendrocyte precursor cells obtained from rat-adipose-tissue-derived stromal cells,[34] and have shown improved functional outcomes after SCI.[33,34]

3.1.2. *Optimal dosage of stem cells in SCI*

Experimental studies in rat models have used intravenous delivery of stem cells in doses ranging from 1×10^6 cells to 3×10^6 cells and have shown improved spinal cord function and recovery.[29,31–34] In the first human pilot study of intravascular delivery of autologous adult bone marrow stem cells, Sykova *et al.* used dose ranges of 20×10^8 to 160×10^8 CD34+ cells in patients with acute ($n = 7$) and chronic ($n = 13$) SCI, and showed improved functional recovery in the patients treated in the acute phase of SCI.[30]

3.1.3. *Optimal timing of stem cell therapy in SCI*

Saporta *et al.* compared the timing of intravascular (intravenous) delivery of human umbilical cord blood stem cells 24 hours versus 5 days after SCI and evaluated the functional outcomes in the two groups.[32] They showed that intravenous delivery of human umbilical cord blood stem cells in the late acute phase, namely 5 days after SCI, was associated with significantly improved functional outcomes compared to transplantation during the hyperacute phase after SCI.[32] In their human pilot study, Sykova *et al.* confirmed these results and showed that intravascular delivery of stem cells in the late acute or subacute phase (10–30 days) is associated with good functional outcomes (6 out of 7 patients had improvement).[30]

In a weight drop impact rat model of SCI, Zurita *et al.* showed that systemic intravenous delivery of bone marrow stem cells (stromal cells) in the chronic phase (>3 months) after SCI is not associated with significant improvement in functional outcomes compared to direct local injection of stem cells in the chronic phase, which does seem to be associated with improved outcomes after SCI.[29] In the human pilot study, Sykova *et al.* treated 13 patients in the chronic phase of SCI (2–17 months) and showed that only 1 out of the 13 patients had improved functional outcomes compared to 6 out of the 7 patients treated in the acute or subacute phase after SCI.[30] However, the majority of the patients in the chronic phase group after SCI were treated with systemic intravenous delivery of bone marrow stem cells and not local intra-arterial therapy.[30] The optimal timing of intravascular delivery of stem cells (intravenous or intra-arterial delivery) seems to be in the late acute or subacute phase (5–30 days) after SCI.

3.1.4. *Optimal intravascular route of delivery in SCI*

Most experimental studies of intravascular stem cell delivery in rat models of SCI have examined the intravenous route of delivery and have shown improved functional outcomes in the late acute phase after SCI.[28,32] In the first human pilot study of intravascular delivery of stem cells, Sykova *et al.* treated the majority of the patients in the acute or subacute phase (*n* = 7) with the intra-arterial route of delivery (6 out of 7 patients had selective catheterization of the vertebral arteries which supply the cervical and upper thoracic spinal cord; one patient had intravenous therapy), and showed improved functional outcomes.[30] Spinal angiography with selective catheterization of spinal intercostal and lumbar arteries is a safe and feasible procedure, and is routinely performed for a variety of disorders (Fig. 13). Future studies need to compare intravenous and intra-arterial delivery of stem cells in the late acute or subacute phase after SCI. In addition, they need to evaluate the role of selective catheterization of the vertebral arteries, segmental intercostal arteries or lumbar arteries based on the location of SCI as a means of focused intravascular delivery of stem cells.

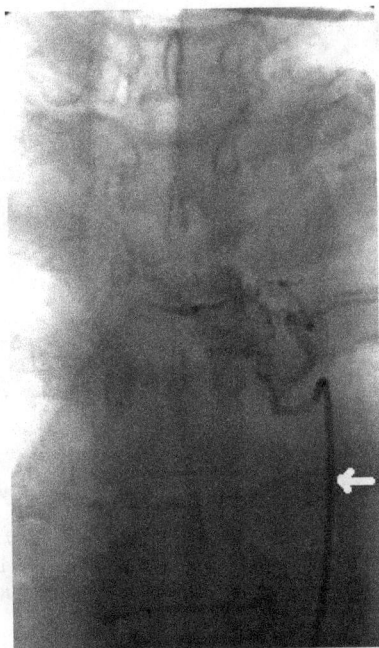

Figure 13. Selective catheterization of a spinal intercostal artery (left T6 thoracic spinal intercostal artery) using a 5 French Cobra-2 catheter, and the unsubtracted fluoroscopic image is shown. (White arrow represents the 5 French Cobra-2 catheter.)

Systemic intravenous delivery of stem cells in the chronic phase (>2 months) after SCI does not seem to be as effective as local injection.[20] Of the 13 patients in the chronic phase after SCI, Sykova *et al.* treated 12 with intravenous delivery of stem cells, but recorded an improvement in only 1.[30] Future studies need to also evaluate the role of intra-arterial delivery of stem cells in the chronic phase after SCI.

4. Ischemic Stroke

Stroke remains the third leading cause of death and a major cause of morbidity in the United States, and about 750,000 people suffer a new or recurrent stroke every year.[35] The estimated direct and indirect costs for stroke in 2006 were US$57.9 billion.[35] Given the enormity of this disease and the morbidity and mortality associated with it, there is clearly a need for new

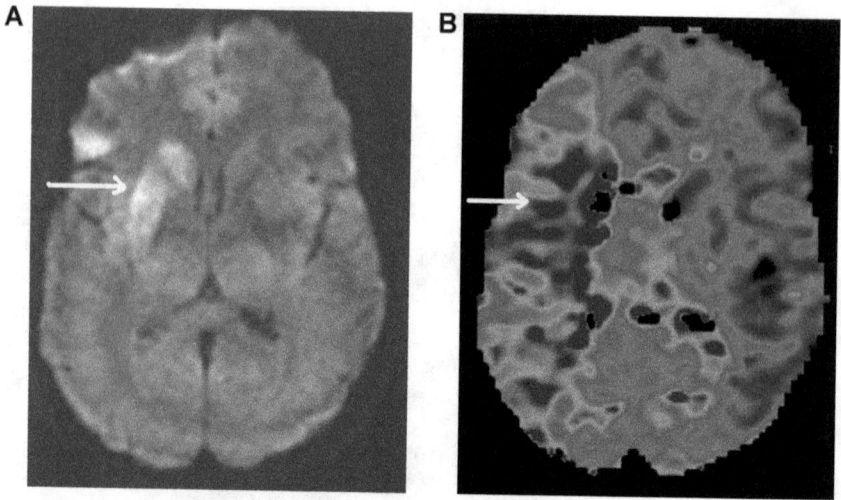

Figure 14. 49-year-old lady with acute right middle cerebral artery (MCA) ischemic stroke. (A) MR diffusion scan showing the ischemic core (infarcted tissue; bright area indicated by (white arrow). (B) MR perfusion scan showing the ischemic penumbra (tissue at risk for infarct; dark area indicated by white arrow).

innovative treatment strategies. Current treatment strategies in acute ischemic stroke are aimed at reopening cerebral vessels and salvaging the ischemic penumbra with intravenous r-tpa (<3 hours from symptom onset), intra-arterial thrombolysis (<6 hours from symptom onset) and/or intra-arterial mechanical clot retrieval (<8 hours from symptom onset).[36]

Despite these advances, clinical success has been limited partly because the current treatments only aim to salvage the ischemic penumbra, without any effort toward salvaging the ischemic core (Fig. 14).[36] Stem cells have the potential to repair organs and tissues that have been damaged as a result of disease, including stroke, and the potential to salvage the ischemic core and offer a new and innovative treatment strategy for patients with acute ischemic stroke.[36]

4.1. Experimental studies of intravascular stem cell therapies in ischemic stroke

4.1.1. *Choice of stem cell type in ischemic stroke*

In a Phase 2 study, Kondziolka *et al.* showed the safety and feasibility of stereotactic intracerebral injection of "neuronal stem cells" (NT2) derived from a human teratocarcinoma cell line ($2\text{–}6 \times 10^6$ cells) in patients with chronic subcortical motor strokes 1–6 years after the stroke.[5] The theoretical risk includes the possibility of these NT2 cells reverting to a neoplastic state. This has increased the interest in intravascular delivery of other types of nonneural stem cells, namely bone marrow stem cells, which have shown promising results in rat models of acute ischemic stroke,[37] and human studies of acute myocardial infarction and stroke patients.[38–40] The additional advantage of bone marrow stem cells is that they are readily available and each patient's own stem cells can be harvested through bone marrow biopsy and used to treat the patient.

4.1.2. *Optimal dosage of stem cells in ischemic stroke*

In a series of 30 patients with MCA ischemic strokes (5 treated and 25 controls), Bang *et al.* showed that intravenous autologous bone marrow mesenchymal stem cell transplantation (a total of 1×10^8 cells given as two doses; the dose was based on body mass extrapolations from doses used in animal studies) is safe, feasible, and may improve functional recovery.[38] Furtado de Mendonca *et al.* reported one patient with acute MCA ischemic stroke who was treated with intra-arterial autologous bone marrow mononuclear stem cell transplantation (1×10^8 cells given as a single dose 4 days after stroke onset), and showed that intra-arterial delivery is safe and feasible.[39] The optimal dose for intravascular delivery of stem cells in humans is not clear, but a dose of 1×10^8 cells seems to be well tolerated.

4.1.3. *Optimal timing of stem cell therapy in ischemic stroke*

Rodent studies have shown that intravascular (intravenous) delivery of human umbilical cord blood stem cells ameliorates neurologic deficits poststroke, with the maximal benefit occurring at 48 hours poststroke.[41] Primate studies have suggested that stem cells survive and can differentiate when transplanted up to 7 days poststroke, but after 7 days, fibrosis and scar formation in the infarct site limits the development of new neuronal connections, rendering stem cell therapy less suitable.[42]

The body has its own endogenous stem cell response to injury, including stroke.[43,44] Peripheral blood mononuclear stem cells ($CD34^+$) have been shown to increase after ischemic stroke and seem to surge at around

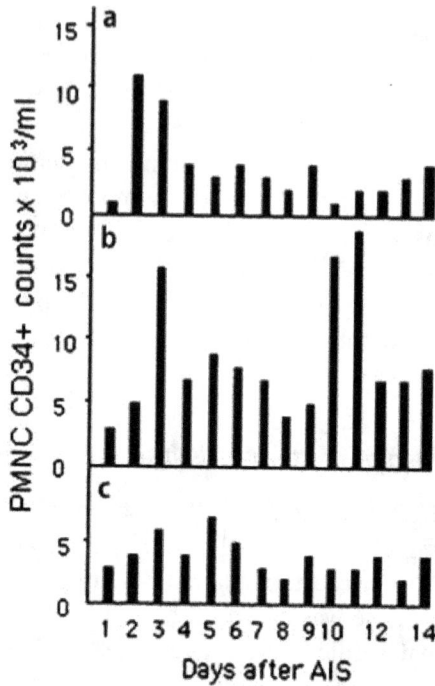

Figure 15. Bursting pattern of peripheral blood mononuclear CD34⁺ stem cell mobilization after acute ischemic stroke (AIS), suggesting a pattern of increase in endogenous stem cell mobilization around days 1–3 and around days 7–10.

2 and 7 days after a stroke (Fig. 15).[45] Circulating endothelial progenitor cells (CD34⁺/CD133⁺), which are a subpopulation of immature CD34⁺ cells, have been shown to be a useful biomarker of cardiovascular disease, and may prove to be a much better biomarker of the body's endogenous stem cell response.[46] Current research at the University of Minnesota is evaluating the levels of circulating endothelial progenitor cells (CD34⁺/CD133⁺) in patients with acute ischemic stroke and trying to understand the optimal timing of exogenous stem cell therapy based on the surge of endogenous stem cell response.

It is thought that some of the effects of these circulating progenitor cells could be a result of trophic mediators.[47] Identifying the trophic mediators, such as granulocyte-colony-stimulating factor (G-CSF), associated with the endogenous stem cell response might help augment the endogenous stem cell response.[47] STEMS (the Stem cell Trial of recovery EnhanceMent after Stroke) is a randomized, controlled pilot study that has shown that G-CSF administration is safe, feasible, and mobilizes bone marrow CD34[+] stem cells after subacute stroke.[47] Researchers at the University of Minnesota are also evaluating the possibility of augmenting the endogenous stem cell and trophic mediator response after acute ischemic stroke to aid in recovery. The optimal time window for stem cell therapy in acute ischemic stroke patients is still not well understood, but is probably between day 2 and day 7.

4.1.4. *Optimal intravascular route of delivery in ischemic stroke*

Intravascular delivery of stem cells has the advantage of being less invasive than direct intracerebral injection. Experimental studies in rat models of acute ischemic stroke have shown that both intravenous and intra-arterial delivery of stem cells have ameliorated neurologic deficits.[37,41] In patients with acute ischemic stroke, Bang *et al.* have demonstrated the safety and feasibility of the intravenous route of delivery of bone marrow stem cells.[38] Intra-arterial delivery of bone marrow stem cells has been studied extensively in acute myocardial infarction patients; more than 8 trials have shown that this route of administration is safe and feasible.[40] Recently Mendonca *et al.* have also shown that intra-arterial delivery of stem cells can be safely performed in a stroke patient.[39] Similar to the thrombolysis trials in acute stroke,[36] both the intravenous and the intra-arterial route seem to be viable options for delivery of stem cells in stroke patients. Clearly, there is growing evidence in the literature suggesting that it is time for a pilot study.

5. Intracerebral Hemorrhage

Intracerebral hemorrhage (ICH) is a deadly form of stroke and has an incidence of nearly 37,000–50,000 new cases every year in the United States (Fig. 16).[48] It is a life-threatening condition, with 6-month mortality rates ranging from 30%–55%.[48] Current therapies to treat ICH include treatments aimed at decreasing hematoma expansion or evacuating the hematoma.[48] There are no current therapies aimed at salvaging the injured brain after hemorrhagic stroke. Several experimental studies have tried to evaluate the role of intravascular delivery of stem cells in improving functional outcome after ICH.

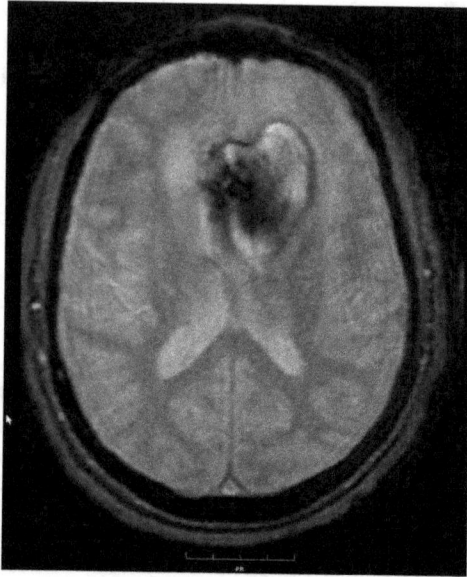

Figure 16. Axial MRI of the brain using gradient echo sequence showing an intracerebral hemorrhage (the hemorrhage is seen as a dark area as a result of the susceptibility artifact on MRI from the blood products).

5.1. Experimental studies of intravascular stem cell therapies in ICH

5.1.1. *Choice of stem cell type in ICH*

Human bone marrow stem cells (stromal cells) delivered by the intravascular (intravenous) route have been shown to improve neurological deficits in a rat model of ICH.[49] In such a model, Zhang *et al.* have used rat mesenchymal bone marrow stem cells for intravascular delivery (intravenous and intracarotid) and have shown improved functional outcomes.[50] Low *et al.* have shown that intravenous delivery of human umbilical cord blood stem cells is associated with significant improvement in neurologic deficits at 2 weeks after experimental ICH in rats.[51] Intravascular delivery, including both intravenous[52] and intra-arterial intracarotid delivery,[53] of neural stem cells has been studied in experimental rat ICH models and has shown differentiation of these cells into neurons in the perihematomal areas and significant improvement in functional recovery after ICH.[52,53]

5.1.2. *Optimal dosage of stem cells in ICH*

Seyfried *et al.* also evaluated dose escalation studies of human adult bone marrow stem cells (stromal cells) in a rat model of ICH (3×10^6 cells versus 5×10^6 cells versus 8×10^6 cells).[49] There were improved functional outcomes in all three dosage groups without any significant difference between the groups, suggesting that there may be a ceiling effect at the 3×10^6 cells dosage level.[49] Zhang *et al.* have shown that rat mesenchymal bone marrow stem cells at a dose of 2×10^6 cells were effective in improving outcomes via the intracarotid and intraventricular routes, but not the intravenous route.[50] Low *et al.* have shown that human umbilical cord blood stem cells at doses of 2.4–3.2×10^6 cells were effective in ameliorating neurological deficits after ICH.[51] The above studies indicate that the optimal dose for intravascular delivery of stem cells in a rat model of ICH probably ranges between 2.4 and 3×10^6 cells.

5.1.3. *Optimal timing of stem cell therapy in ICH*

Zhang *et al.* evaluated the role of intravascular delivery of rat mesenchymal bone marrow stem cells on functional outcome after ICH at different time windows in the acute phase of ICH (days 1, 3, 5 and 7).[50] They have shown that the maximal improvement in functional outcomes in both the intravenous and the intra-arterial route of delivery was at day 3 after ICH.[50] Li *et al.* have evaluated the optimal therapeutic time window for intra-arterial intracarotid delivery of neural stem cells and have shown that intravascular delivery of stem cells in the subacute phase of ICH was associated with maximal improvement in functional outcomes after ICH.[53] Clinical studies of patients have shown that hematoma expansion after ICH usually occurs in the first 48 hours and maximal brain edema around ICH usually occurs between day 3 and day 5 after ICH. The above two studies together indicate a possible bimodel time window for intravascular delivery of stem cells after ICH,[50,53] the first peak being at day 3 (after hematoma expansion has completed and before maximal brain edema occurs), and the second peak from day 7 to day 14 (after maximal brain edema occurs).

5.1.4. *Optimal intravascular route of delivery in ICH*

Zhang *et al.* evaluated the role of intravenous, intracarotid and intraventricular transplantation of rat mesenchymal stem cells at a dose of 2×10^6 cells in an experimental model of ICH.[50] They found that the stem cells delivered via the intracarotid and intraventricular routes but not the intravenous route differentiated into neuronal, asytrocytic and oligodendrocytic cells and had improved functional outcomes.[50] In addition, the stem cells delivered via the intracarotid route were distributed across the cortex, hippocampus and bleeding foci, whereas the stem cells delivered via the intraventricular route were distributed only around the lateral ventricles and bleeding foci.[50] The intra-arterial, intracarotid route seems a feasible and optimal route of delivery of stem cells in ICH (Fig. 17).[50,53] However, further studies are needed to compare the intravenous and intra-arterial routes of delivery at optimal doses and also at optimal therapeutic time windows to identify the most optimal route of delivery.

6. Brain Tumors

Glioblastoma multiforme is one of the most common forms of primary brain tumors and represents around 22–27% of all primary brain tumors.[54] Despite advances in neurosurgical techniques as well as stereotactic radiosurgery and chemotherapeutic regimens, glioblastomas continue to be associated with high mortality rates. Glioblastomas are one of the deadliest forms of primary brain tumors, and despite recent advances, the mean lifetime survival of a patient with such a tumor is only between 32 and 56 weeks from time of detection.[55] One of the reasons for the failure of neurosurgical resection and the dismal survival and high mortality rates is that microscopic tumor cells can migrate distantly from the solid tumors and infiltrate normal brain tissue, leading to recurrent gliomas.[56] Another reason for the dismal survival rates and recurrence of these tumors after radiosurgery is that these tumors are associated with CD133[+] cancer stemlike cells that are radioresistant and cause recurrences (Fig. 18a).[57]

6.1. Molecular therapies for brain tumors

Molecular therapies that can destroy or inhibit the growth of these tumor cells, including the CD133[+] cancer stemlike cells, could hold great promise (Fig. 18b).[57] There are several molecular targets for antitumor therapy. First, experimental studies have shown that "suicide gene" therapy with vectors expressing the thymidine kinase gene can inhibit transcription of

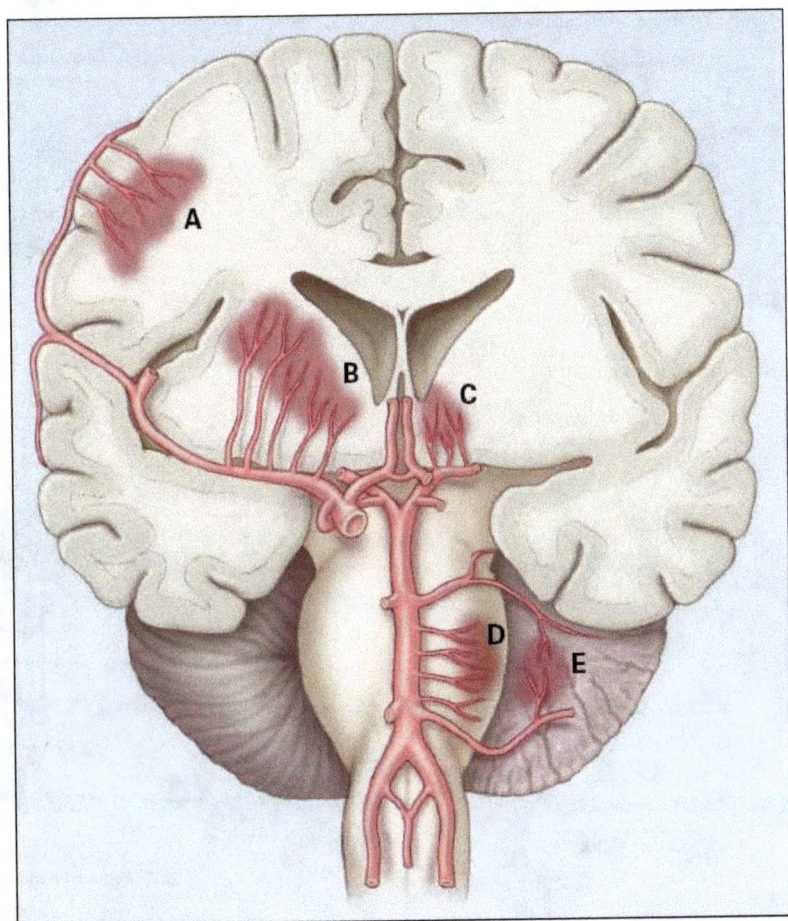

Figure 17. Most common sites and sources of spontaneous intracerebral hemorrhage and associated cerebral arterial distribution: intracerebral hemorrhages most commonly involve cerebral lobes, originating from penetrating cortical branches of the anterior, middle or posterior cerebral arteries (A); basal ganglia, originating from ascending lenticulostriate branches of the middle cerebral artery (B); the thalamus, originating from ascending thalmogeniculate branches of the posterior cerebral artery (C); the pons, originating from paramedian branches of the basilar artery (D); and the cerebellum, originating from penetrating branches of the posterior inferior, anterior inferior or superior cerebellar arteries (E). (Reprinted with permission from *N Engl J Med.* 2001, May 10; **344(19)**: 1450–1460.)

DNA in rapidly dividing cells and thereby arrest tumor growth.[55,56,58] Second, brain tumors can arise from deregulation of signaling pathways that are normally activated during brain development.[59,60] Inhibition of one such signaling pathway, the Hedgehog pathway, using cyclopamine has

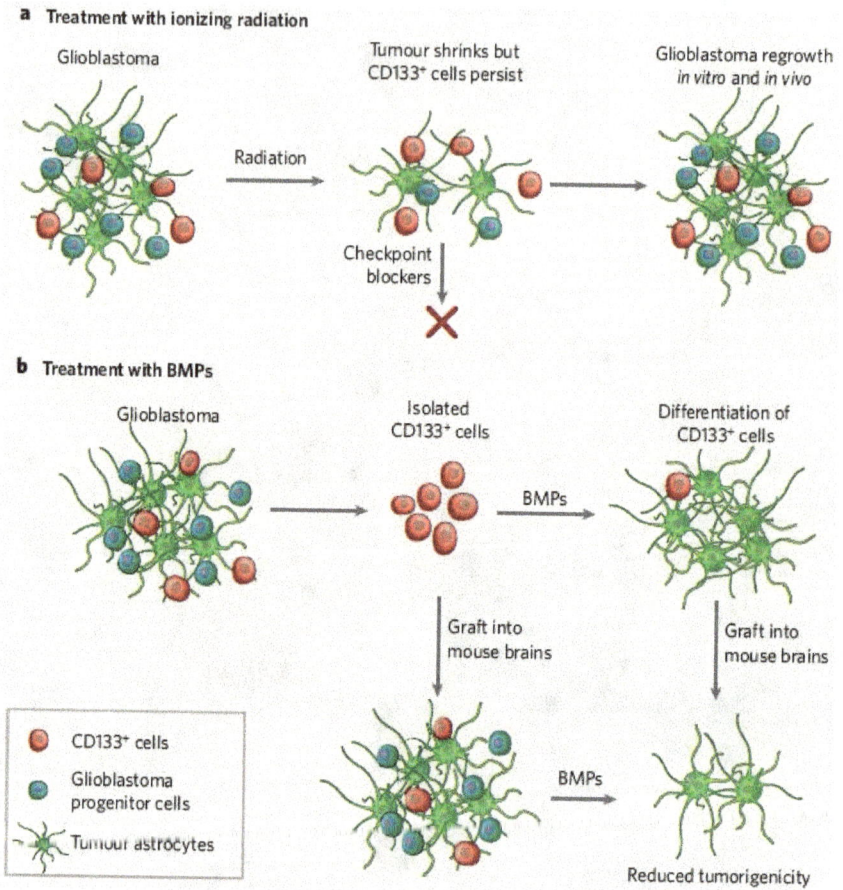

Figure 18. Schematic representation of the response of glioblastomas to ionizing radiation and bone morphogenetic proteins (BMPs). Glioblastomas are heterogeneous tumors that contain a few tumor-initiating CD133-positive stem cells and other, more differentiated, CD133-negative stem cells such as glioblastoma progenitor cells. (a) After radiation treatment the bulk of the glioblastoma responds and the tumor shrinks. However, the CD133+ cells activate checkpoint controls for DNA repair more strongly than the CD133− cells; these CD133+ cells are resistant to radiation and prompt the tumor recurrence. (b) The effect of BMPs on glioblastomas. BMPs normally cause neural stem cells to differentiate into astrocytes; and when used to treat isolated glioblastoma CD133+ cells, they weaken the cells' tumorigenicity. (Reprinted with permission from *Nature*. 2006, Dec 7; **444(7120)**: 687–688.)

been shown to deplete cancer stemlike cells in gliobastomas.[59] Inhibition of another signaling pathway, the NOTCH pathway, using inhibitors of gamma secretase has also been shown to deplete cancer stemlike cells and block tumor growth in embryonal neural tumors such as medulloblastomas.[60]

Third, interferon-beta (IFN-beta) has potent antitumor activity but is limited by systemic side effects.[61] Targeted therapy with IFN-beta could have significant antitumor effects.[61]

Having molecular antitumor therapies is the first step in targeting malignant brain neoplasms. The second step involves identifying novel delivery systems that can take these molecular therapies such that they can reach all the infiltrative microscopic tumor cells that are distant from the solid tumor, as well as reach the CD133+ cancer stemlike cells to prevent recurrence of these tumors.

6.2. Cellular-Vector-mediated molecular therapy ("Bystander killing") for malignant brain neoplasms

Glioblastomas have been shown to actively attract bone marrow stem cells (stromal cells) by secreting angiogenic cytokines, such as interleukin 8, transforming growth factor ss1, neurotrophin 3 and vascular endothelial growth factor A.[62,63] Malignant glioma cells also attract adult hematopoietic progenitor cells via TGF-beta,[64] and irradiation and hypoxia also promote homing of these hematopoietic progenitor cells toward gliobastomas.[65] Components of the extracellular matrix of glioblastomas have also been shown to play an important role in glioma-induced tropism of neural stem cells.[54,66] Glioblastomas are highly vascularized tumors and are also associated with increased mobilization of endothelial progenitor cells that are CD34+ and CD133+, possibly to aid with neoangiogenesis.[67] Given the tendency for glioblastoma cells to actively attract various stem cells, these stem cells can potentially serve as excellent cellular vectors for various molecular antitumor therapies.

Delivery of cellular vectors can be via the intratumoral or the intravascular route. Recently, intratumoral injection of bone marrow stem cells expressing a "suicide gene," namely thymidine kinase, has been evaluated for treatment of glioblastomas.[56] The limitation with intratumoral delivery of stem cells is that the bone marrow stem cells injected into the solid tumor cavity alone may have difficulty reaching other parts of the cerebral hemispheres where there are infiltrative microscopic tumor cells. Endovascular delivery of chemotherapeutic agents for intracranial neoplasms has been shown to be safe and feasible.[68] Endovascular intra-arterial delivery of particulate embolic material such as polyvinyl alcohol particles is also routinely performed. Intravascular (intravenous or intra-arterial) delivery of stem cells is a novel route of delivery that holds great promise.[61,69,70] Nakamizo *et al.* have shown that when delivered via an intracarotid route, human-bone-marrow-derived mesenchymal stem cells were capable of migrating toward glioblastoma cells, and when engineered to release TGF-beta, significantly increased animal survival.[70] Dickson *et al.* have shown that in a murine model of disseminated neuroblastoma, intravenous delivery of tumor tropic neural progenitor cells transduced to express TGF-beta significantly restricted tumor growth.[61] Brown *et al.* have also shown that intravenous delivery of tumor tropic neural stem cells can localize to various tumors of neural and nonneural origin and may play a useful role in cellular-vector-mediated molecular antitumor therapy.[69] Further studies are needed to evaluate the type of molecular therapies these stem cells could deliver, the optimal dosage of stem cells needed, the need for blood–brain barrier disruption as an adjunct to aid in increased cell migration, and the optimal route of intravascular delivery, namely the intravenous or the intra-arterial route.

7. Multiple Sclerosis

Multiple sclerosis is the most common inflammatory demyelinating disease affecting the central nervous system, and is characterized pathologically by multifocal areas of demyelination with relative preservation of axons, loss of oligodendrocytes and astroglial scarring. It is characterized clinically by a primary progressive course, a relapsing remitting (RR) course or a secondary progressive course, and is associated with increasing neurological disability and significantly impaired quality of life.[71,72] The mainstay of current therapies for RR forms of multiple sclerosis includes various disease-modifying immunomodulating drugs, such as IFN-beta or glatiramer acetate for early forms of RR multiple sclerosis, and mitoxantrone or natalizumab for aggressive forms of RR multiple sclerosis.[73] Hematopoietic stem cell transplantation (intravenous delivery) in patients with multiple sclerosis has been attempted with varying degrees of success.[74–94] Autologous hematopoietic stem cell transplantation entails mobilization and preservation of hematopoietic stem cells, followed by myeloablative chemotherapy with or without total body irradiation to arrest the abnormal immune system, followed by reinfusion of hematopoietic stem cells that subsequently helps establish a more tolerant immune system.[73] Although there have been successes in the use of hematopoietic stem cell transplantation in multiple sclerosis patients,[73] it was not felt to be helpful in patients with widespread central nervous system damage with large areas of demyelination, and the transplant-related mortality rates prior to the year 2000 were around 5.3%.[88] The reason for these high transplant-related mortality rates was felt to be the myeloablative chemotherapeutic regimen using busulphan.[88] Current transplant-related mortality rates are <1% with newer chemotherapeutic regimens and are more acceptable.[73] Future studies need to evaluate the role of intravascular delivery of hematopoietic stem cells in selected populations of patients with multiple sclerosis who are identified at an early stage of the disease but have the tendency to progress to a state where there would be widespread damage to the central nervous system.[73]

8. Epilepsy

Epilepsy is very prevalent in the US. Approximately 50 million people have epilepsy, around 40% of whom have temporal lobe epilepsy.[95] Nearly 35% of people with temporal lobe epilepsy are refractory to medical therapy, and most of them also develop learning and memory disturbances as well as depression.[95] Temporal lobe resection is useful when the lesion is unilateral and the epileptogenic site can be localized to the temporal lobe.[96] However, patients with epileptogenic loci in the eloquent cortex or if present in bilateral temporal lobes, are not ideal candidates for surgery.[96] Cell loss or subsequent rewiring due to cell loss in the hippocampus has been associated with lowering of seizure thresholds.[96] Replacing this cell loss with stem cells that could potentially differentiate into inhibitory GABA-ergic interneurons and thereby raise the seizure threshold has been suggested as a possible alternative to surgery.[95] Immortalized mouse neural stem cells engineered to produce GABA under the control of doxycycline have been shown to have seizure-suppressing capabilities after transplantation into the dentate gyrus of rats.[96] Advances in interventional neuroimaging techniques have allowed safe and selective catheterization of intracranial vessels such as the anterior temporal arteries (branches of the middle cerebral artery) as well as the anterior, middle and posterior inferior temporal arteries (branches of the posterior cerebral artery).[13] Further studies are needed to evaluate the role of intravascular delivery of various types of neural and nonneural stem cells in seizure suppression in animal models of refractory epilepsy.

9. Motor Neuron Disease

Amyotrophic lateral sclerosis (ALS) is a debilitating neurodegenerative disease that affects the upper and lower motor neurons and causes diffuse motor neuron degeneration.[97] Garbuzova-Davis *et al.* have shown that intravenous delivery of human umbilical cord blood stem cells was associated with nearly 10–12 weeks of survival of these engrafted stem cells in the brain and spinal cord.[97] These engrafted stem cells also expressed neuronal and glial markers in a mouse model of ALS.[97] Future studies need to further evaluate the therapeutic potential of intravascular delivery of various types of stem cells in animal models of ALS.

10. Muscular Dystrophy

Muscular dystrophies are a group of primary muscle disorders having a hereditary basis characterized by widespread muscle damage due to underlying abnormalities in muscle structural proteins. No effective treatment exists for muscular dystrophies. Bachrach *et al.* have shown the safety and feasibility of intra-arterial transfemoral transplantation of myogenic progenitor cells in a mouse model, and have shown that engraftment rates are 5–8%, compared to 1% with the intravenous route of delivery.[98] Torrente *et al.* have shown that intra-arterial delivery of muscle-derived stem cells results in these cells attaching to the capillaries of muscles and then participating in muscle regeneration by restoring dystrophin in a mouse model of muscular dystrophy.[99] Sampaolesi *et al.* have observed that wild-type mesangioblasts delivered intra-arterially are associated with skeletal muscle regeneration in a mouse model of limb-girdle muscular dystrophy.[100] Sampaolesi *et al.* have also shown that in a large animal model of muscular dystrophy, namely the golden retriever model, intra-arterial delivery of wild-type mesangioblasts significantly ameliorated the symptoms of canine muscular dystrophy compared to intra-arterial delivery of autologous genetically corrected cells.[101] Further studies are needed to evaluate the role of selective catheterization of the muscular branches of the femoral and axillary arteries and intra-arterial delivery of various types of stem cells in functional outcome following muscular dystrophy.

11. Diabetes

Diabetes mellitus affects more than 5 million people in the US. It is a major independent risk factor for heart disease and stroke.[35] Adequate control of blood glucose levels can help decrease the long-term complications of diabetes but is usually difficult to achieve. Zhao *et al.* have shown that when transplanted into the peritoneal cavity of streptozocin-induced diabetic mice, human umbilical cord blood stem cells can differentiate into endoderm-derived insulin-producing cells and eliminate hyperglycemia.[102] Future studies need to evaluate the role of intra-arterial delivery of various types of stem cells into the celiac or superior mesenteric arteries supplying the pancreas to help decrease the long-term risks of hyperglycemia.

The long-term complications of diabetes mellitus include peripheral neuropathy, retinopathy and nephropathy.

11.1. Peripheral neuropathy

Peripheral neuropathies are due to deranged structure and function of peripheral motor, sensory and/or autonomic nerves. Diabetes mellitus is a common cause of peripheral neuropathy and can present as a symmetrical polyneuropathy, an asymmetrical neuropathy, or a combination. Although symptomatic treatments for diabetic peripheral neuropathy exist, there is no current therapy to reverse the damage to the peripheral nerves. Around 10% of people with idiopathic peripheral neuropathy are associated with monoclonal proteins. When there is no underlying disease to account for these monoclonal proteins, it is referred to as monoclonal gammopathy of unknown significance (MGUS). Lee *et al.* have reported a patient with progressive neurologic deficits secondary to MGUS refractory to treatment with chemotherapy, intravenous immunoglobulin and plasma exchange who was treated with chemotherapy followed by intravascular autologous bone marrow stem cell transplantation and showed excellent recovery at 12 months.[103] Siddiqi *et al.* have shown that in 12 patients with peripheral neuropathy associated with Krabbe's disease, intravascular hematopoietic stem cell transplantation was associated with improvement

386   *Stem Cells and Regenerative Medicine*

in peripheral nerve conduction abnormalities, suggesting remyelination of the nerves.[104] Future studies need to evaluate the role of intravascular delivery of various types of stem cells as a treatment for severe progressive diabetic peripheral neuropathy.

11.2. Retinopathy

Retinopathy is a serious microvascular complication of diabetes, and people with diabetes tend to fear blindness even more than premature death.[105] The incidence of blindness is around 50,000 new cases every year in the US, and the prevalence of blindness is greater than 10 million affected people in the US.[105] Purified retinal progenitor cells harvested at the correct stage and transplanted into the subretinal space in a mouse model of retinitis pigmentosa showed that a small portion of the progenitor cells migrated into the retina.[106] These progenitor cells also formed synaptic connections with cells in the inner nuclear layer and were associated with improved papillary light reflexes.[106] Current advances in interventional neuroimaging techniques have made it possible to safely and selectively catheterize intracranial vessels, including the ophthalmic arteries.[13] Selective catheterization of the ophthalmic artery for delivery of chemotherapeutic agents in retinoblastoma has been reported.[107] Similarly, selective catheterization of the ophthalmic artery and intra-arterial delivery of stem cells (into the ophthalmic artery and subsequently the cilioretinal arteries) for improving visual function as a viable alternative to direct injection of stem cells in various conditions, such as retinopathies, retinal strokes, retinitis pigmentosa and age-related macular degeneration, need to be studied in animal models.

12. Sensorineural Deafness

Nearly 28 million people in the US are affected by impaired hearing.[108] Hearing loss is associated with significant impairment in quality of life related to reduced social interactions, isolation, depression and possible memory impairment.[108] Hearing loss can be due to either conductive hearing loss or sensorineural deafness. Although there are treatments for conductive hearing loss, there are no effective treatments for sensorineural deafness. Stem cells could potentially help improve hearing in patients with sensorineural hearing impairment by replacing inner ear neurons and hair cells.[109] Augmenting endogenous stem cell responses as well as exogenous stem cell therapies with embryonic and adult stem cells has shown promise in replacing inner ear neurons and hair cells.[109] Exogenous stem cell therapies using surgical or direct implantation into the cochlea have been limited by the inhibitory effect on stem cell growth due to various factors in the extracellular space of the cochlea.[109] The inner ear is supplied by the labyrinthine artery, which is a distal branch of the anterior–inferior cerebellar artery. The anterior–inferior cerebellar artery is a branch of the basilar artery and can be safely and selectively catheterized using steerable or flow-directed microcatheters. Intra-arterial delivery of stem cells could potentially be another useful route of delivery of stem cells for patients with sensorineural deafness and needs to be studied in animal models.

Summary

Intravascular delivery systems for stem cell transplantation are an exciting new frontier in the field of regenerative medicine. Advances in interventional neurology and neuroimaging techniques have made it possible to safely and selectively catheterize small arteries in previously unapproachable areas, and have opened the door to newer delivery systems for cells, genes, drugs and cellular vectors. Future research should focus on optimizing the choice of stem cell type, the dosage of stem cells, the timing of therapy, the route of delivery, and the appropriate subpopulation with various neurologic and nonneurologic conditions that could benefit from this exciting new frontier in regenerative medicine.

References

1. Thomson JA, Itskovitz-Eldor J, Shapiro SS, *et al.* (1998) Embryonic stem cell lines derived from human blastocytes. *Science* **282**: 1145–1147.
2. Jiang Y, Jahagirdar BN, Reinhardt RL, *et al.* (2002) Pluripotency of mesenchymal stem cells derived from adult marrow. *Nature* **418**: 41–49.
3. Xiao J, Nan Z, Motooka Y, Low WC. (2005) Transplantation of a novel cell line population of umbilical cord blood stem cells ameliorates neurological deficits associated with ischemic brain injury. *Stem Cells Dev* **14**: 722–733.
4. De Coppi P, Bartsch G, Jr, Siddiqui MM, *et al.* (2007) Isolation of amniotic stem cell lines with potential for therapy. *Nat Biotechnol* **25**: 100–106.
5. Kondziolka D, Steinberg GK, Wechsler L, *et al.* (2005) Neurotransplantation for patients with subcortical motor stroke: a phase 2 randomized trial. *J Neurosurg* **103**: 38–45.

6. Jin K, Sun Y, Xie L, *et al.* (2005) Comparison of ischemia-directed migration of neural precursor cells after intrastriatal, intraventricular, or intravenous transplantation in the rat. *Neurobiol Dis* **18**: 366–374.

7. Shi E, Kazui T, Jiang X, *et al.* (2006) Intrathecal injection of bone marrow stromal cells attenuates neurologic injury after spinal cord ischemia. *Ann Thorac Surg* **81**: 2227–2233.

8. Shi E, Kazui T, Jiang X, *et al.* (2007) Therapeutic benefit of intrathecal injection of marrow stromal cells on ischemia-injured spinal cord. *Ann Thorac Surg* **83**: 1484–1490.

9. Willing AE, Lixian J, Milliken M, *et al.* (2003) Intravenous versus intrastriatal cord blood administration in a rodent model of stroke. *J Neurosci Res* **73**: 296–307.

10. Downey GP, Doherty DE, Schwab B, 3rd, *et al.* (1990) Retention of leukocytes in capillaries: role of cell size and deformability. *J Appl Physiol* **69**: 1767–1778.

11. Tolar J, O'Shaughnessy MJ, Panoskaltsis-Mortari A, *et al.* (2006) Host factors that impact the biodistribution and persistence of multipotent adult progenitor cells. *Blood* **107**: 4182–4188.

12. Freyman T, Polin G, Osman H, *et al.* (2006) A quantitative, randomized study evaluating three methods of mesenchymal stem cell delivery following myocardial infarction. *Eur Heart J* **27**: 1114–1122.

13. Janardhan V, Qureshi AI. (2007) Advances in interventional neuroimaging. *Neurotherapeutics* **4**: 414–419.

14. Thurman DJ, Alverson C, Dunn KA, *et al.* (1999) Traumatic brain injury in the United States: a public health perspective. *J Head Trauma Rehabil* **14**: 602–615.

15. Guerrero JL, Thurman DJ, Sniezek JE. (2000) Emergency department visits associated with traumatic brain injury: United States, 1995–1996. *Brain Inj* **14**: 181–186.

16. Lu J, Moochhala S, Moore XL, *et al.* (2007) Adult bone marrow cells differentiate into neural phenotypes and improve functional recovery in rats following traumatic brain injury. *Neurosci Lett* **398**: 12–17.

17. Mahmood A, Lu D, Qu C, *et al.* (2006) Long-term recovery after bone marrow stromal cell treatment of traumatic brain injury in rats. *J Neurosurg* **104**: 272–277.

18. Lu D, Li Y, Wang L, *et al.* (2001) Intraarterial administration of marrow stromal cells in a rat model of traumatic brain injury. *J Neurotrauma* **18**: 813–819.

19. Mahmood A, Lu D, Chopp M. (2004) Marrow stromal cell transplantation after traumatic brain injury promotes cellular proliferation within the brain. *Neurosurgery* **55**: 1185–1193.

20. Mahmood A, Lu D, Chopp M. (2004) Intravenous administration of marrow stromal cells (MSCs) increases the expression of growth factors in rat brain after traumatic brain injury. *J Neurotrauma* 21: 33–39.

21. Mahmood A, Lu D, Qu C, *et al.* (2007) Treatment of traumatic brain injury with a combination therapy of marrow stromal cells and atorvastatin in rats. *Neurosurgery* 60: 546–553.

22. Mahmood A, Lu D, Wang L, *et al.* (2001) Treatment of traumatic brain injury in female rats with intravenous administration of bone marrow stromal cells. *Neurosurgery* 49: 1196–1203.

23. Lu D, Mahmood A, Wang L, *et al.* (2001) Adult bone marrow stromal cells administered intravenously to rats after traumatic brain injury migrate into brain and improve neurological outcome. *Neuroreport* 12: 559–563.

24. Mahmood A, Lu D, Lu M, Chopp M. (2003) Treatment of traumatic brain injury in adult rats with intravenous administration of human bone marrow stromal cells. *Neurosurgery* 53: 697–702.

25. Mahmood A, Lu D, Qu C, *et al.* (2005) Human marrow stromal cell treatment provides long-lasting benefit after traumatic brain injury in rats. *Neurosurgery* 57: 1026–1031.

26. Lu D, Sanberg PR, Mahmood A, *et al.* (2002) Intravenous administration of human umbilical cord blood reduces neurological deficit in the rat after traumatic brain injury. *Cell Transplant* 11: 275–281.

27. National Spinal Cord Injury Statistical Center, University of Alabama at Birmingham. (2006) Spinal cord injury: facts and figures at a glance. http://www.spinalcord.uab.edu/show.asp?durki=21446 Accessed Aug. 1, 2007.

28. Urdzikova L, Jendelova P, Glogarova K, *et al.* (2006) Transplantation of bone marrow stem cells as well as mobilization by granulocyte-colony stimulating factor promotes recovery after spinal cord injury in rats. *J Neurotrauma* 23: 1379–1391.

29. Zurita M, Vaquero J. (2006) Bone marrow stromal cells can achieve cure of chronic paraplegic rats: functional and morphological outcome one year after transplantation. *Neurosci Lett* 402: 51–56.

30. Sykova E, Homola A, Mazanec R, *et al.* (2006) Autologous bone marrow transplantation in patients with subacute and chronic spinal cord injury. *Cell Tranplant* 15: 675–687.

31. Liu Z, Li Y, Qu R, *et al.* (2007) Axonal sprouting into the denervated spinal cord and synaptic and postsynaptic protein expression in the spinal cord after transplantation of bone marrow stromal cells in stroke rats. *Brain Res* 1149: 172–180.

32. Saporta S, Kim JJ, Willing AE, *et al.* (2003) Human umbilical cord blood stem cells infusion in spinal cord injury: engraftment and beneficial influence on behavior. *J Hemotother Stem Cell Res* 12: 271–278.

33. Fujiwara Y, Tanaka N, Ishida O, *et al.* (2004) Intravenously injected neural progenitor cells of transgenic rats can migrate to the injured spinal cord and differentiate into neurons, astrocytes and oligodendrocytes. *Neurosci Lett* **366:** 287–291.

34. Kang SK, Shin MJ, Jung JS, *et al.* (2006) Autologous adipose tissue-derived stromal cells for treatment of spinal cord injury. *Stem Cells Dev* **15:** 583–594.

35. American Hearth Association. (2006) 2006 Heart and Stroke Statistical Update. http://www.american-heart.org/statistics/05stroke.html

36. Janardhan V, Qureshi AI. (2004) Mechanisms of ischemic brain injury. *Curr Cardio Rep* **6:** 117–123.

37. Shen LH, Li Y, Chen J, *et al.* (2006) Intracarotid transplantation of bone marrow stromal cells increases axon-myelin remodeling after stroke. *Neuroscience* **137:** 393–399.

38. Bang OY, Lee JS, Lee PH, Lee G. (2005) Autologous mesenchymal stem cell transplantation in stroke patients. *Ann Neurol* **57:** 874–882.

39. Mendonca ML, Freitas GR, Silva SA, *et al.* (2006) Safety of intra-arterial autologous bone marrow mononuclear cell transplantation for acute ischemic stroke. *Arq Bras Cardiol* **86:** 52–55.

40. Staudacher DL, Preis M, Lewis BS, *et al.* (2006) Cellular and molecular therapeutic modalities for arterial obstructive syndromes. *Pharmacol Ther* **109:** 263–273.

41. Vendrame M, Cassady J, Newcomb J, *et al.* (2004) Infusion of human umbilical cord blood cells in a rat model of stroke dose-dependently rescues behavioral deficits and reduces infarct volume. *Stroke* **35:** 2390–2395.

42. Roitberg BZ, Mangubat E, Chen EY, *et al.* (2006) Survival and early differentiation of human neural stem cells transplanted in a nonhuman primate model of stroke. *J Neurosurg* **105:** 96–102.

43. Machalinski B, Paczkowska E, Koziarska D, Ratajczak MZ. (2006) Mobilization of human hematopoietic stem/progenitor-enriched CD34+ cells into peripheral blood during stress related to ischemic stroke. *Folia Histochem Cytobiol* **44:** 97–101.

44. Paczkowska E, Larysz B, Rzeuski R, *et al.* (2005) Human hematopoietic stem/progenitor-enriched CD34+ cells are mobilized into peripheral blood during stress related to ischemic stroke or acute myocardial infarction. *Eur J Haematol* **75:** 461–467.

45. Dunac A, Frelin C, Popolo-Blondeau M, *et al.* (2007) Neurological and functional recovery in human stroke are associated with peripheral blood CD34+ cell mobilization. *J Neurol* **254:** 327–332.

46. Taguchi A, Matsuyama T, Moriwaki H, *et al.* (2004) Circulating CD34-positive cells provide an index of cerebrovascular function. *Circulation* **109:** 2972–2975.

47. Sprigg N, Bath PM, Zhao L, *et al.* (2006) Granulocyte-colony-stimulating factor mobilizes bone marrow stem cells in patients with subacute ischemic stroke: the Stem cell Trial of recovery EnhanceMent after Stroke (STEMS) pilot randomized, controlled trial (ISRCTN 16784092). *Stroke* 37: 2979–2983.

48. Marietta M, Pedrazzi P, Girardis M, Torelli G. (2007) Intracerebral haemorrhage: an often neglected medical emergency. *Intern Emerg Med* 2: 38–45.

49. Seyfried D, Ding J, Han Y, *et al.* (2006) Effects of intravenous administration of human bone marrow stromal cells after intracerebral hemorrhage in rats. *J Neurosurg* 104: 313–318.

50. Zhang H, Huang Z, Xu Y, Zhang S. (2006) Differentiation and neurological benefit of the mesenchymal stem cells transplanted into the rat brain following intracerebral hemorrhage. *Neurol Res* 28: 104–112.

51. Nan Z, Grande A, Sanberg CD, *et al.* (2005) Infusion of human umbilical cord blood ameliorates neurologic deficits in rats with hemorrhagic brain injury. *Ann N Y Acad Sci* 1049: 84–96.

52. Jeong SW, Chu K, Jung KH, *et al.* (2003) Human neural stem cell transplantation promotes functional recovery in rats with experimental intracerebral hemorrhage. *Stroke* 34: 2258–2263.

53. Li F, Liu Y, Zhu S, *et al.* (2007) Therapeutic time window and effect of intracarotid neural stem cells transplantation for intracerebral hemorrhage. *Neuroreport* 18: 1019–1023.

54. Ziu M, Schmidt NO, Cargioli TG, *et al.* (2006) Glioma-produced extracellular matrix influences brain tumor tropism of human neural stem cells. *J Neurooncol* 79: 125–133.

55. Germano IM, Fable J, Gultekin SH, Silvers A. (2003) Adenovirus/herpes simplex-thymidine kinase/ganciclovir complex: preliminary results of a phase I trial in patients with recurrent malignant gliomas. *J Neurooncol* 65: 279–289.

56. Miletic H, Fischer Y, Litwak S, *et al.* (2007) Bystander killing of malignant gliomas be bone marrow-derived tumor-infiltrating progenitor cells expressing a suicide gene. *Mol Ther* 15: 1373–1381.

57. Dirks PB. (2006) Cancer: stem cells and brain tumours. *Nature* 444: 687–688.

58. Herrlinger U, Woiciechowski C, Sena-Esteves M, *et al.* (2000) Neural precursor cells for delivery of replication-conditional HSV-1 vectors to intracerebral gliomas. *Mol Ther* 1: 347–357.

59. Bar EE, Chaudhry A, Lin A, *et al.* (2007) Cyclopamine-mediated hedgehog pathway inhibition depletes stem-like cancer cells in glioblastoma. *Stem Cells* 25: 2524–2533.

60. Fan X, Matsui W, Khaki L, *et al.* (2006) Nothc pathway inhibition depletes stem-like cells and blocks engraftment in embryonal brain tumors. *Cancer Res* **66:** 7445–7452.

61. Dickson PV, Hamner JB, Burger RA, *et al.* (2007) Intravascular administration of tumor tropic neural progenitor cells permits targeted delivery of interferon-beta and restricts tumor growth in a murine model of disseminated neuroblastoma. *J Pediatr Surg* **42:** 48–53.

62. Birnbaum T, Roider J, Schankin CJ, *et al.* (2007) Malignant gliomas actively recruit bone marrow stromal cells by secreting angiogenic cytokines. *J Neurooncol* **83:** 241–247.

63. Schichor C, Birnbaum T, Etminan N, *et al.* (2006) Vascular endothelial growth factor A contributes to glioma-induced migration of human marrow stromal cells (hMSC). *Exp Neurol* **199:** 301–310.

64. Tabatabai G, Bahr O, Mohle R, *et al.* (2005) Lessons from the bone marrow: how malignant gliomas cells attract adult haematopoietic progenitor cells. *Brain* **128:** 2200–2211.

65. Tabatabai G, Frank B, Mohle R, *et al.* (2006) Irradiation and hypoxia promote homing of haematopoietic progenitor cells towards gliomas by TGF-beta-dependent HIF-1alpha-mediated induction of CXCL12. *Brain* **129:** 2426–2435.

66. Aboody KS, Brown A, Rainov NG, *et al.* (2000) Neural stem cells display extensive tropism for pathology in adult brain: evidence from intracranial gliomas. *Proc Natl Acad Sci USA* **97:** 12846–12851.

67. Zheng PP, Hop WC, Luider TM, *et al.* (2007) Increased levels of circulating endothelial progenitor cells and circulating endothelial nitric oxide synthase in patients with gliomas. *Ann Neurol* **62:** 40–48.

68. Qureshi AI. (2004) Endovascular treatment of cerebrovascular diseases and intracranial neoplasms. *Lancet* **363:** 804–813.

69. Brown AB, Yang W, Schmidt NO, *et al.* (2003) Intravascular delivery of neural stem cells lines to target intracranial and extracranial tumors of neural and non-neural origin. *Hum Gene Ther* **13:** 1777–1785.

70. Nakamizo A, Marini F, Amano T, *et al.* (2005) Human bone marrow-derived mesenchymal stem cells in the treatment of gliomas. *Cancer* **65:** 3307–3318.

71. Janardhan V, Bakshi R. (2000) Quality of life and its relationship to brain lesions and atrophy on magnetic resonance images in 60 patients with multiple sclerosis. *Arch Neurol* **57:** 1485–1491.

72. Janardhan V, Bakshi R. (2002) Quality of life in patients with multiple sclerosis: the impact of fatigue and depression. *J Neurol Sci* **205:** 51–58.

73. Martin R. (2007) Is haematopoietic stem cell transplantation a treatment option for severe MS or not? *Brain* **130:** 1181–1182.

74. Carreras E, Saiz A, Marin P, *et al.* (2003) CD34⁺ selected autologous peripheral blood stem cell transplantation for multiple sclerosis: report of toxicity and treatment results at one year of follow-up in 15 patients. *Haematologica* **88**: 306–314.

75. Espigado I, Marin-Niebla A, Rovira M, *et al.* (2003) Phase I/II trials of autologous peripheral blood stem cell transplantation in autoimmune diseases resistant to conventional therapy: preliminary results from the Spanish experience. *Transplant Proc* **35**: 742–743.

76. Fassas A, Anagnostopoulos A, Kazis A, *et al.* (2000) Autologous stem cell transplantation in progressive multiple sclerosis — an interim analysis of efficacy. *J Clin Immunol* **20**: 24–30.

77. Fasses A, Anagnostopoulos A, Kazis A, *et al.* (1997) Peripheral blood stem cell transplantation in the treatment of progressive multiple sclerosis: first results of a pilot study. *Bone Marrow Transplant* **20**: 631–638.

78. Fassas A, Passweg JR, Anagnostopoulos A, *et al.* (2002) Hematopoietic stem cell transplantation for multiple sclerosis: a retrospective multicenter study. *J Neurol* **249**: 1088–1097.

79. Grigg A, Tubridy NJ, Szer J, *et al.* (2004) Cladribine followed by autologous stem-cell transplantation in progressive multiple sclerosis. *Intern Med J* **34**: 66–69.

80. Kozak T, Havrdova E, Pit'ha J, *et al.* (2001) Immunoablative therapy with autologous stem cell transplantation in the treatment of poor risk multiple sclerosis. *Transplant Proc* **33**: 2179–2181.

81. Kozak T, Havrodova E, Pit'ha J, *et al.* (2000) High-dose immunosuppressive therapy with PBPC support in the treatment of poor risk multiple sclerosis. *Bone Marrow Transplant* **25**: 535–541.

82. Loh Y, Oyama Y, Statkute L, *et al.* (2007) Development of a secondary autoimmune disorder after hematopoietic stem cell transplantation for autoimmune diseases: role of conditioning regimen used. *Blood* **109**: 2643–2648.

83. Mancardi GL, Saccardi R, Filippi M, *et al.* (2001) **57**: 62–68.

84. Metz I, Lucchinetti CH, Openshaw H, *et al.* (2007) Autologous haematopoietic stem cell transplantation fails to stop demyelination and neurodegeneration in multiple sclerosis. *Brain* **130**: 1254–1262.

85. Nash RA, Bowen JD, McSweeney PA, *et al.* (2003) High-dose immunosuppressive therapy and autologous peripheral blood stem cell transplantation for severe multiple sclerosis. *Blood* **102**: 2364–2372.

86. Openshaw H, Lund BT, Kashyap A, *et al.* (2000) Peripheral blood stem cell transplantation in multiple sclerosis with busulfan and cyclophosphamide conditioning: report of toxicity and immunological monitoring. *Biol Blood Marrow Transplant* **6**: 564–575.

87. Papadaki HA, Tsagournisakis M, Mastorodemos V, *et al.* (2005) Normal bone marrow hematopoietic stem cell reserves and normal stromal cell function support the use of autologous stem cell transplantation in patients with multiple sclerosis. *Bone Marrow Transplant* 36: 1053–1063.

88. Saccardi R, Kozak T, Bocelli-Tyndall C, *et al.* (2006) Autologous stem cell transplantation for progressive multiple sclerosis: update of the European Group for Blood and Marrow Transplantation autoimmune diseases working party database. *Mult Scler* 12: 814–823.

89. Saccardi R, Mancardi GL, Solari A, *et al.* (2004) Autologous HSCT for severe progressive multiple sclerosis in a multicenter trial: impact on disease activity and quality of life. *Blood* 105: 2601–2607.

90. Saiz A, Carreras E, Berenguer J, *et al.* (2001) MRI and CSF oligoclonal bands after autologous hematopoietic stem cell transplantation in MS. *Neurology* 56: 1084–1089.

91. Samijn JP, Boekhorst PA, Mondria T, *et al.* (2006) Intense T cell depletion followed by autologous bone marrow transplantation for severe multiple sclerosis. *J Neurol Neurosurg Psychiatry* 77: 46–50.

92. Statkute L, Verda L, Oyama Y, *et al.* (2007) Mobilization, harvesting and selection of peripheral blood stem cells in patients with autoimmune diseases undergoing autologous hematopoietic stem cell transplantation. *Bone Marrow Transplant* 39: 317–329.

93. Su L, Xu J, Ji BX, *et al.* (2006) Autologous peripheral blood stem cell transplantation for severe multiple sclerosis. *Int J Hematol* 84: 276–281.

94. Xu J, Ji BX, Su L, *et al.* (2006) Clinical outcomes after autologous haematopoietic stem cell transplantation in patients with progressive multiple sclerosis. *Chin Med J (Engl)* 119: 1851–1855.

95. Shetty AK, Hattiangady B. (2007) Prospects of stem cell therapy for temporal lone epilepsy. *Stem Cells* 25: 2396–2407.

96. Thompson KW. (2005) Genetically engineered cells with regulatable GABA production can affect after discharges and behavioral seizures after transplantation into the dentate gyrus. *Neuroscience* 133: 1029–1037.

97. Garbuzova-Davis S, Willing AE, Zigova T, *et al.* (2003) Intravenous administration of human umbilical cord blood cells in a mouse model of amyotrophic lateral sclerosis: distribution, migration, and differentiation. *J Hemotother Stem Cell Res* 12: 255–270.

98. Bachrach E, Perez AL, Choi YH, *et al.* (2006) Muscle engraftment of myogenic progenitor cells following intraarterial transplantation. *Muscle Nerve* 34: 44–52.

99. Torrente Y, Tremblay JP, Pisati F, *et al.* (2001) Intraarterial injection of muscle-derived CD34$^+$ Sac-1$^+$ stem cells restores dystrophin in mdx mice. *J Cell Biol* 152: 335–348.

100. Sampaolesi M, Torrente Y, Innocenzi A, *et al.* (2003) Cell therapy of alpha-sarcoglycan null dystrophic mice through intra-arterial delivery of mesoangioblasts. *Science* **301:** 487–492.

101. Sampaolesi M, Blot S, D'Antona G, *et al.* (2006) Mesoangioblast stem cells ameliorate muscle function in dystrophic dogs. *Nature* **444:** 574–579.

102. Zhao Y, Wang H, Mazzone T. (2006) Identification of stem cells from human umbilical cord blood with embryonic and hematopoietic characteristics. *Exp Cell Res* **312:** 2454–2464.

103. Lee YC, Came N, Schwarer A, Day B. (2002) Autologous peripheral blood stem cell transplantation for peripheral neuropathy secondary to monoclonal gammopathy of unknown significance. *Bone Marrow Transplant* **30:** 53–56.

104. Siddiqi ZA, Sanders DB, Massey JM. (2006) Peripheral neuropathy in Krabbe disease: effect of hematopoietic stem cell transplantation. *Neurology* **67:** 268–272.

105. Bennett J. (2007) Retinal progenitor cells — timing is everything. *N Engl J Med* **356:** 1577–1579.

106. MacLaren RE, Pearson RA, MacNeil A, *et al.* (2006) Retinal repair by transplantation of photoreceptor precursors. *Nature* **444:** 203–207.

107. Yamane T, Kaneko A, Mohri M. (2004) The technique of ophthalmic arterial infusion therapy for patients with intraocular retinoblastoma. *Int J Clin Oncol* **9:** 69–73.

108. Daniel E. (2007) Noise and hearing loss: a review. *J Sch Health* **77:** 225–231.

109. Martinez-Monedero R, Edge AS. (2007) Stem cells for the replacement of inner ear neurons and hair cells. *Int J Dev Biol* **51:** 655–661.

15

Stem Cell Strategies for Treating Inner Ear Dysfunction

John H. Anderson* and Steven K. Juhn

1. Introduction

Hearing loss and problems with balance can result from damage to the inner ear due to disease (e.g. Meniere's), toxins, adverse side effects of medications (e.g. aminoglycoside antibiotics), noise, and aging. The health-related impact on the population is significant. Indeed, the most common type of human hearing impairment is due to age-related changes in the inner ear (presbycusis), with 40% of the population over 65 having some form of progressive, bilateral hearing loss.[1] Similarly, a deterioration of inner ear function causes problems with balance,[2] and this has been directly linked to an increase in falls in the aged. Also, there is an age-dependent deterioration of eye–head coordination involving the vestibulo-ocular reflex (VOR).[3,4]

The age-related changes in the cochlea and vestibular labyrinth are multifaceted. An increase in apoptotic cell death has been described for both hair cells and supporting cells in the mouse cochlea.[5] There is a progressive loss of dendrites and cell bodies of spiral ganglion neurons in the cochleas of aged humans and animals.[6,7] Additionally, there is evidence that the spiral ligament and stria vascularis of the cochlear lateral wall are involved in age-related hearing loss.[8,9] In a cross-sectional study of normal human temporal bones spanning ages from birth to 100, Merchant et al.[10] found a progressive (age-dependent) reduction in the number of hair cells in the vestibular sensory epithelia, reaching a maximum of 40%. As with the

*Correspondence: Department of Otolaryngology, University of Minnesota, 396 MMC-420 Delware St SE, Minneapolis, MN 55455, USA. E-mail: anders00@umn.edu

cochlea, Usami *et al.*[5] found evidence for a decrease in both hair cells and supporting cells in the vestibular sacculus.

Unfortunately there is a very limited ability for regeneration in the adult mammalian inner ear. There is no definitive evidence for spontaneous regeneration of hair cells in the cochlea[11] and limited capacity in the vestibular sensory epithelium.[11–13] The inability for regeneration in the adult mammalian cochlea is due to a lack of stem cells.[14] However, Li *et al.*[15] demonstrated that there are stem cells in the adult sensory epithelium of the utriculus that can differentiate *in vitro* and *in vivo* into some cells representing all three dermal lineages. Techniques to stimulate such cells in the adult might form the basis for repairing damage to the vestibular labyrinth and thereby improve balance. Also, there are several reports of cells in the cochlear sensory epithelium in the young, postnatal rat[16] and mouse[17,18] that are capable of proliferation and are multipotent. These cells might be used for future experiments to study the mechanisms of cell differentiation and to help develop strategies for transplantation into the adult cochlea to repair damaged tissue.[19,20]

A complete restoration of inner ear structure and function would require the proliferation and differentiation of several cell types, e.g. supporting cells in the sensory epithelium, different types of hair cells, neurons of the vestibular and auditory nerves, and cells of the lateral wall of the cochlea. On the other hand, in some cases the replacement of a single cell type or a limited number of cell types might still be of benefit in restoring function.

In this review we consider recent studies and results that could form the basis for developing strategies and techniques to stimulate endogenous cells to divide and differentiate or transdifferentiate, and identify stem cells that could be used for transplants into the inner ear. Several reviews and perspectives have been presented by Breuskin *et al.*,[21] Li *et al.*,[20] Hu and Ulfendahl,[19] Nakagawa and Ito,[22] and Martinez-Monedero and Edge.[23] Figure 1 (from Hudspeth[24]; see also Hu and Ulfendahl[19]) shows a schematic that highlights the different parts of the inner ear where stem cells could be injected and the location of the sensory and neural tissues where regeneration and recovery from damage might occur.

2. Endogenous Stem Cells in the Adult Inner Ear

Li *et al.*[15] first reported the existence of cells in the adult inner ear that have characteristics of stem cells, including the formation of clonal spheres and

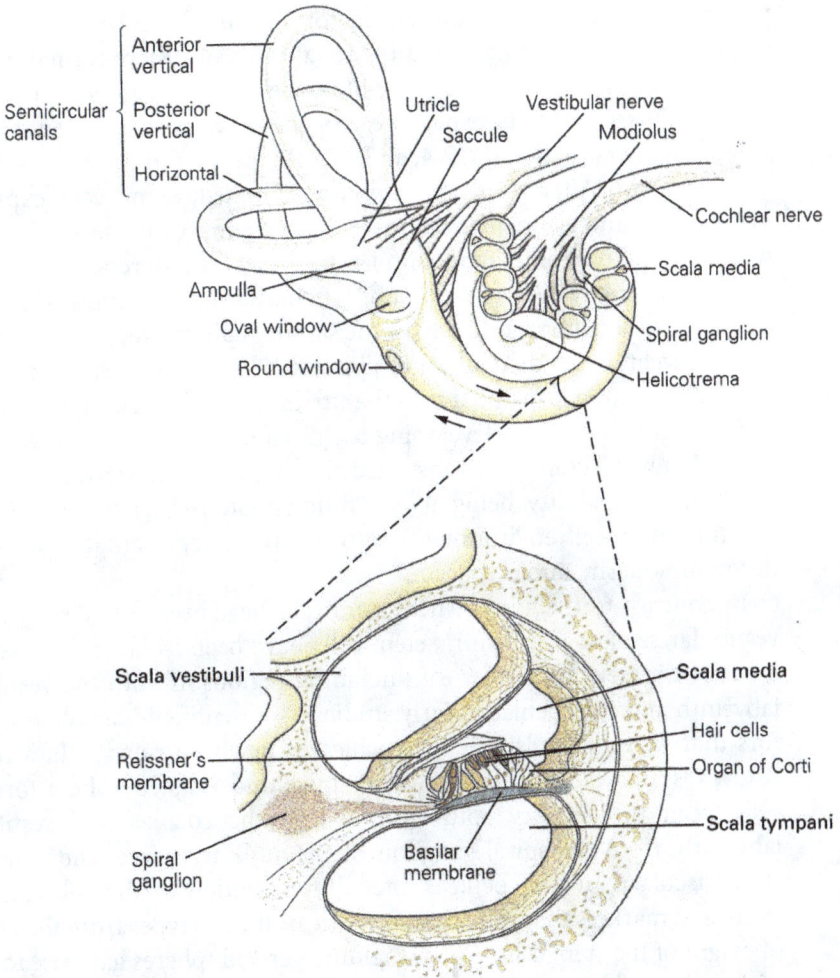

Figure 1. Structures of the inner ear. The upper drawing shows the vestibular labyrinth (semicircular canals, utricle, and saccule) and cochlea. The lower drawing shows a cross-section of the cochlea with its three fluid-filled compartments. Sound pushes the stapes against the oval window, exerting pressure on the perilymph fluid of the scala vestibuli. The fluid pressure wave travels through the connected scala tympami to the round window membrane. The scala media is a separate compartment, filled with endolymph (with a high K$^+$ and low Na$^+$ concentration) which bathes the ciliary part of the hair cells. (Reproduced with permission of the McGraw-Hill Companies. From E. Kandel, *et al.*, *Principles of Neural Science*, 4th ed., p. 592.)

multipotent differentiation potential. In this study,[15] cells were isolated from the sensory epithelium (in the macular region) of the vestibular utriculus of the adult mouse. Some of the isolated cells were mitotically active and formed clonal spheres. The growth factors epidermal growth factor (EGF), insulin-like growth factor 1 (IGF-1), and basic fibroblast growth factor (bEGF) increased the magnitude of sphere formation, and some of the sphere cells were capable of self-renewal. In addition, other sphere cells seemed to have become progenitor cells, expressing nestin and several markers (Pax-2, BMP-4, and BMP-7) present in the early developing inner ear. After differentiation was initiated, some cells expressed myosin VIIa and espin, hair cell markers, and some of those also had protrusions positive for F-actin, similar to a hair cell stereociliary bundle. Another distinctly different subset of differentiated cells expressed markers characteristic of supporting cells of the sensory epithelium, pan-cytokeratin and the cyclin-dependent kinase inhibitor, p27^{Kip1}. When the mouse sphere-derived cells were implanted into the utricle of chicken embryos, the cells were integrated into the developing cochlea and differentiated toward hair cells. Finally, Li *et al.*[15] demonstrated that some sphere-derived cells had pluripotent capability, being able to differentiate (under the influence of an embryonic cell environment) into cell types representative of endoderm, mesoderm, and ectoderm.

In contrast to the adult, where stem cells have been found only in the vestibular sensory epithelium, stem cells have been isolated from several areas of the newborn inner ear, including regions in both the vestibular labyrinth and the cochlea.[17] Early studies had identified hair cell progenitors that were derived from the cochlea of newborn rats.[25,26] In a recent study, Oshima *et al.*[17] used newborn mice and isolated sphere-forming cells from the sensory epithelium within the cochlea and vestibular labyrinth, from the spiral ganglion of the auditory nerve, and from the stria vascularis in the cochlea. After differentiation of the spheres, cells expressed markers that were characteristic of the cell types from the region of origin of the stem cells, e.g. epithelium-derived spheres gave rise to cells with an expression profile similar to that of hair cells and not neurons. At day 21 the greatest capacity for sphere-forming was in the vestibular epithelium, including the utriculus, sacculus, and semicircular canals, compared to the cochlear tissues. Thereafter the cells from the cochlear tissues lost their capacity for sphere formation, whereas there were still a considerable number of sphere-forming cells derived from the vestibular epithelia up to 120 days postnatal.

The decrease in sphere-forming capability could be the basis for the lack of hair cell and auditory nerve regeneration in the adult cochlea. However, Oshima *et al.*[17] hypothesize that there might not have been a loss of stem cells in the cochlea, but rather a change in the state and differentiation of the cells that is reflected in a decrease of developmental and progenitor cell markers. The cells might be the supporting cells in specific regions of the sensory epithelium.[26] With regard to the auditory neurons, Oshima *et al.*[17] further hypothesize that there might not have been a postnatal loss of stem cells in the spiral ganglion. The work of Rask-Andersen *et al.*[27] supports this idea. That study showed that adult human and guinea pig auditory spiral ganglion tissue cultured with the growth factors EGF and bFGF gave rise to spheres of cells. In the presence of the neurotrophic factors GDNF, BDNF, and NT-3, there was differentiation of the spheres into neurons and glial cells.

Future studies could test these hypotheses and explore possibilities for activating the supporting cells. Indeed, recent studies with adult animals have shown that transvection with the *Math1* gene[28,29] results in the formation of new hair cells in the adult organ of Corti. Math1 is a helix-loop-helix transcription factor that is required for the differentiation of hair cells, but not for supporting cells.[30] When the adenovirus vector was injected into the endolymph of the guinea pig cochlea, there was an over-expression of Math1 in nonsensory supporting cells of the cochlear epithelium. This gave rise to new hair cells, and there was some neural axonal growth toward the new cells. Moreover, the study by Izumikawa *et al.*[28] used deaf guinea pigs, and there was a functional improvement in hearing.

Finally, a study by Staecker *et al.*[31] showed that transvection with the *Math1* gene in mice which had received an ototoxic aminoglycoside resulted in regeneration of vestibular hair cells. Also, there was a functional recovery based on an improvement in swimming and an increase in gain of the vestibulo-ocular reflex. Although there was no regeneration in the cochlea in that study, the authors noted that the vector was injected into the perilymph. Kawamoto *et al.*[29] had previously shown that injection (with the Math1 vector) into the endolymphatic space was required for cochlear hair cell regeneration.

3. Stem Cell Transplants into the Inner Ear

The ability of stem cells to differentiate into various cell types, coupled with their propensity to integrate most readily into regions of tissue damage, suggests that stem cells might be uniquely suited to the challenges associated with damage or loss of cells in the sensory regions of the inner ear. However, the environment of the inner ear presents significant challenges for transplantation therapy. Encased in bone, the delicate sensory structures of the inner ear are difficult to access surgically without causing further damage to the tissues. Furthermore, the compartmentalization and unique fluid compositions within the inner ear might limit the ability of stem cells to access damaged regions. Nevertheless, surgical approaches to the semicircular canals, scala tympani and vestibuli, and the lateral wall of the cochlea that could be used for injections have been examined in the mouse.[32–35]

3.1. Neural stem cells migrate within the inner ear and are influenced by the cellular environment

Preliminary inner ear transplantation studies used fetal neural stem cells (NSCs) derived from the hippocampus.[22] Prior studies with NSCs had shown some functional recovery after spinal cord injury and potential for differentiation into neurons and glial cells in the central nervous system and retina. Those results were encouraging and suggested that similar strategies could be used for ameliorating the loss of auditory neurons.[19]

Ito *et al.*[36] reported that NSCs survived up to 4 weeks after being injected into the cochlea of normal neonatal rats, that there was migration within the cochlea, and that a few cells showed differentiation toward cochlear hair cells. However, there was no evidence for integration within the sensory epithelium.

A subsequent study by Parker *et al.*[37] showed that NSCs injected through the round window membrane can survive, migrate, and differentiate within the adult mouse cochlea. These NSCs exhibited a preference for integrating into the spiral ganglion nucleus, and expressed neural markers such as NF-200 and MAP-2. While a few transplanted cells did express myosin VIIa and parvalbumin, markers suggestive of a hair cell fate, these cells were located distal to the sensory epithelium in the spiral prominence.

Iguchi *et al.*[38] injected NSCs into the lateral semicircular canal of 6-week old mice. Transplant-derived cells migrated to the perilymphatic compartments of the cochlea and survived up to 4 weeks. Most cells expressed the glial marker GFAP and were immunoreactive for the neurotrophins GDNF and/or BDNF. Since neurotrophins might help protect auditory neurons,[39] these results suggest potential for NSC treatments to help prevent a progressive loss of auditory neurons, such as during aging or after a cochlear implant.

A similar predominance of differentiation to glial cells was reported by Tamura et al.[40] They injected fetal mouse NSCs into the area of the auditory nerve fibers within the modiolus of 6-week old mice. Fourteen days prior to transplantation, the mice had been treated with cisplatin injected locally into the posterior semicircular canal in one ear. This resulted in a loss of about 60% of the neurons in the spiral ganglion of that ear. Two weeks after transplantation, the NSCs were present within the area of injection and had also migrated to other areas of the modiolus and scala tympani. In addition, about 10% of the NSC-derived cells differentiated into cells expressing the neuron marker TuJ1, and about 80% expressed the glial marker GFAP. Future studies need to address how to increase the number of cells that differentiate into neurons in a damaged cochlea, such as by manipulating transcription factors, and whether (and under what conditions) functional connections can develop within the sensory epithelium and brainstem.[19,22]

Tateya et al.[41] also used an ototoxic animal model. Three days after local injection of neomycin into the posterior semicircular canal of 4-week old mice, NSCs were injected through the lateral wall of the cochlea. Although the majority of stem cells expressed markers for glial cells (GFAP) or neurons (MAP2), a few cells which had migrated to the vestibular epithelia expressed a marker for hair cells, myosin VIIa. These results provide some evidence for the migration of NSCs into a region of tissue damage, the sensory epithelia, and for differentiation into an appropriate cell type.

Although the studies of Tateya et al.[41] and Tamura et al.[40] used mice with damaged inner ear sensory epithelia, there was little differentiation toward cell types other than neurons and glial cells. However, the recent study by Parker et al.[37] provides evidence for the influence of microenvironmental factors in a damaged cochlea. Clonal neural stem cells (cNSCs) derived from the murine fetal cerebellar cells were injected into the scala tympani of adult mice and guinea pigs shortly after the beginning of the noise exposure. The mice were subjected to 2 hours of noise, allowed to recover for 48 hours, and then injected with the cNSCs; the guinea pigs were subjected to 72 hours of noise and then immediately transplanted. At these time points when the cNSCs were injected, the cellular and tissue responses to the noise would still be progressing, eventually causing significant structural damage to the cochlea involving many different cell types. The results showed that the stem cells migrated to the organ of Corti and to regions of the auditory nerve endings. There were several genetic expression profiles. Some cells had characteristics of cochlear hair cells or supporting cells and others had characteristics of auditory neural tissues. The profiles were consistent with the idea that the local cellular environment within the cochlea influenced the differentiation of the stem cells.

The above studies used NSCs derived from fetal tissue. However, it is also possible to use adult NSCs. Hu *et al.*[42] injected adult mouse NSCs into the scala tympani of adult guinea pigs deafened by the ototoxic effects of neomycin. In damaged cochlea, there was survival up to 4 weeks post-transplantation, although the absolute number of surviving NSCs was low (less than 0.7%), similar to other studies.[38] Also, there was migration to the areas of damage along the auditory nerve endings close to the sensory epithelium and to the spiral ganglion. Although some NSCs expressed a neuronal marker, there was a higher percentage (95%) when the NSCs were transduced with ngn2 (an early regulator gene for neuron differentiation) prior to transplantation. This supports the concept of using cell replacement in combination with genetic engineering to help direct subsequent differentiation after transplantation.

Recently, Sekiya *et al.*[43] used conditionally immortal mouse auditory neuroblasts. The cells were surgically implanted into adult rat auditory nerves, survived, and migrated along the auditory nerve. They differentiated into cells that assumed the normal bipolar morphology of auditory neurons and expressed the neuronal cytoskeletal protein beta3-tubulin, consistent with axon growth. On the other hand, there was no evidence for expression of GFAP, a marker for glial cells, or other markers for epithelial and mesenchymal cells. There was only minimal degradation of the auditory brainstem response, indicating that the surgical technique did not damage a large number of host neurons. The results show promise for the successful control of the state of donor cells prior to injection, selective differentiation into cells with an auditory neuron phenotype, and transplantion into the auditory nerve.

3.2. Multipotent adult progenitor cells are candidates for replenishing different cell types

Another possibility for inner ear transplants is to use stem cells derived from adult bone marrow.[22] One type, termed multipotent adult progenitor cells (MAPCs), is capable of differentiating into derivatives of all three germ layers *in vitro* and *in vivo*, similar to embryonic stem (ES) cells.[44]

As such, MAPCs might be ideal candidates for stem cell repair of the damaged inner ear. Because these stem cells are derived from adult bone marrow, they offer the benefits associated with transplantation of autologous tissue while avoiding the technical and ethical difficulties associated with ES cells. Questions that need to be addressed include the following: Can MAPCs survive within the adult inner ear and integrate within the sensory epithelia layer? Do MAPCs differentiate into inner-ear-specific cell types and can they effect changes in auditory and/or vestibular function?

Studies in our laboratories have begun to address these questions.[45] The long-term goal is to investigate cochlear and vestibular hair cell repair in the aged and following exposure to ototoxic drugs. The experiments described below used young, adult Balb/c mice with normal inner ear function. The initial results show that following injection through the round window membrane, equal numbers of MAPCs were found in the bone marrow of both the transplanted and the contralateral ear and the femurs. The stem cells survived over 4 weeks posttransplantation and some remained at the site of injection. Therefore, MAPCs leave the inner ear scala tympani via circulation to reside in the bone marrow throughout the animal. Future studies will investigate different sites of injection and the possibilities for differentiation within the cochlea and vestibular epithelial regions.

3.2.1. *Surgical and injection techniques*

First, injections of dye were made through the round window membrane to demonstrate that the surgical, injection, and histological techniques were accurate. The round window niche was accessed through a small hole in the ventral-inferior aspect of the bulla. A 1.0 μl injection of 10% Evans blue in PBS was made through the round window membrane. The animals were sacrificed 1 hour postinjection, and the temporal bones were dissected, fixed in 4% paraformaldehyde overnight, decalcified in 10% EDTA for 5 days, and processed for cryosectioning. Dye-labeled cells were found throughout the cochlea, including the organ of Corti (Fig. 2). Significantly, no dye was found in other regions of the inner ear.

3.2.2. *MAPC stem cell transplants*

LacZ-containing MAPC stem cells were prepared as a single cell suspension at a concentration of 40,000 cells/μl, with > 95% viability as determined

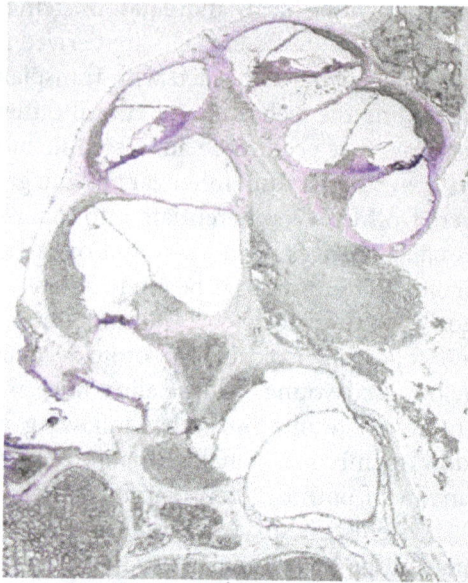

Figure 2. Injection of dye through the round window. A solution of 10% Evans blue was injected through the round window membrane (rwm). The purple dye labeled structures throughout the basal to apical turns of the cochlea, with the heaviest labeling in the walls of the scala tympani and particularly the basal membrane of the organ of Corti. In no cases was dye found in either the vestibular or bone marrow regions of the inner ear (*n* = 6 temporal bones).

by trypan blue exclusion. MAPCs were injected (1.0 μl) into the right cochlea through the round window of 5–6-week old Balb/c mice using a pulled glass pipette attached to a 1.0 μl Hamilton syringe. The animals were sacrificed either 1 or 2 weeks posttransplantation, and the temporal bones were processed for X-gal histochemistry and cryosectioning.

There were two main results. First, MAPCs were found in all animals within the bone marrow of both the transplanted and the contralateral, untreated inner ear (Fig. 3). There was no difference in localization after 1- vs 2-week survival times. The total number of MAPCs in the bone marrow of a representative animal was counted. There were 1,336 cells in the transplanted ear and 1,120 cells in the contralateral, untreated ear. In most of the animals (10 of 12), no MAPCs remained within the cochlea. This suggests that MAPCs responded to their environment by actively migrating out of the scala tympani, entering the vasculature, and homing to the bone marrow throughout the animal, including both temporal bones.

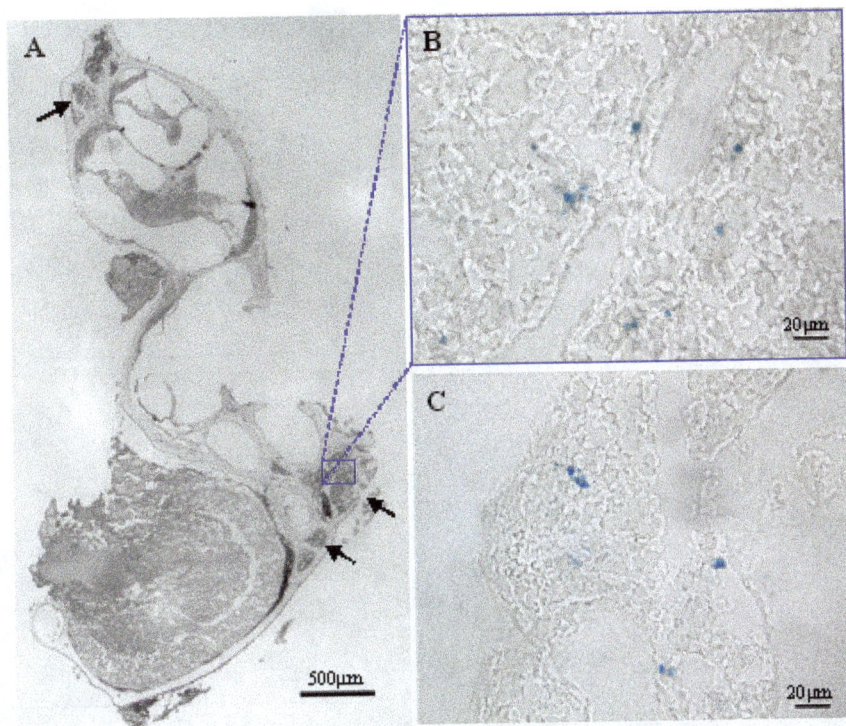

Figure 3. Transplanted MAPCs are preferentially located in the bone marrow. MAPC stem cells were injected through the round window membrane, and animals were sacrificed 1 or 2 weeks later. MAPCs contain the *LacZ* gene, and were identified by staining with X-gal, which forms a blue precipitate. MAPCs were found in the bone marrow of the ear in all transplanted animals, at both 1 week ($n = 8$) and 2 weeks ($n = 4$) posttransplantation. Within the temporal bone encapsulating the inner ear there are several bone marrow regions (**A**; arrows). X-gal stained MAPCs are clearly present in the bone marrow in both the transplanted (**B**) and the contralateral, untreated ear (**C**).

The second finding was that in 2 of the 12 animals, MAPCs had remained within the cochlea (Fig. 4). These MAPCs were located in the lateral portion of the scala tympani at both 1 and 2 weeks posttransplantation, indicating that stem cells derived from adult bone marrow can survive for at least 2 weeks within the cochlea. The lack of significant integration of MAPCs into host tissue was not surprising, as low levels of stem cell integration into undamaged tissue have been well documented in other areas, such as the central nervous system.

The MAPCs in the scala tympani were part of a conglomerate of cells which included nonlabeled host-derived cells. To determine whether the

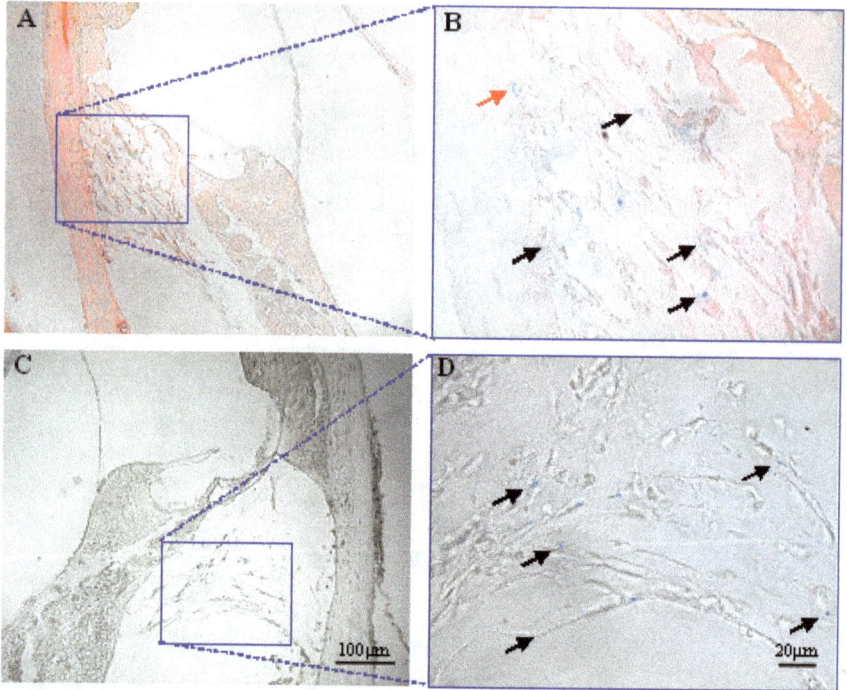

Figure 4. MAPC stem cells were found in the scala tympani at both 1 and 2 weeks posttransplantation. MAPCs were found in the scala tympani 1 week after transplantation (1 of 8 animals). The MAPCs were localized to the lateral portion of the scala tympani, directly beneath the organ of Corti (**A** and **B**; counterstained with Eosin). They were part of a conglomerate of cells, consisting of X-gal-labeled stem cells (arrows) and unlabeled, host-derived cells. In rare cases, it appeared that MAPCs had integrated into the lateral wall of the scala tympani (**B**; red arrow). MAPCs were also found in the scala tympani 2 weeks after transplantation (1 of 4 animals). The MAPCs were localized to the lateral portion of the scala tympani, directly beneath the organ of Corti (**C** and **D**). There was no difference in the appearance of the transplanted stem cells after a 1- or 2-week survival time.

host-derived cells were part of an immune response, we tested for the presence of the CD68 antigen, a marker for macrophages and several types of lymphocytes. There was no positive staining for CD68 within the conglomerate of transplanted MAPCs and host-derived cells (Fig. 5). On the other hand, CD68-positive cells were found in the bone marrow of the inner ear, but were never colocalized with transplanted MAPCs. While the host cell type in the conglomerate is unknown, the absence of CD68 staining suggests that those cells were not part of an immune response.

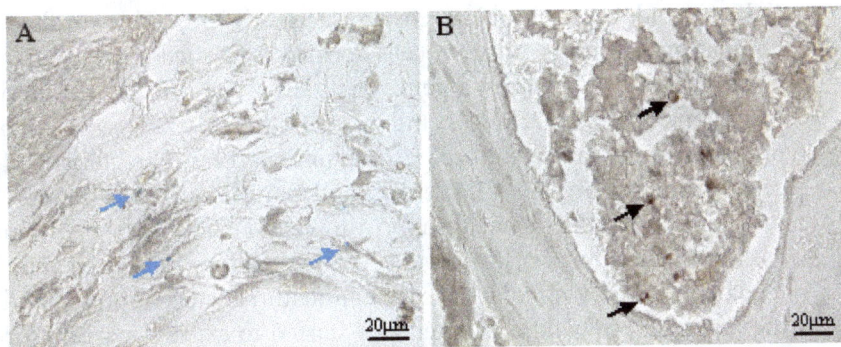

Figure 5. Transplantation of MAPCs did not appear to initiate a host immune response. The presence of an immune response to the MAPCs in the scala tympani was investigated using immunodetection of CD68, a marker for white blood cells and macrophages. There was no positive staining for CD68 among the X-gal stained MAPCs (blue arrows) in the scala tympani (A). CD68-positive cells (black arrows) were found in bone marrow regions of the inner ear (B).

3.2.3. *Summary of MAPC transplant results with normal mice*

Following injections through the round window membrane, equal numbers of MAPCs were found in the bone marrow of both the transplanted and the contralateral ear. Therefore, MAPCs leave the scala tympani via circulation to reside in the bone marrow throughout the animal. This indicates that MAPCs have the ability actively migrate from the environment of the perilymph and adjacent tissues and enter the vasculature. In addition, some MAPCs remained in the scala tympani within a conglomerate of host-derived cells, surviving for at least 2 weeks posttransplantation. During this period there did not appear to be any host immune response. There were no lymphocyte or macrophage infiltrates and no evidence of inflammation in the area of the MAPCs.

3.4. **Embryonic stem cells can be induced to inner ear progenitor cells**

Li *et al.*[46] were able to generate hair cells from a renewable source of progenitors. Murine ES cells were used. Embryoid bodies were cultured with growth factors known to play important roles during development of the inner ear: epidermal growth factor (EGF), insulin-like growth factor 1

(IGF-1), and basic fibroblast growth factor (bFGF). This resulted in a population of nestin-positive cells termed "enriched progenitor cells." These progenitors expressed genes that define the otic placode and otic vesicle, including Pax2, PMP7, and Jagged-1. *In vitro* differentiation of the progenitor cells gave rise to cells expressing markers characteristic of hair cells, including myosin VIIa, espin, and parvalbumin 3, and p27^{Kip1}, a marker found in the inner ear supporting cells. Finally, *in vivo* experiments with chicken embryos showed that the progenitors integrated into both the cochlear and vestibular sensory epithelium at sites of injury, and that the integrated cells differentiated and started to express markers for hair cells and form ciliary bundles.

It has been demonstrated that the inner ear progenitor cells can respond to the normal embryonic cellular environment and differentiate toward hair cells.[46] Furthermore, there is evidence that *Math1*-deficient inner ear progenitor cells can be rescued by normal cells during embryogenesis. A recent study by Du *et al.*[47] showed that *Math1*-null chimeric mice had both normal and mutant hair cells in all inner ear sensory epithelia, indicating that the mutant hair cell precursors, lacking the *Math1* gene, responded to their environment and differentiated into hair cells.

The protocol provided for the generation of inner ear progenitors and their differentiation into hair cells was based on murine ES cell lines.[46] Subsequently, the protocol was applied to human ES cells, and the resulting progenitors also expressed inner ear markers.[48] Shi *et al.*[49] reported that human ES cells could differentiate in a stepwise manner into cells with some characteristics of auditory sensory neurons when the progenitor cells were induced with BMP4. Future studies could use these sources of progenitors for studying the cellular mechanisms for differentiation into different cell types and the efficacy of transplantation into damaged ears.

3.5. Bone marrow hematopoietic stem cells differentiate toward inner ear fibrocytes

In a preliminary study by Naito *et al.*,[50] autologous, multipotent mesenchymal cells from the bone marrow (but not identified as MAPCs) were transplanted into the cochlea of gentamicin treated chinchillas. The stem cells survived up to 3 weeks in the scala vestibuli and tympani, the lateral wall of the cochlea, and the modiolus. It was reported that a few cells expressed NF-200 and might have been differentiating toward neurons. However, a study by Lang *et al.*[51] indicates that whole bone marrow cells as well as the hematopoietic stem cell (HSC) component do not differentiate into neurons or hair cells. When the HSCs were injected into lethally irradiated adult mice, the stem cells survived in the inner ear and were found where fibrocytes and mesenchymal cells are normally located and expressed markers for ion transport fibrocytes. The authors concluded that in the normal adult, HSCs might serve to maintain the population of fibrocytes in the spiral ligament of the inner ear. It is known that inner ear fibrocytes are specialized and play important roles in regulating the ionic environment and fluid homeostasis.[52] After cochlea damage due to noise or drug toxicity, damaged fibrocytes can recover and their loss can be replaced. The HSCs might participate in this repair and regeneration.

Future Challenges

The existence of inner ear stem cells affords the possibility of more fully characterizing the process and regulatory mechanisms by which these stem cells can be activated and the progenitor hair cells differentiate and develop into mature cells. In addition, the hair cell progenitors could be used for *in vitro* studies to identify drugs that could modify or control the intracellular signaling pathways for hair cell differentiation. *In vivo* studies could investigate the process of synapse formation with the afferent neurons and functional recovery from inner ear damage.

Although recent studies of endogenous stem cells and cell transplantation into the inner ear have shown promise for replacing and/or protecting hair cells and associated neurons,[19] there are challenging issues that need to be addressed[53]: the possibilities for transdifferentiation of endogenous cells,[54] the choice and availability of cells for transplantation, the induction and appropriate differentiation of transplanted cells, and the extent of functional recovery of hearing and balance. It is necessary to understand the mechanisms for directing stem cell differentiation along specific pathways,[55] to establish techniques for isolating specific cell types, to keep cell proliferation under appropriate constraints, to control interactions between the host immune system and nonautologous transplanted cells that might initiate a graft rejection, and to assess the long-term viability of transplanted cells.

References

1. Gates GA, Mills JH. (2005) Presbycusis. *Lancet* **366:** 1111–1120.
2. Rauch SD, Velazquez-Villasenor L, Dimitri PS, Merchant SN. (2001) Decreasing hair cell counts in aging humans. *Ann N Y Acad Sci* **942:** 220–227.
3. Paige GD. (1992) Senescence of human visual–vestibular interactions. 1. Vestibulo-ocular reflex and adaptive plasticity with aging. *J Vestib Res* **2:** 133–151.
4. Baloh RW, Enrietto J, Jacobson KM, Lin A. (2001) Age-related changes in vestibular function: a longitudinal study. *Ann N Y Acad Sci* **942:** 210–219.
5. Usami S, Takumi Y, Fujita S, *et al.* (1997) Cell death in the inner ear associated with aging is apoptosis? *Brain Res* **747:** 147–150.
6. Spoendlin H, Schrott A. (1989) Analysis of the human auditory nerve. *Hear Res* **43:** 25–38.
7. White JA, Burgess BJ, Hall RD, Nadol JB. (2000) Pattern of degeneration of the spiral ganglion cell and its processes in the C57BL/6J mouse. *Hear Res* **141:** 12–18.
8. Ichimiya I, Suzuki M, Mogi G. (2000) Age-related changes in the murine cochlear lateral wall. *Hear Res* **139:** 116–122.
9. Hequembourg S, Liberman MC. (2001) Spiral ligament pathology: a major aspect of age-related cochlear degeneration in C57BL/6 mice. *J Assoc Res Otolaryngol* **2:** 118–129.
10. Merchant SN, Velazquez-Villasenor L, Tsuji K, *et al.* (2000) Temporal bone studies of the human peripheral vestibular system. Normative vestibular hair cell data. *Ann Otol Rhinol Laryngol Suppl* **181:** 3–13.
11. Zheng JL, Keller G, Gao WQ. (1999) Immunocytochemical and morphological evidence for intracellular self-repair as an important contributor to mammalian hair cell recovery. *J Neurosci* **19:** 2161–2170.
12. Forge A, Li L, Corwin JT, Nevill G. (1993) Ultrastructural evidence for hair cell regeneration in the mammalian inner ear. *Science* **259:** 1616–1619.
13. Warchol ME, Lambert PR, Goldstein BJ, *et al.* (1993) Regenerative proliferation in inner ear sensory epithelia from adult guinea pigs and humans. *Science* **259:** 1619–1622.
14. Martinez-Monedero R, Oshima K, Heller S, Edge AS. (2007) The potential role of endogenous stem cells in regeneration of the inner ear. *Hear Res* **227:** 48–52.
15. Li H, Liu H, Heller S. (2003) Pluripotent stem cells from the adult mouse inner ear. *Nat Med* **9:** 1293–1299.
16. Lou X, Zhang Y, Yuan C. (2007) Multipotent stem cells from the young rat inner ear. *Neurosci Lett* **416:** 28–33.
17. Oshima K, Grimm CM, Corrales CE, *et al.* (2007) Differential distribution of stem cells in the auditory and vestibular organs of the inner ear. *J Assoc Res Otolaryngol* **8:** 18–31.

18. Savary E, Hugnot JP, Chassigneux Y, *et al.* (2007) Distinct population of hair cell progenitors can be isolated from the postnatal mouse cochlea using side population analysis. *Stem Cells* **25:** 332–339.
19. Hu Z, Ulfendahl M. (2006) Cell replacement therapy in the inner ear. *Stem Cells Dev* **15:** 449–459.
20. Li H, Corrales CE, Edge A, Heller S. (2004) Stem cells as therapy for hearing loss. *Trends Mol Med* **10:** 309–315.
21. Breuskin I, Bodson M, Thelen N, *et al.* (2007) Strategies to regenerate hair cells: identification of progenitors and critical genes. *Hear Res.* [EPUB ahead of print].
22. Nakagawa T, Ito J. (2004) Application of cell therapy to inner ear diseases. *Acta Otolaryngol Suppl* **551:** 6–9.
23. Martinez-Monedero R, Edge AS. (2007) Stem cells for the replacement of inner ear neurons and hair cells. *Int J Dev Biol* **51:** 655–661.
24. Hudspeth AJ. (2000) Hearing. In: Kandel ER, Schwartz JH, Jessell TM (eds.), *Principles of Neural Science*, New York, McGraw-Hill, pp. 590–613.
25. Malgrange B, Thiry M, Van De Water TR, *et al.* (2002) Epithelial supporting cells can differentiate into outer hair cells and Deiters' cells in the cultured organ of Corti. *Cell Mol Life Sci* **59:** 1744–1757.
26. Zhai S, Shi L, Wang BE, *et al.* (2005) Isolation and culture of hair cell progenitors from postnatal rat cochleae. *J Neurobiol* **65:** 282–293.
27. Rask-Andersen H, Bostrom M, Gerdin B, *et al.* (2005) Regeneration of human auditory nerve. *In vitro/in video* demonstration of neural progenitor cells in adult human and guinea pig spiral ganglion. *Hear Res* **203:** 180–191.
28. Izumikawa M, Minoda R, Kawamoto K, *et al.* (2005) Auditory hair cell replacement and hearing improvement by Atoh1 gene therapy in deaf mammals. *Nat Med* **11:** 271–276.
29. Kawamoto K, Ishimoto S, Minoda R, *et al.* (2003) *Math1* gene transfer generates new cochlear hair cells in mature guinea pigs *in vivo*. *J Neurosci* **23:** 4395–4400.
30. Bermingham NA, Hassan BA, Price SD, *et al.* (1999) *Math1*: an essential gene for the generation of inner ear hair cells. *Science* **284:** 1837–1841.
31. Staecker H, Praetorius M, Baker K, Brough DE. (2007) Vestibular hair cell regeneration and restoration of balance function induced by *Math1* gene transfer. *Otol Neurotol* **28:** 223–231.
32. Iguchi F, Nakagawa T, Tateya I, *et al.* (2004) Surgical techniques for cell transplantation into the mouse cochlea. *Acta Otolaryngol Suppl* **551:** 43–47.
33. Jero J, Tseng CJ, Mhatre AN, Lalwani AK. (2001) A surgical approach appropriate for targeted cochlear gene therapy in the mouse. *Hear Res* **151:** 106–114.
34. Kawamoto K, Oh SH, Kanzaki S, *et al.* (2001) The functional and structural outcome of inner ear gene transfer via the vestibular and cochlear fluids in mice. *Mol Ther* **4:** 575–585.

35. Staecker H, Li D, O'Malley BW, Jr, Van De Water TR. (2001) Gene expression in the mammalian cochlea: a study of multiple vector systems. *Acta Otolaryngol* **121**: 157–163.

36. Ito J, Kojima K, Kawaguchi S. (2001) Survival of neural stem cells in the cochlea. *Acta Otolaryngol* **121**: 140–142.

37. Parker MA, Corliss DA, Gray B, *et al.* (2007) Neural stem cells injected into the sound-damaged cochlea migrate throughout the cochlea and express markers of hair cells, supporting cells, and spiral ganglion cells. *Hear Res* **232**: 29–43.

38. Iguchi F, Nakagawa T, Tateya I, *et al.* (2003) Trophic support of mouse inner ear by neural stem cell transplantation. *Neuroreport* **14**: 77–80.

39. Shinohara T, Bredberg G, Ulfendahl M, *et al.* (2002) Neurotrophic factor intervention restores auditory function in deafened animals. *Proc Natl Acad Sci USA* **99**: 1657–1660.

40. Tamura T, Nakagawa T, Iguchi F, *et al.* (2004) Transplantation of neural stem cells into the modiolus of mouse cochleae injured by cisplatin. *Acta Otolaryngol Suppl* **551**: 65–68.

41. Tateya I, Nakagawa T, Iguchi F, *et al.* (2003) Fate of neural stem cells grafted into injured inner ears of mice. *Neuroreport* **14**: 1677–1681.

42. Hu Z, Wei D, Johansson CB, *et al.* (2005) Survival and neural differentiation of adult neural stem cells transplanted into the mature inner ear. *Exp Cell Res* **302**: 40–47.

43. Sekiya T, Holley MC, Kojima K, *et al.* (2007) Transplantation of conditionally immortal auditory neuroblasts to the auditory nerve. *Eur J Neurosci* **25**: 2307–2318.

44. Jiang Y, Jahagirdar BN, Reinhardt RL, *et al.* (2002) Pluripotency of mesenchymal stem cells derived from adult marrow. *Nature* **418**: 41–49.

45. Oh S, Jurney WM, Ortiz-Gonzalez XR, *et al.* (2003) Survival and distribution of adult-derived stem cells transplanted into the adult mouse inner ear. *Soc Neurosci Abstr.*

46. Li H, Roblin G, Liu H, Heller S. (2003) Generation of hair cells by stepwise differentiation of embryonic stem cells. *Proc Natl Acad Sci USA* **100**: 13495–13500.

47. Du X, Jensen P, Goldowitz D, Hamre KM. (2007) Wild-type cells rescue genotypically Math1-null hair cells in the inner ears of chimeric mice. *Dev Biol* **305**: 430–438.

48. Rivolta MN, Li H, Heller S. (2006) Generation of inner ear cell types from embryonic stem cells. *Methods Mol Biol* **330**: 71–92.

49. Shi F, Corrales E, Edge A. (2007) Sensory neurons produced by induction of human ES cells with BMP4: engraftment in the organ of Corti. *Assoc Res Otolaryngol Abstr.*

50. Naito Y, Nakamura T, Iguchi F, *et al.* (2003) Transplantation of autologous mesenchymal stem cells into the cochlea of the chinchilla. *Assoc Res Otolaryngol Abstr.*

51. Lang H, Ebihara Y, Schmiedt RA, *et al.* (2006) Contribution of bone marrow hematopoietic stem cells to adult mouse inner ear: mesenchymal cells and fibrocytes. *J Comp Neurol* **496:** 187–201.
52. Delprat B, Ruel J, Guitton MJ, *et al.* (2005) Deafness and cochlear fibrocyte alterations in mice deficient for the inner ear protein otospiralin. *Mol Cell Biol* **25:** 847–853.
53. Battey JF. (2007) Stem cells: current challenges and future promise. *Dev Dyn* [EPUB ahead of print].
54. Batts SA, Raphael Y. (2007) Transdifferentiation and its applicability for inner ear therapy. *Hear Res* **227:** 41–47.
55. Fritzsch B, Beisel KW, Hansen LA. (2006) The molecular basis of neurosensory cell formation in ear development: a blueprint for hair cell and sensory neuron regeneration? *Bioessays* **28:** 1181–1193.

STEM CELLS FOR LIVER AND KIDNEY

16

Renal Stem Cells

Sandeep Gupta* and Mark E. Rosenberg

1. Introduction

Stem cells participate in tissue formation, maintenance, and restoration. Imbalance between loss of mature cells and replacement by functional cells is potentially responsible for many degenerative diseases and may be an important pathophysiologic factor in the partial recovery following organ injury. Stem cells are viewed as potential therapeutic tools for tilting this balance between supply and demand of cells in favor of regeneration.

As reviewed in this book, stem cells and their niches have been defined in many organs. Considerable controversy still exists as to whether there are stem cells present in the adult kidney. Interest in adult kidney stem cells has been kindled by a number of factors, including the quest for understanding and manipulating the regenerative response of the kidney following acute injury, and the recent successes in cellular therapy of tubular and glomerular diseases. In this article we will review the current state of knowledge regarding adult kidney stem cells, beginning with lessons learned from kidney development and progressing through current attempts to isolate and characterize adult kidney stem cells and the use of these and other stem cells in the therapy of kidney disease.

2. Stem Cells and Kidney Development

The kidney, as well as much of the urogenital system, is derived from the intermediate mesoderm of the embryo. Kidney development proceeds

*Correspondence: Division of Renal Diseases and Hypertension, Department of Medicine, University of Minnesota, 516 Delaware Street SE, Box 736, Minneapolis, MN 55455, USA. Tel: (612) 624-9444; Fax: (612) 626-3840; E-mail: gupta024@umn.edu

through three stages of development named the pronephros, mesonephros, and metanephros. The permanent metanephric kidney is formed through a series of reciprocal interactions between the ureteric bud and the metanephric mesenchyme that induces mesenchymal-to-epithelial transformation (MET) and eventually nephron formation. The molecular mechanisms controlling these interaction have been well characterized and the subject of excellent reviews.[1,2]

Stem cells exist in the metanephric mesenchyme, which is the tissue of origin for most of the structures of the mature kidney including glomeruli, tubules except for the collecting duct, the interstitium, and vasculature.[3,4] The ureteric bud gives rise to the collecting duct and ureter. Nephrogenesis is complete in humans by 36 weeks of gestation. In the rodent kidney nephrogenesis persists for several weeks after birth in the more peripheral nephrogenic zone of the kidney. Renal stem cells persist in the adult kidneys of other organisms, such as the skate and the fresh water teleost. These cells can participate in new nephron formation following partial nephrectomy.[5-7] The existence of stem cells in the adult mammalian kidney remains controversial, and there is no evidence to suggest that new nephron formation can occur in adult life.

Knowledge derived from kidney development and lower organisms is contributing in a number of ways to the study of adult kidney stem cells. The molecular signature of developing renal cells is being used to define stem cell markers that can be used to identify and track adult stem cells For example, CD24 and cadherin 11 are two cell surface proteins that are strongly expressed by uninduced metanephric mesenchyme.[8] Pathways involved in cell lineage progression in the metanephric kidney have provided important clues as to how to manipulate embryonic and adult stem cells to undergo renal cell lineage progression. *In vitro* nephrogenesis can occur when metanephric kidneys are grown in organ culture. This culture system has been used as a readout in defining the differentiation potential of injected stem cells. Finally, both the metanephric mesenchyme and intact metanephric kidneys are being used in experimental studies for the treatment of adult kidney diseases.[9,10]

Embryonic stem (ES) cells can form renal structures under certain circumstances. For example, ES cells can form mesonephric ducts or ureteric buds in teratomas.[11] ES cells can form renal structures when injected into cultured metanephric kidneys or into the kidneys of cultured mouse embryos.[12,13] Preconditioning strategies have been used to enhance the renal developmental potential of ES cells. Treatment of murine ES cells with activin A, retinoic acid, and BMP-7 results in close to 100% incorporation of cells into renal tubules in kidney culture experiments.[14] ES cells engineered to express Wnt4 can develop into mature tubules when injected into cultured metanephric kidneys.[15] However, a major issue even beyond ethics in the use of ES cells for therapeutic purposes is the risk of teratoma formation.

3. Normal Cell Turnover in the Kidney

The mature kidney, composed of at least 26 differentiated cell types, is generally viewed as an organ with low cell turnover. The percentage of proliferating cells (proliferation index), estimated by counting mitotic figures, was found to be 0.1% for proximal tubular cells.[16] Normal cell turnover has also been estimated by counting the number of cells in the adult kidney stained with such proliferation markers as proliferating cell nuclear antigen (PCNA) or Ki-67.[17,18] In these studies the proliferative index ranges between 0.22% for proximal tubular epithelium and 0.33% for distal tubular epithelium. Mesangial cell turnover is more rapid, with an estimated 1% of cells being renewed daily.[19] Despite what may seem like low cell turnover, studies have estimated that approximately 70,000 renal tubular epithelial cells are lost into the urine every hour.[20] The cell source for replacing lost kidney cells remains to be defined. In contrast to the low cell turnover seen in the uninjured kidney, a robust regenerative response occurs following acute kidney injury.

4. Acute Renal Failure

Acute renal failure (ARF), a common complication of many medical conditions, results in significant morbidity and mortality. Development of ARF is associated with increased short term complications and may contribute to long term loss of renal function.[21,22] The incidence of ARF increases substantially as the severity of illness during hospitalization increases. Outcomes after an episode of severe ARF are poor. Hospital mortality is overall greater than 50% in individuals who develop postoperative ARF requiring dialysis.[23] Much of this increased mortality is attributable to confounding factors. However, even after multivariate adjustment, ARF is an independent risk factor for mortality.[21] Less severe episodes of ARF are also associated with an increased short term adverse event rate.[24-27]

There is a growing body of evidence suggesting that episodes of ARF are not fully reversible and may lead to the development of chronic and progressive renal dysfunction.[23,28-31] Part of the reason for this poor outcome may be related to an inadequate regenerative response.

4.1. Renal regeneration following acute renal failure

Toxic and ischemic insults to the kidney lead to ARF, most often manifested as acute tubular necrosis (ATN). Loss of injured or dead cells leads to denudation of tubular basement membrane with sloughed cells and cellular debris filling tubular lumens (Fig. 1). Following injury the kidney undergoes a robust regenerative response leading to recovery of renal function. New cells are required to replace damaged cells. Three possible sources of new tubular cells are adjacent less damaged tubular cells, extrarenal cells presumably of bone marrow origin that home to the injured kidney, and resident renal stem cells (Fig. 2). There is evidence to

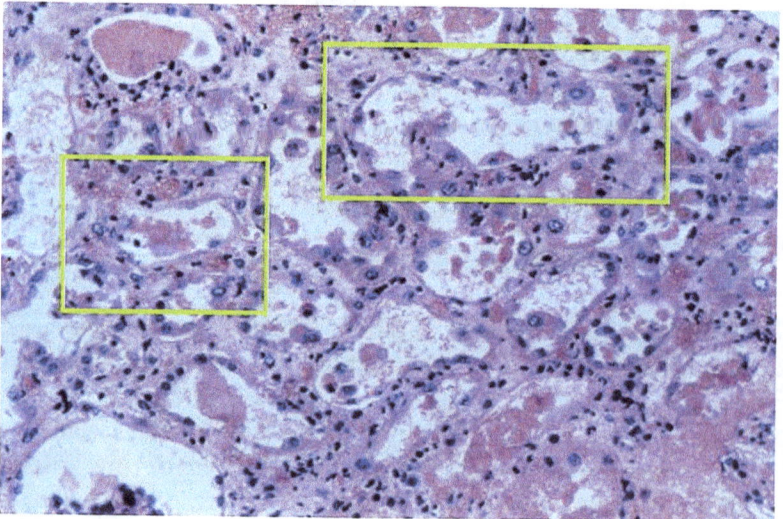

Figure 1. Histology of the kidney following acute renal failure, showing large areas of denuded basement membrane secondary to epithelial cell loss (boxes). Tubular lumens contain sloughed cells and cell debris. The source of cells that replace these lost cells has been a major stimulus for the study of renal stem cells.

Possible cellular sources of
regenerating tubules

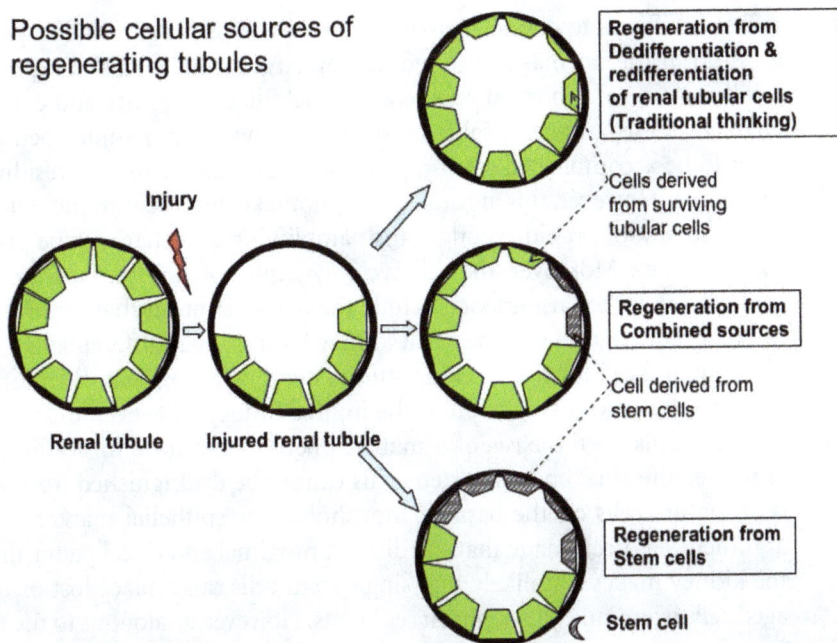

Figure 2. Schematic diagram showing possible cellular sources for regenerating renal tubules.

support a role for less injured tubular cells. Recapitulating developmental paradigms, these cells dedifferentiate, proliferate, and eventually reline denuded tubules, restoring the structural and functional integrity of the kidney.[17,32–36] Molecular events defining this regenerative response have been characterized, and strategies to accelerate the repair process tested both in experimental models and in humans.[17,32–38]

There is evidence that bone marrow-derived cells can migrate to the kidney and form tubular epithelial cells following acute renal injury (discussed below).[39–43] However, the *in vivo* contribution of extrarenal cells to kidney regeneration is minimal.[44–46]

5. Adult Renal Stem Cells

There is ongoing debate and controversy as to whether stem cells exist in the adult kidney, and if they do, what role stem cells play in the regenerative response following ARF. The lack of a definitive marker for kidney stem cells makes it difficult to define a renal stem cell niche, or follow cell lineage progression in the normal or injured kidney. Arguments against a significant role for stem cells in renal regeneration include the rapidity and extent of DNA synthesis and cell proliferation that follows kidney injury being too great to be accomplished by a small number of adult stem cells residing in the kidney. However, this argument does not take into account the potential presence of more rapid cycling transit amplifying cells that may be present in the kidney. Moreover, the cell cycle dynamics of stem cells is far from understood. Other arguments include the demonstration that proliferating or label-retaining cells in the adult kidney have normal differentiated morphology, and cell lineage tracking studies demonstrating that differentiated cells can undergo cell division in the injured kidney. However, differentiating stem cells on the basis of a mature phenotype can be misleading. For example, intestinal epithelial stem cells cannot be distinguished from adjacent mature cells on the basis of morphology or epithelial markers.[47] The argument has been made that the distinct proximal and distal patterning of the kidney makes it unlikely that single stem cells can replace lost or damaged cells from multiple nephron segments. However, analogous to the findings for the hematopoietic system, the existence of multiple committed progenitor cells in the adult kidney is a likely possibility. The concept of multiple stem cells is supported by the number of different cells isolated by different groups using the selection methods discussed below.

Arguments favoring the presence of adult renal stem cells include the regenerative capacity of the kidney, the presence of renal stem cells in lower organisms, the ability of label-retaining cells to proliferate, and the isolation of stem cells from the adult kidney. These latter studies will be summarized below focusing on isolation methods, the phenotype of isolated cells, the differentiation potential of the cells, and the therapeutic efficacy of the cells in models of renal injury. Tables 1 and 2 provide a summary of this information.

Candidate adult renal stem cells have been isolated using four different selection strategies that have been used to isolate stem cells from other organs. The first method takes advantage of the fact that stem cells are slow cycling cells. Therefore, when the DNA of the cells is labeled with a marker such as bromodeoyuridine (BrdU), the cells retain the label for a long period of time. This label retention can be used to identify and isolate putative stem cells. The second method of isolating side population (SP) cells takes advantage of the fact that SP cells extrude Hoechst dye through the activity of multidrug resistance proteins on the cell surface that are part of the ATP-binding cassette (ABC) transporter superfamily.

Table 1. Summary of Studies Isolating Renal Stem Cells

Study	Isolation Method	Species	Cells Cultured	Stem Cell Markers	Other Markers	Location	Differentiation into Kidney Tubules	Differentiation into Other Lineages
Oliver[50]	Label retaining	Rat	Yes			Papilla	Yes	Neuron
Maeshima[52]	Label retaining	Rat	Yes			Proximal tubule	Yes	Not tested
Iwatani[53]	Side population	Rat	No	Sca-1	C-kit, CD45	Proximal tubule	No	Heme, muscle, liver
Hishikawa[54]	Side population	Mouse	No	Sca-1	Musculin/MyoR	Interstitium	Not tested	Not tested
Challen[55]	Side population	Mouse	No	Sca-1, CD24	Edoglin/CD105	Tubule	Yes	Osteocyte, adipocyte
Bussolati[56]	Marker	Human	Yes	CD133	Pax-2, CD44	Interstitium	Yes	Endothelium
Sagrinati[57]	Marker	Human	Yes	Oct4, CD24, CD133, Bml1		Glomerular parietal epithelium	Yes	Osteocyte, adipocyte, neuron
Dekel[58]	Marker	Mouse	Yes	Sca1		Interstitium of papilla	Not tested	Osteocyte, adipocyte, neuron
Kitamura[59]	Culture	Rat	Yes	Sca1, Musashi-1	Pax-2, Wt-1, Wnt-4	Proximal tubule	Yes	Not tested
Gupta[60]	Culture	Rat	Yes	Oct4, Rex1	Pax-2, CD44	Proximal tubule	Yes	Neuron, liver endothelium

Table 2. Studies of Renal Stem Cells in Acute Kidney Injury

Study	Injury Model	Incorporation of Cells into Tubules	Improved Function
Oliver[50]	Ischemia/reperfusion	Yes	Not tested
Maeshima[52]		Not tested	Not tested
Iwatani[53]	Glomerulonephritis, gentamicin	No	Not tested
Hishikawa[54]	Cisplatin	No	Yes
Challen[55]	Glomerulonephritis	Possibly	Yes
Bussolati[56]	Rhabdomyolysis	Yes	
Sagrinati[57]	Rhabdomyolysis	Yes	Yes
Dekel[58]	Ischemia/reperfusion	Yes	Not tested
Kitamura[59]	Ischemia/reperfusion	Yes	Yes
Gupta[60]	Ischemia/reperfusion	Yes	Yes

SP cells (Hoechst low cells) isolated from many different organs contain multipotent stem cells.[48] The third method for isolating renal stem cells uses specific cell surface markers that have been used to identify stem cells from other organs or the metanephric kidney. The markers used to isolate renal stem cells include CD133, stem cell antigen-1 (Sca-1), and CD24. The fourth method uses culture conditions that have been successful in selecting stem cells in other organ systems.

5.1. Label-retaining cells

Oliver and colleagues have isolated cells from the renal papilla of young mice and rats that are slow-cycling cells and have characteristics of renal stem cells.[49] When grown in culture these papillary cells express epithelial and mesenchymal markers, form cellular spheres, and display some evidence of plasticity with differentiation into neurons under appropriate culture conditions. Following *in vivo* ischemic renal injury, these cells proliferate.

Maeshima *et al.* identified a population of cells in the adult rat kidney scattered among renal tubular cells.[50] These cells were identified as label-retaining cells, and were found predominantly in proximal tubules. Following renal ischemia, these label-retaining cells undergo proliferation, and progeny of these cells initially express vimentin, a mesenchymal cell marker, and later become positive for E-cadherin, an epithelial cell marker. The cells that have been subsequently isolated demonstrate plasticity, and can be integrated into the developing kidney.[51]

5.2. **Side population cells**

Three different groups isolated and characterized kidney SP cells. Iwatani *et al.* isolated SP cells from adult rat kidneys that comprised 0.03–0.1% of the cells of the digested kidney.[52] These cells do not participate in kidney regeneration following experimental glomerulonephritis or tubular injury induced by gentamicin. Hishikawa *et al.* isolated SP cells from adult mouse kidneys that expressed Musculin/MyoR, a transcription factor found in skeletal muscle precursors.[53] These cells localize to the renal interstitium. Number of cells decreases following acute renal failure, and infusion of the cells is associated with expression of renal protective factors and improved kidney function in an acute model of renal injury, but not in chronic renal disease. Challen *et al.* isolated SP cells from adult mouse kidneys that comprised 0.14% of the kidney cells.[54] These cells do not express Musculin/MyoR, have multilineage differentiation potential, and are heterogenous, including the presence of a monocytic component. Interestingly, they express genes involved in Notch signaling. Renal protective paracrine effects are observed following infusion of the cells in ARF.

5.3. **Candidate markers used to isolate renal stem cells**

Bussolati *et al.* isolated and cultured a population of cells from adult human kidneys using CD133 as a selection marker.[55] These cells can be differentiated *in vitro* and *in vivo* into epithelial and endothelial cells, can form tubules and vessels, and express early and late nephron markers. Injection of the cells three days after myoglobinuric ARF results in some incorporation of these human cells into mouse tubules. These cells have limited self-renewal properties and express HLA Class I antigens.

Sagrinati *et al.* used CD 24 and CD133 to select a subset of parietal glomerular epithelial cells that have a high self-renewal potential and can be differentiated into renal tubular cells, as well as osteocytes and adipocytes.[56,57] Injection of these cells into SCID mice with myoglobinuric ARF results in incorporation of injected cells into regenerating tubules and improved renal function.

Dekel *et al.* used Sca-1 magnetic cell sorting to isolate nontubular Sca-1 positive cells from mouse kidneys.[57] The cells are present in the interstitium of the kidney, and can differentiate into myogenic, osteogenic, adipogenic, and neural lineages. When injected into the renal parenchyma in a model of ARF, the cells adopt a renal phenotype.

5.4. Selective culture conditions

Kitamura isolated a population of rapidly proliferating cells from microdissected proximal tubules that express the stem cell markers Sca-1 and Musashi-1 as well as early nephron markers.[58] The cells can differentiate into mature tubular cells in culture. These cells have a triploid karyotype, although they do not undergo tumor formation in nude mice.

Gupta *et al.* isolated a unique population of cells from rat kidneys that were called multipotent renal progenitor cells (MRPCs).[59] Features of these cells include: spindle-shaped morphology; self-renewal for over 200 population doublings without evidence for senescence; normal karyotype and DNA analysis; expression of vimentin, CD90 (thy1.1), Pax-2, and Oct4 but not cytokeratin, MHC class I or II, or other markers of more differentiated cells (Fig. 3). MRPC exhibit plasticity, demonstrated by the ability of the cells to be induced to express endothelial, hepatocyte, and neural markers by Q-RT-PCR and immunohistochemistry. The cells can differentiate into renal tubules when injected under the capsule of an

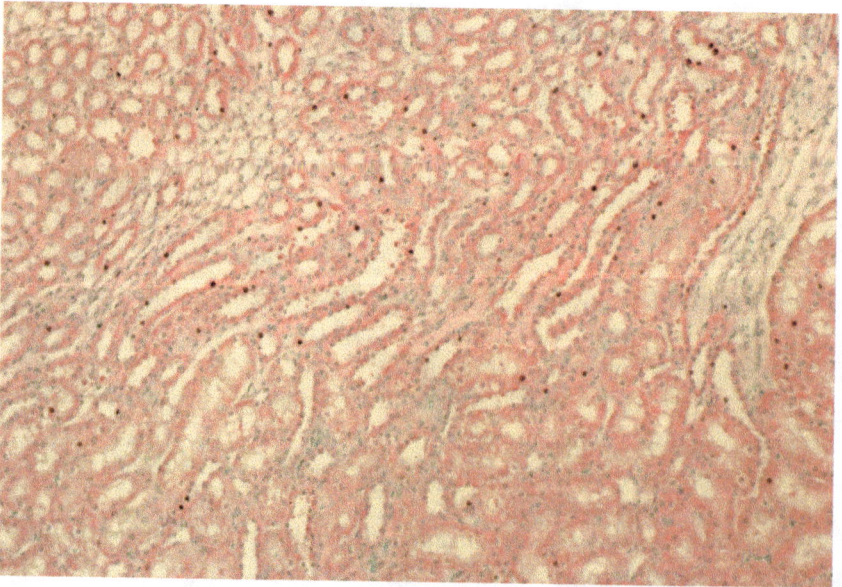

Figure 3. Expression of Oct4 in the normal kidney. Oct4 immunostaining (brown nuclear staining) was observed in occasional cells associated primarily with renal proximal tubules. The section is taken from the cortical-medullary junction where staining was most prominent.

uninjured kidney or intra-arterially following renal ischemia-reperfusion injury.

5.5. Summary

Differences in the cells isolated in these studies may be due to different selection markers, species of origin, age of the kidneys, and culture conditions. The different cells have been localized to multiple sites in the kidney (Table 1) including tubular and interstitial cells of the renal papilla, the proximal tubule, the cortical interstitium, and the parietal epithelial cell of the glomerulus. Different markers have been used to confirm these localizations, although in many cases it remains a challenge to confirm that the cells isolated are the same as the *in vivo* cells. Despite this, the proximal tubule was a site for the stem cells in many of the studies. Interestingly, the proximal tubule is the site of greatest injury in ARF. Hence, it is logical to house stem cells at the site of maximal demand.

The isolated cells express a number of stem cell markers, including Sca-1, Oct4, Rex1, CD24, and CD133. The marker expressed by many of the cells was Sca1. Sca-1 is a member of the Ly-6 family and is one of the earliest cell surface markers of hematopoietic stem cells, but has since been found to be expressed in bone marrow-derived multipotential cells, and skeletal muscle satellite cells, and is also present on mature cells. Oct4 (also referred to as POU5F1) is a POU (Pit-Oct-Unc) domain transcription factor expressed in embryonic stem cells, as well as in primordial germ cells and adult gonads.[60–63] It plays a critical role in maintaining the pluripotency of embryonic stem cells and the viability of primordial germ cells.[61,63–65] Differentiation of embryonic stem cells is associated with downregulation of Oct4.[66,67] Oct4 expression has been demonstrated in stem cells isolated from umbilical cord blood, bone marrow, hair follicles, muscle, skin, breast, pancreas, liver, and amniotic fluid, as well as endothelial progenitor cells and neural stem cells.[60,68–81] These markers will be useful for localizing a renal stem cell niche and can be used to study cell lineage progression in the normal and the injured kidney.

6. Cellular Therapy of Renal Disease

Use of cellular units for therapeutic purposes is referred to as "cellular therapy." Using stem cells for cellular therapy has generated renewed excitement and enthusiasm in the medical community caring for patients with limited therapeutic options, such as in kidney, brain, cardiac, and degenerative diseases. Two cardinal characteristics of stem cells, namely self-renewal and plasticity, make them particularly attractive agents for cellular therapy. Self-renewal makes it possible to expand and maintain stem cells as cell banks for future therapeutic use. Plasticity raises the therapeutic possibility of using stem cells derived from one source for treatment of disease in another tissue type.

Both renal and extrarenal stem cells can repopulate existing renal scaffolds vacated by lost cells. However, generation of new nephrons, as seen in other organisms, likely requires the presence of the primitive kidney specific stem cells. An extreme example of cellular therapy would be formation of complete kidneys utilizing multipotent stem cells. However, this is not foreseeable in the near future, as our understanding of stem cell biology is in its infancy. However, based on current knowledge, stem cells can be potentially utilized for replacement of damaged cells, delivery of genes and other bioactive substances, and as a cell source for bioartificial dialyzers. Exogenous administration of stem cells leads to homing to the site of injury, making these cells potentially useful for targeted delivery of cellular units, genes, and proteins.

6.1. Therapy of acute renal failure

Stem cells derived from both renal and extrarenal sources have been used for the cellular therapy of kidney diseases. Initial proof-of-principle evidence for participation of extrarenal cells in kidney regeneration came from sex mismatched bone marrow and kidney transplant recipients demonstrating renal differentiation of male cells in female kidneys.[39,40,82] Stem cells of extrarenal origin can be incorporated into tubules,[40–42] glomeruli,[39,83–85] interstitium,[82] and vasculature[82] of the kidney. To further characterize the subpopulation of extrarenal cells participating in renal repair, fractionated bone marrow and pure populations of bone marrow-derived stem cells have been utilized for renal repair with conflicting outcomes. Significant incorporation of whole[39,85] and fractionated bone marrow[42,86] hematopoietic stem cells (HSCs),[41] and mesenchymal stem cells (MSCs)[87] into injured kidneys was demonstrated by earlier investigators. Kale *et al.* demonstrated incorporation of Lin$^-$ Sca-1$^+$ c-kit$^+$ cells in 2–5% of renal tubules following ischemic injury.[42] Similarly, Lin *et al.* demonstrated significant incorporation of HSCs in renal tubules following ischemic injury.[41] Morigi *et al.*, on the contrary, did not find incorporation or functional benefit following HSC administration, while MSCs were incorporated into renal tubules and led to significant functional improvement.[87]

These initial studies showing incorporation of extrarenal cells into the kidney were encouraging. However, many of the subsequent studies did not support these results. Variable outcomes could be due to differences in the quality of stem cells, the stringency of evaluation, the renal injury model, the method and timing of stem cell administration, and the species of the animal studied. The best of the studies show no, or at best minimum, incorporation of extrarenal stem cells into the kidney following injury.[43,44,46,88] Apart from occasional cells in peritubular capillaries, Duffield *et al.* did not find incorporation of bone marrow-derived stem cells in ischemia-reperfusion or unilateral ureteric obstruction models of renal injury despite multiple methods of detection.[44] Bone marrow-derived cells were identified either by LacZ, GFP, or by the presence of the Y chromosome in a female recipient of male bone marrow transplantation. Lin *et al.* detected a small number of bone marrow-derived epithelial cells 28 days after ischemia reperfusion injury.[46] The majority of cells repairing the kidney were derived from intrarenal cells. Wagers *et al.* did not find significant incorporation of lin⁻Sca-1⁺Thy1.1ˡᵒ GFP-HSC in the uninjured kidneys of lethally irradiated mice even after 9 months of follow-up.[88] Although injury is proposed to increase the likelihood of stem cell incorporation, Szczypka *et al.* did not see incorporation of bone-derived cells in a folic acid model of acute renal failure.[43]

Endothelial progenitor cells (EPCs) are bone marrow-derived cells that play an important role in the pathophysiology of kidney diseases and associated cardiovascular morbidity.[89,90] Endothelial dysfunction is directly responsible for many kidney diseases, such as radiation nephritis, HUS-TTP, calcineurin inhibitor toxicity, vaculitis, and acute ischemic renal injury. Apart from their role in kidney disease, EPCs are involved in the pathophysiology of such systemic diseases as type 1 diabetes.[91] They exert significant paracrine effects following exogenous administration by upregulating cytokines such as IL-8 that improve proliferation of adjacent mature endothelial cells during recovery.[92] Acute kidney injury results in damage to the tubular epithelial and vascular endothelial cells. Blood supply to the injured kidneys is further compromised due to endothelial cell swelling. EPCs have been shown to provide protection in models of ARF. These cells are rapidly mobilized from the bone marrow and home to the injured kidney.[93] Furthermore, exogenous administration of EPC provides renal protection.[94]

The unique microenvironment present in the kidney, or the need for kidney-specific stem cells, might be responsible for the limited incorporation of extrarenal stem cells seen in adult kidneys. The kidney is one of the few organs that undergo mesenchymal-to-epithelial transformation (MET) during the process of nephrogenesis, which may pose an additional challenge for the incorporation of extrarenal stem cells. Kidney-specific stem cells might be at an advantage in accomplishing MET. In support of this concept, many investigators have shown incorporation of exogenously administered kidney specific stem cells into the adult kidney[49,55,56,58,59] and developing metanephric mesenchyme.[51] Many of these studies also showed functional and structural improvement following administration of cells in models of acute kidney injury.[56,59]

6.1.1. *Functional improvement*

Stem cells can potentially improve organ function by replacement of damaged cells, or by facilitating recovery of endogenous cells. The latter phenomenon is often referred to as "paracrine effect." Either transdifferentiation or cell fusion is a likely mechanism for stem cell incorporation. During transdifferentiation, stem cells mature into a functional cell type, while during fusion stem cells fuse and acquire the phenotype of a mature cell. The transdifferentiation potential of stem cells might be more limited than was initially proposed. Many investigators have now demonstrated fusion as the likely explanation for stem cell plasticity. Fusion, though a compromise, cannot be entirely discounted as a way of improving organ function. In a model of tyrosinaemia type I, mice with mutations in the fumarylacetoacetate hydrolase gene ($Fah^{-/-}$) regained normal liver structure and function by fusion with $Fah^{+/+}$ bone marrow cells.[95,96] Although fusion is an attractive alternate mechanism for plasticity, all cell types and organs are not equally suited to undergo fusion. For example, stem cells are more suited than fibroblast cells, and the liver, brain, and heart are more suited than the kidney for fusion.[97]

Many investigators have shown the functional benefit of stem cell therapy in models of renal injury such as ischemia-reperfusion and cisplatin-induced acute renal failure.[42,87,98] Stem cells have also been demonstrated to facilitate tissue regeneration by paracrine mechanisms.[92] Togel *et al.* demonstrated improvement in renal function and histology by paracrine effects of MSC infusion in a rat model of ischemia reperfusion.[99] MSC infusion resulted in upregulation of anti-inflammatory cytokines and downregulation of proinflammatory cytokines. Given the more rigorous recent studies showing minimal incorporation of infused stem cells, the current view is that it is the paracrine effects of infused cells that are primarily responsible for the functional improvement seen in renal injury models.

6.2. Therapy of glomerular disease

Mesangial cells are one of the three cell types residing in the glomerulus of the kidney. These cells are involved in the pathophysiology of many glomerular disorders, including diabetic nephropathy, IgA nephropathy, and membranoproliferative glomerulonephritis. Early studies demonstrated the ability of bone marrow cells to reconstitute mesangial cells.[86] Masuya *et al.* further defined the origin of the bone marrow cell responsible for this effect by demonstrating that clonally expanded hematopoietic stem cells (Lin$^-$ Sca-1$^+$, c-kit$^+$, CD34$^-$) can differentiate into mesangial cells in lethally irradiated mice.[83] Pathophysiologic relevance for the bone marrow-to-mesangial cell transformation comes from studies demonstrating that bone marrow transplantation can attenuate glomerular injury in mouse models of IgA nephropathy.[100]

The role of bone marrow-derived cells has been studied in other models of glomerular disease. For example, glomerulosclerosis can be transmitted by mesangial cell progenitors derived from bone marrow.[102] In an anti-Thy1 model of glomerulonephritis, bone marrow cells contributed to the early proliferating mesangial cell population and persisted in the mesangium for at least eight weeks following induction of mesangial injury.[102] Injection of human bone marrow cells into diabetic mice resulted in the presence of injected cells in the glomerulus, particularly the glomerular endothelium.[103] This therapy was associated with decreased mesangial expansion and a decrease in macrophage infiltration. Alport syndrome is a human disease characterized by progressive kidney failure and caused by mutations in type IV collagen genes. In a mouse model of Alport syndrome due to deficiency of the COL4A3 chain of type IV collagen, transplantation of wild type bone marrow into irradiated COL4A3 knockout mice led to restoration of expression of the α3 chain of type IV collagen and improvement in the structure and function of the kidney.[104] In the same mouse model of Alport syndrome, Ninichuk and colleagues demonstrated that five-weekly injections of mesenchymal stem cells decreased interstitial damage but did not result in cell incorporation or renal functional improvement.[105]

6.3. Safety concerns

Autologous stem cells can be genetically engineered to rectify defective genes, or for delivery of molecules of interest. The safety of modified stem cells is far from established. Transplantation of autologous hematopoietic stem cells that were genetically engineered to correct adenosine deaminase deficiency resulted in the development of acute leukemia due to genetic integration of the vector.[106] Abnormal electrical integration of the stem cells in organs such as the brain and the heart may result in a fatal outcome.[107,108] Such catastrophic outcomes due to abnormal electrical integration are fortunately unlikely in the kidney. However, engineered stem cells may lose control over the regulation of delivered molecules such as erythropoietin and vitamin D, resulting in an undesirable outcome. Autologous stem cells are not always available, or may not be the most suitable stem cells for the required therapeutic purpose. In such situations allogenic stem cells can be used. However, these cells may be immunogenic and require immunosuppressant medications. Although not substantiated by evidence so far, the use of adult stem cells can theoretically lead to cancer or teratoma formation, as has been seen with embryonic stem cells.

7. Future Directions

Despite significant advances in the field of renal stem cells, significant major issues remain unresolved. The fundamental question of whether stem cells exist in the adult kidney remains controversial. Many groups have isolated cells from the kidney with properties of renal progenitor cells, although the phenotype of many of these cells may be influenced by *in vitro* culture conditions. Nonetheless, the cells have provided important clues as to location, markers, and differentiation and therapeutic potential. Defining renal stem cell markers is critical for the field to move forward. Such markers will allow for the characterization of the kidney stem cell niche and the signaling pathways that control stem cell behavior in the normal and the injured kidney. The presence of stem cells in the adult kidney has important implications for our understanding of normal cell turnover in the kidney and the source of regenerating cells following ARF. Identification of renal stem cell markers will set the stage for genetic strategies to define such renal cell lineage progression.

The *in vitro* differentiation of renal stem cells can provide an important model system for studying the cell biology involved in nephron formation and for dissecting the role of specific factors in renal cell lineage progression. Understanding the process of mesenchymal-to-epithelial transformation (MET) may also lead to strategies for limiting or reversing renal interstitial fibrosis. During the process of fibrosis, epithelial cells undergo epithelial-to-mesenchymal transformation (EMT) leading to migration of cells from the tubule to the interstitial space and subsequent transformation into fibroblasts.[109–112] The potential to reverse this process through MET has exciting therapeutic potential for the treatment of chronic kidney disease. Stem cell depletion could also be a pathophysiologic factor in many chronic progressive renal diseases that have a prominent tubulointerstitial component (e.g. radiation nephritis).

ARF is associated with a poor short term prognosis and adverse long term effects on renal function. The lack of specific beneficial therapies has led investigators to explore strategies for both protecting against renal injury and accelerating the repair process.[113–117] In most cases, strategies targeting renal protection are dependent on knowing ahead of time when the injury event will occur. Therapeutic strategies directed at enhancing renal repair have the potential to limit the duration of ARF and to better preserve long term kidney function. Pharmacologic therapies directed at enhancing renal regeneration are limited in their scope and have been unsuccessful when applied to the treatment of human ARF.[118,119] On the other hand, cellular therapies can affect a number of different mechanisms, both autocrine and paracrine, and therefore have greater potential to enhance repair. Understanding the *in vivo* control of stem cell behavior may also lead to pharmacologic approaches to mobilizing renal stem cells and accelerating the replacement of injured or dying cells and thereby enhance the renal regenerative response to kidney injury.

References

1. Dressler GR. (2006) The cellular basis of kidney development. *Annu Rev Cell Dev Biol* 22: 509–529.
2. Schedl A, Hastie ND. (2000) Cross-talk in kidney development. *Curr Opin Genet Dev* 10: 543–549.

3. Herzlinger D, Koseki C, Mikawa T, al-Awqati Q. (1992) Metanephric mesenchyme contains multipotent stem cells whose fate is restricted after induction. *Development* **114**: 565–572.
4. Oliver JA, Barasch J, Yang J, *et al.* (2002) Metanephric mesenchyme contains embryonic renal stem cells. *Am J Physiol Renal Physiol* **283**: F799–809.
5. Drummond IA, Mukhopadhyay D, Sukhatme VP. (1998) Expression of fetal kidney growth factors in a kidney tumor line: role of FGF2 in kidney development. *Exp Nephrol* **6**: 522–533.
6. Elger M, Hentschel H, Litteral J, *et al.* (2003) Nephrogenesis is induced by partial nephrectomy in the elasmobranch *Leucoraja erinacea*. *J Am Soc Nephrol* **14**: 1506–1518.
7. Salice CJ, Rokous JS, Kane AS, Reimschuessel R. (2001) New nephron development in goldfish (*Carassius auratus*) kidneys following repeated gentamicin-induced nephrotoxicosis. *Comp Med* **51**: 56–59.
8. Challen GA, Martinez G, Davis MJ, *et al.* (2004) Identifying the molecular phenotype of renal progenitor cells. *J Am Soc Nephrol* **15**: 2344–2357.
9. Hammerman MR. (2004) Organogenesis of kidneys following transplantation of renal progenitor cells. *Transpl Immunol* **12**: 229–239.
10. Dekel B, Burakova T, Arditti FD, *et al.* (2003) Human and porcine early kidney precursors as a new source for transplantation. *Nat Med* **9**: 53–60.
11. Yamamoto M, Cui L, Johkura K, *et al.* (2006) Branching ducts similar to mesonephric ducts or ureteric buds in teratomas originating from mouse embryonic stem cells. *Am J Physiol Renal Physiol* **290**: F52–F60.
12. Steenhard BM, Isom KS, Cazcarro P, *et al.* (2005) Integration of embryonic stem cells in metanephric kidney organ culture. *J Am Soc Nephrol* **16**: 1623–1631.
13. Yokoo T, Ohashi T, Shen JS, *et al.* (2005) Human mesenchymal stem cells in rodent whole-embryo culture are reprogrammed to contribute to kidney tissues. *Proc Natl Acad Sci USA* **102**: 3296–3300.
14. Kim D, Dressler GR. (2005) Nephrogenic factors promote differentiation of mouse embryonic stem cells into renal epithelia. *J Am Soc Nephrol* **16**: 3527–3534.
15. Kobayashi T, Tanaka H, Kuwana H, *et al.* (2005) Wnt4-transformed mouse embryonic stem cells differentiate into renal tubular cells. *Biochem Biophys Res Commun* **336**: 585–595.
16. McCreight CE, Sulkin NM. (1959) Cellular proliferation in kidneys of young and senile rats following unilateral nephrectomy. *J Gerontol* **14**: 440–443.
17. Witzgall R, Brown D, Schwarz C, Bonventre JV. (1994) Localization of proliferating cell nuclear antigen, vimentin, c-Fos, and clusterin in the postischemic kidney, evidence for a heterogenous genetic response among

nephron segments, and a large pool of mitotically active and dedifferentiated cells. *J Clin Invest* **93**: 2175–2188.

18. Nadasdy T, Laszik Z, Blick KE, *et al.* (1994) Proliferative activity of intrinsic cell populations in the normal human kidney. *J Am Soc Nephrol* **4**: 2032–2039.
19. Pabst R, Sterzel RB. (1983) Cell renewal of glomerular cell types in normal rats: an autoradiographic analysis. *Kidney Int* **24**: 626–631.
20. Prescott LF. (1966) The normal urinary excretion rates of renal tubular cells, leucocytes and red blood cells. *Clin Sci* **31**: 425–435.
21. Chertow GM, Levy EM, Hammermeister KE, *et al.* (1998) Independent association between acute renal failure and mortality following cardiac surgery. *Am J Med* **104**: 343–348.
22. Nash K, Hafeez A, Hou S. (2002) Hospital-acquired renal insufficiency. *Am J Kidney Dis* **39**: 930–936.
23. Augustine JJ, Sandy D, Seifert TH, Paganini EP. (2004) A randomized controlled trial comparing intermittent with continuous dialysis in patients with ARF. *Am J Kidney Dis* **44**: 1000–1007.
24. Levy EM, Viscoli CM, Horwitz RI. (1996) The effect of acute renal failure on mortality: a cohort analysis. *JAMA* **275**: 1489–1494.
25. McCullough PA, Wolyn R, Rocher LL, *et al.* (1997) Acute renal failure after coronary intervention: incidence, risk factors, and relationship to mortality. *Am J Med* **103**: 368–375.
26. Thakar CV, Worley S, Arrigain S, *et al.* (2005) Influence of renal dysfunction on mortality after cardiac surgery: modifying effect of preoperative renal function. *Kidney Int* **67**: 1112–1119.
27. Thakar CV, Yared JP, Worley S, *et al.* (2003) Renal dysfunction and serious infections after open-heart surgery. *Kidney Int* **64**: 239–246.
28. Basile DP, Donohoe D, Roethe K, Osborn JL. (2001) Renal ischemic injury results in permanent damage to peritubular capillaries and influences long-term function. *Am J Physiol Renal Physiol* **281**: F887–F899.
29. Basile DP, Fredrich K, Alausa M, *et al.* (2005) Identification of persistently altered gene expression in the kidney after functional recovery from ischemic acute renal failure. *Am J Physiol Renal Physiol* **288**: F953–F963.
30. Pagtalunan ME, Olson JL, Tilney NL, Meyer TW. (1999) Late consequences of acute ischemic injury to a solitary kidney. *J Am Soc Nephrol* **10**: 366–373.
31. Bhandari S, Turney JH. (1996) Survivors of acute renal failure who do not recover renal function. *Quart J Med* **89**: 415–421.
32. Safirstein R, Price PM, Saggi SJ, Harris RC. (1990) Changes in gene expression after temporary renal ischemia. *Kidney Int* **37**: 1515–1521.
33. Safirstein R. (1994) Gene expression in nephrotoxic and ischemic acute renal failure. *J Am Soc Nephrol* **4**: 1387–1395.

34. Bacallao R, Fine LG. (1989) Molecular events in the organization of renal tubular epithelium: from nephrogenesis to regeneration. *Am J Physiol* **257**: F913–F924.

35. Safirstein R. (1999) Renal regeneration: reiterating a developmental paradigm. *Kidney Int* **56**: 1599.

36. Maeshima A, Maeshima K, Nojima Y, Kojima I. (2002) Involvement of Pax-2 in the action of activin A on tubular cell regeneration. *J Am Soc Nephrol* **13**: 2850–2859.

37. Imgrund M, Grone E, Grone HJ, *et al.* (1999) Re-expression of the developmental gene Pax-2 during experimental acute tubular necrosis in mice. *Kidney Int* **56**: 1423–1431.

38. Devarajan P, Mishra J, Supavekin S, *et al.* (2003) Gene expression in early ischemic renal injury: clues towards pathogenesis, biomarker discovery, and novel therapeutics. *Mol Genet Metab* **80**: 365–376.

39. Poulsom R, Forbes SJ, Hodivala-Dilke K, *et al.* (2001) Bone marrow contributes to renal parenchymal turnover and regeneration. *J Pathol* **195**: 229–235.

40. Gupta S, Verfaille C, Chmielewski D, *et al.* (2002) A role for extrarenal cells in the regeneration following acute renal failure. *Kidney Int* **62**: 1285–1290.

41. Lin F, Cordes K, Li L, *et al.* (2003) Hematopoietic cells contribute to the regeneration of renal tubules after ischemia-reperfusion injury in mice. *J Am Soc Nephrol* **14**: 1188–1199.

42. Kale S, Karihaloo A, Clark PR, *et al.* (2003) Bone marrow stem cells contribute to repair of the ischemically injured renal tubule. *J Clin Invest* **112**: 42–49.

43. Szczypka MS, Westover AJ, Clouthier SG, *et al.* (2005) Rare incorporation of bone marrow-derived cells into kidney after folic acid-induced injury. *Stem Cells* **23**: 44–54.

44. Duffield JS, Park KM, Hsiao LL, *et al.* (2005) Restoration of tubular epithelial cells during repair of the postischemic kidney occurs independently of bone marrow-derived stem cells. *J Clin Invest* **115**: 1743–1755.

45. Krause D, Cantley LG. (2005) Bone marrow plasticity revisited: protection or differentiation in the kidney tubule? *J Clin Invest* **115**: 1705–1708.

46. Lin F, Moran A, Igarashi P. (2005) Intrarenal cells, not bone marrow-derived cells, are the major source for regeneration in postischemic kidney. *J Clin Invest* **115**: 1756–1764.

47. Marshman E, Booth C, Potten CS. (2002) The intestinal epithelial stem cell. *Bioessays* **24**: 91–98.

48. Challen GA, Little MH. (2006) A side order of stem cells: the SP phenotype. *Stem Cells* **24**: 3–12.

49. Oliver JA, Maarouf O, Cheema FH, *et al.* (2004) The renal papilla is a niche for adult kidney stem cells. *J Clin Invest* **114**: 795–804.

50. Maeshima A, Yamashita S, Nojima Y. (2003) Identification of renal progenitor-like tubular cells that participate in the regeneration processes of the kidney. *J Am Soc Nephrol* **14:** 3138–3146.

51. Maeshima A, Sakurai H, Nigam SK. (2006) Adult kidney tubular cell population showing phenotypic plasticity, tubulogenic capacity, and integration capability into developing kidney. *J Am Soc Nephrol* **17:** 188–198.

52. Iwatani H, Ito T, Imai E, *et al.* (2004) Hematopoietic and nonhematopoietic potentials of Hoechst(low)/side population cells isolated from adult rat kidney. *Kidney Int* **65:** 1604–1614.

53. Hishikawa K, Marumo T, Miura S, *et al.* (2005) Musculin/MyoR is expressed in kidney side population cells and can regulate their function. *J Cell Biol* **169:** 921–928.

54. Challen GA, Bertoncello I, Deane JA, *et al.* (2006) Kidney side population reveals multilineage potential and renal functional capacity but also cellular heterogeneity. *J Am Soc Nephrol* **17:** 1896–1912.

55. Bussolati B, Bruno S, Grange C, *et al.* (2005) Isolation of renal progenitor cells from adult human kidney. *Am J Pathol* **166:** 545–555.

56. Sagrinati C, Netti GS, Mazzinghi B, *et al.* (2006) Isolation and characterization of multipotent progenitor cells from the Bowman's capsule of adult human kidneys. *J Am Soc Nephrol* **17:** 2443–2456.

57. Dekel B, Zangi L, Shezen E, *et al.* (2006) Isolation and characterization of nontubular sca-1⁺lin⁻ multipotent stem/progenitor cells from adult mouse kidney. *J Am Soc Nephrol* **17:** 3300–3314.

58. Kitamura S, Yamasaki Y, Kinomura M, *et al.* (2005) Establishment and characterization of renal progenitor like cells from S3 segment of nephron in rat adult kidney. *FASEB J* **19:** 1789–1797.

59. Gupta S, Verfaillie CM, Chmielewski DH, *et al.* (2006) Isolation and characterization of kidney-derived stem cells. *J Am Soc Nephrol* **17:** 3028–3040.

60. Dyce PW, Zhu H, Craig J, Li J. (2004) Stem cells with multilineage potential derived from porcine skin. *Biochem Biophys Res Commun* **316:** 651–658.

61. Yeom YI, Fuhrmann G, Ovitt CE, *et al.* (1996) Germline regulatory element of Oct-4 specific for the totipotent cycle of embryonal cells. *Development* **122:** 881–894.

62. Mitalipov SM, Kuo HC, Hennebold JD, Wolf DP. (2003) Oct-4 expression in pluripotent cells of the rhesus monkey. *Biol Reprod* **69:** 1785–1792.

63. Kehler J, Tolkunova E, Koschorz B, *et al.* (2004) Oct4 is required for primordial germ cell survival. *EMBO Rep* **5:** 1078–1083.

64. Pesce M, Scholer HR. (2000) Oct-4: control of totipotency and germline determination. *Mol Reprod Dev* **55:** 452–457.

65. Rosner MH, Vigano MA, Ozato K, *et al.* (1990) A POU-domain transcription factor in early stem cells and germ cells of the mammalian embryo. *Nature* **345:** 686–692.

66. Reubinoff BE, Pera MF, Fong CY, *et al.* (2000) Embryonic stem cell lines from human blastocysts: somatic differentiation *in vitro*. *Nat Biotechnol* **18:** 399–404.

67. Niwa H, Miyazaki J, Smith AG. (2000) Quantitative expression of Oct-3/4 defines differentiation, dedifferentiation or self-renewal of ES cells. *Nat Genet* **24:** 372–376.

68. Jiang Y, Vaessen B, Lenvik T, *et al.* (2002) Multipotent progenitor cells can be isolated from postnatal murine bone marrow, muscle, and brain. *Exp Hematol* **30:** 896–904.

69. Jiang Y, Jahagirdar BN, Reinhardt RL, *et al.* (2002) Pluripotency of mesenchymal stem cells derived from adult marrow. *Nature* **418:** 41–49.

70. Schwartz RE, Reyes M, Koodie L, *et al.* (2002) Multipotent adult progenitor cells from bone marrow differentiate into functional hepatocyte-like cells. *J Clin Invest* **109:** 1291–1302.

71. Reyes M, Dudek A, Jahagirdar B, *et al.* (2002) Origin of endothelial progenitors in human postnatal bone marrow. *J Clin Invest* **109:** 337–346.

72. Yu H, Fang D, Kumar SM, *et al.* (2006) Isolation of a novel population of multipotent adult stem cells from human hair follicles. *Am J Pathol* **168:** 1879–1888.

73. Baal N, Reisinger K, Jahr H, *et al.* (2004) Expression of transcription factor Oct-4 and other embryonic genes in CD133 positive cells from human umbilical cord blood. *Thromb Haemost* **92:** 767–775.

74. D'Ippolito G, Diabira S, Howard GA, *et al.* (2004) Marrow-isolated adult multilineage inducible (MIAMI) cells, a unique population of postnatal young and old human cells with extensive expansion and differentiation potential. *J Cell Sci* **117:** 2971–2981.

75. Davis SF, Hood J, Thomas A, Brunnell BA. (2006) Isolation of adult rhesus neural stem and progenitor cells and differentiation into immature oligodendrocytes. *Stem Cells Dev* **15:** 191–199.

76. Romagnani P, Annunziato F, Liotta F, *et al.* (2005) CD14+CD34 low cells with stem cell phenotypic and functional features are the major source of circulating endothelial progenitors. *Circ Res* **97:** 314–322.

77. Romero-Ramos M, Vourc'h P, Young HE, *et al.* (2002) Neuronal differentiation of stem cells isolated from adult muscle. *J Neurosci Res* **69:** 894–907.

78. Trosko JE, Tai MH. (2006) Adult stem cell theory of the multi-stage, multi-mechanism theory of carcinogenesis: role of inflammation on the promotion of initiated stem cells. *Contrib Microbiol* **13:** 45–65.

79. Tsai MS, Hwang SM, Tsai YL, *et al.* (2006) Clonal amniotic fluid-derived stem cells express characteristics of both mesenchymal and neural stem cells. *Biol Reprod* **74:** 545–551.

80. Xiao J, Nan Z, Motooka Y, Low WC. (2005) Transplantation of a novel cell line population of umbilical cord blood stem cells ameliorates neurological deficits associated with ischemic brain injury. *Stem Cells Dev* **14:** 722–733.

81. Zhou YF, Fang F, Fu JR, *et al.* (2005) An experimental study on astrocytes promoting production of neural stem cells derived from mouse embryonic stem cells. *Chin Med J (Engl)* **118**: 1994–1999.

82. Grimm PC, Nickerson P, Jeffrey J, *et al.* (2001) Neointimal and tubulointerstitial infiltration by recipient mesenchymal cells in chronic renal-allograft rejection. *N Engl J Med* **345**: 93–97.

83. Masuya M, Drake CJ, Fleming PA, *et al.* (2003) Hematopoietic origin of glomerular mesangial cells. *Blood* **101**: 2215–2218.

84. Prodromidi EI, Poulsom R, Jeffery R, *et al.* (2006) Bone marrow-derived cells contribute to podocyte regeneration and amelioration of renal disease in a mouse model of Alport syndrome. *Stem Cells* **24**: 2448–2455.

85. Rookmaaker MB, Smits AM, Tolboom H, *et al.* (2003) Bone marrow-derived cells contribute to glomerular endothelial repair in experimental glomerulonephritis. *Am J Pathol* **163**: 553–562.

86. Imasawa T, Utsunomiya Y, Kawamura T, *et al.* (2001) The potential of bone marrow-derived cells to differentiate to glomerular mesangial cells. *J Am Soc Nephrol* **12**: 1401–1409.

87. Morigi M, Imberti B, Zoja C, *et al.* (2004) Mesenchymal stem cells are renotropic, helping to repair the kidney and improve function in acute renal failure. *J Am Soc Nephrol* **15**: 1794–1804.

88. Wagers AJ, Sherwood RI, Christensen JL, Weissman IL. (2002) Little evidence for developmental plasticity of adult hematopoietic stem cells. *Science.* **297**: 2256–2259.

89. Choi JH, Kim KL, Huh W, *et al.* (2004) Decreased number and impaired angiogenic function of endothelial progenitor cells in patients with chronic renal failure. *Arterioscler Thromb Vasc Biol* **24**: 1246–1252.

90. Rodriguez-Ayala E, Yao Q, Holmen C, *et al.* (2006) Imbalance between detached circulating endothelial cells and endothelial progenitor cells in chronic kidney disease. *Blood Purif* **24**: 196–202.

91. Loomans CJ, de Koning EJ, Staal FJ, *et al.* (2004) Endothelial progenitor cell dysfunction: a novel concept in the pathogenesis of vascular complications of type 1 diabetes. *Diabetes* **53**: 195–199.

92. He T, Peterson TE, Katusic ZS. (2005) Paracrine mitogenic effect of human endothelial progenitor cells: role of interleukin-8. *Am J Physiol Heart Circ Physiol* **289**: H968–972.

93. Patschan D, Plotkin M, Goligorsky MS. (2006) Therapeutic use of stem and endothelial progenitor cells in acute renal injury: ca ira. *Curr Opin Pharmacol* **6**: 176–183.

94. Patschan D, Krupincza K, Patschan S, *et al.* (2006) Dynamics of mobilization and homing of endothelial progenitor cells after acute renal ischemia: modulation by ischemic preconditioning. *Am J Physiol Renal Physiol* **291**: F176–F185.

95. Vassilopoulos G, Wang PR, Russell DW. (2003) Transplanted bone marrow regenerates liver by cell fusion. *Nature* **422:** 901–904.
96. Wang X, Willenbring H, Akkari Y, *et al.* (2003) Cell fusion is the principal source of bone marrow-derived hepatocytes. *Nature* **422:** 897–901.
97. Alvarez-Dolado M, Pardal R, Garcia-Verdugo JM, *et al.* (2003) Fusion of bone marrow-derived cells with Purkinje neurons, cardiomyocytes and hepatocytes. *Nature* **425:** 968–973.
98. Lange C, Togel F, Ittrich H, *et al.* (2005) Administered mesenchymal stem cells enhance recovery from ischemia/reperfusion-induced acute renal failure in rats. *Kidney Int* **68:** 1613–1617.
99. Togel F, Hu Z, Weiss K, *et al.* (2005) Administered mesenchymal stem cells protect against ischemic acute renal failure through differentiation-independent mechanisms. *Am J Physiol Renal Physiol* **289:** F31–F42.
100. Imasawa T, Nagasawa R, Utsunomiya Y, *et al.* (1999) Bone marrow transplantation attenuates murine IgA nephropathy: role of a stem cell disorder. *Kidney Int.* **56:** 1809.
101. Cornacchia F, Fornoni A, Plati AR, *et al.* (2001) Glomerulosclerosis is transmitted by bone marrow-derived mesangial cell progenitors. *J Clin Invest* **108:** 1649–1656.
102. Ito T, Suzuki A, Imai E, *et al.* (2001) Bone marrow is a reservoir of repopulating mesangial cells during glomerular remodeling. *J Am Soc Nephrol* **12:** 2625–2635.
103. Lee RH, Seo MJ, Reger RL, *et al.* (2006) Multipotent stromal cells from human marrow home to and promote repair of pancreatic islets and renal glomeruli in diabetic NOD/scid mice. *Proc Natl Acad Sci USA* **103:** 17438–17443.
104. Sugimoto H, Mundel TM, Sund M, *et al.* (2006) Bone marrow-derived stem cells repair basement membrane collagen defects and reverse genetic kidney disease. *Proc Natl Acad Sci USA* **103:** 7321–7326.
105. Ninichuk V, Gross O, Segerer S, *et al.* (2006) Multipotent mesenchymal stem cells reduce interstitial fibrosis but do not delay progression of chronic kidney disease in collagen4A3-deficient mice. *Kidney Int* **70:** 121–129.
106. Hacein-Bey-Abina S, Von Kalle C, Schmidt M, *et al.* (2003) LMO2-associated clonal T cell proliferation in two patients after gene therapy for SCID-X1. *Science* **302:** 415–419.
107. Dunnett SB, Bjorklund A, Lindvall O. (2001) Cell therapy in Parkinson's disease — stop or go? *Nat Rev Neurosci* **2:** 365–369.
108. Menasche P, Hagege AA, Vilquin JT, *et al.* (2003) Autologous skeletal myoblast transplantation for severe postinfarction left ventricular dysfunction. *J Am Coll Cardiol* **41:** 1078–1083.
109. Iwano M, Plieth D, Danoff TM, *et al.* (2002) Evidence that fibroblasts derive from epithelium during tissue fibrosis. *J Clin Invest* **110:** 341–350.

110. Zeisberg M, Hanai J, Sugimoto H, *et al.* (2003) BMP-7 counteracts TGF-beta1-induced epithelial-to-mesenchymal transition and reverses chronic renal injury. *Nat Med* 9: 964–968.
111. Zeisberg M, Kalluri R. (2004) The role of epithelial-to-mesenchymal transition in renal fibrosis. *J Mol Med* 82: 175–181.
112. Zeisberg M, Shah AA, Kalluri R. (2005) Bone morphogenic protein-7 induces mesenchymal to epithelial transition in adult renal fibroblasts and facilitates regeneration of injured kidney. *J Biol Chem* 280: 8094–8100.
113. Zager RA, Johnson AC, Lund S, Hanson S. (2006) Acute renal failure: determinants and characteristics of the injury induced hyper-inflammatory response. *Am J Physiol Renal Physiol* 291: F546–F556.
114. Zager RA, Johnson AC, Hanson SY, Lund S. (2006) Acute nephrotoxic and obstructive injury primes the kidney to endotoxin-driven cytokine/chemokine production. *Kidney Int* 69: 1181–1188.
115. Nath KA, Norby SM. (2000) Reactive oxygen species and acute renal failure. *Am J Med* 109: 665–678.
116. Nath KA. (2006) Heme oxygenase-1: a provenance for cytoprotective pathways in the kidney and other tissues. *Kidney Int* 70: 432–443.
117. Nath KA. (1996) Adaptation to the nephrotoxicity of heme proteins. *Exp Nephrol* 4: 139–143.
118. Hammerman MR, Miller SB. (1994) Therapeutic use of growth factors in renal failure. *J Am Soc Nephrol* 5: 1–11.
119. Hirschberg R, Kopple J, Lipsett P, *et al.* (1999) Multicenter trial of recombinant human insulin-like growth factor 1 in patients with acute renal failure. *Kidney Int* 55: 2423–2432.

17

Liver Stem Cells in Regenerative Medicine

Xin Wang*, Yiping Hu, Zongyu Chen and Xiaoyan Ding

1. Introduction

The liver, unlike many other vital organs, can regenerate itself after injury. The phenomenon of liver regeneration has been known since antiquity and was illustrated in the ancient Greek legend of Prometheus. Many mammals can survive after surgical removal of up to 75% of their liver mass. In experiments, 2/3 liver partial hepatectomy (PH) is performed as acute injury to study liver regeneration. The original number of liver cells can be restored within 1 week, and the liver tissue can reach the original mass within 2–3 weeks. Liver regeneration can be undertaken several times after serial PHs. Because of this special property, stem cells were once considered to exist in the liver organ. For a long time, liver stem cells have been referred to as different types of cells.[1] At least 6 types of cells have been regarded as liver stem cells: (1) embryonic cells that become hepatocytes and bile duct epithelial cells (BDECs) during liver organogenesis; (2) cells responsible for normal *in vivo* cell turnovers of hepatocytes and BDECs in adults; (3) cells that could result in the phenotypes of hepatocytes and BDECs *in vitro*; (4) cells responsible for liver regeneration after chronic injury, such as PH; (5) cells responsible for progenitor-dependent regeneration, such as oval cells; and (6) transplanted cells responsible for liver repopulation. These definitions were only used in the special situations in previous studies. A generally agreeable definition of liver stem cells has not been established because of the insufficient knowledge about the nature of liver stem cells.

Recently, liver stem cells have been studied for clinical application. They are expected to become an endless cell source for making liver tissues and

*Correpondence: Stem Cell Institute, University of Minnesota, Minneapolis, MN 55455, USA. Tel: (612) 625-4478; Fax: (612) 624-2436; E-mail: wangx336@umn.edu

artificial liver organs in the future. Liver stem cells may also be used directly as donor cells in transplantation for the therapies of some liver diseases. This chapter will mainly cover the information closely related to liver stem cells in regenerative medicine. Other information on liver stem cells can be found in previous reviews.[1,2] Here, only mammal liver stem cells will be discussed. Sometimes, the term liver stem cells is used in reference to both liver precursor cells and liver progenitor cells because the lineage commitment of liver stem cells has not been well defined. In addition, it is commonly agreed that liver stem cells are the stem cells for precursors of two types of epithelial cells in liver, hepatocytes and BDECs.

2. Liver Stem Cells Inside the Liver Organ

2.1. Liver stem cells at embryonic stages

Like other cells with stemness, liver stem cells should have two basic capacities: self-renewal and the ability to differentiate into either hepatocytes or BDECs. During embryonic development, the progenitor cells of liver come from their ancestors, endodermal stem cells. They may expend themselves and differentiate into the progenitor cells at several differentiation levels. The differentiated cells gradually build themselves into the liver organ.

2.1.1. *Fetal liver progenitor cells*

Research on experimental embryology in mice has supplied important information about liver development. Previous results indicated that the liver developed from ventral foregut endoderm beginning at embryonic day (ED) 8 of gestation.[3] The first sign of hepatic differentiation was the expression of albumin and α-fetoprotein mRNA in the precursor of the hepatocyte among endodermal cells, even prior to the appearance of cells with hepatocytic morphology.[4] Between ED 8.5 and ED 9.5, the precursor of the hepatocyte proliferated significantly. Late on ED 9.5, these cells migrated toward cardiac mesoderm in the septum transversum.[5] Signals from the cardiac mesoderm could induce the cells to increase the expression levels of albumin and α-fetoprotein mRNAs and to form the tissue structure of the liver bud. In an early study, some specific signals produced by mesenchyme were reported.[6] It was known that fibroblast growth factors (FGFs) 1, 2, and 8 could induce the liver gene expression program in the isolated foregut endodermal tissues of the mouse.[6] The receptors for FGFs 1 and 4 were expressed on foregut endoderm cells and were essential for this induction.

At ED 10.5, the vascularization of the liver bud began, followed by a large increase in liver mass. The early cells in the liver bud were mostly positive for both albumin and α-fetoprotein. The additionally differentiated phenotypes of both hepatocytes and bile duct epithelium emerged in mid-gestation. Definitive lineage of liver stem cells in different differentiation stages is not yet fully understood, but it is generally agreed that bile ducts and hepatocytes emerged from common precursors, the hepatoblasts.[3] It was not known whether there were only hepatoblasts or whether there was a hierarchy of lineage progression consisting of more primitive liver stem cells and more committed bipotential progenitors, followed by the more differentiated hepatocytic progenitors and bile epithelial progenitors.

2.1.2. *Cell surface markers of fetal liver progenitor cells*

Cell surface markers are generally used to define stem cells. The search for related surface markers for liver stem or progenitor cells has only started recently. The special markers for separating hepatoblasts from other early liver stem cells were introduced by Schmelzer *et al.*[7] Fetal liver progenitor cells were used in the studies. The neuronal cell adhesion molecule (NCAM), the epithelial cell adhesion molecule (EpCAM), and claudin-3 (CLDN-3) were suggested as the key markers for identifying liver stem cells. They were used to identify several groups of liver stem or progenitor cells in the fetal liver. The low level of albumin and a complete absence of expression of α-fetoprotein and adult liver-specific proteins were known to be properties of liver stem cells, but not of liver progenitor cells and differentiated liver cells. Here, hepatoblasts were defined as the relatively differentiated liver progenitor cells that expressed high levels of α-fetoprotein and albumin, low levels of adult liver-specific proteins and cytokeratin (CK) 19, and none of NCAM and CLDN-3. Mature hepatocytes did not express EpCAM, NCAM, CLDN-3, CK19, or α-fetoprotein, but expressed with well-known adult-specific marker profiles, including expression of high levels of albumin, cytochrome

P4503A4, connexins, phosphoenolpyruvate carboxykinase, and transferrin. However, these definitions of several types of liver stem and progenitor cells have not been proven by rigorous assays, such as the assay using a colony of single cells.

In another study, liver stem cells were defined by the CD antibodies used in fluorescent activated cell sorting (FACS) followed by *in vitro* characterization during single cell-based culture. The cells with c-Met$^+$ CD49f$^+$/low c-Kit$^-$ CD45$^-$ TER119$^-$ were isolated from mouse fetal livers at ED 11.5.[8] These CD-marker-defined cells were regarded as liver stem cells for their ability to expand from single cells into colonies in culture, and could differentiate *in vitro* into some epithelia phenotypes within the liver, pancreas, and intestine at detectable levels. However, the cell origin and functionality of these stem-cell-like cells have not yet been proven.

2.2. Liver stem cells in the adult liver

Studies of adult liver stem cells can be traced back to the early 20th century. It was once hypothesized that liver regeneration after acute injury depended on a group of stem or progenitor cells in the adult liver, which could differentiate into hepatocytes and BDECs. However, there is still no evidence to support the hypothesis that liver stem cells are activated during acute injuries such as PH. In contrast, the existence of adult liver stem cells in chronic liver injury has been supported by many studies. Among the candidates for liver stem cells in the adult liver, oval cells and small hepatocyte-like progenitor cells (SHLPs) were two major groups of liver stem cells suggested from most previous studies. These two types of cells may be the progeny of the natural liver stem cells that were suggested as a potential niche at the canal of Hering (Fig. 1).

2.2.1. *Oval cells*

Oval cells as liver stem cells were first found during the process of 2-acetylaminofluorene (AAF) chemical induced liver cancer in the rat.[2,9] Oval cells were generally regarded as small epithelial cells that could proliferate tremendously to repair liver injury. Because these cells have oval nuclei and a high ratio of nuclei to cytoplasm, they were termed oval cells to distinguish them from other adult liver cells. After this finding, the studies of

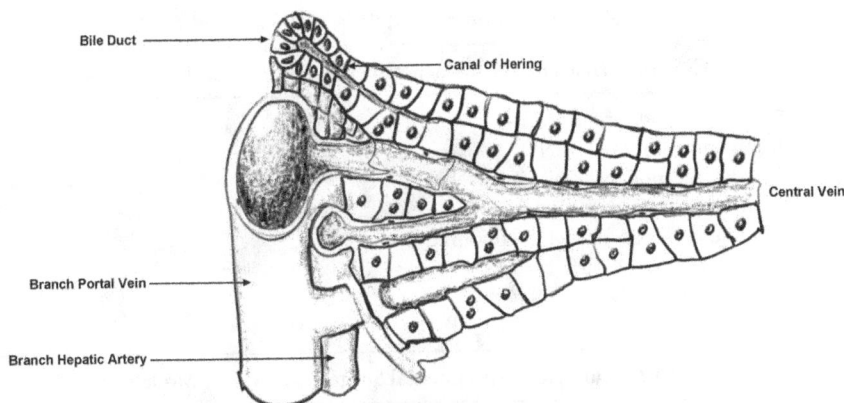

Figure 1. Diagram of the sinusoidal structure of the liver lobulaer. Blood flow enters hepatic sinusoids from the branch hepatic artery and the branch portal vein. The liver plate is a single layer extending from the portal space to the central vein. The canal of Hering is the region that connects between the BDECs in the bile duct and the last hepatocyte extended from the liver plate, which is regarded as a site of the niche for liver stem cells and the originating site for oval cells.

both normal liver stem cell biology and liver tumoregenesis from potential liver cancer stem cells were greatly facilitated. Several animal models used to study the oval cells have been developed (Table 1). The oval cells in mouse livers have also been induced with chemical treatments. However, the morphology of mouse oval cells is not the same as that of rat oval cells. Mouse oval cells are not oval in shape and do not have oval nuclei. They are sometimes called atypical ductular proliferation (ADP) cells.[10] The reason for the specific occurrence of oval cells in chronic liver injury instead of acute injury has been discovered gradually. In chemical-induced or viral infection-induced injuries, the death of mature hepatocytes results in the loss of cells' proliferation capacity, which exists in acute liver injury such as PH. A ductular reaction is generally induced in the injured liver. Oval cells start to emerge at the canal of Hering. In this specific region, oval cells neighbor newly generated hepatocytes and BDECs, which could be easily interpreted as the oval cells differentiating into both the new hepatocytes and BDECs. Collective results of *in vitro* assays have also shown that oval cells differentiate into cells with the markers for both hepatocytes and BDECs.[11] Oval cells might be related to hepatoblasts, because the typical proteins expressed in hepatoblasts, such as HNFs, GATA box, and CAATP,

Table 1. Induction of Oval Cells in the Rat and the Mouse

Chemical Treatments Plus Experimental Surgeries in Rat Models[91,110–116]
2-acetylaminofluorene (AAF)
Diethylnitrosomine (DEN)
Solt–Farber model (DEN + AAF + PH)
Modified Solt–Farber model (AAF + PH)
Choline-deficient diet + DL-ethionine
D-galactosomine + PH
Losiocarpine + PH
Retrorsine + PH

Chemicals Plus Experimental Surgeries in Mouse Models[10,117–119]
Dipin
3,5-dielhoxycorbonyl-1.4-dihydrocollidine (DDC)
Phenobarbital + cocaine + PH
Choline-deficient diet + DL-ethionine

are found in oval cells. In addition, oval cells express the common markers for stem cells, such as c-Kit, CD34, and Flt-3 receptor, which further supports the theory of oval cells as stem cells in the liver.[1]

Undifferentiated liver stem cells that are preserved at the canal of Hering as a stem cell niche are not activated by the signals released during acute liver injury. When signals such as growth factors/cytokines are released during chronic liver injury, the liver stem cells could be stimulated in silence to activate cell proliferation and differentiation. Some transcription factors found in fetal liver cells are activated in oval cells.[2] After activation, proliferating oval cells invade the parenchyma of liver tissue. The new liver parenchyma cells differentiated from oval cells are accompanied by many Ito cells that express growth factors for cell proliferation. However, definitive evidence of a direct relationship between oval cells and natural liver stem cells is still lacking.

2.2.2. *Small hepatocyte-like progenitor cells (SHLPs)*

Small hepatocyte-like progenitor cells (SHLPs) may be another type of liver progenitor cells.[12] In the retrorsine/PH rat model, SHLPs are found to have some different characteristics from the oval cells induced in other oval cell models. SHLPs share markers with fetal liver hepatocytes, mature hepatocytes, and oval cells. They express the hepatocyte markers albumin and trasthyretin. However, they do not express the biliary markers glutathione S-transferase (GST) and BD.1 membrane protein for BDECs, suggesting that they are only hepatocytic progenitors instead of bipotent liver progenitors. It is also possible that SHLPs are the progeny cells of oval cells during chronic injury.

2.3. Liver cancer stem cells

In accordance with the general definition of stem cells, cancer stem cells should be in a primitive state with the capacity to self-renew and differentiate into "mature" cancer cells. The existence of liver cancer stem cells was proposed a long time ago,[2] but liver cancer stem cells have not yet been isolated. Normal liver stem cells are probably related to liver cancer stem cells, and a close relationship between liver cancer stem cells and oval cells might exist. In the studies of the 1950s, the discovery of oval cells was made from experiments aimed at understanding the mechanism of liver cancer cells.[2] At that early time, there were two hypotheses proposed from these studies. One is dedifferentiation: liver tumors arise from the dedifferentiation of mature cells that have not terminally differentiated. The other theory is maturation arrest, which hypothesizes that liver tumors arise from maturation arrest of immature liver progenitor/stem cells. In the second hypothesis, oval cells were regarded as liver cancer stem cells that originated from immature liver progenitor/stem cells and were studied as the model of maturation arrest. Under experimental conditions, hepatocelular carcinoma was found from oval cells in several oval-cell-inducing models. Oval cells colocalized with "mature" liver cancer cells in the late stage of the chronic liver injury.[2] These liver cancer cells expressed similar markers of oval cells. Additionally, cultured oval cells became liver cancer cells after transplantation in mice and rats.[2] However, there was still no evidence to support the theory that endogenous oval cells could directly become liver cancer cells *in vivo*. In addition, because oval cells are defined on the morphological level, oval cells may be composed of several kinds of cells, including normal liver stem cells and liver cancer stem cells. Regarding the application of liver stem cells in regenerative medicine, liver cancer stem cells must be avoided for cell isolation and transplantation in liver cell therapy.

3. Multiple Sources of Stem Cells Used to Derive Hepatic Cells

There are several types of cells with the characteristics of liver stem cells for generating hepatocytes and BDECs. These stem cell-like cells have been used as model systems to study the nature of liver stem cells, and tested for significant hepatic differentiation under induction in order to establish potential protocols for deriving hepatocytes for regenerative medicine.

3.1. Liver stem cell lines

Most liver stem cell lines were derived from the adult liver and the fetal liver of animals. They have been used as model systems for studying the biological activities of liver stem cells and the mechanisms underlining these activities. Recently, liver stem cell lines have also been derived from human fetal livers. These human liver cell lines might have the potential to be used directly in cell transplantation therapy in the future. Some representative liver stem cell lines are briefly summarized in the following:

3.1.1. *Liver epithelial progenitor cells (LEPCs)*

These cells as liver progenitor cells were derived from livers of adult mice that were treated with retrorsine and partial hepatectomy.[13] LEPCs have dual expressions of both albumin and CK19 and stem cell markers of c-Kit and Thy-1, but not hematopoietic markers of CD34 and CD45. They can differentiate into either hepatocyte-like cells or BDEC-like cells under different stimulations in culture. Upon transplantation in the CCl4-injured liver, LEPCs engraft into liver parenchyma and differentiate into hepatocytes.

3.1.2. *WB-F344 epithelial cells*

WB-F344 epithelial cells were derived from normal rat cells.[12] They can be cultured as normal cell line with many passages. WB-F344 cells can also be induced to differentiate into hepatocytes and BDECs. Under different manipulations, WB-F344 cells can become a variety of tissue-specific cells, including skeletal muscle cells, cardiac muscle cells, insulin-producing β-cells, prostate glandular cells, hematopoietic cells, and liver tumor cells.[14,15]

3.1.3. *Cyto-Met hepatic precursor cells*

Cyto-Met hepatic precursor cells, as hepatic precursor cells with genetic modification of cyto-Met overexpression, were derived from transgenic mice.[16] Cyto-met was represented as a protein with a constitutively active form of c-Met (the receptor of hepatocyte growth factor, HGF). Two morphologically distinct types of progenitor cells were found in the culture of this mouse liver cell line. The clonal cells with epithelial morphology resembled hepatocytes and could differentiate into only the hepatocytic phenotype. In contrast, the clonal cells with palmate morphology could differentiate into two types of cells, namely hepatocyte-like cells and BDEC-like cells, which suggested that the capacity of bipotential existed only in a subpopulation of cells with palmate morphology.

3.1.4. *HBC-3 cells*

HBC-3 cells are the bipotential cells derived from ED 9.5 of the mouse embryonic liver.[17] This line of clonal cells could be induced to differentiate into either hepatocytes or BDECs with formation of duct structures under different stimulations in culture.

3.1.5. *Human liver cell lines*

AKN-1 cells were the first reported human liver cell line[18] from a normal liver. They carried liver progenitor characteristics. Recently, additional human fetal hepatocyte (HFH) lines were reported. These human liver progenitor cell-like cell lines were derived from second-trimester human fetal livers and cultured in serum medium.[19] They were successfully passed for 9–12 months. Cell populations with liver stem cells were thought to contain these HFH cell lines. Based on the collected experiences, the same group further derived human fetal multipotent progenitor cells (hFLMPCs).[20] The single cell clones were isolated and passed for up to 100 population doublings. Interestingly, the cells can be induced to differentiate into hepatocytes and BDECs *in vitro*, as well as fat, bone, cartilage, and endothelial cells. hFLMPCs can differentiate into functional hepatocytes *in vivo* when transplanted into animal models of liver disease.[20]

3.2. Embryonic stem cells becoming hepatocytes

Mammalian embryonic stem (ES) cells are pluripotent stem cells derived from the inner cell mass of blastocysts.[21,22] Mouse ES cells in culture are known to differentiate into many types of adult somatic cells,[23] including hepatocyte-like cells.

3.2.1. *Differentiation of mouse ES cells into hepatic cells*

Several methods of inducing ES cells to differentiate into hepatocyte-like cells have been established. In early reports, mouse ES cells differentiated into hepatocyte-like cells through a process of embryoid body (EB) formation.[4–34] EB formation was regarded as an uncontrolled process of ES cell differentiation in culture with activation of differentiation factors supplied from fetal bovine serum, after removing leukemia inhibitory factor (LIF), which prevents mouse ES cells from differentiation. The structure of the EB is three-dimensional cell aggregates containing cells representing all three germ layers. Cells with endoderm and liver properties were found in EBs. These cells expressed makers for definitive endoderm or hepatic cells (summarized in Table 2). After formation of the EB, a continuous differentiation of hepatic cells could be further induced by the growth factors known from the studies of embryonic liver development, including acidic fibroblast growth factor (aFGF), basic FGF (bFGF), HGF, oncostatin M, and insulin. One major problem for most previous studies was that the frequency of ES cell-derived hepatic cells generated in the process via EB formation was extremely low. Therefore, the readout of ES cell-derived hepatic differentiation depended mainly on the detection of hepatic gene transcripts and proteins in the assays of RT-PCR or immunohistochemistry on several cells.

Other methods of inducing hepatic cells from ES cells were either with a partial process of EB formation at early stages or without EB formation.[35,36] Retinoic acid as a cell differentiation factor was used to initiate the process of ES cell differentiation.[35,36] More ES cell-derived cells with albumin[36] or α-fetoprotein[35] could be obtained in these experiments in comparison with the albumin- or α-fetoprotein positive cells induced in other methods.[35,36] However, these reports did not address whether the cells expressing albumin and α-fetoprotein were the derived hepatic cells or the derived cells of primitive or visceral endoderm, because both albumin and α-fetoprotein are not specific markers for hepatic cells at the early stage of ES cell differentiation. Recently, the protocol for induction and

Table 2. Parameters Used for Analysis of ES Cells Differentiating into Hepatic Cells

I. Markers Used for ES cells Differentiating into Hepatic Cells[24–34,105,106]

α1-antitrypsin (AAT)
Alpha fetal protein (AFP)
Albumin (ALB)
Aldolase B
Apolipoprotein A1
Apolipoprotein A2
Apolipoprotein E (ApoE)
Asialoglycoprotein receptor 1 (Asgr1)
Cytokeratin 8 (CK8)
Cytokeratin 18 (CK18)
Cytokeratin 19 (CK19)
Complement C3 (C3)
Cytochrome P450 (P450)
Carbamoyl phosphate synthetase I (CPSase I)
Dipeptidyl-peptidase IV (DPP IV)
Markers for hepatic cells
α1-antitrypsin (AAT)
Glucose-6-phosphatase (G-6-P)
Glutathione S-transferase (GST)
Hepatocyte nuclear factor 3α (HNF3α)
Hepatocyte nuclear factor 3β (HNF3β)
Hepatocyte nuclear factor 3γ (HNF3γ)
Hepatocyte nuclear factor 4α (HNF4α)
Liver-specific organic anion transporter 1 (LST-1)
N-system amino acid transporter (NAT)
Phosphaenolpyruvate carboxykinase (PEPCK)
Peroxisomal membrane protein 1–like protein (PXMPL-L)
Phenylalanine hydroxylase (PAH)
Transcription factor Sox 17α
Transthyretin (TTR)
Transcription factor GATA 4
Tyrosine aminotransferase (TAT)
Tryptophan-2,3-dioxygenase (TDO)
Transferrin (TFN)

II. *In Vitro* Functional Assays for Stem Cell-derived Hepatocyte-like Cells[24,29,48,49]

Urea synthesis assay
Pentoxyresorufin assay
Uptake of a-LDL
Periodic acid–Schiff for glycogen assay
Uptake of Indocyanine Green assay
Triacylglycerol synthesis assay
G-6-P activity

isolation of ES cell-derived definitive endoderm cells was reported.[37] In this report, the process of EB formation as short as 2.5 days in culture was used to initiate the early differentiation of ES cells. Mesoendodermal cells, which were the cells with bipotential differentiation to either mesoderm or endoderm, were labeled by the promoter trap of Brachyury gene expression and further purified by FACS. Subsequently, fetal bovine serum was removed from the induction medium because it enhanced the differentiation of ES cells into the directions of ectoderm and mesoderm cells, which negatively influenced the direction of differentiation into definitive endoderm cells. The important information obtained from this report was that activin A could be used to induce ES cell-derived Brachyury positive cells to differentiate into committed definitive endodermal cells and finally cells expressing many hepatic markers. Activin A as a regulator was originally known for its definitive endoderm-inducing potential in the embryonic development of *Xenopus laevis*.[37,38] With the protocol of using activin A at early induction, the ES cell-derived cells with expressions of albumin and α-fetoprotein, and the derived cells with expression of tyrosine aminotransferase (TAT) and carbamol phosphate synthesis (CPS1), could be induced to significantly high frequencies.[37]

More recently, two reports indicated that activin A could be directly used to induce the definitive endoderm differentiation from both human ES cells[39] and mouse ES cells.[40] The whole process of EB formation was not required for these induction processes. Together, the results from these reports suggest that induction of hepatic differentiation from ES cells can be conducted under controllable processes, either via a short time of EB formation or totally without EB formation.

Compared with the research on mouse ES cells, there have been fewer studies of human ES cell-derived hepatocytes. Human ES cell-derived hepatocytes might be directly used for cell transplantation therapy in the future. Hepatic differentiation from human ES cells was first reported with detectable message signals of albumin, $\alpha 1$ anti-trypsin, and α-fetoprotein.[41] In a later report, human ES cells in culture were shown to differentiate into hepatocyte-like cells after treatment with sodium butyrate.[42] Human ES cell-derived hepatocyte-like cells expressed proteins including albumin, $\alpha 1$ anti-trypsin, CK18, and CK8, accumulated glycogen, and had inducible cytochrome P450 activity. During the induction, α-fetoprotein was not expressed, suggesting that the mechanism of differentiation might be different from naturally occurring developmental differentiation. In addition, biodegradable polymer scaffolds were once used to promote human

ES cells differentiating into hepatocyte-like cells.[43] EBs from human ES cells were treated with activin A and insulin-like growth factor 1 (IGF-1). After 8 days of induced differentiation in culture, the cells were seeded onto the scaffolds coated with matrigel or fibronectin. Expression of both albumin and α-fetoprotein was found in the induced cells.

In addition, a recent report indicated that activin A could be used directly to induce the differentiation of definitive endoderm from human ES cells.[39] Therefore, the induced definitive endoderm cells could be further used to generate hepatocytes in a more efficient manner.[39]

3.2.2. *Enrichment of ES cell-derived liver progenitor cells*

Because the cell surface marker for liver progenitor cells and hepatocytes is yet unknown, the strategy of the promoter trap was used for labeling the hepatic cells differentiated from ES cells. There were two approaches to achieving the promoter trap. The first was with knock-in of a reporter gene, such as the enhanced green fluorescent protein (EGFP), which resulted in the expression of the reporter gene under the control of a specific promoter, such as the promoter of α-fetoprotein for fetal hepatic cells,[26] or the promoter of Brachyury for mesoendodermal cells[37] and the promoters of the genes goosecoid (Gsc) and Sox17 for primitive streak cells.[40] The second approach was to deliver DNA plasmid vectors carrying a fragment of the promoter driven-reporter gene.[35,36] Both of these approaches have been successfully used to label the ES cell-derived cells at different stages of endodermal differentiation or hepatic differentiation.

Recently, the cell surface markers CXCR4 and E-cadherin (ECD) have been found on ES cell-derived mesoendodermal cells and definitive endoderm cells, which resulted in successful purification of these two type of cells in FACS.[40] Cell surface markers specific for liver stem or progenitor cells are expected to be known in the future, which will be useful for the purification of ES cell-derived liver progenitor cells.

3.2.3. *In vitro and in vivo function of ES cell-derived hepatic cells*

For proving the hepatic function, ES cell-derived hepatocyte-like cells were studied *in vitro* for the hepatic functions known from previous studies of cultured hepatocytes, including the capacity for synthesis of urea[24,29] and triacylglucerol,[29] uptake of indocyamine green,[27] metabolic activities of glucose collection and ammonia elimination (Table 2).[30] The *in vivo* function of ES cell-derived hepatocyte-like cells was also studied after the cells were transplanted into livers with induced injury. In many reports, ES cell-derived hepatocyte-like cells could be detected in homing of livers of recipient mice. Recently, ES cell-derived hepatocyte-like cells have been successfully proven to have the capacity for liver repopulation. This finding will be further discussed later in this article.

3.3. Hepatic differentiation from the stem cells isolated from the hematopoietic system

Several types of cells in bone marrow have been considered as the stem cells for hepatic cells. In addition, the existence of particular liver stem cells in bone marrow was once suggested from several studies *in vivo* and *in vitro*.[44]

3.3.1. *Mesenchymal stem cells differentiating into liver cells*

Mesenchymal stem cells (MSCs) are typically isolated from bone marrow and cultured in typical plastic dishes used for tissue culture.[45] The lifespan of MSCs is known within the limitation of 15–50 population doublings. Under various stimulations both *in vitro* and *in vivo*, MSCs can become several different kinds of differentiated cells belonging to all three embryonic lineages.[45] Compared to the early findings of other cells that differentiated from MSCs under inductions, it has only recently been discovered that MSCs can also be induced to become hepatocyte-like cells. MSCs can be isolated and expanded from two sources, namely bone marrow and umbilical cord blood. The induced differentiation to let MSCs become hepatocytes has been successfully undertaken *in vitro* for both types of MSCs from human donors.[46] Similar results were almost found for rat bone marrow-derived MSCs.[47] The MSC-derived hepatocyte-like cells had the typical morphology of hepatocytes and expressed proteins specific for hepatocytes. However, the *in vivo* hepatic functions of MSC-derived hepatocyte-like cells after transplantation assay have not been fully proven yet.

3.3.2. *Multipotent adult progenitor cells differentiating into liver cells*

As another type of progenitor cells, multipotent adult progenitor cells (MAPCs) have been isolated from postnatal bone marrow tissues of the human, mouse, rat, and swine.[48,49] In contrast to MSCs, MAPCs demonstrate more primitive characters. MAPCs are CD44-negative and major histocompatibility complex class I (MHC I)–negative, unlike MSCs. In addition, MAPCs can be expanded under defined low-serum conditions for more than 100 population doublings in humans and swine or 400 in mice and rats without telomere shortening or karyotypic abnormalities. They can be successfully induced to differentiate into many types of cells from three embryonic germ layers, including hepatocytes. Several reports have shown that MAPCs can differentiate into hepatocytes and BDECs both *in vivo* and *in vitro*. Transplanted mouse MAPCs with hepatic differentiation can be traced in the livers of recipient mice. MAPCs replaced the endogenous hepatocytes up to the level of 20%. For *in vitro* induction, human, mouse, and rat MAPCs, which were cultured on Matrigel with FGF-4 and HGF, could differentiate into epithelioid cells that expressed the typical markers in hepatocytes and BDECs, including hepatocyte nuclear factor 3β (HNF 3β), GATA4, CK19, transthyretin, α-fetoprotein, CK18, HNF-4, and HNF-1. These cells also acquired functional characteristics of hepatocytes. They secreted urea and albumin, had phenobarbital-inducible cytochrome p450, could take up low-density lipoprotein (LDL), and stored glycogen. These results suggested that MAPCs could differentiate into cells with morphological, phenotypic, and functional characteristics of hepatocytes.

3.3.3. *Fusion between hepatocytes and bone marrow-derived cells*

Bone marrow-derived liver cells may originate from the differentiation of MAPCs or MSCs *in vivo*. However, there is still no evidence to prove that MAPCs or MSCs can significantly differentiate into hepatocytes after transplantation. Their capacities for liver repopulation have not yet been proven; whether liver stem cells exist in bone marrow is still controversial. Recently, results from several experiments showed that cell fusion happens between bone marrow-derived cells and hepatocytes.[50] Subsequent studies indicated that the macrophages from bone marrow fused with hepatocytes to become new hepatocytes.[51] However, research to isolate some special liver stem cells in bone marrow is ongoing, because bone marrow-derived hepatocytes without cell fusion have also been observed in other experiments.[52]

3.4. Liver stem cells and transdifferentiation

Transdifferentiation has recently become a subject of interest in the field of stem cell research. Until recently, the term has not been well defined.

It could be used to suggest the capacity of liver stem cells to become cells other than hepatocytes and BDECs, or to suggest that nonliver stem cells become hepatocytes or BDECs. If transdifferentiation can be proven, the range of application for stem cells will be greatly increased. The experimental results of transdifferentiation have been reported in many studies, but unfortunately, many results were from *in vitro* studies without known molecular mechanisms. In addition, the frequency of suggested transdifferentiation is relatively low. Some findings on transdifferentiation related to liver cells are introduced in the following.

3.4.1. *Liver stem cell lines to become other cells outside of the liver*

WB-F344 epithelial cells as liver stem cells derived from normal rat liver cells[15] have been known to exhibit multipotent properties *in vitro*. In addition to becoming hepatocytes and BDECs, WB-F344 cells can also become a variety of other cells, including skeletal muscle cells, cardiac muscle cells, insulin-producing β cells, prostate glandular cells, hematopoietic cells, and liver tumor cells under different manipulations.[14,15]

3.4.2. *Noncharacterized liver stem cells in the pancreas*

During embryogenesis, the main pancreatic cells, including exocrine acinar and duct cells, and endocrine α, β, and δ cells, develop from a common definitive endodermal precursor located in the ventral foregut.[53,54] Importantly, the main epithelial cells of the liver, hepatocytes and BDECs, are from the same region of the foregut endoderm.[55] The hepatic anlage develops ventrally toward the cardiac mesenchyme, which induces the hepatoblast differentiation pathway. The pancreas buds from the same region, with its ventral lobe growing anteriorly in the same direction as the liver and its larger dorsal lobe growing posteriorly. Thus, the ventral lobe of the pancreas is particularly closely related anatomically to the liver.

The tight relationship between the liver and the pancreas in embryonic development has raised the possibility that a common hepatopancreatic precursor/stem cell may persist in adult life in both organs. Indeed, several independent lines of evidence suggest that the adult pancreas contains cells that can give rise to hepatocytes. The best-known example is the emergence of hepatocytes in copper-depleted rats after refeeding of copper.[56] In this system, weanling rats are fed a copper-free diet for 8 weeks, which leads to complete acinar atrophy, and then are re-fed with copper. Within weeks, cells with multiple hepatocellular characteristics emerge from the remaining pancreatic ducts. This work has been interpreted to suggest the presence of a pancreatic liver stem cell.[57] This notion is supported by the appearance of hepatocellular markers in human pancreatic cancers, which suggests common ancestors of cancer stem cells for the liver and the pancreas.[58] Recently, a specific cytokine has been identified as a candidate for driving this process. Transgenic mice in which the keratinocyte growth factor (KGF) gene is driven by an insulin promoter consistently develop pancreatic hepatocytes.[59] Thus, the existence of pancreatic liver precursors has been shown in several different mammalian species and under multiple experimental conditions. Both the adult liver and the adult pancreas may continue to harbor a small population of primitive hepatopancreatic stem cells with the potential to give rise to the same differentiated progeny as during embryogenesis.

3.4.3. *Peripheral blood monocytes to become hepatocytes*

Unlike stem cells such as MSCs and MAPCs, monocytes as differentiated cells from the hematopoietic system were shown with hepatocytic differentiation. The differentiation of human peripheral blood monocytes into hepatocyte-like and pancreatic islet-like cells has been reported.[60] Monocytes were treated with macrophage colony-stimulating factor and interleukin 3, followed by incubation with hepatocyte- and pancreatic islet-specific differentiation media. In response to macrophage colony-stimulating factor and interleukin 3, monocytes resumed cell division and reprogrammed to be capable of differentiating into new hepatocytes which closely resemble primary human hepatocytes with respect to morphology, expression of hepatocyte markers, and specific metabolic functions. After transplantation into the liver of mice with severe immunodeficiency disease — nonobese diabetic mice, new hepatocytes integrated well into the liver tissue and showed a morphology and albumin expression similar to that of primary human hepatocytes transplanted under identical conditions. The ability to reprogram, expand, and differentiate peripheral blood monocytes into hepatocytes opens a real possibility of clinical application of programmable cells of monocytic origin in liver tissue repair and organ regeneration.

3.4.4. *Salivary gland cells to become hepatocytes*

Salivary gland progenitor (SGP) cells could be isolated from the rat[61] and the mouse[62] after the experimental injury of salivary gland duct ligation. The cells expressed common stem cell markers such as Sca-1 and c-Kit. In culture, SGP cells have the capacity for cell expansion from a single cell clone, which suggested the stemness of these cells. In addition, they have shown a capacity for differentiations of pancreatic-cell-like cells and hepatic oval cell-like cells.[62] When transplanted into mice with standard PH, the partially differentiated SGP cells could further differentiate into cells with mature hepatocytic markers and SGP cell-derived CK19-positive cells composed into the standard duct structures.

4. Liver Regeneration

There are three types of liver regeneration under different mechanisms: (1) liver tissue turnover under normal conditions; (2) liver regeneration after acute liver injury; and (3) progenitor-dependent liver regeneration under chronic injury. However, it is still not fully understood how liver stem cells take part in all of these liver regeneration processes. Characterization of liver stem cells during liver regeneration will be helpful for manipulating the activities of liver stem cells for application of liver regeneration in clinical therapy.

4.1. Liver tissue turnover under normal conditions

Adult hepatocytes have an average lifespan of 200–300 days, after which they are replaced by new hepatocytes.[63] One of the mechanisms for this activity of normal liver tissue turnover was suggested in the "streaming liver" model.[64] The stream is consistent with the main bloodstream in the liver from the portal vein to the central vein. Such liver regeneration is similar to the regeneration in the intestine. Young hepatocytes appeared from a site near the portal vein, which might be from stem cells, and migrated toward the site of the central vein. The gene expression patterns along the stream supported the turnover process during the young cell migration in liver zones 1 and 2 and from the portal vein to the central vein (Fig. 2). This movement was suggested as a typical lineage progression in the adult liver. Furthermore, the ploidy and size of hepatocytes showed the gradients in the liver zones from the portal vein to the central vein. Central hepatocytes tend to be larger cells, with more ploidy than their periportal counterparts.

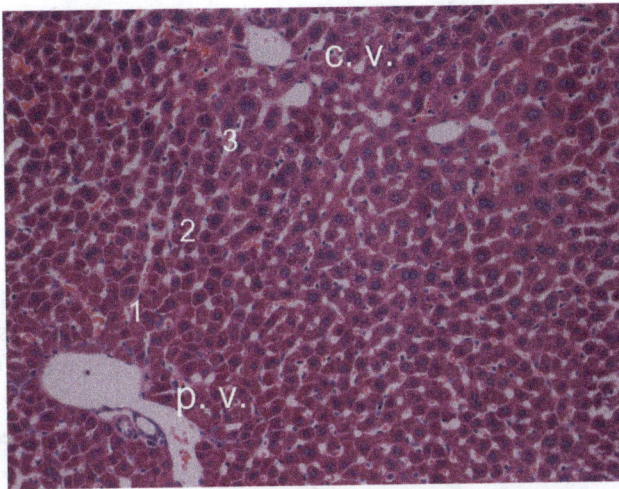

Figure 2. Matabolic zonation of the liver lobule. A histological section of the mouse liver was stained by HE. The blood flow is from the portal vein (p.v.) through zones 1, 2, and 3 to the central vein (c.v.).

Another possible explanation is that normal liver turnover in adult animals is mediated by *in situ* cell division of hepatocytes, instead of the turnover in streaming. The liver stem cells are not related to normal liver turnover.[65,66] The patterns of cells and gene expression could be reversed by the reverse of the main blood flow in the liver. Therefore, normal liver turnover does not depend on the suspected liver stem cells in the zone of the periportal vein. The experimental results from retroviral marking on hepatocytes support this new hypothesis. Retroviral-marked hepatocytes formed small clones, remained largely coherent, and were equally distributed everywhere, instead of a gradient distribution.[66] Additional results to support the second mechanism come from experiments utilizing the mosaic pattern of X inactivation in female mice to analyze patterns of hepatocytes in growth.[67,68]

4.2. Liver regeneration after acute liver injury

The liver has a tremendous capacity to regenerate after injury. Because of the inherent difficulty in studying liver regeneration in human subjects, many animal models have been established for studying liver regeneration after injury. These include surgical models such as PH and portal branch ligation, pharmacological models using a variety of hepatotoxic agents, and pre-existing liver injury models such as hereditary metabolic or anatomic defects of the liver (reviewed in Ref. 69). Our current understanding of liver regeneration after acute injury comes largely from the surgical and pharmacological models, whereas pre-existing liver injury models provide valuable information about liver regeneration after chronic injury.

A lot of information came from the rat model of 2/3 PH established originally by Higgins and Anderson many years ago.[2] In this classic model, the rat liver restored its original mass in about three days after surgical removal of 70% of the liver mass, including the right and left lobes. Strictly speaking, regeneration was a biological misnomer, since the resected lobes did not grow back. Instead, liver mass restoration occurred through the compensatory hyperplasia of the remaining lobes. Interestingly, the restoration process stopped after the original mass was gained, even after repeated PHs. This clearly indicates that liver regeneration is a highly coordinated process with tight biological control.

The molecular mechanisms underlying the liver regeneration process have recently been studied. Three stages were proposed for the liver regeneration after PH: initiation, proliferation, and termination. The initiation stage occurred within 4 hours of PH and was characterized by priming of quiescent hepatocytes by factors such as TNF-α, IL-6, and nitric oxide released by injured hepatocytes and nonhepatocytes upon stimulation of possible gut-derived cytokines. The priming led to the expression in the hepatocytes of immediate early genes such as c-fos, c-myc, jun B, NF-κB, STAT3, AP-1, and C/EBPβ.[15,70] These transcription factors in turn caused secondary activation of multiple genes, including growth factor and receptor genes, resulting in primed hepatocytes that were more competent to replicate. The proliferation stage was characterized by proliferation of primed hepatocytes stimulated by potent mitogens such as HGF, TGF-α, and EGF together with comitogens such as norepinephrine and insulin, which induced cyclins and cyclin-dependent kinases critical for cell cycle progression.[70] Indeed, hepatocyte DNA synthesis was shown to start 12–16 hours after the hepatectomy and peaked at 24–48 hours. The onset of mitosis followed 6–8 hours later, reaching its maximum 48 hours after surgery.[71] Although numerous growth factors had been implicated in regulating liver regeneration, two receptor–ligand and growth factor signaling systems appeared to be mainly involved: HGF and its receptor (Met) and the EGF receptor and its large family of ligands and coreceptors.[72,73] At the termination stage, a stop signal kept the regenerated liver to an appropriate functional size. Little is known about the stop signal but TGF-β was implicated because of its inhibitory effects on DNA synthesis in the regenerating hepatocytes.

It is important to note that although the liver mass was restored 3 days after surgery, the liver histology at this time was very different from normal and was characterized by desinusoidalized hepatocyte clusters of 12–15 cells with little extracellular matrix. Stellate cells then began to migrate into the clusters and new sinusoidal vasculatures were formed. The normal liver histology was not re-established until 7–10 days after the surgery.[71,74]

4.3. Progenitor-dependent liver regeneration under chronic injury

In the Higgins–Anderson model of rat PH, the regeneration process involved the parenchymal hepatocytes almost exclusively. Indeed, differentiated hepatocytes were considered functional stem cells because of their maintained potential for repeated division, even though this potential might gradually subside with aging.[75] However, hepatocytes are certainly not the only cell type capable of liver regeneration. This has been clearly illustrated in a modified rat PH model where hepatocyte proliferation was inhibited by the hepatocarcinogen AAF.[76] In the presence of AAF, oval cells, rather than hepatocytes, became activated and proliferated.[75] The differentiated hepatocytes served as functional liver stem cells after mild and even severe liver injuries such as 2/3 PH in the rat, whereas progenitor cells such as oval cells played a role only if the differentiated hepatocytes failed to bear the brunt of regeneration as in the case of inhibition by toxins such as AAF. The two-compartment system was validated by other models of liver injury as well as recent clinicopathological correlation in human studies.

Compared with that in acute injury, liver regeneration in long-term chronic injury has been less thoroughly studied. Both animal models and human studies implicated the involvement of hepatocytes as well as progenitor cells. Increased numbers of mitotically active hepatocytes were observed in chronic liver diseases.[77,78] Also noted was proliferative bile ductular reaction that corresponded to progenitor cell activation and proliferation.[79] Furthermore, the progenitor response correlated well with the severity of the chronic liver injury.[80] There was evidence that the time course of liver regeneration was delayed and sometimes regeneration was inhibited in chronically injured livers. Cell cycle arrest was implicated in the delay and inhibition.[81] In addition, liver regeneration in chronic injury was often associated with development of cirrhosis and hepatocellular carcinoma.[82] The molecular mechanism underlying these processes is still under investigation.

5. Liver Repopulation

Liver repopulation is a process of significant cell replacement in liver parenchyma. When a small number of donor cells are transplanted through the portal vein or spleen, the transplanted cells can migrate into the liver along with blood flow in the portal vein. These cells can be blocked in the microstructure of the liver from further migration, or could engraft in the recipient liver and expand to occupy the whole recipient liver [EKH1]. Under special experimental conditions, the donor hepatocytes could replace up to 90% of endogenous hepatocytes, or up to > 50% of the liver mass.[83] For these situations, the process of liver repopulation is analogous to that of bone marrow transplanted into the lethally irradiated host to expand and replace the hematopoietic system of the host. However, the efficiency of engraftment after cell transplantation was less than 1% under conventional conditions. Although transplanted hepatocytes can engraft, the expected therapeutic liver repopulation following hepatocyte transplantation is difficult to achieve.

Liver repopulation has special significance for regenerative medicine. Orthotopic liver transplantation is usually required to treat end stage liver diseases. However, because there are not enough donor liver organs, most patients with end stage liver diseases die waiting. For the last 20 years, hepatocyte transplantation has been considered an alternative transplantation strategy for whole liver organs in curing some liver diseases.[84] Partial successes in clinical trials highlight the potential benefits.[84] The first attempt at human hepatocyte-based treatment of an inherited liver-based metabolic disorder was made in the context of *ex vivo* gene therapy of familial hypercholesterolemia.[85] Many investigators have transplanted allogeneic hepatocytes to correct metabolic disorders, including ornithine transcarbamylase (OTC) deficiency, alpha-1-antitrypsin deficiency, glycogen storage disease type Ia, infantile Refsum disease and Crigler–Najjar syndrome type I. However, hepatocyte transplantation is still in experimental stages and therapeutic liver repopulation following hepatocyte transplantation is still difficult to achieve. This is largely due to the limited availability of donor hepatocytes and the inability to efficiently expand the cells in the recipient liver to achieve therapeutic liver repopulation. Developing protocols that allow more efficient *in vivo* expansion of hepatocytes would render hepatocytes more suitable for liver repopulation in therapeutic application.

5.1. Animal models for hepatocyte transplantation assay

With special genetic modification of gene expression, or chemical and surgical treatments, several animal models were successfully generated for studying hepatocyte transplantation and liver repopulation (Table 3). Three of these models will be discussed in the following section. These animal models were used more frequently than the others because they carried advantages in comparison with the others in manipulation of experiments and the obviously strong selection pressures for donor cells.

Table 3. Animal Models for Liver Repopulation

Genetically Modified Model Systems[86,90,120–122]	Selective Pressure
Albumin–urokinase transgenic mouse	Urokinase-mediated hepatocyte injury
FAH knockout mouse	Accumulation of toxic tyrosine metabolites
Albumin–HSVTK transgenic mouse	HSVTK-mediated conversion of ganciclovir to toxin
Bcl2 expression mouse donor hepatcytes	Donor cells resistant to Fas-ligand-induced apoptosis
Cinnamon rat (Wilson's disease)	Accumulation of copper

Treatment Conditioning Model Systems[95,116,123,124]	Selective Pressure
Retrorsine pretreated rat/mouse	DNA of host hepatocytes is damaged by retrorsine
Irradiation pretreated rat	DNA of host hepatocytes is damaged by X-rays
Monocrotaline pretreated rat/mouse	DNA of host hepatocytes is damaged by monocrotaline

5.1.1. *Urokinase plasminogen-activator transgenic mouse*

Albumin–urokinase plasminogen activator (Alb–uPA) transgenic mice were the first model to achieve liver repopulation.[86] In uPA transgenic mice, the expression of the uPA gene was under the control of the albumin promoter, which is expressed in a liver-specific manner. Expression of the uPA gene in hepatocytes induced cell death. These animals displayed sudden loss of hepatocytes and liver inflammation with necrosis. Most animals died, but some could survive and showed spontaneous development of cell nodules with normal hepatocytes. These nodules were all found in clones that contained cells originating from a spontaneous loss of the uPA transgene. The reverted hepatocytes had a selective growth advantage and thus resulted in repopulation of the liver of uPA transgenic mice. When hepatocytes isolated from wild-type mice were transplanted into young uPA mice, the donor hepatocytes could act as endogenous hepatocytes and have a powerful effect on growth and repopulate the liver of recipients.[87]

5.1.2. *Fumarylacetoacetate hydrolase knocked-out mouse*

Mice in which the gene of the tyrosine catabolic enzyme fumarylacetoacetate hydrolase (FAH) is knocked out[1] represent a robust model for liver repopulation. In humans, FAH gene deficiency causes the disease hepatorenal tyrosinemia or hereditary tyrosinemia type 1 (HT1).[88] FAH mutant (FAH$^{-/-}$) mice develop severe liver injury, which is invariably lethal. The FAH-mutant-derived liver injury can be rescued by blocking the tyrosine pathway upstream using the drug 2-(2-nitro-4-trifluoro-methylbenzyol)-1,3 cyclohexanedione (NTBC).[89] Thus, FAH$^{-/-}$ mice can develop to adulthood, breed, and remain healthy until NTBC is removed. After NTBC withdrawal, liver failure occurs and the mice die within 4–6 weeks. When FAH$^{-/-}$ mice are transplanted with single cell suspension of syngeneic wild-type hepatocytes, the wild-type FAH-positive hepatocytes migrate to the liver and divide until > 95% of the liver cell mass is replaced by donor cells,[90] and animals with repopulated livers are healthy (Fig. 3). Moreover, repopulating cells can be serially transplanted to measure the cell division capacity and to generate large numbers of cells for genetic and cytological studies.[50,78]

5.1.3. *Retrorsine-treated animals*

Laconi *et al.* (1995) developed an approach to liver repopulation by chemically blocking the regenerative capacity of host cells using retrorsine,

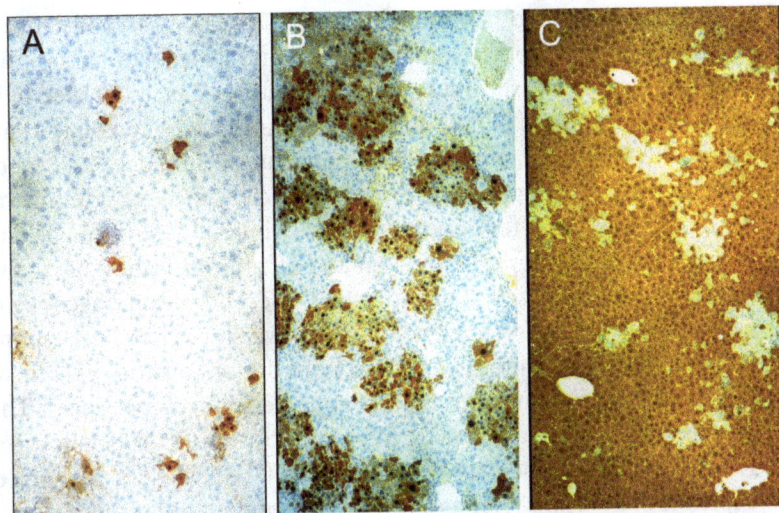

Figure 3. Liver repopulation in the FAH$^{-/-}$ mouse after wild-type hepatocyte transplantation. Wild-type hepatocytes can engraft and replace up to 95% of endogenous hepatocytes in the recipient. (A) A FAH$^{-/-}$ liver sample was harvested after 2 days of hepatocyte transplantation and a section was stained by anti-FAH antibody. (B) A FAH$^{-/-}$ liver sample was harvested after 3 weeks of hepatocyte transplantation and a section was stained by anti-FAH antibody. The donor cells started to repopulate into nodules. (B) A FAH$^{-/-}$ liver sample was harvested after 8 weeks of hepatocyte transplantation and a section was stained by anti-FAH antibody. The complete liver repopulation reached the final level of > 95%.

structurally similar pyrrolizidine alkaloids.[91] The retrorsine chemical compound was selectively metabolized to the active form by hepatocytes, which alkylated cellular DNA and caused proliferation arrest of hepatocytes in the G2 phase of the cell cycle.[92,93] Gordon *et al.* (2000) showed recently that retrorsine-treated animals with PH induced proliferation of endogenous, small, hepatocyte-like progenitor cells.[94] Retrorsine treatment followed by PH can also be used to achieve near-total liver repopulation by transplanted donor cells. Typically, genetically marked rat hepatocytes that were positive for the bile canalicular membrane protein dipeptidyl peptidase IV (DPPIV) were injected into the spleen of a congenic strain of mutant rats not expressing DPPIV enzyme activity. Within 2 months, there was 40–60% replacement by transplanted hepatocytes in female rats and > 95% replacement in male rats. Similar to the finding in rats, retrorsine treatment plus PH in mice has resulted in selection effects for donor hepatocytes in liver repopulation.[95]

Compared with other models, the FAH$^{-/-}$ mouse model has several advantages in hepatocyte transplantation and liver repopulation. First, the time and degree of liver injury in FAH$^{-/-}$ mice can be controlled by switching administration of NTBC on and off.[89] This makes it possible to maintain the health of FAH$^{-/-}$ mice so that they can be raised to adulthood. The breeding of homozygous FAH$^{-/-}$ mice is efficient with the administration of NTBC. The selection for donor hepatocytes in FAH$^{-/-}$ mice can be done flexibly, and repeatedly by switching on and off administration of NTBC based on the overall health of the recipients.[50,96] Second, the endogenous hepatocytes of FAH$^{-/-}$ mice do not undergo autoreversion. Thus, liver repopulation in FAH$^{-/-}$ mice is completely derived from donor hepatocytes, which makes the serial transplantation of repopulating hepatocytes possible.[78] Third, the detection of repopulated wild-type FAH-positive donor hepatocytes in the FAH$^{-/-}$ mouse is based on the existence of FAH DNA and FAH enzyme expression. By using FAH as a marker, we can quantitatively determine the donor-derived hepatocytes with FAH expression in the background of FAH-expressing negative cells in the recipient's liver.

In addition to testing liver repopulation in animal models with strong selection for donor hepatocytes such as in FAH$^{-/-}$ mice, models without significant selection of donor hepatocytes, such as PH of the normal liver, have also been used to test liver repopulation.[97,98] Because there is no selective pressure for donor hepatocytes in the liver of PH, liver repopulation is significantly difficult. Based on this property in a PH model, cells with strong proliferating capacity, such as fetal liver progenitor cells, could be studied in liver repopulation.[99] These experiments have elucidated the advantage of liver progenitor cells in liver repopulation. In addition, it will be possible to study the enhanced liver population capacity of hepatocytes that are modified with sustained expression of a cell cycle enhancer such as FoxM1. The genetically modified hepatocytes with overexpression of the cell cycle regulator are expected to overcome the difficulty of using normal hepatocytes as donor cells for liver repopulation just after PH.

5.2. Liver repopulations from mature hepatocytes

Several experiments have strongly suggested that fully differentiated mature hepatocytes, which constitute the majority of liver cells, are responsible for liver repopulation and have a stem-cell-like capacity for cell division. One study used the transplantation assay model of FAG$^{-/-}$ mouse to serially transplant wild-type hepatocytes at limiting dilution through seven rounds.

Complete repopulation was achieved in each round, and it was estimated that the repopulating cells in the seventh-round recipients had undergone at least 100 cell doublings, similar to what has been seen in serial transplantation of hematopoietic stem cells.[78] Interestingly, the only donor-derived cells in this experiment were hepatocytes. No BDECs or other cell types of donor origin were found, thus raising the possibility of a "unipotential" stem cell. In another report, three sets of experiments were performed to address whether differentiated hepatocytes or putative stem cells were responsible for repopulation.[77] First, cell fractionation by centrifugal elutriation was used to identify and purify three major-size fractions of hepatocytes (cell diameters of 16 μm, 21 μm, and 27 μm). Each fraction was transplanted in competition with unfractionated liver cells carrying a distinct genetic marker, which served as a baseline reference for liver repopulation. The larger polyploid hepatocytes, which represented about 70% of the hepatocyte population, were primarily responsible for liver repopulation. In contrast, small diploid hepatocytes were inferior to the larger cells in competitive repopulation experiments. Second, competitive repopulation was performed between naive liver cells and those that had been serially transplanted up to seven times. Importantly, serial transplantation neither enhanced nor diminished the liver repopulation capacity. If serial liver repopulation were stem cell-dependent, this result would suggest that the ratio of progenitors to differentiated hepatocytes was kept constant during a 10–20-fold cell expansion. More likely, this result means that virtually all the original input cells (> 95% hepatocytes) were capable of serial transplantation. The third set of experiments involved retroviral marking of donor hepatocytes *in vitro* and *in vivo*. Again, no evidence for a rare stem cell responsible for liver repopulation was detected. Together, these experiments strongly suggest that fully differentiated hepatocytes, which constitute the majority of liver cells, are responsible for liver repopulation and have a stem-cell-like capacity for cell division.

5.3. Liver repopulations from fetal liver progenitor cells

Several recent reports have indicated that liver progenitor cells isolated from rat or mouse fetal livers can differentiate into hepatocytes after being transplanted into the liver of the adult rat or mouse.[100–102] Unlike adult hepatocytes, fetal liver progenitor cells might become not only mature hepatocytes during liver repopulation, but also BDECs for reconstitution of the biliary system.[102] In addition, fetal liver progenitor cells also showed enhanced proliferating capacity during liver repopulation in comparison with mature hepatocytes,[99] which suggested the advantage of using progenitor cells instead of mature hepatocytes for therapeutic liver repopulation.

One early report using fetal rat liver cells in the retrorsine model indicated that there were at least three distinct subpopulations of hepatoblasts at ED 12–14.[102] One population appeared to be bipotential on the basis of the histochemical markers albumin and CK19, and the other two had either the unipotent hepatocyte marker albumin or the biliary epithelial cell marker CK19. After transplantation, the bipotential cells were able to proliferate in retrorsine-treated recipients, whereas the unipotent cells proliferated even in the liver of the rat without retrorsine treatment. However, none of the fetal liver cell populations proliferated spontaneously. Either PH or thyroid hormone treatment was required to induce proliferation of the transplanted cells. Nonetheless, fetal liver cells proliferated more readily than adult cells. The transplanted fetal cells gave rise to both mature hepatocyte cords and mature bile duct structures. This result suggested that new hepatocytes from transplanted fetal cells could undertake liver repopulation in the same manner as adult hepatocytes. The reasons for how fetal liver progenitor cells undertook hepatic differentiation to become mature hepatocytes in the circumstance of the adult liver have not been elucidated yet. However, the capacity of liver progenitor cells to finish differentiation in recipient livers will enable us to use them directly in cell transplantation therapy in the future.

In a similar transplantation assay, rat fetal liver cells as donor cells showed differences in cell proliferation and cell apoptosis in comparison with endogenous hepatocytes.[99] The "cell–cell competition" between donor fetal liver cells and endogenous hepatocytes was suggested during fetal liver cell repopulation. Liver repopulation resulted from the stronger proliferating activity of transplanted fetal liver cells and reduced apoptosis of their progeny compared with host hepatocytes, coupled with increased apoptosis of host hepatocytes immediately adjacent to transplanted cells.

5.4. Liver repopulations from various liver progenitor cells

5.4.1. *Liver repopulations from oval cells*

Oval cells have been proven capable of liver repopulation after transplantation.[100,103] In a study with a rat model system,[100] oval cells in the pancreas were induced by feeding rats a copper-depletion diet. Oval cell fractions enriched with candidate epithelial progenitor cells from the rat pancreas were isolated and transplanted into the liver of an inbred strain of Fischer rats. Using a dipeptidyl dipeptidase IV genetic marker system to follow the fate of transplanted oval cells with expression of the albumin gene, pancreatic oval cells differentiate into hepatocytes after transplantation to the liver, express liver-specific proteins, and become fully integrated into the liver parenchymal structure. In another study on the mouse, oval cells were induced and isolated from the adult mouse liver by 3,5-diethoxycarbonyl-1,4-dihydrocollidine (DDC).[103] Transplantation into FAH$^{-/-}$ mice was used to determine their capacity for liver repopulation. In competitive repopulation experiments, hepatic oval cells were at least as efficient as mature hepatocytes in repopulating the liver. In addition, oval cells were derived not from hepatocytes, but from liver nonparenchymal cells. This finding supported a model in which intrahepatic progenitors differentiate into hepatocytes irreversibly. Other results indicate that oval cells do not originate in bone marrow but in the liver itself. Most importantly, results from transplantation models of both the rat and the mouse indicate that oval cells have valuable properties for therapeutic liver repopulation.

5.4.2. *Liver repopulations from pancreatic liver progenitor cells*

Pancreatic oval cells induced by copper depletion were shown to give rise to morphologically normal hepatocytes *in vivo*.[100] Other experiments also proved that pancreatic hepatocyte precursors existed in the normal pancreas without the use of toxic induction regimens and copper depletion. Liver repopulation experiments were performed in the FAH$^{-/-}$ mouse model. Pancreatic cell suspensions from adult wild-type mice without any pretreatment were transplanted into FAH$^{-/-}$ recipients, and selection was induced by NTBC withdrawal. Extensive liver repopulation (> 50%) was observed in about 10% of transplant recipients, and another 35% had histological evidence for donor-derived hepatocyte nodules.[104] The experimental results indicated that the adult mouse pancreas contained hepatocyte precursors, even under normal, nonpathologic conditions. These pancreatic liver stem cells may have other differentiation potential, particularly toward the pancreatic endocrine lineages.[104]

5.4.3. *Liver repopulations from ES cell-derived hepatic cells*

The transplantation assay has become one of the experiments necessary for proving the functionality of ES cell-derived hepatocytes. Previously, the *in vivo* hepatic function of ES cell-derived hepatocytes could not be well studied with a robust model. The mouse ES cell-derived hepatocyte-like cells were only known to be capable of homing in mouse recipients that were pretreated with PH only,[26] PH + CCl4 injection,[24] or dimethyl-nitroamine (DMN) injection.[36] Recently, two reports have shown the results of liver repopulation from mouse-ES cell-derived cells using mouse models with selectable pressure of donor hepatic cells.[105,106] In one report, mouse ES cells were first transfected with the EGFP fluorescent reporter gene regulated by a albumin enhancer/promoter.[106] Then, the genetically modified ES cells were induced to differentiate into hepatocytes in a protocol of a serum-free chemically defined medium. The embryoid bodies and differentiation of hepatic lineage cells occurred in the absence of exogenous growth factors or feeder cell layers. EGFP+ cells could increase in number and acquire hepatocyte-like morphology and hepatocyte-specific markers. When transplanted into Alb–uPA/SCID mice, the FACS-enriched EGFP+ cells developed into functional hepatocytes without evidence of cell fusion and participated in the repair of the diseased liver. In the late treatment of the same recipients with CCl4-induced liver injury, ES cell-derived hepatocytes were responsive to normal growth regulation and proliferated at the same rate as the host hepatocytes. In addition, the transplanted GFP+ cells also differentiated into BDECs. In another report, mouse ES cells were treated with activin A in serum-free conditions.[105] The cells could be induced into definitive endoderm progenitor cells defined by the coexpression of either a combination of Brachyury, Foxa2, and c-Kit, or a combination of c-Kit and CXCR4. With additional treatment of bone morphogenetic protein 4 in combination with basic fibroblast growth factor and activin A, induced definitive endoderm progenitor cells could become hepatic cells expressing hepatocyte marker proteins, secreting albumin, storing glycogen, and showing ultrastructural characteristics of mature hepatocytes. When transplanted into FAH−/− mice, the ES cell-derived hepatocyte-like cells repopulated in the recipients of the FAH−/− mouse with injured livers and matured into hepatocytes expressing FAH. The functionality of ES cell-derived hepatocyte-like cells was proven with the capacity for liver repopulation.

Human-ES cell-derived hepatocytes will become a potential source for generating hepatocytes in the future. A transplantation assay model suitable for human hepatic cells can be used for proving the *in vivo* functionality of human-ES cell-derived hepatocytes. Previous reports indicated that xenogeneic engraftment of human hepatocytes in the mouse liver

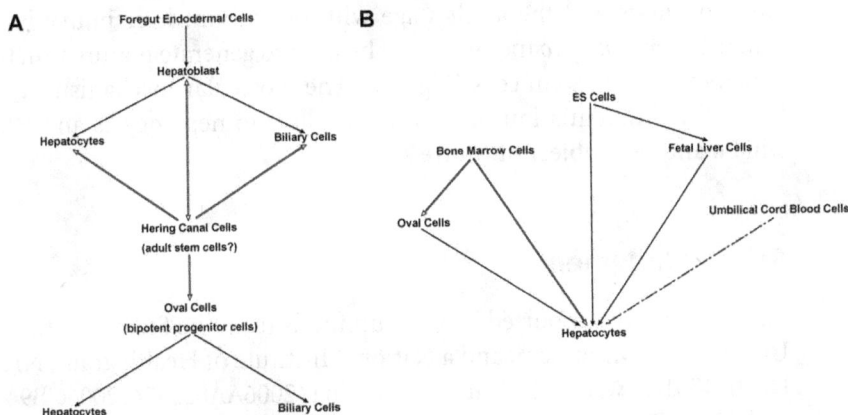

Figure 4. Hepatic lineage relationships in adult and embryonic stages (**A**) and during *in vitro* induction (**B**). Solid arrows indicate the known differentiation. Unfilled lines supply the information on possible relationships. Dashed lines mark the weak results without *in vivo* proving.

could be achieved in the Alb–uPA/Rag-2$^{-/-}$ mouse[107] and the Alb–uPA/SCID mouse.[108,109] These model systems are expected to be useful in analyzing the *in vivo* function of human-ES cell-derived hepatocyte-like cells.

Conclusion

The subject of liver stem cells is of interest for both scientific research and medical application. Basic study of liver developmental biology and liver organogenesis has provided important knowledge in understanding the nature of liver stem cells in embryos. In addition, liver injuries induced under experimental conditions supplied additional information for understanding the activities of liver stem cells in adults under special conditions. In the future, the special properties of liver regeneration and liver repopulation will enable us to use liver stem cells in a special manner for the purpose of clinical therapy. Although liver stem cells have not been fully characterized, tremendous evidence has indicated that several types of stem cells and stem cell-like cells can be used to generate hepatocytes and BDECs. These cells will become valuable sources of donor cells for liver-regenerative medicine. They may be used for generating liver tissues or used directly for liver repopulation in the future. In Fig. 4A, our current knowledge about the basic lineage relationships of hepatocytes during embryogenesis and the adult stage with or without liver injury is illustrated. Several *in vitro* methods may be used to generate mature functional hepatocytes from stem cells (Fig. 4B). The molecular mechanisms underlining the differentiation of liver stem cells into hepatocytes and BDECs will be another subject of future studies.

Acknowledgments

The authors are supported by start-up funds from the Stem Cell Institute, University of Minnesota, and a National Institute of Health grant, NIDDK DK-074561, as well as the National Funding (2006AA022474, 2006CB943901 and 2006CB947900).

References

1. Grompe M, Finegold M. (2001) Liver stem cells. In: Marshak DR, Gardner R, Godbout MJ (eds.), *Stem Cell Biology,* Cold Spring Harbor Laboratory Press, pp. 455–497.
2. Sell S, Ilic Z. (1997) *Liver Stem Cells: Medical Intelligence Unit,* 1st edn. Springer-Verlag, New York.
3. Zaret KS. (2000) Liver specification and early morphogenesis. *Mech Dev* 92: 83–88.
4. Gualdi R, Bossard P, Zheng M, *et al.* (1996) Hepatic specification of the gut endoderm *in vitro*: cell signaling and transcriptional control. *Genes Dev* 10: 1670–1682.
5. Duncan SA. (2003) Mechanisms controlling early development of the liver. *Mech Dev* 120: 19–33.
6. Jung J, Zheng M, Goldfarb M, Zaret KS. (1999) Initiation of mammalian liver development from endoderm by fibroblast growth factors. *Science* 284: 1998–2003.
7. Schmelzer E, Wauthier E, Reid LM. (2006) The phenotypes of pluripotent human hepatic progenitors. *Stem Cells* 24: 1852–1858.
8. Zheng YW, Taniguchi H. (2003) Diversity of hepatic stem cells in the fetal and adult liver. *Semin Liver Dis* 23: 337–348.
9. Farber E. (1956) Similarities in the sequence of early histological changes induced in the liver of the rat by ethionine, 2-acetylamino-fluorene, and 3′-methyl-4-dimethylaminoazobenzene. *Cancer Res* 16: 142–148.
10. Preisegger KH, Factor VM, Fuchsbichler A, *et al.* (1999) Atypical ductular proliferation and its inhibition by transforming growth factor beta1 in the

3,5-diethoxycarbonyl-1,4-dihydrocollidine mouse model for chronic alcoholic liver disease. *Lab Invest* **79:** 103–109.

11. Wilson JW, Leduc EH. (1958) Role of cholangioles in restoration of the liver of the mouse after dietary injury. *J Pathol Bacteriol* **76:** 441–449.

12. Walkup MH, Gerber DA. (2006) Hepatic stem cells: in search of. *Stem Cells* **24:** 1833–1840.

13. Li WL, Su J, Yao YC, *et al.* (2006) Isolation and characterization of bipotent liver progenitor cells from adult mouse. *Stem Cells* **24:** 322–332.

14. Cao LZ, Tang DQ, Horb ME, *et al.* (2004) High glucose is necessary for complete maturation of Pdx1-VP16-expressing hepatic cells into functional insulin-producing cells. *Diabetes* **53:** 3168–3178.

15. Fausto N, Campbell JS. (2003) The role of hepatocytes and oval cells in liver regeneration and repopulation. *Mech Dev* **120:** 117–130.

16. Spagnoli FM, Amicone L, Tripodi M, Weiss MC. (1998) Identification of a bipotential precursor cell in hepatic cell lines derived from transgenic mice expressing cyto-Met in the liver. *J Cell Biol* **143:** 1101–1112.

17. Rogler LE. (1997) Selective bipotential differentiation of mouse embryonic hepatoblasts *in vitro*. *Am J Pathol* **150:** 591–602.

18. Nussler AK, Vergani G, Gollin SM, *et al.* (1999) Isolation and characterization of a human hepatic epithelial-like cells line (AKN-1) from a normal liver. *In Vitro Cell Dev Biol Anim* **35:** 190–197.

19. Lazaro CA, Croager EJ, Mitchell C, *et al.* (2003) Establishment, characterization and long-term maintenance of cultures of human fetal hepatocytes. *Hepatology* **38:** 1095–1106.

20. Dan YY, Rhiele KJ, Lazaro C, *et al.* (2006) Isolation of multipotent progenitor cells from human fetal liver capable of differentiating into liver and mesenchymal lineages. *Proc Natl Acad Sci USA* **103:** 9912–9917.

21. Evans MJ, Kaufman MH. (1981) Establishment in culture of pluripotential cells from mouse embryos. *Nature* **292:** 154–156.

22. Martin GR. (1981) Isolation of a pluripotent cell line from early mouse embryos cultured in medium conditioned by teratocarcinoma stem cells. *Proc Natl Acad Sci USA* **78:** 7634–7638.

23. Odorico JS, Kaufman DS, Thomson JA. (2001) Multilineage differentiation from human embryonic stem cell lines. *Stem Cells* **19:** 193–204.

24. Chinzei R, Tanaka Y, Shimizu-Saito K, *et al.* (2002) Embryoid-body cells derived from a mouse embryonic stem cell line show differentiation into functional hepatocytes. *Hepatology* **36:** 22–29.

25. Hamazaki T, Iiboshi Y, Oka M, *et al.* (2001) Hepatic maturation in differentiating embryonic stem cells *in vitro*. *FEBS Lett* **497:** 15–19.

26. Yin Y, Lim YK, Salto-Tellez M, *et al.* (2002) AFP[(+)], ESC-derived cells engraft and differentiate into hepatocytes *in vivo*. *Stem Cells* **20:** 338–346.

27. Yamada T, Yoshikawa M, Kanda S, *et al.* (2002) *In vitro* differentiation of embryonic stem cells into hepatocyte-like cells identified by cellular uptake of indcyanine green. *Stem Cells* 20: 146–154.

28. Jones EA, Tosh D, Wilson DI, *et al.* (2002) Hepatic differentiation of murine embryonic stem cells. *Exp Cell Res* 272: 15–22.

29. Ishizaka S, Shiroi A, Kanda S, *et al.* (2002) Development of hepatocytes from ES cells after transfection with the HNF-3beta gene. *FASEB J* 16: 1444–1446.

30. Yamamoto H, Quinn G, Asari A, *et al.* (2003) Differentiation of embryonic stem cells into hepatocytes: biological functions and therapeutic application. *Hepatology* 37: 983–993.

31. Kuai XL, Cong XQ, Li XL, Xiao SD. (2003) Generation of hepatocytes from cultured mouse embryonic stem cells. *Liver Transpl* 9: 1094–1099.

32. Fair JH, Cairns BA, Lapaglia M, *et al.* (2003) Induction of hepatic differentiation in embryonic stem cells by co-culture with embryonic cardiac mesoderm. *Surgery* 134: 189–196.

33. Tanaka Y, Yoshikawa M, Kobayashi Y, *et al.* (2003) Expressions of hepatobiliary organic anion transporters and bilirubin-conjugating enzyme in differentiating embryonic stem cells. *Biochem Biophys Res Commun* 309: 324–330.

34. Lebkowski JS, Gold J, Xu C, *et al.* (2001) Human embryonic stem cells: culture, differentiation, and genetic modification for regenerative medicine applications. *Cancer J* 7 **Suppl 2:** S83–93.

35. Ishii T, Yasuchika K, Fujii H, *et al.* (2005) *In vitro* differentiation and maturation of mouse embryonic stem cells into hepatocytes. *Exp Cell Res* 309: 68–77.

36. Teratoni T, Yamamoto H, Aoyagi K, *et al.* (2005) Direct hepatic fate specification from mouse embryonic stem cells. *Hepatology* 41: 836–846.

37. Kubo A, Shinozaki K, Shannon JM, *et al.* (2004) Development of definitive endoderm from embryonic stem cells in culture. *Development* 131: 1651–1662.

38. Keller G. (2005) Embryonic stem cell differentiation: emergence of a new era in biology and medicine. *Genes Dev* 19: 1129–1155.

39. D'Amour KA, Agulnick AD, Eliazer S, *et al.* (2005) Efficient differentiation of human embryonic stem cells to definitive endoderm. *Nat Biotechnol* 23: 1534–1541.

40. Yasunaga M, Tada S, Torikai-Nishikawa S, *et al.* (2005) Induction and monitoring of definitive and visceral endoderm differentiation of mouse ES cells. *Nat Biotechnol* 23: 1542–1550.

41. Schuldiner M, Yanuka O, Itskovitz-Eldor J, *et al.* (2000) Effects of eight growth factors on the differentiation of cells derived from human embryonic stem cells. *Proc Natl Acad Sci USA* 97: 11307–11312.

42. Rambhatla L, Chiu CP, Kundu P, *et al.* (2003) Generation of hepatocyte-like cells from human embryonic stem cells. *Cell Transplant* 12: 1–11.

43. Levenberg S, Huang NF, Lavik E, *et al.* (2003) Differentiation of human embryonic stem cells on three-dimensional polymer scaffolds. *Proc Natl Acad Sci USA* **100:** 12741–12746.

44. Grompe M. (2003) The role of bone marrow stem cells in liver regeneration. *Semin Liver Dis* **23:** 363–372.

45. Pittenger MF, Mackay AM, Beck SC, *et al.* (1999) Multilineage potential of adult human mesenchymal stem cells. *Science* **284:** 143–147.

46. Lee KD, Kuo TK, Whang-Peng J, *et al.* (2004) *In vitro* hepatic differentiation of human mesenchymal stem cells. *Hepatology* **40:** 1275–1284.

47. Kang XQ, Zang WJ, Song TS, *et al.* (2005) Rat bone marrow mesenchymal stem cells differentiate into hepatocytes *in vitro*. *World J Gastroenterol* **11:** 2479–2484.

48. Schwartz RE, Reyes M, Koodie L, *et al.* (2002) Multipotent adult progenitor cells from bone marrow differentiate into functional hepatocyte-like cells. *J Clin Invest* **109:** 1291–1301.

49. Zeng L, Rahrmann E, Hu Q, *et al.* (2006) Multipotent adult progenitor cells from swine bone marrow. *Stem Cells* **24:** 2355–2366.

50. Wang X, Willenbring H, Akkari Y, *et al.* (2003) Cell fusion is the principal source of bone marrow-derived hepatocytes. *Nature* **422:** 897–901.

51. Willenbring H, Bailey AS, Foster M, *et al.* (2004) Myelomonocytic cells are sufficient for therapeutic cell fusion in liver. *Nat Med* **10:** 744–748.

52. Harris RG, Herzog EL, Bruscia EM, *et al.* (2004) Lack of a fusion requirement for development of bone marrow-derived epithelia. *Science* **305:** 90–93.

53. Spooner BS, Walther BT, Rutter WJ. (1970) The development of the dorsal and ventral mammalian pancreas *in vivo* and *in vitro*. *J Cell Biol* **47:** 235–246.

54. Rutter WJ. (1980) The development of the endocrine and exocrine pancreas. *Monogr Pathol* **21:** 30–38.

55. Shiojiri N, Lemire JM, Fausto N. (1991) Cell lineages and oval cell progenitors in rat liver development. *Cancer Res* **51:** 2611–2620.

56. Rao MS, Dwivedi RS, Yeldandi AV, *et al.* (1989) Role of periductal and ductular epithelial cells of the adult rat pancreas in pancreatic hepatocyte lineage: a change in the differentiation commitment. *Am J Pathol* **134:** 1069–1086.

57. Reddy JK, Rao MS, Yeldani AV, *et al.* (1991) Pancreatic hepatocytes: an *in vivo* model for cell lineage in pancreas of adult rat. *Dig Dis Sci* **36:** 502–509.

58. Hruban RH, Molina JM, Reddy MN, Boitnott JK. (1987) A neoplasm with pancreatic and hepatocellular differentiation presenting with subcutaneous fat necrosis. *Am J Clin Pathol* **88:** 639–645.

59. Krakowski ML, Kritzik MR, Jones EM, *et al.* (1999) Pancreatic expression of keratinocyte growth factor leads to differentiation of islet hepatocytes and proliferation of duct cells. *Am J Pathol* **154:** 638–691.

60. Ruhnke M, Ungefroren H, Nussler A, *et al.* (2005) Differentiation of *in vitro*-modified human peripheral blood monocytes into hepatocyte-like and pancreatic islet-like cells. *Gastroenterology* **128**: 1774–1786.
61. Okumura K, Nakamura K, Hisatomi Y, *et al.* (2003) Salivary gland progenitor cells induced by duct ligation differentiate into hepatic and pancreatic lineages. *Hepatology* **38**: 104–113.
62. Hisatomi Y, Okumura K, Nakamura K, *et al.* (2004) Flow cytometric isolation of endodermal progenitors from mouse salivary gland differentiate into hepatic and pancreatic lineages. *Hepatology* **39**: 667–675.
63. Fausto N, Webber EM. (1994) Liver regeneration. In: Arias IM (ed.), *The Liver — Biology and Pathology* (Raven, New York.), pp. 1059–1084.
64. Zajicek G, Oren R, Weinreb Jr M. (1985) The streaming liver. *Liver* **5**: 293–300.
65. Kennedy S, Rettinger S, Flye MW, Ponder KP. (1995) Experiments in transgenic mice show that hepatocytes are the source for postnatal liver growth and do not stream. *Hepatology* **22**: 160–168.
66. Bralet MP, Branchereau S, Brechot C, Ferry N. (1994) Cell lineage study in the liver using retroviral mediated gene transfer: evidence against the streaming of hepatocytes in normal liver. *Am J Pathol* **144**: 896–905.
67. Shiojiri N, Imai H, Goto S, *et al.* (1997) Mosaic patterns of ornithine transcarbamylase expression in spfash mouse liver. *Am J Pathol* **151**: 413–421.
68. Shiojiri N, Inujima S, Ishikawa K, *et al.* (2001) Cell lineage analysis during liver development using the spf(ash)-heterozygous mouse. *Lab Invest* **81**: 17–25.
69. Palmes D, Spiegel HU. (2004) Animal models of liver regeneration. *Biomaterials* **25**: 1601–1611.
70. Fausto N. (2000) Liver regeneration. *J Hepatol* **32**: 19–31.
71. Kountouras J, Boura P, Lygidakis NJ. (2001) Liver regeneration after hepatectomy. *Hepatogastroenterology* **48**: 556–562.
72. Pahlavan PS, Feldmann RE Jr, Zavos S, Kountouras J. (2006) Prometheus' challenge: molecular, cellular and systemic aspects of liver regeneration. *J Surg Res* **134**: 238–251.
73. Michalopoulos GK, DeFrances MC. (1997) Liver regeneration. *Science* **276**: 60–66.
74. Court FG, Wemyss-Holden SA, Dennison AR, Maddern GJ. (2002) The mystery of liver regeneration. *Br J Surg* **89**: 1089–1095.
75. Alison M. (1998) Liver stem cells: a two compartment system. *Curr Opin Cell Biol* **10**: 710–715.
76. Alison M, Sarraf C. (1998) Hepatic stem cells. *J Hepatol* **29**: 676–682.
77. Overturf K, al-Dhalimy M, Finegold M, Grompe M. (1999) The repopulation potential of hepatocyte populations differing in size and prior mitotic expansion. *Am J Pathol* **155**: 2135–2143.

78. Overturf K, al-Dhalimy M, Ou CN, *et al.* (1997) Serial transplantation reveals the stem-cell-like regenerative potential of adult mouse hepatocytes. *Am J Pathol* 151: 1273–1280.

79. Tan J, Hytiroglou P, Wieczorek R, *et al.* (2002) Immunohistochemical evidence for hepatic progenitor cells in liver disease. *Liver* 22: 365–373.

80. Eleazar JA, Memeo L, Jhang JS, *et al.* (2004) Progenitor cell expansion: an important source of hepatocyte regeneration in chronic hepatits. *J Hepatol* 41: 983–991.

81. Marshall A, Rushbrook S, Davies SE, *et al.* (2005) Relation between hepatocyte G1 arrest, impaired hepatic regeneration, and fibrosis in chronic hepatitis C virus infection. *Gastroenterology* 128: 33–42.

82. Bisgaard HC, Thorgeirsson SS. (1996) Hepatic regeneration: the role of regeneration in pathogenesis of chronic liver diseases. *Clin Lab Med* 16: 325–339.

83. Grompe M. (2006) Principles of therapeutic liver repopulation. *J Inherit Metab Dis* 29: 421–425.

84. Grompe M. (1999) Therapeutic liver repopulation for the treatment of metabolic liver diseases. *Hum Cell* 12: 171–180.

85. Fox IJ, Roy-Chowdhury J. (2004) Hepatocyte transplantation. *J Hepatol* 40: 878–886.

86. Rhim JA, Sandgren EP, Degen JL, *et al.* (1994) Replacement of disease mouse liver by hepatic cell transplantation. *Science* 263: 1149–1152.

87. Sandgren EP, Palmiter RD, Heckel JL, *et al.* (1991) Complete hepatic regeneration after somatic deletion of an albumin-plasminogen activator transgene. *Cell* 66: 245–256.

88. Mitchell GA, Grompe M, Lambert M, Tanguay RM. (2001) Hypertyrosinemia. In: Scriver CR (ed.), *The Metabolic and Molecular Basis of Inherited Disease*, McGraw-Hill, New York, pp. 1777–1805.

89. Grompe M, Lindstedt S, al-Dhalimy M, *et al.* (1995) Pharmacological correction of neonatal lethal hepatic dysfunction in a murine model of hereditary tyrosinaemia type I. *Nat Genet* 10: 453–460.

90. Overturf K, al-Dhalimy M, Tanguay R, *et al.* (1996) Hepatocytes corrected by gene therapy are selected *in vivo* in a murine model of hereditary tyrosinaemia type I. *Nat Genet* 12: 266–273.

91. Laconi E, Sarma DS, Pani P. (1995) Transplantation of normal hepatocytes modulates the development of chronic liver lesions induced by a pyrrolizidine alkaloid, lasiocarpine. *Carcinogenesis* 16: 139–142.

92. Samuel A, Jago MV. (1975) Localization in the cell cycle of the antimitotic action of the pyrrolizidine alkaloid, lasiocarpine and of its metabolite, dehydroheliotridine. *Chem Biol Interact* 10: 185–197.

93. Mattocks AR, Driver HE, Barbour RH, Robins DJ. (1986) Metabolism and toxicity of synthetic analogues of macrocyclic diester pyrrolizidine alkaloids. *Chem Biol Interact* 58: 95–108.

94. Gordon GH, Coleman WB, Hixson DC, Grisham JW. (2000) Liver regeneration in rats with retrorsine-induced hepatocellular injury proceeds through a novel cellular response. *Am J Pathol* **156**: 607–619.
95. Guo D, Fu T, Nelson JA, *et al.* (2002) Liver repopulation after cell transplantation in mice treated with retrorsine and carbon tetrachloride. *Transplantation* **73**: 1818–1824.
96. Lagasse E, Connors H, al-Dhalimy M, *et al.* (2000) Purified hematopoietic stem cells can differentiate into hepatocytes *in vivo*. *Nat Med* **6**: 1229–1234.
97. Yuan RH, Ogawa A, Ogawa E, *et al.* (2003) p27Kip1 inactivation provides a proliferative advantage to transplanted hepatocytes in DPPIV/Rag2 double knockout mice after repeated host liver injury. *Cell Transplant* **12**: 907–919.
98. Sandhu JS, Petkov PM, Dabeva MD, Shafritz DA. (2001) Stem cell properties and repopulation of the rat liver by fetal liver epithelial progenitor cells. *Am J Pathol* **159**: 1323–1334.
99. Oertel M, Menthena A, Dabeva MD, Schafritz DA. (2006) Cell competition leads to a high level of normal liver reconstitution by transplanted fetal liver stem/progenitor cells. *Gastroenterology* **130**: 507–520.
100. Dabeva MD, Hwang SG, Vasa SR, *et al.* (1997) Differentiation of pancreatic epithelial progenitor cells into hepatocytes following transplantation into rat liver. *Proc Natl Acad Sci USA* **94**: 7356–7361.
101. Nierhoff D, Ogawa A, Oertel M, *et al.* (2005) Purification and characterization of mouse fetal liver epithelial cells with high *in vivo* repopulation capacity. *Hepatology* **42**: 130–139.
102. Dabeva MD, Petkov PM, Sandhu J, *et al.* (2000) Proliferation and differentiation of fetal liver epithelial progenitor cells after transplantation into adult rat liver. *Am J Pathol* **156**: 2017–2031.
103. Wang X, Foster M, al-Dhalimy M, *et al.* (2003) The origin and liver repopulating capacity of murine oval cells. *Proc Natl Acad Sci USA* **100 Suppl 1**: 11881–11888.
104. Wang X, al-Dhalimy M, Lagasse E, *et al.* (2001) Liver repopulation and correction of metabolic liver disease by transplanted adult mouse pancreatic cells. *Am J Pathol* **158**: 571–579.
105. Gouon-Evans V, Boussemart L, Gadue P, *et al.* (2006) BMP-4 is required for hepatic specification of mouse embryonic stem cell-delivered definitive endoderm. *Nat Biotechnol* **24**: 1402–1411.
106. Heo J, Factor VM, Uren T, *et al.* (2006) Hepatic precursors derived from murine embryonic stem cells contribute to regeneration of injured liver. *Hepatology* **44**: 1478–1486.
107. Dandri M, Burda MR, Torok E, *et al.* (2001) Repopulation of mouse liver with human hepatocytes and *in vivo* infection with hepatitis B virus. *Hepatology* **33**: 981–988.

108. Mercer DF, Schiller DE, Elliott JF, *et al.* (2001) Hepatitis C virus replication in mice with chimeric human livers. *Nat Med* 7: 927–933.

109. Tateno C, Yoshizane Y, Saito N, *et al.* (2004) Near completely humanized liver in mice shows human-type metabolic responses to drugs. *Am J Pathol* 165: 901–912.

110. Teebor GW, Becker FF. (1971) Regression and persistence of hyperplastic hepatic nodules induced by N-2-fluorenylacetamide and their relationships to hepatocarcinogenesis. *Cancer Res* 31: 1–3.

111. Schwarze PE, Pettersen EO, Shoaib MC, Seglen PO. (1984) Emergence of a population of small, diploid hepatocytes during hepatocarcinogenesis. *Carcinogenesis* 5: 1267–1275.

112. Solt DB, Medline A, Farber E. (1977) Rapid emergence of carcinogen-induced hyperplastic lesions in a new model for the sequential analysis of liver carcinogenesis. *Am J Pathol* 88: 595–618.

113. Evarts RP, Nakasukasa H, Marsden ER, *et al.* (1990) Cellular and molecular changes in the early stages of chemical hepatocarcinogenesis in the rat. *Cancer Res* 50: 3430–3444.

114. Shinozuka H, Lombardi B, Sell S, Iammarino RM. (1978) Early histological and functional alterations of ethionine liver carcinogenesis in rats fed a choline-deficient diet. *Cancer Res* 38: 1092–1098.

115. Lemire JM, Shiojiri N, Fausto N. (1991) Oval cell proliferation and the origin of small hepatocytes in liver injury induced by D-galactosamine. *Am J Pathol* 139: 535–552.

116. Laconi E, Oren R, Mukhopadhyay DK, *et al.* (1998) Long-term, near-total liver replacement by transplantation of isolated hepatocytes in rats treated with retrorsine. *Am J Pathol* 153: 319–329.

117. Factor VM, Radaeva SA, Thorgeirsson SS. (1994) Origin and fate of oval cells in dipin-induced hepatocarcinogenesis in the mouse. *Am J Pathol* 145: 409–422.

118. Rosenberg D, Ilic Z, Yin L, Sell S. (2000) Proliferation of hepatic lineage cells of normal C57BL and interleukin-6 knockout mice after cocaine-induced periportal injury. *Hepatology* 31: 948–955.

119. Knight B, Yeoh GC, Husk KL, *et al.* (2000) Impaired preneoplastic changes and liver tumor formation in tumor necrosis factor receptor type 1 knock-out mice. *J Exp Med* 192: 1809–1818.

120. Braun KM, Degen JL, Sandgren EP. (2000) Hepatocyte transplantation in a model of toxin-induced liver disease: variable therapeutic effect during replacement of damaged parenchyma by donor cells. *Nat Med* 6: 320–326.

121. Yoshida Y, Tokusashi Y, Lee GH, Ogawa K. (1996) Intrahepatic transplantation of normal hepatocytes prevents Wilson's disease in Long–Evans cinnamon rats. *Gastroenterology* 111: 1654–1660.

122. Mignon A, Guidotti JE, Mitchell C, *et al.* (1998) Selective repopulation of normal mouse liver by Fas/CD95-resistant hepatocytes. *Nat Med* **4:** 1185–1188.

123. Guha C, Sharma A, Gupta S, *et al.* (1999) Amelioration of radiation-induced liver damage in partially hepatectomized rats by hepatocyte transplantation. *Cancer Res* **59:** 5871–5874.

124. Witek RP, Fisher SH, Petersen BE. (2005) Monocrotaline, an alternative to retrorsine-based hepatocyte transplantation in rodents. *Cell Transplant* **14:** 41–47.

TECHNOLOGY FOR PRODUCTION
OF STEM CELLS

Clinical Manufacture of Stem Cells and Derivative Cell Populations

David H. McKenna* and John Wagner[†]

1. Introduction

Over the past decade there have been numerous major advances in stem cell biology, transplantation, and immunology. Prior to clinical testing, the US Food and Drug Administration (FDA) requires that investigational biological therapies be produced in accordance with current Good Manufacturing Practices (cGMP) and current Good Tissue Practices (cGTP). Development and production of novel biological therapies, including stem and progenitor cells, represent major research efforts. In general, the development of a novel biotherapeutic reagent starts with a proposal by an investigator who presents detailed results of preclinical studies performed in a basic research laboratory. A clinical trial outlining the method of testing the biological reagents' safety and potential efficacy (if appropriate) must subsequently be developed. From this vantage point, a detailed laboratory procedure can be developed for the large-scale clinical manufacture of stem cells or their derivative cell populations. The general strategy for the "translation" of a basic laboratory procedure into a clinical manufacturing procedure is shown in Fig. 1.

2. Preclinical Studies

A principal goal of ongoing research in regenerative medicine is the delivery of potentially curative cellular therapies to patients with a multitude of

*Correspondence: Clinical Cell Therapy Laboratory, University of Minnesota Medical Center, Minneapolis, MN 55455, USA. E-mail: mcken020@umn.edu
[†]Molecular and Celluar Therapeutics, 1900 Fitch Avenue, Saint Paul, MN 55108, USA.

Strategy

Identification of clinical-
grade reagents and methods
and scale-up

Test potency of
product in preclinical
model systems

Establish clinical model
system for testing biological
reagent

cGMP manufacture
of reagents

Validation

File IND

IRB, CPRC, FDA approval

Product Release for Phase I/II
Clinical Trials

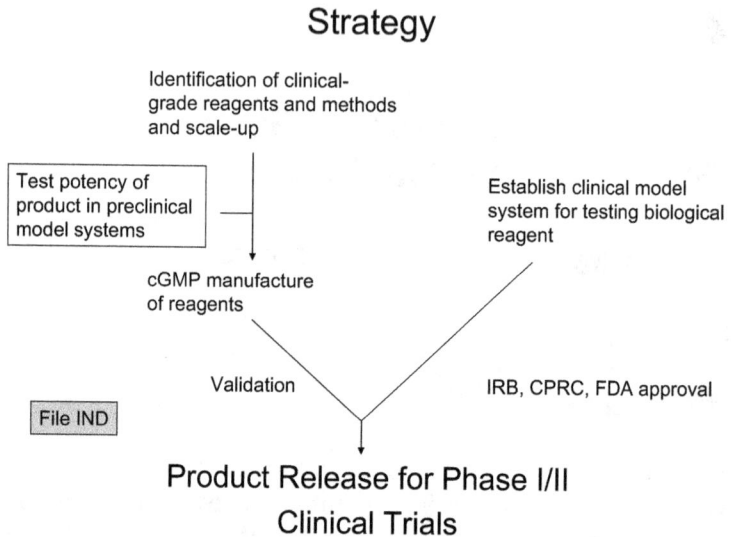

Figure 1. General strategy for the "translation" of a basic laboratory procedure into a clinical manufacturing procedure.

illnesses. The first step is the "proof of concept" typically demonstrated in an animal species and physiologic state that is most relevant for the clinical indication. Studies performed in animal models should support a rationale for conducting a human clinical trial. A relevant animal species would be one in which the biological response to the proposed therapy would be expected to mimic the human response. At a minimum, a dose response relationship and toxicology profile should be obtained. Treated animals should be monitored for general health status, with serum chemistry and hematology profiles. Target and nontarget tissues should be examined microscopically for histopathology changes supporting both safety and efficacy (see Table 1). In addition, the basic laboratory must maintain adequate documentation of the stem cell manufacturing, and manipulations, and should follow current Good Laboratory Practices (cGLP).

Importantly, the investigator must address the limitations of the animal models employed, demonstrate the genetic stability of stem cell products derived from long-term cultures, develop a rationale for the route of cell administration/delivery system, and acquire the analytical tools/experimental techniques for assessing the comparability between stem cell products derived from different starting source materials.

Table 1. Toxicological and Pharmacological Assessment

(1) Evidence of implantation reaction
(2) Evidence of any inflammatory response in target and nontarget tissues
(3) Evidence of host immune response to injected cells
(4) Cellular fate-plasticity (differentiation, transdifferentiation, fusion)
(5) Evidence of anatomic/functional tissue integration into host physiology
(6) Evidence of morphological alterations in target and nontarget tissues
(7) Determination of cell survival after transplantation (duration of therapeutic effect)
(8) Biological distribution/cell migration
(9) Evidence of cell differentiation (cell autonomy versus competence to respond to local host signals)
(10) Determination of cell proliferative potential *in vivo*
(11) Evidence of tumorigenicity (hyperplastic or unregulated growth)

3. Translational Development

Following the demonstration that a cell preparation potentially corrects an underlying disorder, repairs damaged tissues or ameliorates the expected disease course in animal models (i.e. proof of concept), and the completion of safety and efficacy studies (i.e. toxicology and pharmacology), the stem cell product enters the next phase — translational development. Translational development is the preclinical phase that involves adaptation of research methods into practical, clinically applicable methods that eventually allow for the large-scale manufacture of cell products intended for human use. Many technical, medical, quality, and regulatory issues affecting clinical operations will likely be encountered and are most appropriately addressed at this stage. Once methods for large-scale production have been developed and validated and all quality and regulatory requirements have been implemented, product manufacture for clinical use may commence. The general approach that culminates in the clinical-scale manufacture of stem cells and derivatives, including the necessary infrastructure (e.g. facility and quality systems requirements) and technical, regulatory, and medical considerations, is described in detail.

4. Infrastructure

4.1. Facility

Stem cell therapies intended for human infusion must be produced using cGMP and cGTP, the regulatory requirements promulgated by the FDA.

cGMP and cGTP are discussed in greater detail within "Regulatory Issues" below. Here we discuss considerations for the design and operation of a cell processing laboratory to facilitate cGMP/cGTP stem cell therapy production.

Title 21 of the Code of Federal Regulations (CFR), Part 211(21 CFR 211), Subpart C covers "Buildings and Facilities." Within this subpart of the CFR are eight sections: Design and Construction Features; Lighting; Ventilation, Air Filtration, Air Heating and Cooling; Plumbing; Sewage and Refuse; Washing and Toilet Facilities; Sanitation; and Maintenance. These regulations provide the minimal framework for the establishment of a production facility. Similar but abbreviated regulations are found in 21 CFR 1271.190, "Facilities," of the cGTP.

The measure of air quality of the production space is described in 21 CFR 209. Clean rooms are classified according to the number of particles larger than 0.5 microns per cubic foot of air. For early phase clinical studies there are no specific requirements for production area classification. The standard practice, however, is to perform all open manipulations under class 100 (i.e. less than 100 particles larger than 0.5 microns per cubic foot of air, or ISO 5) conditions (i.e. a biosafety cabinet). The area surrounding the class 100 work space is typically class 10,000 (ISO 7). Air of this quality provides "bioburden control" for the class 100 space and may be used for closed-system manipulations. Additionally, larger equipment that does not typically fit within a biosafety cabinet (e.g. ultracentrifuge, incubators) may be operated in class 10,000 space. Further information on aseptic processing is available as an FDA Guidance.[1]

There are several additional considerations for the cell processing facility, including process flow (i.e. well-planned and unidirectional), flexibility in design to allow for production of multiple products, and the concept of monitoring (environment, equipment, product, and personnel). A detailed discussion on these matters is beyond the scope of this article; in-depth discussions have been published.[2,3]

The often-used term "cGMP facility" suggests that compliance with federal regulations is dependent upon a physical structure. As noted above, some physical amenities are necessary for clinical-grade production of cellular therapies; however, it is primarily the quality system essentials (e.g. appropriate policies, personnel, and procedures) that allow for compliance. In addition to the quality system essentials there is a requirement for personnel with expertise in the technical, medical/scientific, and regulatory aspects of cellular therapy.

4.2. Quality systems

Quality assurance (QA) is the sum of activities planned and performed to provide confidence that all systems that influence product quality are reliable and functioning as expected.[4] The QA program defines the policies and environment necessary for attaining minimum quality and safety standards.[5] The basic components of a program include standard operating procedures (SOPs), documentation/record-keeping and traceability requirements, personnel qualifications and training (including a continuing education program), building and facilities/equipment quality control (QC), process control, auditing and investigation, and error and accident system/management.[6,7] At a cell processing laboratory, the quality program is truly the means by which the regulations are practiced. The FDA, realizing the importance of QA, has placed major emphasis within cGTP on the establishment of an effective quality program. It is the expectation of the FDA that the quality program will ensure a laboratory's compliance with regulations.

In addition to the guidance and oversight provided by the FDA, standards developed by professional organizations such as the American Association of Blood Banks (AABB) and the Foundation for the Accreditation of Cellular Therapy (FACT) exist to assist the laboratory in its effort to consistently maintain a quality program.[8,9] AABB and FACT standards are very similar to the FDA regulations, being founded on the same quality system essentials. The FDA, FACT, and AABB along with the individual transplant center and cell processing laboratory strive toward a common goal — to protect patient (and donor) health and ensure quality care while providing a cellular therapy product that can be used confidently with the hope of improving quality of life.

5. Technical Expertise

Personnel with training in cell processing, including cell isolation/selection, culture and expansion, and final preparation, are of course a necessity. Technologists will perform a variety of cell processing methods, including routine blood bank-based methods as well as more specialized, cell therapy-specific procedures and technologies. Several of these will be discussed under "Production" below. With experience in sterile technique and universal (standard) precautions, a medical technologist (with a four-year degree), possibly with blood banking experience, is well-suited for the clinical cell processing laboratory. Technologists with backgrounds in specialties other than blood banking, such as hematology, flow cytometry, or molecular diagnostics, may be an asset to the lab, considering the variety of quality control tests for stem cell therapies. Having a laboratory supervisor or manager with broad experience in cell processing and exceptional organizational skills is essential to facilitating movement of therapies from "bench to bedside." Oversight of technical aspects within the laboratory is ultimately a responsibility of the laboratory and/or medical director.[8,9]

6. Regulatory Support

Experts in regulatory affairs facilitate interactions with the FDA and may serve as valuable advisors to the investigators. Due to the complexity of FDA submissions many institutions have developed programs that solely function to provide assistance for investigational new drug (IND) submissions. An outline of the regulatory steps involved in stem cell therapies follows.

Prior to initiating clinical trials with stem cells, an IND application must be filed with the FDA, specifically the Center for Biologics Evaluation and Research (CBER). The IND application form (Form FDA 1571) is submitted with the following: a table of contents, an introductory statement, a general investigational plan, an investigator's brochure (if not a sponsor–investigator IND application), the "protocols" (including the study protocol, investigator data, facilities data, and institutional review board data), the chemistry, manufacturing, and control data, the pharmacology and toxicology data, the previous human experience, and any other additional information. The FDA will review the application based upon the best available science, possibly seeking input from an advisory committee.[10] Following review they will respond within 30 days, acknowledging receipt of the application and assigning a number to the IND. The clinical trial may commence upon notification of approval from the FDA or once 30 days have passed from the date of receipt of the application at the FDA. If the FDA requests more information from the investigators, the study may be placed on "clinical hold" until the FDA receives a satisfactory response. Once approved, the study must be conducted appropriately under informed consent and a clinical monitoring program; adequate records must be maintained, and adverse events must be reported to the

FDA. Any amendments to cell processing or the clinical trial must be submitted to the FDA. If the amendment is considered major (e.g. dose change), the modification should not be implemented prior to receipt and acknowledgement by the FDA. A detailed annual report must be submitted to the FDA on the anniversary date of the IND. This report typically includes the protocol schemas, cell processing details including any deviations, and clinical results to date. Adverse events and amendments reported earlier in the year should be summarized.

7. Medical Oversight

Medical expertise and oversight are certainly requisite for stem cell-based clinical trials. All physicians involved in a trial must have the appropriate training and knowledge of the stem cell therapy and the particular illness/disease being treated. The medical director of the laboratory oversees product safety and serves as a consultant to the rest of the clinical team.[8,9] Physicians providing direct care for enrolled patients manage administration of the cells and postadministration monitoring in addition to the standard patient care.

8. Production

A description of the steps involved in the transfer of technology from the research laboratory to the clinical cell processing laboratory is followed by a very broad discussion on current methods and available technologies commonly used in blood- and marrow-based stem cell processing.

8.1. Technology transfer

Prior to initiation of clinical production, the procedures for cell isolation and culture must be "translated" into large-scale, clinically suitable methods. This task requires a close collaboration between the research laboratory and the clinical cell processing laboratory. One possible strategy is a tiered approach, as outlined in Table 2.[3] The initial step (tier 1) typically involves an evaluation (i.e. scientific, medical, technical, regulatory) for feasibility and a review of the small-scale production (research) model where procedural modifications are identified. As the research activity is

Table 2. Tiered Approach to Product Development

	Tier-1	Tier-2	Tier-3
Location	Basic/Translational Laboratory	Translational Laboratory	Clinical Cell Therapy Laboratory
Function	Identification of required laboratory procedure modification; evaluation of modifications on small scale Performance of all laboratory readout assays	Transfer of technology and scale-up, initial development of SOPs, IND preparation, QA/QC plan Development of CLIA-approved readout assays	Final SOPs, validation runs, clinical product manufacture and product release
QA Oversight	No	Yes	Yes
Criteria to Move to Next Tier	Must have a workable method in place for large-scale testing	Large-scale method using clinical-grade reagents and methods must be reproducible	Successful validation runs and sign-off by QA Program prior to production of products for clinical testing
Product to Patients	No	No	Yes

often carried out in open systems (e.g. wells of a culture tray), one obvious change is modification of the culture system. Optimally, a closed system could be constructed for the manufacture of the stem cells or cell derivatives to avoid the potential for microbial contamination. There are several available cell selection technologies that are considered closed systems, some of which are discussed below. However, closed systems are not always possible. Use of alternative strategies (e.g. flasks, cell factories) should be justified. Beyond cell isolation and culture, there must be a detailed plan for final preparation with a possible need for validated transportation to distant clinical sites.

Once a feasible method for large-scale production has been demonstrated, development of the standard operating procedures (SOPs), the batch production records (BPRs), and the QA/QC plan is initiated (tier 2).

Modifications to production, including further assessment of clinical-grade reagents and supplies, continue at this step in the process. With reproducible manufacturing methods using appropriate reagents and supplies, the stem cell product moves forward (tier 3), where SOPs/BPRs, the QA/QC plan, and regulatory documents are finalized and validation runs occur prior to clinical product manufacture.

8.2. Cell isolation

There are several techniques of cell isolation or enrichment, and the specific method or combination of methods will be chosen based upon many variables, including source and type of stem cell. Approaches may yield heterogeneous (mixed) or homogeneous (pure) populations of cells. Counterflow centrifugal elutriation is a cell processing method that separates cell populations based upon two physical characteristics — size and density (sedimentation coefficient). Within the instrument a centrifuge separates a cell suspension based upon density; a medium is then passed through the cell collection chamber in the direction opposite to the centrifugal force. Adjustment of the flow rate of the medium and/or the speed of centrifugation to enable the counterflow rate to balance centrifugal force allows for alignment of cells based upon sedimentation characteristics (size and density). Fairly specific cell populations can then be diverted as a fraction of the initial product; blood- (including umbilical cord blood) and marrow-derived stem cells are generally considered to have the size and density of lymphoid cells, suggesting elutriation as a potential technology for isolation. Instruments with large-scale capacity are available,[11,12] and one is considered a closed system.[12]

Density gradient separation is another cell processing method that, like elutriation, provides a mixed cell population. This technique involves using a solution that is composed of particles of varying sizes that form a gradient according to the sedimentation rate. Introduction of the cell sample into this solution followed by centrifugation allows for separation of cell populations based upon size or density. This method is widely used with blood and marrow to enrich the mononuclear cell fraction while depleting red cells and granulocytes. Clinically acceptable solutions[13] and large-scale capacity instruments with functionally closed systems[14] exist. One obvious advantage of these less selective isolation methods is the unmodified condition of the cells (i.e. no attached reagents) in the end product.

Systems incorporating monoclonal antibody-based technologies that target cell surface antigens are often used for stem cell isolation or enrichment.[15-17] These methods may involve magnetic capture[16,17] of the cell type of interest by either positive selection (i.e. target cells retained) or negative selection (i.e. target cells depleted). Alternatively, high levels of purity may be attained with cells collected using a fluorescence-activated, high-speed cell sorter. This technology offers much opportunity for stem cell research, as selection of cells can simultaneously be based upon size, surface antigen expression, and cell granularity.[18] As a consequence of relative purity, the yield may be reduced. Furthermore, different methods may impact cell function. Today, most have enriched these very rare stem cell populations using a depletion approach to reduce undesirable, non-stem cell populations. Depending upon the approach to selection, it is necessary to address a number of technical and regulatory issues related to the specific antibodies or antibody/magnetic particles used.

8.3. Culture/expansion

The culture vessel, medium and culture supplements, and possibly feeder cells are the primary components of the stem cell culture system, where there may be technical and regulatory concerns. Culture bag sets offer the advantage of a closed system, limiting microbial contamination during culture. However, culture bags may not provide the functional end population for a variety of reasons. Alternatively, plastic devices may be preferable to culture bags for cell manufacture. If plastic devices are considered, surface area constraints may necessitate the use of wells, or T25, T75, or T150 flasks, particularly at the earliest time points when few cells are isolated for culture initiation. However, if appropriate, cell factories should be considered over multiple flasks so as to limit the number of manipulations to the culture. Despite attempts to scale up, there may be circumstances where certain stem cell types are best expanded in wells; for initial (i.e. phase I) clinical studies this may be allowed by the FDA, particularly if studies supporting this need are included with the IND application.

Medium and culture supplements are a major focus of process modification, as preclinical studies typically utilize research-grade medium, cytokines, growth factors, and vessel-coating materials/matrices, as well as fetal calf serum (FCS). Early phase trials may be permitted to proceed using the research-grade material. However, it will be necessary to demonstrate that alternatives, such as serum-free conditions or use of human sera rather than FCS, are not possible.

Several types of cGMPs-grade medium, cytokines, growth factors, and vessel-coating materials/matrices are currently available[19–21] and should be considered for clinical manufacture. Great effort should be invested to eliminate FCS from processing so as to avoid concerns for disease transmission as well as possibilities for immune sensitization and reactions.[22] Human AB serum or serum albumin from plasma collected from countries considered free of risk for bovine spongiform encephalopathy/variant Creutzfeldt–Jakob disease and processed under cGMPs with appropriate infectious disease testing is a possible alternative to FCS. When experiments demonstrate that FCS remains a requirement for successful cell culture, only serum from qualified herds with complete infectious disease testing should be used.

Oftentimes stem cell cultures rely upon feeder cells to allow the culture to proceed in an undifferentiated manner. These feeder cells may be of human or nonhuman (e.g. murine) origin. Use of nonhuman feeder cells does fit the definition of xenotransplantation as defined by CBER.[23] However, the FDA does not intend for the xenotransplantation requirements to preclude use of human embryonic stem cells in clinical trials.[10] Regardless of the source of the feeder cell layer, adventitious agent testing of the stem cells must be performed to demonstrate that the culture is free from infectious agents that may be transmitted to the recipient. If the source of feeder cells is nonhuman, this testing, of course, will include an extensive list of relevant microorganisms known to infect the given animal source. Plans to eliminate the need for feeder cells have already begun[24]; concerns surrounding the demonstration of immunogenic nonhuman sialic acid reinforce the rationale for proceeding without nonhuman feeder cells.[25]

8.4. Cell banks

A cell banking strategy may be useful in the manufacture of stem cells and/or stem cell derivatives, particularly in the allogeneic, "universal" donor setting. Additionally, cell banks may be developed for feeder or accessory cells necessary for culture of the final product. Establishment of a cell bank involves large-scale manufacturing and cryopreservation of cells for storage typically in specified aliquots for patient dosage or subsequent culture/expansion. There are two types of cell banks — master cell

banks (MCBs) and working cell banks (WCBs), each with unique logistic
and regulatory considerations.

8.5. Master cell banks

An MCB may be defined as a homogeneous culture of cells that have been
manufactured together and fully characterized by various test methods to
determine safety, identity, purity, potency, and stability. Aliquots of cells
from the MCB are typically placed in cryovials and stored in the vapor
phase of liquid nitrogen. Because the MCB is the ultimate source from
which the final product is produced, testing requirements are extensive.
Table 3 lists possible characterization tests for a human-derived MCB.

Table 3. Sample List of Characterization Testing for a Human-Derived MCB

MCB Qualification Testing
Cell identification by isoenzymes
Transmission electron microscopy
USP bacteriostasis/fungistasis by direct transfer method
USP sterility by direct transfer method
Mycoplasma assay (points to consider test)
In vivo detection of adventitious viruses in adult and suckling mice and eggs
In vitro detection of adventitious virus on three cell lines
HIV-I viral nucleic acid by PCR
HIV-II viral nucleic acid by PCR
HTLV-I viral nucleic acid by PCR
HTLV-II viral nucleic acid by PCR
HAV viral nucleic acid by PCR
HBV viral nucleic acid by PCR
HCV viral nucleic acid by PCR
CMV viral nucleic acid by PCR
EBV viral nucleic acid by PCR
HSV1 viral nucleic acid by PCR
HSV2 viral nucleic acid by PCR
Human herpes virus 6 nucleic acid by PCR
Human herpes virus 7 nucleic acid by PCR
Human herpes virus 8 nucleic acid by PCR
Human parvovirus B19 nucleic acid by PCR
Bovine viruses*
Porcine viruses*
Murine viruses*

*Non-human species-related testing may be required depending upon culture conditions.

Nonhuman species viral and other testing may be necessary, depending upon the culture conditions. Final testing requirements are determined through discussion with the FDA. With novel stem cell-based therapies, decisions for testing will likely be on a case-by-case basis.

8.6. Working cell banks

A WCB is derived from an MCB. Like an MCB, a WCB may be defined as a homogeneous culture of cells that have been manufactured together. However, because of the level of characterization testing of the MCB, testing of the WCB is generally more limited. Testing may include sterility (bacterial/fungal culture), mycoplasma testing, and limited purity and identity testing (e.g. cell phenotype by flow cytometry). As with the MCB, however, final testing requirements are decided by the FDA.

8.7. Final preparation

Final preparation is somewhat dependent upon whether the final stem cell-therapy product is fresh or cryopreserved. If cryopreservation is practical, all product testing results may be available prior to thaw and infusion. However, stability studies may indicate that the freezing process harms the stem cells, supporting infusion of cells freshly harvested from the culture. Regardless of whether the product is administered as fresh or thawed cells, attempts should be made to eliminate noninfusible reagents (e.g. medium, fetal calf serum, cryoprotectant) from the final product. This is typically accomplished with a wash step which involves a series of centrifugations and dilutions with an infusible, cell-compatible solution such as 5% human serum albumin. The wash and resuspension may appear straightforward and inconsequential as compared to the more technical areas of cell isolation and culture; however, excessive washing or centrifuge speeds may affect cell viability and/or recovery as well as function. Attention to details when planning the validation of final preparation will help to avoid detrimental effects on the cells at this late stage of production.

9. Other Issues Related to Production

A description of considerations for determination of donor eligibility and product testing, including lot release testing, additional critical elements in the clinical manufacture of stem cells and stem cell derivatives, follows.

9.1. Qualification of donors: Evaluation of human stem cell sources

Whether the source of the stem cell population is derived from adult donors, fetal/neonatal donors (e.g. umbilical cord blood, amniotic fluid, or fetal tissues), or the inner cell mass of 5–10-day old embryos, donor qualification criteria need to be determined. It is critical that the tissue donors or gamete donors (in the case of embryonic stem cells) be screened for communicable infectious and genetic diseases (where possible). Allogeneic donors need to have a medical history with specific testing for infectious disease agents when possible. Licensed screening tests routinely used for allogeneic, hematopoietic stem cell products include: HIV and HCV (viral antigen or viral nucleic acid by PCR), anti-HIV I/II, anti-HCV, anti-HTLV I/II, anti-HBc, HBsAg, anti-CMV and EBV, and *Treponema pallidum* (by serology). Such testing may not be possible for fetal or gamete donors. In such cases, the FDA must be consulted and an exemption may be permitted. Testing of the cultured cell product itself, however, will likely be required. Donor eligibility for human cells, tissues, and cell- and tissue-based products is covered under 21 CFR 1271, Subpart C.

9.2. Lot release/quality control testing

The QA program defines QC procedures, specific tasks to be used to monitor product quality.[5] QC is an essential component of the QA system and is designed to evaluate the manufacturing process from start (i.e. raw materials) to finish (i.e. final product).[26] A robust system of QC is essential in a clinical cell processing laboratory, as it serves two major purposes: (1) determination of the suitability of a product for a given patient and (2) assessment of overall production and laboratory practices. The importance of determination of the suitability of a patient-specific cell therapy product is self-evident, with safety being of paramount importance in early phase trials. Monitoring of overall production and laboratory practices allows for process improvement and facilitates standardization of production.

Product characterization and release testing is specific testing performed on the final product to ensure safety, purity, identity, potency, and stability.

It is based on requirements of the General Biological Products Standards (21 CFR 610). Briefly, safety testing serves to confirm that the product is free of adventitious agents, bacteria, fungi, and mycoplasma. Purity testing is used to demonstrate that the product contains no endotoxin and no extraneous material other than that which is unavoidable in the manufacturing process. Identity testing confirms that the product within the bag/container is that which is identified on the label. Potency testing measures the functional or biological activity, and stability testing determines retention of this function after short- and long-term storage. Table 4 summarizes product testing and provides examples of the main categories of testing.

Upon licensure the FDA states "No lot of any licensed product shall be released by the manufacturer prior to the completion of tests for conformity with standards applicable to such a product."[27] However, prior to licensure, the FDA focuses primarily upon safety, as is evident

Table 4. Product Testing

Characterization/Release Testing for...	Examples
Safety	• Serologic markers of infectious disease • Gram stain • Sterility cultures (USP/clinical laboratory) • Mycoplasma detection (culture, PCR) • Testing for adventitious agents
Purity	• Endotoxin • Assays for assessment of in-process reagent removal (e.g. antibodies/beads, vectors, cytokines) • Flow-cytometric analysis
Identity	• Flow-cytometric analysis • Cytochemistry • Molecular diagnostic testing (e.g. quant. RT-PCR) • HLA typing
Potency	• Viability (microscopy/flow cytometry) • Clonogenic assays (LTC-IC, CFU, SRC) • Proliferative potential (e.g. G-banding, Southern blot for telomere length) • Pluripotency (*in vivo* differentiation potential in NOD-SCID mice)
Stability	• As noted for potency (and possibly other testing above) after short- and long-term storage

in the Guidance for Industry: INDs — Approaches to Complying with cGMP during Phase I. Per the FDA: "For known safety-related concerns, specifications should be established and met. For some product attributes, all relevant acceptance criteria may not be known at this stage of product development. This information will be reviewed in the IND submission."[28]

This "graded" approach to product characterization allows the investigator to gather additional product manufacturing data to develop and validate testing and/or to show equivalency to methods currently accepted by the FDA. The benefits of this approach are clearly evident with safety and potency testing. The FDA requires that equivalency to FDA-accepted methods for sterility, mycoplasma, and endotoxin testing be established by licensure,[27] and establishment of a truly functionally predictive potency assay may be the most difficult aspect of full product characterization. With the FDA model for earlier phase studies, these challenging tasks are more readily accomplished.

10. Product Administration

With physician approval and following standard operating procedures for product and patient identification, the stem cell therapy may be administered. Although hematopoietic stem/progenitor cells are typically infused by intravenous drip into a central line without a needle, pump, or filter, the route of administration of other stem cell therapies will be dependent upon the cell type, intended use, or a combination of these and other considerations. It is anticipated that specific stem cell populations and cell derivatives will be delivered by routes other than intravenously (e.g. intra-arterially or direct injection into a specific organ or tissue). Regardless of the method of administration, at a minimum, vital signs should be checked before administration, immediately, and one hour after administration, and at regular intervals thereafter through 24-hour postadministration. The FDA may request that vital signs and other monitoring (dependent upon the cells and anticipated localization) continue for 48 hours or longer. If any adverse reactions are associated with the stem cell product, more frequent monitoring may be required. The patient's physician, the principal investigator, and the medical director of the cell processing laboratory should be notified immediately of any signs or symptoms of an unexpected or severe adverse reaction. As the stem cell therapy is investigational, the QA unit should be notified for review and potential reporting to the FDA.

The extensive experience with infusion-related reactions or toxicities gained from hematopoietic stem/progenitor cell transplantation and blood transfusion may be of benefit.[29–33] Depending upon processing methods, allergic, hemolytic, and febrile reactions may occur. With open systems processing and/or highly manipulated stem cell products, microbial contamination is always a concern. A final wash step removing the majority of the cellular debris from long-term culture or cell lysis due to the freezing/thawing process (e.g. red cells and granulocytes) may serve to limit toxicity. In the event of an unexpected or severe adverse reaction, an investigation including any appropriate laboratory testing should be initiated. If the product has red cells or red cell fragments, direct antiglobulin testing and antibody titers may be necessary in order to rule out hemolytic reactions. Gram stain and blood culture should be ordered if any signs or symptoms suggest microbial contamination. If the product was cryopreserved in dimethylsulfoxide (DMSO) prior to administration, procedures to avoid toxicity to the patient and thawed cells should be in place. Patients should not receive more than 1 mg/kg of DMSO; thawed stem cells should be infused as soon as possible following thaw, or a wash step should be considered and postthaw product stability should be established. Depending upon the route of administration, hydration and use of prophylactic antiemetics, antipyretics, and antihistamines may be advised.

The laboratory responsible for the stem cell-therapy product production and/or distribution should include an infusion form for patient care staff to document receipt and infusion of the product. At a minimum the completed form should contain the name and unique identifier of the recipient, the proper product type/name and product identifier, and the names or identifiers of the medical staff receiving and verifying the product, the date and time of issue, and the product condition by visual inspection.[8] It is recommended that the form have space designated for additional information, including cell dose, unit volume, proper identification procedure, start time and duration of administration, description of patient status before and after administration, and any complications associated with the procedure. Return of the completed form to the laboratory should be a requirement.

11. Regulatory Issues

There are many regulatory issues to consider with the manufacture of a stem cell therapy for human clinical trials. Below we have outlined and provided an historical reference of the FDA regulations that govern production of cellular therapies, cGMP and cGTP; we also have included a discussion on the Chemistry, Manufacturing, and Controls (CMC) section of the Investigational New Drug (IND) Application. An excellent, more thorough review of the regulatory considerations for the manufacture of stem cells has recently been published.[34]

11.1. cGMP/cGTP

Biological products have been under regulation by the federal government since 1902, when the Biologics Control Act was passed. Legislative action was taken in response to the tragic deaths of 13 children in Saint Louis, Missouri, following administration of tetanus-infected diphtheria antitoxin. The Act required manufacturers to obtain both establishment and product licensure. The unfortunate occurrence of other disasters prompted additional regulations that supplemented the Biologics Control Act. The Food, Drug, and Cosmetic Act was passed in 1938, after 100 people died following ingestion of an elixir of sulfanilamide that contained diethylene glycol. This Act replaced the Pure Food and Drug Act of 1906, and it required proof of safety in addition to purity. Amendments to the Food, Drug, and Cosmetic Act in 1962 in response to the observed congenital defects due to thalidomide included a requirement of proof of product efficacy, thus completing the now well-known licensure triad: safety, purity, and efficacy.[35–37] By this time many of the elements of cGMP, scientifically sound methods, practices, or procedures that are followed and documented throughout product development and manufacturing, were well established in the pharmaceutical industry.[38]

The aim of cGMP is to ensure product safety, purity, and efficacy. Current GMPs have their origin in the Food, Drug, and Cosmetic Act, and the original application was to the pharmaceutical industry, as outlined above. However, the approach was adapted by the Food and Drug Administration (FDA) for blood center operations[39] and has more recently been applied to the manufacture of cellular therapies outside of the blood bank setting. It soon became evident, however, that the unique and diverse nature of cellular therapies required special consideration with regard to regulation.

The FDA has since established a tiered, risk-based system to regulate cellular therapies. The framework of this system consists of cGMP and the more recently finalized cGTP. The concept of cGTP is similar to that of cGMP, though more narrow in focus and aimed primarily at prevention at introduction, transmission, and spread of communicable disease. The cGTP Final Rule (21 CFR 1271, subparts D–F) is part of 21 CFR Part 1271 — Human Cells, Tissues, and Cellular and Tissue-Based Products; subparts A and B are the Registrations Final Rule, and subpart C covers the Donor Eligibility Final Rule. cGTP serve to supplement cGMP. In brief, minimally manipulated, "standard of care" (i.e. non-IND) cell therapy products require adherence to cGTP (Section 361, Public Health Service Act,; 21 CFR Part 1271 — Human Cells, Tissues, and Cellular and Tissue-Based Products); products undergoing more than minimal manipulation for novel applications (i.e. requiring IND) will be held to cGTP as well as the more rigorous cGMP, requiring approval prior to manufacture for clinical use [Section 351, Public Health Service Act (Biologic); Section 505, Food, Drug, and Cosmetic Act (Drug); Investigational New Drug Application — 21 CFR Part 312]. As noted earlier, AABB and FACT have written standards to serve as guidance for compliance with these regulations.[8,9] As the field of cell therapy continues to grow, it is anticipated that the regulations will expand and/or adapt as indicated.

11.2. Investigational new drug (IND) application: Chemistry, manufacturing, and controls (CMC)

The section within the IND application that covers production of the stem cell therapy is the Chemistry, Manufacturing, and Controls, or CMC. The CMC is, of course, a critical component of both the IND submission and the clinical trial. Product manufacturing, characterization, and testing information, including lot release specifications, are included here. The amount of information to be included is dependent upon the phase of the clinical trial. The graded nature of the CMC allows the researcher to accumulate product information in early phase trials while assuring the safety and rights of the subjects. As such, information sufficient to allow proper evaluation of subject safety by the FDA should be included in the CMC. Ultimately, sufficient information should be submitted to assure proper identification, quality, purity, and strength of the stem cell product.

The team approach to writing the CMC involving the laboratory and medical director(s), principal investigator(s), laboratory/technical staff, and QA/regulatory staff works well. All parties bring valuable contributions to the effort. Once acceptable production methods have been established, the CMC section can be submitted. Amendments to the CMC may require immediate notification to the FDA depending upon the nature of the amendment. Minor changes may be reported in the annual report.

There are many FDA documents that offer guidance for writing CMC sections for stem cell therapies allowing for fulfillment of the regulations. Several are listed below:

- Points to Consider in the Characterization of Cell Lines Used to Produce Biologicals, 1993.
- Content and Format of Investigational New Drug Applications (INDs) for Phase 1 Studies of Drugs, Including Well-Characterized, Therapeutic, Biotechnology-Derived Products, 1995.
- Guidance for Industry: Guidance for Human Somatic Cell Therapy and Gene Therapy, 1998.
- Guidance for Reviewers: Instructions and Template for Chemistry, Manufacturing, and Control (CMC) Reviewers of Human Somatic Cell Therapy Investigational New Drug Applications (INDs) [Draft Guidance], 2003.
- Guidance for Industry: Gene Therapy Clinical Trials — Observing Participants for Delayed Adverse Events [Draft Guidance], 2005.
- Guidance for Industry: INDs — Approaches to Complying with cGMP during Phase I [Draft Guidance], 2006.
- Potency Measurements for Cellular and Gene Therapy Products, Cellular, Tissue and Gene Therapies Advisory Committee, FDA Briefing Document, 2006. Available at http://www.fda.gov/ohrms/dockets/ac/06/briefing/2006-4205B1-index.htm

The 2003 Guidance for Reviewers is particularly helpful, as it outlines the approach recommended to FDA reviewers of CMC sections of cell therapy-related submissions. As such, the 2003 Guidance for Reviewers may be used as the framework for the CMC section, with the additional documents serving an informational and supportive role. Table 5 outlines this approach to writing the CMC.

Table 5. Sample Outline for CMC Writing*

I. **Product Manufacturing and Characterization Information**

A. General
- Product type, derivation
- Where processed?
  - Relevant accreditations (FACT, AABB, CAP, CLIA)
  - Reference Type V Master File (MF) for facility (if filed)

B. Procurement
- Starting material
  - Blood, UCB, marrow, other
- Where collected?
- Process description

C. Infectious Disease Testing and Prevention of Cross-Contamination
- Donor suitability per cGTP
- Medical history
- List of testing
- Quarantine if positive result
- Process if product with positive result to be infused

D. Cell Processing
- Description of processing methods
- Flow diagram outlining processing and testing

E. Reagents
- Table indicating reagent, manufacturer, status [reference to FDA approval, certificates of analysis (COA), MFs, existent IND]
- HSA (from countries considered free of vCJD risk?)
- Reference and include COA for reagents not approved for human infusion

II. **Product Testing**

A. Microbiological Testing
- Sterility testing
  - Method/test
  - When tested? Final product? In process? Days held?
  - Indicate sample will not be washed or manipulated before testing
- Mycoplasma testing
  - Method/test (e.g. PCR, culture)
  - When tested? Final product? In process? Days held?
  - Indicate sample will not be washed or manipulated before testing
- Gram stain
- Reference to donor infectious disease testing (Section I.C.)

B. Identity
- Labeling, segregation, any test methods employed

C. Purity
- Make-up of final suspension (e.g. washed cells in 5% HSA)
- Analysis (e.g. flow cytometry, endotoxin)

(*Continued*)

Table 5. (*Continued*)

D. Potency
   - Analysis (e.g. flow cytometry as *in vitro* surrogate, other *in vitro* functional assays, *in vivo* clinical assessment)
E. Additional Testing
   i.   Viability
        – Method (e.g. microscopy, flow cytometry)
        – When?
   ii. Cell Dose
        – Method (e.g. hematology analyzer)
        – Actual doses (range)
        – Minimum dose to allow for infusion?
   iii. Other
        – Retain aliquot?

III. **Product Release Criteria Testing and Additional Testing**
   - One table with lot release testing (assay, method, where tested, specification)
     – E.g. endotoxin, LAL method (manufacturer of kit), cell therapy lab, $\leq$ 5 EU/kg; viability, flow cytometry (7-AAD), clinical flow cytometry lab, $\geq$ 70%
   - Second table with additional (not lot release) testing
     – E.g. sterility/mycoplasma testing on final product that will not be available prior to release; research-type assays

IV. **Product Stability**
   - Stability testing to support postproduction clinical use
   - Fresh? Cryopreserved? Transit time/conditions

V. **Other Issues**

A. Product Tracking
   - Labeling per standards/regulations
   - Unique identifiers (#, name)
   - Confirmation prior to administration
   - Segregation system
B. Labeling
   - Per standards/regulations
   - Additional items on label
   - Include "Caution: New Drug — Limited by Federal Law to Investigational Use" per 21 CFR 312.6
   - Attach sample label/hangtag
C. Container/Closure
   - Bags, tubing sets, flasks, etc.
   - Indicate compatibility with cells

(*Continued*)

Table 5. (*Continued*)

D. Environmental Impact
  - "The sponsor claims categorical exclusion [under 21 CFR 25.31(e)] for the study under this IND. To the sponsor's knowledge, no extraordinary circumstances exist."
E. Validation and Qualification of the Manufacturing Process and Facility
  - Indicate process validation performed prior to clinical use
  - Reference facility MF if on file

*This approach is based upon the FDA Guidance for Reviewers: Instructions and Template for Chemistry, Manufacturing, and Control (CMC) Reviewers of Human Somatic Cell Therapy Investigational New Drug Applications (INDs) [Draft Guidance], 2003.

Summary

Stem cells and their cell derivatives hold great promise, with the potential of revolutionizing medical practice. However, these cells are also high-risk, in that they have an unparalleled proliferative and differentiation potential. This (along with the fact that the "approved" embryonic stem cell lines have limited "histories" and prior exposure to nonhuman species feeder cells) only serves to increase the potential risk of untoward events in human recipients. Therefore, it is imperative that the manufacture of these new and exciting stem cell therapies be held to the highest standards so that it will be possible to dissect out why the first generation of cell therapies derived from stem cells either succeed or fail. This article has served to introduce the issues facing the translation of stem cells from the basic laboratory reagent to the clinical product.

References

1. US Food and Drug Administration. (2004) Guidance for industry sterile drug products produced by aseptic processing — current good manufacturing practice.
2. Burger SR. (2000) Design and operation of a current good manufacturing practices cell-engineering laboratory. *Cytotherapy* 2: 111–122.
3. McKenna DH, Kadidlo DM, Miller JS, *et al.* (2005) The Minnesota Molecular and Cellular Therapeutics Facility: a state-of-the-art biotherapeutics engineering laboratory. *Transfus Med Rev* 19: 217–228.

4. US Food and Drug Administration. (1995) Guideline for quality assurance in blood establishments. Docket no. 91N-0450.

5. Newman-Gage H. (1995) Application of quality assurance practices in processing cells and tissues for transplantation. *Cell Transplant* 4: 447–454.

6. McCullough J. (1995) Quality assurance and good manufacturing practices for processing hematopoietic progenitor processing. *J Hematother* 4: 493–501.

7. Bennett ST, Johnson NL, Lasky LL. (1992) Quality assurance and standards in hematopoietic progenitor processing. *J Clin Apher* 7: 138–144.

8. American Association of Blood Banks. (2007) Standards for cellular therapy product services. 2nd edn.

9. Foundation for the Accreditation of Cellular Therapy (FACT). (2006) International standards for cellular therapy product collection, processing, and administration. 3rd edn.

10. Fink DW. Embryonic stem cell-based therapies: US-FDA regulatory expectations. Presented at the Stem Cell Research Forum of India, February 1, 2007. Available at (http://www.fda.gov/cber/genetherapy/stemcell012907df.htm)

11. Gao IK, Noga SJ, Wagner JE, *et al.* (1987) Implementation of a semi-closed large scale counter-flow centrifugal elutriation system. *J Clin Apher* 3:154–160.

12. Rouard H, Leon A, De Reys S, *et al.* (2003) A closed and single-use system for monocyte enrichment: potential for dendritic cell generation for clinical applications. *Transfusion* 43: 481–487.

13. Seeger FH, Tonn T, Krzossok N, *et al.* (2007) Cell isolation procedures matter: a comparison of different isolation protocols of bone marrow mononuclear cells used for cell therapy in patients with acute myocardial infarction. *E Heart J* 28: 766–772.

14. Jin NR, Hill R, Segal G, *et al.* (1987). Preparation of red-blood-cell depleted marrow for ABO-incompatible marrow transplantation by density-gradient separation using the IBM 2991 blood cell processor. *Exp Hematol* 15: 93–98.

15. Berger M, Adams S, Tigges B, *et al.* (2006) Differentiation of umbilical cord blood-derived multi-lineage progenitor cells into respiratory epithelial cells. *Cytotherapy* 8(5): 480–487.

16. Miltenyi S, Muller W, Weichel W, Radbruch A. (1990) High gradient magnetic cell separation with MACS. *Cytometry* 11: 231–238.

17. Rowley SD, Loken M, Radich J, *et al.* (1998) Isolation of CD34[+] cells from blood stem cell components using the Baxter Isolex system. *Bone Marrow Transplantation* 21: 1253–1262.

18. Gee AP and Durett AG. (2002) Cell sorting for therapeutic applications — points to consider. *Cytotherapy* 4(1): 91–92.

19. Mu LJ, Lazarova P, Gaudernack G, *et al.* (2004) Development of a clinical grade procedure for generation of mRNA transfected dendritic cells from

purified frozen CD34$^{(+)}$ blood progenitor cells. *Int J Immunopathol Pharmacol* **17**(3): 255–263.

20. Shaw PH, Gilligan D, Wang X-W, *et al.* (2003) *Ex vivo* expansion of megakaryocyte precursors from umbilical cord blood CD34$^+$ cells in a closed liquid culture system. *Biology of Blood and Marrow Transplantation* **9**: 151–156.

21. Hanenberg H, Xiao XL, Dilloo D, *et al.* (1996) Colocalization of retrovirus and target cells on specific fibronectin fragments increases genetic transduction of mammalian cells. *Nat Med* **2**(8): 876–882.

22. Ortel TL, Mercer MC, Thames EH, *et al.* (2001) Immunologic impact and clinical outcomes after surgical exposure to bovine thrombin. *Annals of Surgery* **233**(1): 88–96.

23. US Food and Drug Administration (2003) Guidance for industry: source animal, product, preclinical, and clinical issues concerning the use of xenotransplantation products in humans.

24. Xu R-H, Peck RM, Li DS, *et al.* (2005) Basic FGF and suppression of BMP signaling sustain undifferentiated proliferation of human ES cells. *Nature Methods* **2**(3): 185–190.

25. Martin MJ, Muotri A, Gage F, Varki A. (2005) Human embryonic stem cells express an immunogenic nonhuman sialic acid. *Nature Medicine* **11**(2): 228–232.

26. Christiansen GD. Quality control and quality assurance issues in biopharmaceutical processing. In: Avis KE, Wagner CM, WU VL (eds.), *Biotechnology: Quality Assurance and Validation*, Englewood, CO: HIS Health Group, Drug Manufacturing Technology Series, Vol. 4, 1999, pp. 33–43.

27. US Food and Drug Administration. General biological products standards. 21 CFR 610.

28. US Food and Drug Administration. (2006) Guidance for industry: INDs — approaches to complying with cGMP during phase I [draft guidance].

29. Davis JM, Rowley SD, Braine HG, *et al.* (1990) Clinical toxicity of cryopreserved bone marrow graft infusion. *Blood* **75**: 781–786.

30. Okamoto Y, Takaue Y, Saito S, *et al.* (1993) Toxicities associated with cryopreserved and thawed peripheral blood stem cell autografts in children with active cancer. *Transfusion* **33**: 578–581.

31. Davenport RD. (2005) Pathophysiology of hemolytic transfusion reactions. *Semin Hematol* **42**: 165–168.

32. Kopko PM, Holland PV. (2001) Mechanisms of severe transfusion reactions. *Transfus Clin Biol* **8**: 278–281.

33. Gilstad CW. (2003) Anaphylactic transfusion reactions. *Curr Opin Hematol* **10**(6): 419–423.

34. Weber DJ. (2006) Manufacturing considerations for clinical uses of therapies derived from stem cells. *Methods Enzymol* **420**: 410–430.

35. Solomon JM. (1994). The evolution of the current blood banking regulatory climate. *Transfusion* **34:** 272–277.
36. Zuck TF. (1995) Current good manufacturing practices. *Transfusion* **35:** 955–966.
37. Center for Biologics Evaluation and Research website: http://www.fda.gov/cber/
38. Gee AP. (2002) The impact of regulatory policy on the development of somatic cell therapies in the United States. *Transpl Immunol* **9:** 295–300.
39. US Food and Drug Administration. Current good manufacturing practice for blood and blood components. 21 CFR 606.

Stem Cell Culture Engineering

Kartik Subramanian and Wei-Shou Hu*

The stem cells from embryos as well as a variety of tissue sources hold promise for regenerative medicine. When one is embarking on the transformation of stem cell science to technologies benefitting patients in need, it is prudent to reflect on the roles of engineering in propelling such transformation. While many areas of engineering, including innovative fabrication of cell habitats for cultivation and implantation and scale-up, are also crucial, we emphasize the critical need for introducing better quantification of stem cell culture as a first step of converting it to a process. Quantitative characterization of stem cells, their potency and the process of differentiation allows for objective comparison of cells of different sources, protocols, and enhances process robustness. Herein we discuss the key variables to be characterized for stem cells in culture and possible formulas that may serve as a starting point for establishing a sound description of the stem cell culture process.

1. Introduction

The past century was marked by accelerated advances in many fields of science, particularly biology. The elucidation of DNA structure, genetic codes, cellular metabolism, development of cellular structure and their regulation etc. all occurred in the short span of a few decades. These advances propelled new technologies, which in turn furthered biological science. Antibiotics in general, and penicillin in particular, heralded the new technology nurtured by the advances of biological science in the

*Correspondence: 421 Washington Avenue SE, Minneapolis, MN 55455-0132, USA. Tel: (612) 626-7630; Fax: (612) 626-7246. E-mail: wshu@cems.umn.edu

1950s.[1] Advances in synthetic chemistry and enzymology in the following decades sustained the rapid expansion of biotechnology, especially in the pharmaceutical industry as propelled by the semisynthetic penicillins.[2] The arrival of recombinant DNA technology created a new biotechnology, as we know today. A mere quarter century ago, recombinant DNA technology was first applied to produce human growth hormone and insulin. This was followed by the introduction of mammalian (primarily Chinese hamster ovary) cells for therapeutic protein production.[3] Today, the market for therapeutic proteins exceeds US$20 billion in sales in the United States alone. In the third millennium, all hope is focused on genomics-based molecular medicine and stem cell-based regenerative medicine.

As we continue to plow the road to the next technology on the horizon, we are reminded of the nature of technological advances; while revolutionizing technologies have been enabled by scientific discoveries, it was the subsequent engineering innovations that brought the technology to fruition. In pursuing stem cell technology, it is not too early to ask whether engineering should play some role in developing this new technology and in defining that role. Although the term "stem cell engineering" has been used for a number of years, the word "engineering" is often used in the same way as in "genetic engineering." It evokes surgical modifications rather than a discipline-based analysis or synthesis. Even when the word is used to imply engineering in a true sense, the referrer is often focused on tissue engineering or scale-up. Undoubtedly, the transformation of laboratory discoveries into biotechnological products could not happen without successful scale-up. It is also well acknowledged that in the past two decades, advances in biomaterials have brightened the prospects of employing engineering tissue analogs for reversing functional or structural impairment of tissues or organs. However, more fundamentally, engineering's most critical role does not begin with scale-up, but long before scale-up.

The first step in transforming a laboratory protocol into a process is the quantification of the process by describing the stochiometric relationship and the kinetic behavior involved in the transformation of the starting materials or cells into a product. It is such quantification that enables engineering analysis to examine the procedures involved for their robustness and efficiency. A procedure prone to error, poor reproducibility, inadequate scalability or low yield and efficiency is replaced by an alternative procedure or otherwise improved. After all, the most important distinction between an elaborate laboratory protocol and a process is that the latter must be very robust and achievable by reasonably skilled workers rather than specialists. This aspect of engineering contribution is important for stem cell culture because of its somewhat unpredictable nature in cell isolation, maintenance and differentiation. For stem cell technology to become readily applicable for clinical use, these procedures will need to be transformed into robust processes. This article will address the quantitative description that should be applied to stem cell culture.

2. Engineering Characterization of Stem Cells

In the third quarter of the last century, rapid advances were seen in descriptions of microbial growth kinetics. The need for working models controlling cell growth in various fermentation processes, including antibiotics and amino acids, spurred much of the effort in kinetic model development. For bioprocess technologists, engineers and applied biologists alike, using those descriptors becomes virtually a norm when dealing with processes involving cell expansion. Such practice has been adapted for the new wave of mammalian cell based bioprocess in the last quarter of the 20th century. Although the growth requirements for mammalian cells are much more complex, the general kinetic description developed for microbial systems has been deemed adequate. In the following section those practices and the possible ways of extending them for stem cells will be described and discussed.

2.1. Important variables defining the "state" of stem cells

In engineering terms, the "state" of a system is the collection of all the characteristics of the system that allow the system to be completely defined in any condition. The minimum set of variables that can define those characteristics are called state variables. "Minimum set" is invoked because redundant variables, which can be reduced from each other, are eliminated. For a cellular system, the conditions we are interested in are the amount (e.g. number or mass) of cells, their physiological conditions or other biological "activities." For stem cells, the biological activities of particular interest are their self-renewal capabilities, or upon initiation of differentiation, their differentiation state. To describe a stem cell system, one needs to denote the number of cells capable of self-renewal, or any particular differentiated state, and capable of further differentiating to particular lineages. From an engineering perspective, all these variables need to be described quantitatively. To enable the design of a robust bioprocess one needs to further describe how the "state" of the system will be altered by the environment. However, the relationship between cells and the environment is not static; the environment causes cells to proliferate or differentiate while cells consume nutrients and even growth factors, signaling molecules, causing the environment to change. Thus, along the "process" of growth and differentiation, cell and environment alter each other. A quantitative description of how cell and environment affect each other is a second task upon attaining a quantitative description of the state of a stem cell system.

2.2. Cell characterization

Traditional means of characterizing cultured mammalian cells rely on cell numbers, cell size, cell mass and viability. Their physiological state is often judged by how fast they are growing (increase in cell mass) or proliferating (increase in cell number). The description is often written as

$$\frac{dn}{dt} = \mu n - \mu_d n,$$

$$\frac{dn_d}{dt} = \mu_d n,$$

where n is a viable cell concentration (number of cells per unit volume) and n_d is a dead cell concentration, and μ and μ_d are referred to as the specific growth and the death rate respectively. In some cases cell size is also considered. However, it is common to use an average cell size, r, rather than considering cell size distribution in a population. In most cases, the mass of the cell is considered to be directly related to the size.

$$nr^3 = \alpha x,$$

where x is the cell concentration in terms of mass. The population doubling time and the death rate are useful in characterizing stem cell culture. The death rate is small and the doubling time is rather constant. A change of the doubling time of cultured stem cells may be indicative of the onset of senescence due to an adverse environment.

2.3. **Phenotype characterization**

Many phenotypic attributes can be used to characterize a stem cell population. Most important among them is probably their capacity for self-renewal and differentiation. Although direct measurement of the potency of differentiation toward particular lineages is possible, its routine use in describing a cell population is not practical. In general, the characterization of stem cells relies on the presence or absence of certain surface antigens, or the level of transcript for some marker genes that are associated with stem cell potency. The presence of surface antigens is typically quantified by immunophenotyping using flow cytometry, while the level of transcript is by quantitative PCR. Although other than cell concentration the most important state variable is the potency of the cell, the nonavailability of a direct and quantitative description of potency leads to the use of surface markers or transcript levels as descriptors of the potency, and essentially the use of these correlated attributes as state variables. For many stem cell types a set of surface and transcript markers that can completely define the potency of the cell is not available. Instead, the markers are only used to separate the subpopulations. Strictly speaking, such a set of markers is not sufficient in defining the state of cells, since it is likely that only a subset of cells with the given markers have the designated potency. Nevertheless, for the convenience of discussion, we will refer to the markers as state variables with the understanding that in an ideal world the state variables should define the potency of a stem cell.

In Fig. 1, the "state" of multipotent adult progenitor cells (MAPCs) is represented. In terms of some of their state variables, MAPCs are adult stem cells with multilineage differentiation capability derived from rodent, swine and human bone marrow.[4-6] CD31 and CD44 surface marker expression distinguishes MAPC clones expressing high and low levels of transcription factor Oct4 (Fig. 1). Immunohistochemistry confirmation of Oct4 expression of the nuclear localization in high-Oct4 MAPC clones is shown in Fig. 2. Although immunohistochemistry is also important for relating cell/tissue morphology/architecture and phenotype, its results are typically less quantitative and are used in a binary way for verifying the presence or the absence of markers of individual cells. Although there has also been progress in the use of image analysis methods for measuring the amount and strength of immunohistochemical markers, the use of these methods is still in its early stage.[7] Even though flow cytometry provides some quantitative measure of the level of proteins on or in the cells, the

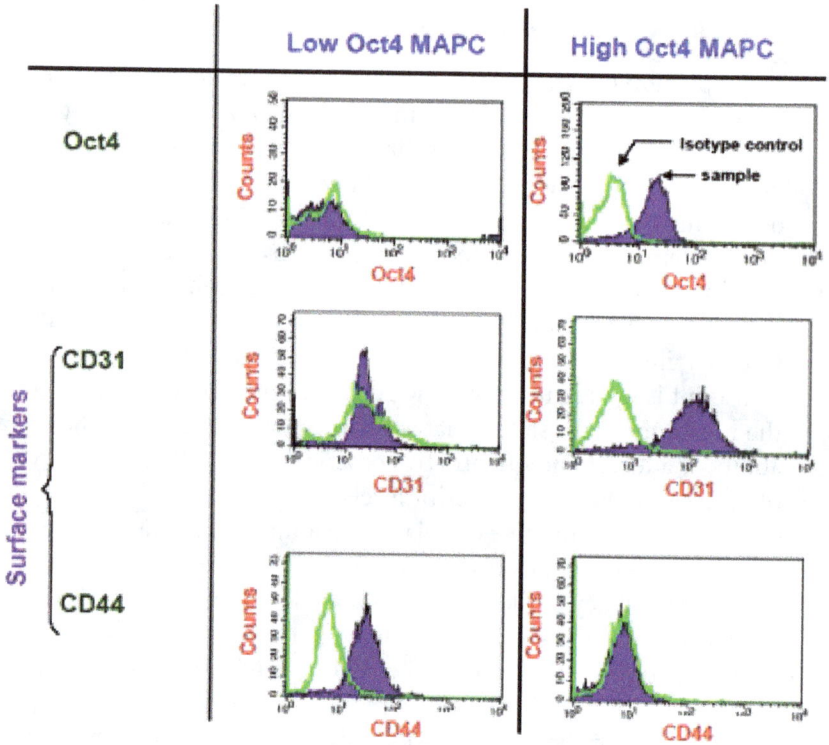

Figure 1. Comparison of the characterization of low Oct4 MAPCs (left) and high Oct4 MAPCs (right). Oct4 and CD31 are expressed in high Oct4 MAPCs but not in low Oct4 MAPCs, while CD44 is expressed in low Oct4 MAPCs but not in high Oct4 MAPCs.

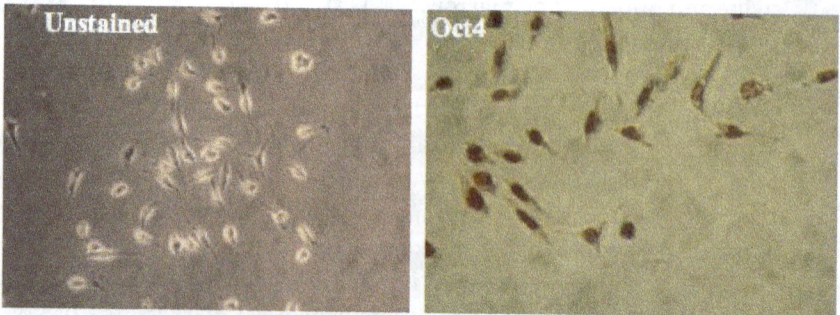

Figure 2. Immunohistochemistry; staining for Oct4 in the nucleus of high Oct4 cells. Unstained cells (left) and stained cells (right).

data it generates are commonly used only semiquantitatively. For example, $CD45^+$ or $CD45^-$, or the presence and absence of $CD45^+$ on the cell surface, is used to distinguish the hematopoietic cell population from the non-hematopoietic cell population. The extent of expression of CD45 cell surface antigens on different cell populations is seldom quantified. Rarely is the fluorescence intensity calibrated with respect to a positive control and reported on a continuous scale — say, from 0 to 1.

In addition, flow-cytometric characterization of stem cells and their differentiation can be greatly aided by employing analytical tools for transcriptome and proteome. In one study, the transcriptomes of several stem cells, including mouse embryonic stem cells (mESCs), neural stem cells (NSCs) and hematopoietic stem cells (HSCs), were compared to identify a core set of "stemness genes" after subtracting the transcripts expressed in the differentiated cell types from the brain and bone marrow. Although a total of 216 genes were shared between the three populations, only 4 were present in all three stem cell populations.[8] If indeed such a set of genes can be identified that confers the property of "stemness," the transcript level of those genes can be used as state variables. On elucidating the transcriptome of multipotent adult progenitor cells (MAPCs) along with mESCs and mesenchymal stem cells (MSCs), a unique signature of MAPCs distinct from other stem cells was established.[9] Identification of cell surface markers is important for guiding and monitoring differentiation toward specific cell types. In one study, mouse ES cells derived $Gsc+Sox17^+$ definitive endoderm and $Gsc-Sox17^+$ visceral endoderm gene expression profiles were compared and seven surface markers were identified. Among them, CXCR4 was used to select and monitor Gsc^+ cells from unmanipulated ES cells under culture conditions that select for definitive endoderm and enrich the cell population.[10] Temporal gene expression analysis of stem cell differentiation has also been employed for the study of differentiation pathways in finding novel transcriptional regulators or targets of known regulators that can help understand and guide differentiation to a particular cell type. For example, cDNA microarray analysis was used to study osteoblast differentiation from mesodermal progenitor cells (MPCs) over a period of 7 days. Forty-one key transcription factors were identified that play an important role in onset of differentiation to an osteoblast lineage in a time-dependent manner.[11] Such studies, along with *in vivo* studies on development of organs like the liver, would enable identification of transcription factors during differentiation with the potential to track the kinetics of differentiation.[12]

The use of proteomics in stem cells has been limited because of the challenges in identification of low abundance proteins and the lack of accessibility to sufficient numbers of purified cells representative of the phenotype of interest. However, the study of the proteome can facilitate the identification of surface markers, as was demonstrated in a study with human multipotent adult progenitor cells (hMAPCs) where a four-stage fractionation system that coupled liquid chromatography and mass-spectrometric methods was employed (LC-MALDI-MS/MS).[13] Out of the 2,151 proteins identified, there were several potential surface marker proteins, two tyrosine kinases previously related to self-renewal potential of stem cells and proteins of the ubiquitination/proteosome machinery potentially involved in directing proliferation and/or differentiation. In another study, iTRAQ was used to quantitatively compare the proteome of hematopoietic stem cells and of hematopoietic progenitor cells. A total of 948 proteins were identified and several hypoxia-related changes in controlling metabolism and oxidative protection were observed, indicating the adaptability of hematopoietic stem cells to anaerobic environments. This study also compared the transcriptome and proteome analysis and demonstrated that the proteomics approach complements the microarray analysis by taking protein turnover and posttranslational modifications into consideration.[14,15] We are likely to see more applications of transcriptome and proteome analysis to characterize stem cells and identify the signature of the defining stem cell properties. Such signatures are essentially state variables.

2.4. Averaged properties and distributed properties hyperproducing cell line

Due to the heterogeneity of cell types commonly observed in stem cell populations, it is important to include both averaged and distributed properties in characterizing the state of stem cell cultures. The heterogeneity or distribution of a population occurs at two different levels. First of all, the entire population may consist of subpopulations of defined and not well-defined characteristics. For example, a population may consist of well-defined stem cells capable of self-renewal, another subpopulation with well-defined progenitor cells and a third subpopulation with poorly characterized properties. On a second level, even within a pure population, not every individual cell will have an equal value of the characteristics being examined, be it a state variable or just some other properties. A complexity in describing stem cell culture populations is that not only are populations usually heterogeneous, consisting of many subpopulations, but also many important characteristics for each subpopulation have their levels distributed over a range.

Many stem cell characterizations rely on quantitative PCR (qRTPCR) to quantify the level of some lineage-specific transcripts. It provides the needed sensitivity for evaluating even transcripts at low levels and does not rely on the availability of surface markers and specific antibodies. However, traditional qRTPCR measurement only gives the averaged values of the population being assessed, not the level in each individual cell or even subpopulations, unless the cells have been sorted. Recently, single-cell PCR has become available for characterizing cell types appearing at low frequency in a population, e.g. differentiated cells of a particular lineage in a stem cell population.[16] On the other hand, cytometry-based measurements give the levels of multiple markers on individual cells and readily provide information on the frequency distribution of different types of cells. Although flow cytometry is highly effective in characterizing isolated different cell types in a population, the requirement of a large number of cells limits the characterization of infrequent, rare stem cells in a heterogeneous population of several cell types.

It is customary to present flow-cytometric data as a histogram on a single measured variable or on a contour plot on two variables. The histogram is basically the number of cells in each bin (step range of the measured variable). Population distribution data can also be expressed in terms of the population density function. This essentially takes an extremely small bin size (dr) and uses the density function to describe the population distribution. By taking the bin size to be infinitesimally small, the number of cells with properties between r_1 and r_2 can be calculated by integrating the population density function over the range of r.

$$S = \int_{r_1}^{r_2} f(r)dr \cdot N,$$

when N is the total number of cells in the entire population. The integral of $f(r)$ by itself does not have a unit. Integration of $f(r)$ from 0 to ∞ gives unity:

$$\int_0^\infty f(r)dr = 1.$$

Given a large-enough sample size, one can obtain the density function from fitting a curve to the histogram. Alternatively, one can plot a cumulative distribution curve and take a derivative function to describe the population density.

2.5. Multiparametric characterization of stem cells

Typically stem cells are characterized by the presence or absence of multiple surface markers. Often they are measured simultaneously by multivariate flow cytometry. The population density function is thus dependent on multiple independent property variables. The fraction of the subpopulation of a particular characteristic is thus the integral over the ranges of those independent variables. For the case of two variables written as

$$S = \sum_0 \varepsilon_i w_i \int_{a_1}^{a_2} \int_{b_1}^{b_2} f(x, y) dx dy.$$

An example of such two-dimensional property distributions is the combination of cell size and DNA content into one graph. The typical graph illustrates the density of a population by presenting each count as a dot. This can be transformed into a two-dimensional histogram and then a two-dimensional population density distribution curve. The two peaks represent two regions in which a large number of cells fall. Those are G2 and G1 cells respectively. It is a common practice to draw isoclines in the same way as contours on a map. The line which cuts through a plane parallel to the bottom plane enclosing a region which includes a certain top percentage of the population creates an isocline. A series of concentric isoclines can be drawn on a two-dimensional plot to represent the population distribution, as shown in Fig. 3. In the case where three independent variables are involved, a three-dimensional plot can be used, with each cell represented by a dotted point in the space. The density of those dots in the space illustrates the density of the subpopulation.

Figure 4 gives an example of three different types of cells in a three-dimensional space. Two of the three subpopulations, which on a two-dimensional plot overlap significantly, can be readily differentiated in the three-dimensional space. Additional features increase the power of differentiating the subpopulations. Hypersurfaces can be created so that a certain fraction of the subpopulation is enclosed inside each layer of the

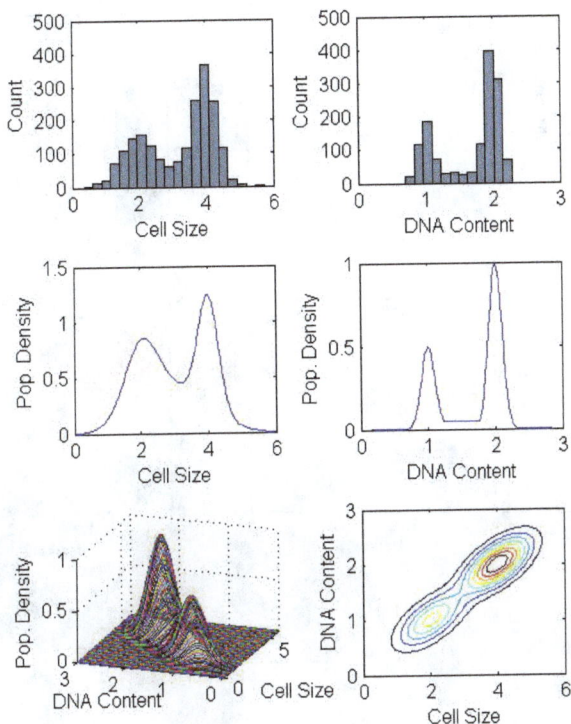

Figure 3. Cell size and DNA content effects on population density. Idealized histograms (top) of cell counts from measurements of cell size (left) and DNA content (right). The bimodal distribution is due to cell cycle phases corresponding to one copy of DNA (after division) or two copies of DNA (just prior to division). Idealized population densities (middle) provide a continuous function to model cell counts. Contour plots (bottom; left – 3D, right – 2D) demonstrating the combined population density as a function of both cell size and DNA content.

hypersurface. One can imagine that if the three sets of features define the "matureness" of a differentiated population, it could be possible to create a hypersurface, within the enclosed surface of which the probability of a cell being mature and differentiated is above some value. As the hypersurface expands to an outer layer, the probability decreases.

In many cases, more than three surface markers are needed. Table 1 gives a summary of some of the multipotent adult stem cells recently identified and the markers that have been measured on them. Comparing the phenotypes of these different stem cells, it is clear that although some markers are common among different stem cells, there is none unique to a particular cell type. Hence, characterization of such stem cells, with subtle phenotype

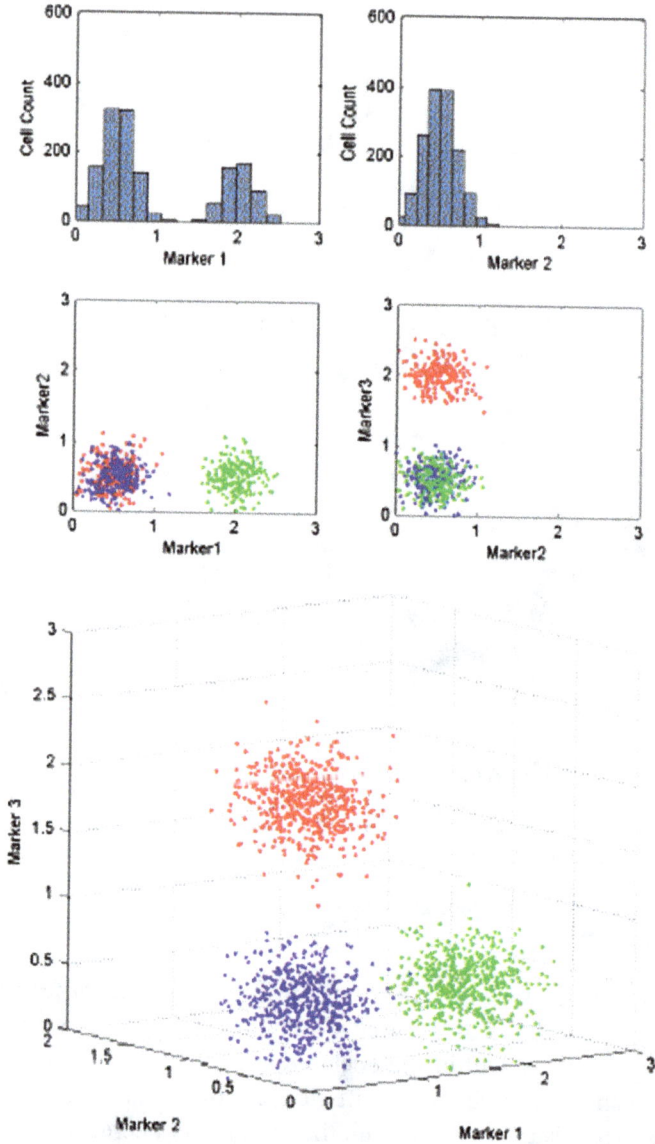

Figure 4. Multiparametric characterization of stem cells. Quantitative measurements for *N* markers should be visualized in an *N*-dimensional space to differentiate between the various populations. For this example, *N* = 3. Histograms (top) for marker 1 and marker 2 reveal only two populations (left) or one population (right). (*N*–1)-dimensional graphs (2D, middle) reveal only two populations when comparing two markers at a time. *N*-dimensional visualization (3D, bottom) reveals three distinct populations. Populations can be deduced from all possible (*N*–1)-dimensional plots, but are easily distinguished in *N*-dimensional plots.

Table 1 Characteristics of Some Adult Stem Cells — Surface Immunophenotype and Differentiation Capability. Since the Different Cell Types Do Not have a Unique Surface Phenotype, Multiparametric Characterization is Useful for Easily Distinguishing Between Various Cell Phenotypes.

Cell Type	Immunophenotype	Differentiation Capability	Reference
MIAMI (Marrow isolated adult multilineage inducible cells)	$CD29^+$, $CD49e^+$, $CD90^+$, $CD34^-$, $CD45^-$, c-kit$^-$	Neural, pancreatic, osteogenic	D'Ippolito *et al.*, *JCS* (2004)
SSEA-1$^+$ BM cells	Sca-1$^+$, $CD44^+$, $CD45^-$, $CD31^-$	Astrocyte, hepatocyte, endothelial	Afonso *et al.*, *Blood* (2007)
VSEL (Very small embryon-like cells)	$CXCR4^+$, SSEA-1$^+$, c-met$^+$	Neural, pancreatic, cardiomyocyte	Kucia *et al.*, Leukemia (2006)
USSCs (Unrestricted somatic stem cells)	$CD13^+$, $CD29^+$, $CD90^+$, $CD34^-$, $CD45^-$, c-kit$^-$	Neural, hematopoietic, hepatic	Kogler *et al.*, *J Exp Med* (2004)
MRPCs (Multilineage renal progenitor cells)	$CD90^+$, $CD44^+$, SSEA-1$^-$, $CD45^-$	Neural, endothelial, hepatic	Gupta *et al.*, *JASN* (2006)

differences between them, makes the utilization of multiparametric characterization space critical. However, visualizing the overlap and separation of different subpopulations in such multidimensional space is nontrivial. We foresee the applicability of dimensionality reduction techniques like principal component analysis (PCA)[17] being used to reduce the multidimensional space to a three-dimensional space that best characterizes the difference between most of the cells belonging to a particular stem cell population from most of the cells of another population. Further, distance measurement methods like *k*-means nearest neighbor can be used to quantify the separation of populations in the feature space.[18]

2.6. Functional characterization

The surface markers and specific transcript expression provide valuable tools for characterizing various stem cells. However, for many types of stem cells that have been isolated in the past few years, no reliable markers are truly definitive. Furthermore, the ultimate test of the potency of "stemness" is the cell's capability of self-renewal and differentiation to particular lineages. A quantitative means of assessing potency and differentiation capability is called for. In principle, potency and differentiation capability are related properties, since potency is defined by a cell's ability to differentiate to the designated lineages. However, the pathway of differentiation is a time-dependent event; after initiation and commitment to a lineage, cells pass through the various states of changing characteristics before reaching "maturation." The course of differentiation is often characterized by changing surface markers or transcript expression, a process that is continuous in nature and evolves over time. However, our analytical mind often grasps discrete events better than an evolving continuum. It is thus customary to express the differentiation process in stages. One can possibly characterize a population of differentiating stem cells by the fraction of cells in each stage of differentiation:

$$1 = \varepsilon_0 + \varepsilon_1 + \cdots + \varepsilon_m + (\varepsilon_w),$$

where ε_0 is the fraction of cells in each stage of differentiation, including those that are not differentiating (ε_i) and others that might have differentiated toward the "wrong" or unintended lineage (ε_w). One can propose that a score be established to describe the degree or capability of differentiation. As a weighted sum of ε_i,

$$S = \sum_0 \varepsilon_i w_i,$$

where w_i is the weighting factor, with $w_m = 1$. Completely differentiated cells have a full score, all others incompletely differentiated with $w_i < 1$, while those differentiating to the "wrong" lineage bear a penalty, and thus $w_w < 0$. In this case, if all cells differentiate to maturation, $\varepsilon_m = 1$, then the score is 1.0. For example, differentiation of embryonic stem cells to liver endoderm can be categorized into three stages: (a) formation of definitive endoderm represented by expression of CXCR4, (b) specification of the hepatic endoderm represented by expression of α-fetoprotein (AFP), and (c) maturation by expression of albumin or cytochrome P450. A differentiation score is possible by assigning weighting factors to be 0.3 for

CXCR4, 0.5 for AFP and 0.9 for albumin to represent the early, mid- and mature marker of hepatic differentiation. In addition, a penalty weighting factor of −0.5 can be assigned for neuroectoderm marker N-CAM or NF200, and −0.5 for a mesoderm marker like Brachyury. Fractions of cells expressing each of these markers could be quantified at different stages of differentiation and the corresponding differentiation score could be computed.

The calculation of such a score would enable measurement of the extent of differentiation to the lineage of choice under some culture environment and provide a value to compare the differentiation abilities of different stem cells.

For pluripotent or multipotent stem cells, differentiation toward multiple lineages will need to be tested. The current definition of stem cell potency is rather discrete and binary. For example, pluripotency refers to a cell's capability to differentiate to lineages of all three germ layers. Failure to differentiate to one of them disqualifies the cell from being considered pluripotent. Thus, a cell in a mixed population of different potencies can be either pluripotent or not. The population is thus characterized by the fraction of pluripotent cells.

Again, the progression of stem cell differentiation and the stem cell "losing" its "stemness" is continuous rather than abrupt and discreet. It might be better to employ a grading or scoring system using some type of weighing factor. Although stem cells are capable of unlimited renewal, it is known that under some laboratory conditions they do change and alter their potency. With a scoring system one can better assess the properties of cells that no longer possess the complete potency. Many of these cells are still valuable for research, and perhaps in clinical application in the future as well. A refined characterization of the potency of these cells will potentially increase their utility.

3. Kinetic Description of Population Dynamics

In the evaluation of a time-evolving system, it is important to describe the speed at which the system is changing. That description allows one to begin to modulate and control the rate and direction of change of the system. In the previous section we discussed quantitative description of stem cell growth. Once quantitative description for "stemness" and differentiation is in place, one also seeks to describe quantitatively how quickly the process is progressing.

3.1. Specific rates

The two categories of rates most important in stem cell culture are the growth rate and the rate of differentiation to a particular lineage or "degeneration" to lose "stemness" in the case of cell maintenance/expansion. The quantity used to describe growth, the specific growth rate (or the doubling time), has been described. The specific rate of differentiation or degeneration can be written simplistically as

$$\frac{dn}{dt} = \mu n - \mu' n,$$

$$\frac{dn'}{dt} = \mu' n,$$

where μ' is the rate constant (often described as the specific rate) of differentiation or losing "stemness." Many differentiation processes are carried out at confluent density. Under such conditions, the increase in the number of cells is negligible and the growth term can be neglected. The differentiation term treats differentiation as a discrete event. A sharp transition from undifferentiated (n) to differentiated (n') cells presumably divides the population into two categories. As discussed in the previous section, the cell's differentiated state is typically characterized by multiple variables and can be visualized in a multidimensional space. Cells of different differentiated states reside in different regions in the space. As cells differentiate, they begin to change the value of those markers and move in the space. A centroid of the population can be defined, and the trajectory of the centroid can be connected. The initial population may have different directions, as depicted in Fig. 5; for example, one may be desired and the other abnormal. They may arise from every region in the same initial population randomly. It is also possible that cells in the subspace in the initial population give rise to its abnormal subpopulation. One can connect the centroids to create a trajectory (a vector) to illustrate the path of differentiation and even calculate the specific rate of differentiation. In Fig. 6, we illustrate a 3D-differentiation trajectory comparison of the influence of three different extracellular matrices, including matrigel, collagen and fibronectin, on MAPCs' ability to differentiate to the hepatic lineage. The x, y and z axes correspond to mRNA expression levels of the early, mid- and late marker of hepatic differentiation, respectively. It is

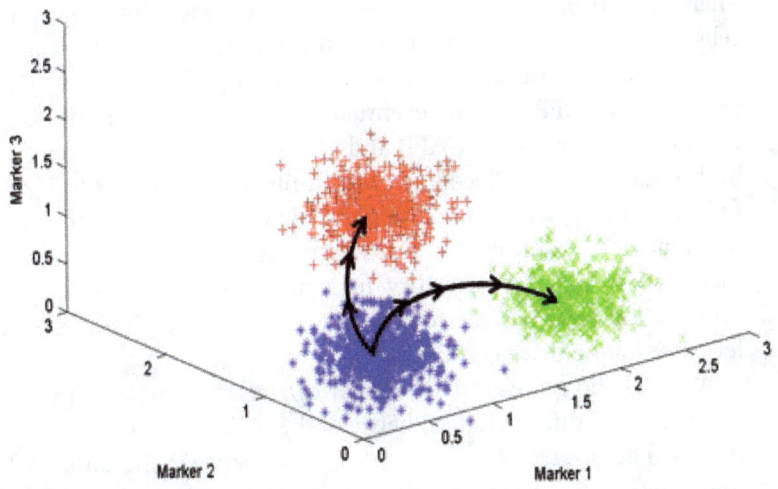

Figure 5. Developmental paths along a three-dimensional space. By connecting the trajectory of the cartroid of the population, vectors of differentiation can be depicted and the specific rate of differentiation can be defined.

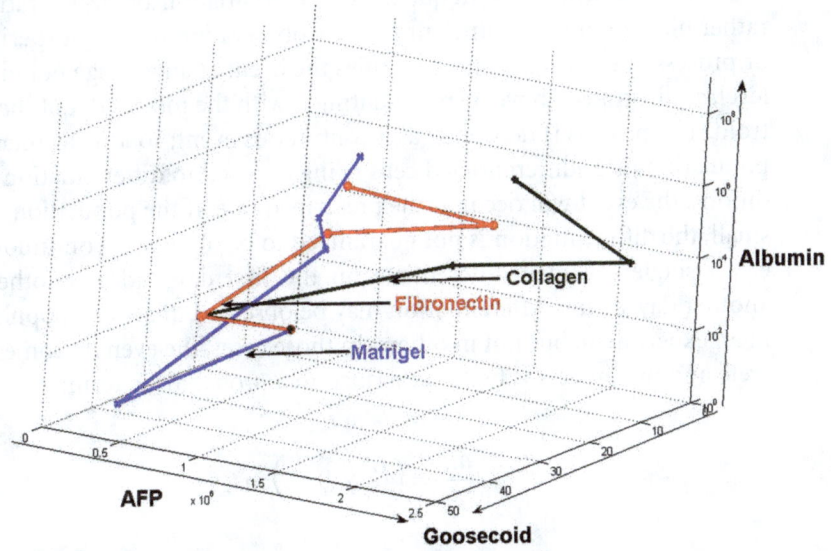

Figure 6. 3D differentiation trajectory. The differentiation of a stem cell can be mapped through N space based on quantitative measurements of N differentiation markers. Data shown is MAPC differentiation toward a hepatocyte fate on three different substrates (fibronectin, collagen and matrigel). Markers shown consist of one early marker (goosecoid), one specification marker (AFP) and one maturation marker (albumin).

interesting to observe the differences in the population trajectory of the cells under the different conditions. In embryonic development, cells begin by expressing Goosecoid, an early definitive endoderm marker, followed by expression of AFP and the eventual expression of albumin along with a decrease in expression of AFP and Goosecoid, indicating maturation. In both matrigel and collagen but not in fibronectin, the cells appear to be following the path similar to embryonic development, albeit with some differences. In addition to indicating the similarity to embryonic development, observing the trajectory in a multidimensional space also gives information on when and to what extent the differentiation begins to differ between the three conditions.

A simplistic mathematical model can be constructed to quantify the parameters of differentiation between the three stages of hepatic differentiation. The inverse of differentiation time between the different stages is the specific rate for that transition (or rate constant). Such quantification would provide insights into the kinetics of differentiation and lay the design principles for understanding and relieving the bottlenecks in differentiation.

In the case where the frequency of differentiation or degeneration is rather high, or at least sufficient for easy observation in every experiment or process, the constant rate quantifies the event at an average population level, as illustrated above by representing it with the movement of the centroid. The process is described as a continuous event; so long as there is a population of undifferentiated cells being exposed to differentiation conditions, the event will occur at the prescribed rate. If the population is too small, the differentiation is not guaranteed to be observed "continuously," e.g. a single cell may differentiate on the first day and/or another on another day, or the differentiation may be observed in a small population in an experiment, but not in others. In those cases, the event becomes discrete and the process is better described in a probabilistic term:

$$\frac{dn'}{dt} = \left(P\big|_{n \to n'^{(n)}} \right) \cdot n.$$

Many processes involving stem cells are likely to fall into this category. The probability function P is affected by the presence of growth factors and signaling molecules.

4. Quantitative Description of the Environment

4.1. Environment

For the development of stem cell bioprocesses, quantitative understanding of the influence of parameters on self-renewal and differentiation is essential. Concentrations of glucose serum, cytokine/growth factor, matrix components, oxygen, pH and temperature have been shown to play a critical role in defining the state of the stem cell culture system.

Even though the bulk of the work has been fairly recent and the conditions may not be optimal, the environment for culturing stem cells is generally well defined. The important medium components fall into three categories: nutrients, growth/differentiation factors, and the bulk salts sustaining the electrolyte homeostasis. Additionally, a buffer (typically carbon-dioxide-based) is present.

Excluding serum, growth factors and signaling molecules, the other components, including nutrients, trace salts and balancing bulk salts, are generally lumped together and called basal medium. The composition of basal medium for stem cell culture is generally not different from other types of cell culture. Under specialized conditions, the concentration of some nutrients may not be typical. For example, higher glucose concentrations may be used in the differentiation to insulin-producing β-cells.

The chemical composition of the environment is often based on the understanding of the signaling pathways and gene regulation of the behavior of the stem cells. In other cases, it is derived empirically. However, to date, a mechanistic understanding of the regulatory networks related to stem cells is rather limited. Optimization methods, such as composite and factorial design, have been employed to devise the composition of the cytokine and growth factors.

In addition to these chemical environments, many other factors, both chemical and physical, may affect differentiation and renewal. Coating of surfaces with extracellular matrix molecules, such as fibronectin or collagen matrigel, is a common practice. Reported evidence suggests that surface stiffness may affect differentiation.[19] Whether hydrostatic pressure or fluid flow affects differentiation (as seen in the response of endothelial cells) has yet to be investigated.

The most complex environmental factor in stem cell culture is probably the feeder layer. For the derivation and cultivation of human embryonic stem cells, it is still a common practice to employ mouse embryonic fibroblasts (MEFs) as feeder layers. These nonproliferating cells (since they have been gamma-ray-irradiated or treated with growth inhibitory antibiotics) provide essential and uncharacterized factor(s) for sustaining the capability of self-renewal. Replacement with other chemical factors or other cells has not always been successful. The role of the feeder layer possibly involves presentation of unknown growth-stimulating molecules on the feeder cell surface or secretion of some stimulatory molecules. It is also possible that a complex process of cross-talking between the growing (stem cell) and nongrowing MEFs modulates the fate of stem cells.[20]

4.2. Characterization of the environment

To characterize the environment of stem cell culture, it is important that one specifies the concentrations of major components. It is also important to characterize the changes in the concentration of those key components. In general, two classes of components are present in terms of their consumption rate: the major components whose consumption is easily quantifiable (such as glucose) and others which are consumed at a much slower rate (such as transferin and lipids) or are hardly consumed at all (such as sodium and calcium). For the former, one typically defines a specific consumption rate, q_s:

$$\frac{ds}{dt} = -q_s x.$$

As cells differentiate, the specific consumption rate may change. In many differentiation protocols, a step change of medium composition is practiced. For example, in stem cell differentiation to liver cells, the medium is changed to remove and replace growth factors. The change in the growth factor composition has a profound effect on the outcome.

There has been an increasing reliance on design of experiment (DOE) statistical tools to cover a wide range of parameter space with an optimized number of experiments. In one study, thrombopoeitin (TPO), flt-3 ligand (FL), Steel factor (SF) and interleukin 11 (IL-11) were optimized for long-term repopulation of hematopoietic stem cells (HSCs) by factorial experimental design and response-surface analysis. The study led to the finding that IL-11 and SF were the most significant stimulators of murine HSC expansion. The results of the factorial design analysis were represented in terms of modified Monod kinetic models, including terms for inhibitory substrates, which allowed state outputs like HSC expansion to be predicted as a function of concentration of the cytokines.[21] In another, related study, two-level factorial design combined with the steepest ascent method was used to optimize serum-free culture for expansion of CD34+ human hematopoietic stem cells derived from cord blood. An HSC medium with lower cytokine concentrations than previously used was employed, with superior or comparable HSC growth.[22] Such statistical methods have also been employed in embryonic stem cell differentiation to the endoderm using automated fluorescence microscopy.[23]

The cellular microenvironment where the stem cell resides, or the "niche," is a complex combination of soluble signals, cellular matrix, physical stimuli, and possibly cell–cell interactions. Many of these factors interact in a nonlinear way. Examining the effect of each factor independently is not adequate for identification of optimal conditions.

Even with an optimized statistical design of experiments, combinatorial screening of multiple parameters at the tissue culture scale can be overwhelming in labor and expensive in cost. Recently, there have been efforts to develop miniaturized platforms for screening multiple parameters using a glass array of "molecular microenvironments" consisting of different combinations of matrix proteins or signaling molecules in a high-throughput manner, as even quantitative analysis of the phenotype can be obtained at a single-cell resolution. An extracellular matrix microarray consisting of 32 different combinations of five different extracellular matrix molecules probed the effects of the differentiation of ES cells.[24] In another study, the regulation of human neural precursor cells was studied upon treatment with morphogens, signaling proteins and extracellular matrix components. Using quantitative single-cell analysis, it was found that costimulation with Wnt and Notch was required to maintain cells in an undifferentiated state.[25] Recently, the increased use of microfluidic technology has opened up new possibilities in perfusion-based culture systems on a miniaturized scale. A microfluidic system can maintain steady medium composition and minimize spatial gradients in the cellular microenvironment.[26] A micro-bioreactor array consisting of 12 independent bioreactors coupled with *in situ* quantitative image analysis was used to study the effect of flow-dependent differences in smooth muscle differentiation of human embryonic stem cells.[27]

High-throughput flow-based miniature systems will become valuable tools for optimizing the environmental conditions for stem cells.

5. Relating Environment to Process

The sections above describe the road to a well-characterized stem cell culture. It is the first step in transforming a laboratory protocol to a process. There are occasions when one may wish to be able to predict the effect of process change on cell behavior. If there is a predictive model, one may first explore the consequences by using computer simulations before setting out to conduct the experiment. Employing an advanced model that incorporates much biological insight, the model can also be used to control the process. For such purposes, a description relating the environment to the behavior of cells is needed.

5.1. Traditional Monod model

Traditionally, the main interest in describing microbial processes is to depict the growth, i.e. the change of cell concentrations over time and how it responds to changes in environmental conditions. Two general approaches are taken to develop a mathematical model that describes the relationship between the environment and cell growth: an empirical model and a mechanistic model. The empirical model is usually obtained by fitting experimental data to functionality. The mechanistic model attempts to depict the response of the system (in this case the system is the cell, and the response is growth) to the environmental factors through the intermediate steps of cellular biochemical reactions and regulations. Except in a few rare cases, the mechanism of growth control is too complex and involves too many steps from the environmental factors to facilitate growth response. The models that have been shown to have practical application are essentially empirical models.

Classical empirical models for cell growth all follow the original one conceived by Monod. This model basically states that as the nutrient concentration increases from nil, the growth rate increases until it reaches a saturation value (maximum growth rate) asymptotically (Fig. 7a):

$$\mu = \frac{\mu_{max} s}{K_s + s},$$

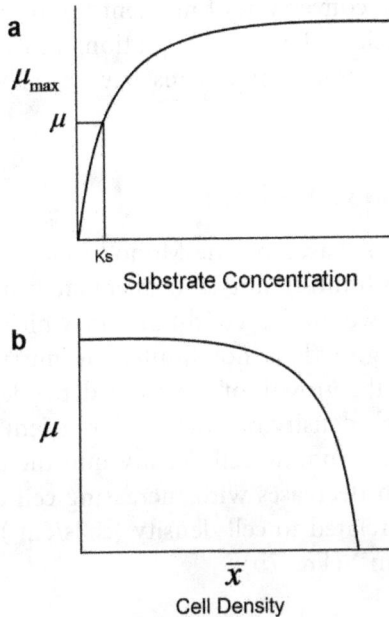

Figure 7. (a) Classic Monod model for cell growth based on one limiting substrate (nutrient). The growth rate increases until it reaches a saturation value (maximum growth rate) asymptotically. (b) Growth rate as a function of cell density; as the cell density increases, the growth rate decreases.

where μ_{max} is the maximum growth rate, s is the substrate concentration and K_s is the substrate concentration where $\mu = \mu_{max}/2$.

The model has a form identical to Michaelis–Menten enzyme kinetics. It describes the growth rate response of the *E. coli* growth rate to glucose concentration reasonably well. Subsequent applications to other micro-organisms and even mammalian cells were reported to be satisfactory. A variety of somewhat modified models are widely used in many processes.

To employ the model, one combines the equations of growth and environmental factors:

$$\frac{dx}{dt} = \frac{\mu_{max}s}{K_s + s}x,$$

$$\frac{ds}{dt} = \frac{\mu_{max}s}{K_s + s}\frac{x}{y_{x/s}},$$

where $y_{x/s}$ is the conversion of nutrient (s) to cell (x) and is often called the yield coefficient. This set of equations and similarly formulated ones for a variety of micro-organisms have been widely applied in many processes.

5.2. Modeling stem cell culture

Although models based on the Monod model describe the growth rate dependence on limiting nutrients adequately, most mammalian cells in culture are grown under conditions in which the nutrient supply is abundant and growth is not limited by nutrient concentration. One notable case is the growth of adhesion-dependent cells. Their growth is restricted by cell density but rarely by nutrient availability. A common practice is to incorporate cell density into the growth rate equation so that the growth decreases with increasing cell density. In that case the growth rate is related to cell density (cells/cm^2) instead of cell concentration (cells/cm^3) (Fig. 7b):

$$\mu = \frac{\mu_{max} \cdot \overline{x}}{k_x + \overline{x}}.$$

In describing stem cell culture, one aims to establish a relationship between the concentrations of cells at a particular state (potency or differentiation) and the environmental factors. Considering only two populations, stem cells $[n(p_1)]$ differentiating to another cell type $[n(p_2)]$,

$$\frac{dn(p_1)}{dt} = \mu \cdot n(p_1) - \delta \cdot n(p_1),$$

$$\frac{dn(p_2)}{dt} = \delta \cdot n(p_1).$$

Basically the equation describes the change of stem cell concentration $[n(p_1)]$ as the result of its own expansion (governed by specific growth rate μ) and the rate of its conversion to differentiated cells at a specific rate δ.

One seeks to define how δ is affected by the environmental factors. For example, if the differentiation rate is affected by molecule g positively at a low concentration level but negatively at a high concentration level, a possible form of the equation is

$$\delta = \frac{\delta_{max}\, g}{K_g + g + k'_g g^2}.$$

A balance equation is also needed for g. One possible scenario is that factor g is taken up or degraded by cells and added through medium addition at rate F and concentration g_0.

$$\frac{dg}{dt} = -\delta_d \cdot g \cdot n + F \cdot g_0.$$

In a practical sense, the difficulty in stem cell culture is in establishing function $\delta(g)$, i.e. the relationship between the rate of differentiation and the growth factor. The example equation shown is a deterministic type. Such deterministic models assume that the rate of differentiation is completely predictable by the relationship specified. In addition, it is also assumed that they imply that all cells have an equal chance to differentiate; given enough time the undifferentiated cell gradually differentiates. In most stem cell differentiations studies, only a fraction of cells differentiate in the desired direction and reach varying degrees of maturation. While it is possible that a heterogeneous starting population is responsible for the failure of the entire population to differentiate, it is also possible that the event of differentiation initiation is stochastic. The presence of the determinant factor merely increases the probability of cell differentiation rather than committing the entire population in a definitive manner. More studies on the stochastic aspect of stem cell culture processes will give more insight into the differentiation commitment.

Another aspect that renders the modeling stem cell process a complex task is the cross-talking among subpopulations. Cell–cell interaction may occur not only between the feeder layer cells and stem cells, but also among the subpopulation of stem cells differentiating otherwise. For example, cells differentiating in different lineages may cross-feed factors and modulate the differentiation path. The cross-talking may be one-way or both-ways. It is often found that cells in the course of differentiation cluster in seemingly random spatial locations in a population and that the differentiating cells trigger their neighbors to differentiate as well. The existence of signals that cells communicate to each other has recently been observed.[18]

In many cases, the mechanism of the transcription regulation event for differentiation is at least partially known. One can possibly develop a mechanistic model for such a gene regulation event and translate it to functionality for S to the differentiation determinant factor. Several recent studies have indicated the propensity of signal transduction cascades and gene regulatory networks to function as biological "switches," where stem cells adopt alternative fates as a function of input signals in a nonlinear feedback-dependent manner.[28] Thus, along with incorporation of deterministic elements in recent modeling efforts, the concept of signaling thresholds is becoming increasingly popular. For example, it was recently shown that the influence of a signaling factor sonic hedgehog (Shh) on adult neural stem cell differentiation was based on a threshold response mechanism.[29,30] In another work, the signaling threshold levels of LIF were demonstrated, both theoretically and empirically, to influence self-renewal vs differentiation decisions.[31] Eventually, such threshold models can be used to design cytokine supplementation strategies as well as modulation of expression of receptors on the cell surface such that ligand-receptor interaction dynamics favorably increase the probability of desired fate choice (self-renewal or directed differentiation) of the cells.

Conclusion

In the past scores of years we have seen many biological discoveries and innovations successfully transform into technologies that have been largely responsible for drastically improved healthcare and nutrition. The most recent ones are new biotechnology based on mammalian cell culture. The first step in such a transformation has always been applying process analysis to the procedure of product manufacture. The success of product manufacture always relies on a robust process. Quantifying the variables and rates involved and assessing the factors that affect those variables and rates quantitatively are the keys to process analysis.

Stem cell cultivation and differentiation are processes with a much greater degree of complexity than what the biotechnology community has encountered in the past. Because of that complexity the need for quantification is actually greater, not smaller. With the promise of stem cells and the rapid advances in stem cell science, the urgency of introducing better quantification into the field to facilitate its transformation into technology should be noted.

References

1. Macfarlane G. (1985) *Alexander Fleming: The Man and the Myth* (Oxford University Press).
2. Rolinson GN, Geddes AM. (2007) The 50th anniversary of the discovery of 6-aminpenicillanic acid (6APA). *Int J Antimicrob Agents* **29**: 3–8.
3. Lubiniecki AS. (1998) Historical reflections on cell culture engineering. *Cytotechnology* **28**: 139–145.
4. Jiang Y, Jahagirdar BN, Reinhardt RL, *et al.* (2002) Pluripotency of mesenchymal stem cells derived from adult marrow. *Nature* **418**: 41–49.
5. Zeng L, Rahrmann E, Hu Q, *et al.* (2006) Multipotent adult progenitor cells from swine bone marrow. *Stem Cells* **24**: 2355–2366.
6. Reyes M, Lund T, Lenvik T, *et al.* (2001) Purification and *ex vivo* expansion of postnatal human marrow mesodermal progenitor cells. *Blood* **98**: 2615–2625.
7. Kaczmarke E, Gorna A, Majewski P. (2004) Techniques of image analysis for quantitative immunohistochemistry. *Rocz Akad Med Bialymst* **49 Suppl 1**: 155–158.
8. Ramalho-Santos M, Yoon S, Matsuzaki Y, *et al.* (2002) "Stemness": transcriptional profiling of embryonic and adult stem cells. *Science* **298**: 597–600.
9. Ulloa-Montoya F, Kidder BL, Pauwelyn KA, *et al.* (2007) Comparative transcriptome analysis of embryonic and adult stem cells with extended and limited differentiation capacity. *Genome Biol* **8**: R163.
10. Yasunaga M, Tada S, Torikai-Nishikawa S, *et al.* (2005) Induction and monitoring of definitive and visceral endoderm differentiation of mouse ES cells. *Nat Biotechnol* **23**: 1542–1550.
11. Li Y, Zhang R, Qiao H, *et al.* (2007) Generation of insulin-producing cells from PDX-1 gene-modified human mesenchymal stem cells. *J Cell Physiol* **211**: 36–44.
12. Jocheim-Richter A, Rudrich U, Koczan D, *et al.* (2006) Gene expression analysis identifies novel genes participating in early murine liver development and adult liver regeneration. *Differentiation* **74**: 167–173.
13. Gevaert K, Pinxteren J, Demol H, *et al.* (2006) Four-stage liquid chromatographic selection of methionyl peptides for peptide-centric proteome analysis: the proteome of human multipotent adult progenitor cells. *J Proteome Res* **5**: 1415–1428.

14. Unwin RD, Smith DL, Blinco D, *et al.* (2006) Quantitative proteomics reveals posttranslational control as a regulatory factor in primary hematopoietic stem cells. *Blood* **107:** 4687–4694.
15. Xu C, Police S, Rao N, Carpenter MK. (2002) Characterization and enrichment of cardiomyocytes derives from human embryonic stem cells. *Circ Res* **91:** 501–508.
16. Aubin JE, Liu F, Candeliere GA. (2002) Single-cell PCR methods for studying stem cells and progenitors. *Methods Mol Biol* **185:** 403–415.
17. Sharov AA, Piao Y, Matoba R, *et al.* (2003) Transcriptome analysis of mouse stem cells and early embryos. *PLoS Biol* **1:** D74.
18. Vits H, Chi CM, Hu WS, *et al.* (1994) Characterizing patterns in plant somatic embryo cultures: its morphology and development. *AIChE J* **40:** 1728–1740.
19. Engler AJ, Sen S, Sweeney HL, Disher DE. (2006) Matrix elasticity directs stem cell lineage specification. *Cell* **126:** 677–689.
20. Bendall SC, Stewart MH, Menendez P, *et al.* (2007) IGF and FGF cooperatively establish the regulatory stem cell niche of pluripotency human cells *in vitro. Nature* **448:** 1015–1021.
21. Audet J, Miller CL, Eaves CJ, Piret JM. (2002) Common and distinct features of cytokine effects of hematopoietic stem and progenitor cells revealed by dose-response surface analysis. *Biotechnol Bioeng* **80:** 383–404.
22. Yao CL, Chu IM, Hsieh TB, Hwang SM. (2004) A systematic strategy to optimize *ex vivo* expansion medium for human hematopoietic stem cells derived from umbilical cord blood mononuclear cells. *Exp Hematol* **32:** 720–727.
23. Chang KH, Zandstra PW. (2004) Quantitative screening of embryonic stem cell differentiation: endoderm formation as a model. *Biotechnol Bioeng* **88:** 287–298.
24. Flaim CJ, Chien S, Bhatia SN. (2005) An extracellular matrix microarray for probing cellular differentiation. *Nat Methods* **2:** 119–125.
25. Soen Y, Mori A, Palmer TD, Brown PO. (2006) Exploring the regulation of human neural precursor cell differentiation using arrays of signaling microenvironments. *Mol Sys Biol* **2:** 37.
26. Yu H, Meyvantsson I, Shkel IA, Beebe DJ. (2005) Diffusion-dependent cell behavior in microenvironments. *Lab Chip* **5:** 1089–1095.
27. Figallo E, Cannizzaro C, Gerecht S, *et al.* (2007) Micro-bioreactor array for controlling cellular microenvironments. *Lab Chip* **7:** 710–719.
28. O'Neill A, Schaffer DV. (2004) The biology and engineering of stem-cell control. *Biotechnol Appl Biochem* **40:** 5–16.
29. Lai K, Robertson MJ, Schaffer DV. (2004) The sonic hedgehog signaling system as a bistable genetic switch. *Biophys J* **86:** 2748–2757.

30. Saha K, Schaffer DV. (2006) Signal dynamics in Sonic hedgehog tissue patterning. *Development* **133:** 889–900.
31. Mahdavi A, Davey RE, Bhola P, *et al.* (2007) Sensitivity analysis of intracellular signaling pathway kinetics predicts targets for stem cell fate control. *PLoS Comput Biol* **3:** e130.

Index

α-fetoprotein   446, 447, 454, 456, 457, 459

A2B5   313, 314
ACE inhibitor   191, 193
activity   300
acute myocardial infarction   191, 192, 201
acute renal failure   421, 422, 427, 430–432
adipose-derived   193
adrenalectomy   302
adult stem cell   64, 66, 75, 78
age   302, 312
akinesia   202
albumin   446, 447, 451, 452, 454–457, 459, 461, 467, 468, 472–474
allogeneic   209
Alzheimer's disease   316, 318
amyotrophic lateral sclerosis (ALS)   316
angiogenesis   316
anti-inflammatory   205
anti-ischemic   207
apolipoprotein E   204–206
apoptosis   192, 197
apoptotic degeneration   312
arrhythmia   195, 196, 198, 202
astrocytes   294
asymmetric differentiation   64, 65, 77

asymmetric division   64, 65, 77, 292
atheroprotection   205
atherosclerosis   193, 195, 202, 203, 206
atherosclerotic   192, 203–206
atypical ductular proliferation   449
auditory   398, 400–405, 410
autologous   198, 200, 201, 209, 210
average properties   520

balance   397, 398, 411
BDNF   310
bedside   196, 202, 210
bench   193, 196, 202, 210
Bergman-glia   292
beta cell   28, 29, 32, 34, 35
beta-blocker   191, 193
bFGF   292, 295
bile duct epithelial cells   445
biocompatibility   209
biodistribution   201
biomarker   213
biorepository   210, 211
bi-potential progenitor   447
blastocyst   294, 305
BLBP   296
blood   305
blood flow   191, 195, 197, 213
BMPs   314

bone marrow   204, 209, 210, 306, 317

bone marrow mononuclear cells 192, 205, 206

brain tumors   378, 379

BrdU   297, 305

BRG1   55, 56

CA1   309

canal of Hering   448–450

cardiac precursor cells   207

cardiac stem cell   166, 167

cardiomyocyte   194, 207

cardiomyopathy   201

cardiomyoplasty   194, 197, 199

catheter   196, 200, 212

CD34   179

CDC2   80, 83, 84

CDC25   69, 70

cdc28   68

CDK1   72

CDK2   71, 83–85

CDK4   69–71, 83, 84, 86

CDK6   69–71, 83, 84, 86

CDK-activating kinase (CAK)   69, 70

cell cycle   63–65, 67–69, 71, 75, 76, 78–80, 82–88

cell fusion   45–50, 52

cell growth   68, 79, 86

cell proliferation   64, 66, 72, 75, 79, 83–85, 87

cell replacement   317

cell therapy   192, 193, 195, 196, 198–200, 202, 206, 210–215

cellular therapy   234, 237, 419, 429, 430

checkpoint   72, 79, 82

chemokine   204, 205

chemotherapeutic   303

chemotherapeutic agents   303, 316

chimera   294

cholesterol   204

cholinergic   296

chromatin   45, 46, 50, 53–58

chronic liver injury   448–451, 466

clinical manufacture   487, 497, 500

clinical trial   191–194, 196, 198, 210–214, 313

clonal   296

c-myc   76, 77

CNPase   313

cochlea   397–407, 410, 411

cochlear stem cell   327–349

collagen   209

comparison   194–196, 207, 208, 211, 213

confocal microscopy   304

connexin-43   195, 196

consensus   213, 214

contractility   193, 194, 200, 202, 213

corneal epithelium   113, 114, 117, 122, 125, 127, 131

coronary artery bypass grafting 194

coronary heart disease   191, 192

cortical pyramidal neurons   312

corticospinal neuronal   312

cortisol   302

covariate   211

cyclin A   71

cyclin B   72, 80, 83

cyclin D   69–71, 86

cyclin dependent kinase (CDK) 69, 70, 80, 81, 83–87

cyclin E   71, 84, 85

cyclin-dependent kinase inhibitor (CKI)   69, 70, 86

cytokeratin (CK)19   447, 452, 455, 459, 462, 472

cytokine   203–205, 302, 307, 309, 311

cytoplasmic beads   304

damage   306
database   210, 211
degenerative   316
dementia   302
demyelinating lesions   314
dentate gyrus   301
depression   302
detergent   209
development   64, 66, 72, 73, 77, 78,
    80, 84, 85
DG   298, 310
diabetes   28, 35
diastolic   193, 194, 197, 199, 201,
    207
differentiate   207
differentiation   27, 29–40, 64–66,
    72–75, 77, 78, 80, 82, 86, 87,
    193, 197, 207–210, 265–269,
    272–279, 283, 284, 328–330,
    332, 334, 335, 337, 338,
    340–348
differentiation trajectory   528, 529
diphtheria toxin   317
distributed properties   520
DNA replication   63, 67, 68, 71,
    80, 81, 83, 85
dobutamine   202
dopamine   298
dopaminergic   296
dose   202, 207, 214
Dp1/2   69, 71
drug-eluting stent   192
drugs   193, 214

E2F   69, 71, 79, 80, 83–85
ectoderm   30
ectodermal   79
ectopy   198
edema   198
efficacy   196, 198, 200, 212, 215
EGF   294
EGF-receptors   292

electrophysiological   309
embryoid body (EB)   5, 6, 454, 456
embryonic   207–210
embryonic stem cell   27, 31, 40,
    45, 64, 66, 72–75, 294, 316
endocrine   27, 29, 31, 32, 35–37, 39
endoderm   28, 30, 31, 35, 36, 38–40
endodermal   79
endodermal stem cells   446
endogenous repair   193, 204
endolymph   399, 401
endothelial cells   306
endothelial dysfunction   203
endothelial progenitor cells   192,
    202, 205
endothelium   203, 204
endovascular   362, 381
end-point   193, 202, 213, 214
engraftment   193, 195–197, 212, 214
enriched environment   302
environment   516, 527, 531, 532,
    534
epiblast   291
epilepsy   302
erythrocytes   5, 7, 9
ES cell-derived hepatocytes   456,
    473, 474
espin   400, 410
estrogen   302
excessive proliferation   313
exercise   302, 308
exocrine   27–29
exogenous cells   315
extracellular matrix molecules   306
extraocular muscle   177, 180, 185

feeder layer   531, 532, 537
fetal human brain tissue   315
FGF   291
FGF-2   302
fibrillation   198–200
fibroblasts   194

flow-limiting lesion   203
foregut   39, 40
fractional shortening   207
FRGY2a/b   55
fumarylacetoacetate hydrolase
       (FAH)   468–470, 473, 474
functional properties   301
functional recovery   312, 316
functionally integrate   309
fuse   305
fusion   305

GABAergic   296, 315
gap phases (G0, G1, G2)   63, 68, 69,
       71, 72, 79–81, 87
gastrulation   291
GFAP   292, 298, 300
GLAST   296
glia   292
glial scar   314
glioblastomas   313
glucagon   29, 37
glucokinase (GK)   31, 33, 35, 36, 38,
       39
glut-2   31, 37, 38
glutamatergic   296
Gp130   73
granule neurons   298
gray matter   314
growth factor   192, 215

hair cell   397–405, 409–411
hearing   397, 401, 411
heart   143, 145, 152, 158, 159,
       164–167, 192, 193, 201, 207,
       213, 221–227, 229, 230, 232,
       234–238, 242, 244–250
heart failure   191–194, 221, 237,
       238, 244, 248
hematopoiesis   3–5, 7, 8, 10, 14–16,
       18

hematopoietic   202, 206, 207
hematopoietic stem cell (HSC)
       67, 75–77, 81–83, 87
hepatoblast   447, 449, 460, 472
hepatocytes   445–454, 456–476
high-fat diet   203–205
hippocampal   301
hippocampal neurogenesis   302
hippocampus   298
histological   197
homing   200, 201
hospitalization   192
human   312
human embryonic stem cells   3
Huntington's disease   309, 313,
       316
hydrogel   209
hyperglycemia   195
hypoxia   197, 304, 309

IGF-1   302
inflammation   203, 209, 302, 308,
       312, 316
inflammatory   197, 203–205
infusion   313
inhibitor of differentiation (ID)
       74, 75
injury   306, 311
inner ear   397–411
insulin   27–40, 195
integration   303, 306, 309
integrins   75
interleukin-6   73
intracardiac   212
intracoronary   196, 212
intramyocardial   196
intravascular delivery   355, 358, 361,
       367, 368–370, 373, 375–377,
       382–384, 386, 387
intravenous   196, 200, 201, 212
invasive   313

irradiation   303
ischemia   194, 195, 203, 304, 308
ischemic   312
islet   27–29, 31, 32, 34, 38
ISWI   54–56

Kaplan-Meier   202
kinetics   515, 519, 530, 535
k-means   525

learning   302
learning and memory   302
left ventricular   192, 193, 197, 199,
      200, 202
leukemia inhibitor factor   73
leukocyte   209
limbal deficiency   125, 127–134
limbal stem cells   114, 116, 118,
      121, 124, 125, 128–132, 134
liver   445–455, 457–476
liver parenchyma cells   450
liver regeneration   445, 448,
      462–466, 475
liver repopulation   445, 458, 459,
      466–475
liver stem cells   445–453, 458–460,
      462, 463, 465, 473, 475,
      476
long term potentiation   302
lymphocytes   4, 5, 11
lysosomal storage disorders   317

M phase   63, 68, 72
macrodepot   196, 197
macrophage   209
magnetic resonance imaging   198,
      213
mammalian development   27
Mash1   315
Math1 gene   401, 410
mathematical model   530, 534

matrix   209
maturation promoting factor
      (MPF)   68
mechanism   193, 204, 207, 209, 211,
      212, 214
medium   197
megakaryocytes   5, 10
membrane dyes   304
memory   300, 309
mesenchymal cells   192, 202
mesenchymal stem cells   458
mesoderm   30
mesodermal   79
methods scale-up   488, 489, 494
microdepot   196, 197
microvascular   209
migrate   306, 312, 316
migrating neuroblasts   300
mitogen activated protein kinase
      (MAPK)   75, 76, 86
mitogens   69, 85, 306, 313
mitosis   68, 304
mitotic inhibitors   303
mobilization   306, 312
monocyte   202, 203
monod model   534–536
monolayer   296
morbidity   191
mortality   191
motorneurons   296
mouse   328, 346
Müller glia   292
multiple sclerosis   314, 316
multipotent   193, 207, 209, 293
multipotent adult progenitor cells
      (MAPCs)   95–107, 404–410,
      458, 459, 461
muscle   305
muscle regeneration   145, 147, 150,
      151, 154, 159, 161, 166, 178,
      179

muscular dystrophy   147, 151, 162,
    164, 167
myocardial infarction   221, 229,
    232, 239, 243, 244
myocardial perfusion   194, 213
myocardium   191, 193, 194, 196,
    199, 200, 207, 212–214
myocyte   194
MyoD   146, 147, 151–153, 155, 157,
    161, 179, 180
myogenesis   146–149, 151, 154, 155,
    161, 167
myogenic precursor cell   177–180,
    183–185
myogenin   194
myosin VIIa   400, 402, 403, 410
Myt1   70

nanog   74, 75
N-cadherin   77
necrosis   192, 197
neoangiogenesis   197
nestin   296
neural stem cell   266, 279, 283
neural transplantation   357
neuroendocrine   296
neuroepithelial   291
neurogenesis   307, 317
Neurogenin2   315
neurohormonal   193
neurons   294
neuroprotective molecules   316
neuroregenerative   315
neurospheres   294
neurotransmitter   316
neurotrophic   316
neurotrophic molecules   316
New York Heart Association   191
NG2   313, 314
niche   75–77, 87, 306
NMDA-receptor   302
Noggin   306
notch   73

nuclear reprogramming   45–50,
    52–54, 56, 57, 59
nucleoplasmin   56–58
nucleotides   305

Oct3/4   73–75, 79
Oct4   425, 428, 429
ocular surface reconstruction   126,
    128, 132, 134
olfactory bulb   298
olfactory discrimination   299
olig2   313, 315
oligodendrocyte precursor cells
    (OPCs)   295, 313
oligodendrocyte progenitor cells
    295
oligodendrocytes   294, 296
opaminergic   315
organ of Corti   401, 403, 405, 406,
    408
osteoblast   76, 77
ototoxic   401, 403–405
oval cells   445, 448–451, 465, 472,
    473

$P15^{INK4b}$ 69
$P16^{INK4a}$ 69, 71
$P18^{INK4c}$ 69, 83
$P19^{INK4d}$ 69
$P21^{CIP1}$ 69, 80–83
$P27^{KIP1}$ 69, 71, 80–83, 85–87
$P57^{KIP2}$ 69
pancreas   27–35, 37–40
pancreatic polypeptide   29
paracrine   298
Parkinson's disease   303, 316
partial hepatectomy   445, 452
pathophysiology   203, 206
Pax-2   400
pax6   296, 315
Pax7   146–149, 151, 153–156, 158,
    164
PDGF   313, 314

PDGFRα 313
pdx-1   33–36
percutaneous   213
periglomerular neurons   298
perilymph   399, 401, 409
PI3K   76, 86
PKC   76
plaque   203–206
plasticity   294, 306
pluripotency   66, 73–75, 87
pluripotent   293
population density   521–523
population dynamics   527
preclinical   200, 201, 206, 214
precursor   293
principal   525
probability   523, 530, 537, 538
progenitor   293
progenitor cell   192, 193, 209, 214,
        265–269, 278, 279, 281, 282,
        284, 301
proinflammatory   204, 205
proliferation   307, 328, 329, 331,
        332, 334–338, 340–343, 347
proliferation in   306
prospectively labeled   312
protein Rb (pRb)   69, 71, 80,
        83–85
proteoglycan   313
pyramindal neurons   309

quantification   513, 514, 530, 539
quiescence   65, 75, 85, 87
quiescent   313

radial glial cells   292
Ramon y Cajal   303
RAS   75, 86
RC2   296
reendothelialization   203
regeneration   193, 207, 211, 311,
        419, 422–424, 427, 430, 432,
        435

regenerative medicine   266, 445,
        446, 451, 452, 466, 475
regenerative therapies   72
regulation   306
regulations and quality   491
rehabilitative   306
re-infarction   191, 193
remodeling   192, 193, 198, 199,
        201–203, 207, 213
repair   193, 195, 198, 203, 204,
        206–209, 211, 212, 214, 215
reproducibility   213
reprogrammed   315
restricted   305
retina   265–271, 273, 278–281, 283,
        284, 298
retroviral   315
retroviral tracing studies   314
risks   313
round window   399, 402, 405–407,
        409

S phase   63, 79–81, 83, 84, 87
sacculus   398, 400
safety   196, 198, 202
satellite cell   145–161, 163, 164,
        166, 167, 178–185, 304
scaffold   209
scala tympani   402–409
scala vestibuli   399, 410
scar tissue   195
schizophrenia   302
self-renewal   64, 65, 72–75, 80, 82,
        83
self-renewing   293, 306
semicircular canal   399, 400, 402,
        403
senescence   65, 209, 293
sensorineural hearing loss   327
sensory epithelium   398–404, 410
serum   314
sex-based   204, 206
side population   145, 147, 148, 157

skeletal muscle   145–148, 150, 152, 155, 158–162, 164–166, 177–180, 183–185
skeletal myoblasts   192, 198, 200
Smad   74
somatic cell nuclear cloning   45, 46, 52
somatic stem cells   305
somatostatin   29, 34, 37
sox2   296
spatial memory   300, 302
specific rates   528
specification   306
spiral ganglion   397, 400–404
spontaneous neural replacement   312
STAT3   73–75
state variable   515, 517, 519, 520
stem cell plasticity   49, 317, 430
stem cell transplantation   355, 363, 366, 373, 382, 383, 385, 387
stem cells   143, 145–150, 152, 155–163, 165–167, 221, 222, 224–226, 229–232, 234–236, 238, 241, 242, 245–247, 293, 417, 419, 420, 422–435
streptozoticin (STZ)   32, 34–36, 38
stress   302
stroke   355, 357, 358, 369, 371–375, 385, 386
stromal cells   5, 8, 13
subcallosal zone (SCZ)   297, 309
subgranular layer   300
subventricular zone (SVZ)   292, 296, 299, 307
surface stiffness   531
survival   192, 195, 196, 202, 212, 300, 307
survive   302

surviving   312
symmetric division   65, 77, 293
synaptic connections   309
synaptic integration   317
synaptically   301
systolic   193–195, 199, 201

telomerase   73, 79
teratoma   207, 316
TGF-β 77
Th1-type   204
Th2-type   205
thymidine analogs   304
totipotency   66
transcription factors   297
transdifferentiation   305
transformation   305
transgenic labels   305
transient amplifying (TA)   78
transit amplifying cells   294, 299, 313
transit time   75, 78–80, 87
transplantation   4, 12–18, 192–196, 198–200, 202, 212–214, 303, 316
treatment   194–196, 202, 204–206, 209, 214
trophic   316
tumors   313
type 2 astrocytes   314
type B cells   298
type C cells   299
type D   300

umbilical cord blood   3, 13
urokinase plasminogen activator (uPA)   468, 474, 475
utriculus   398–400

vascular dysfunction   203
vascular remodeling   203

VEGF   302
ventricular tachycardia   195, 199
vestibular   397–401, 403, 405, 406, 410
viability   194, 202
Vimentin   296

white matter   314

Xenopus eggs   53

zygote   294

www.ingramcontent.com/pod-product-compliance
Lightning Source LLC
Chambersburg PA
CBHW052010230326
41598CB00078B/2232

9 789813 203457